A Guide to Our Solar System

Compiled by

Maci Valerio

Scribbles

Year of Publication 2018

ISBN : 9789386874993

Book Published by

Scribbles

(An Imprint of Alpha Editions)

email - alphaedis@gmail.com

Produced by: PediaPress GmbH
Limburg an der Lahn
Germany
http://pediapress.com/

Contents

Discovery and exploration of the Solar System

Discovery and exploration of the Solar System is observation, visitation, and increase in knowledge and understanding of Earth's "cosmic neighborhood". This includes the Sun, Earth and the Moon, the major planets including Mercury, Venus, Mars, Jupiter, Saturn, Uranus, and Neptune, their satellites, as well as smaller bodies including comets, asteroids, and dust.

> *The Solar System — our Sun's system of planets, moons, and smaller debris — is humankind's cosmic backyard. Small by factors of millions compared to interstellar distances, the spaces between the planets are daunting, but technologically surmountable*
>
> —*NASA*[1]

History

For many thousands of years, astronomers maintained a geocentric world view and did not recognize the existence of a Solar System. Most believed Earth was stationary at the center of the Universe and categorically different from the divine or ethereal objects that moved through the sky. Although the Greek philosopher Aristarchus of Samos had speculated on a heliocentric reordering of the cosmos, Nicolaus Copernicus in the 16th century developed a mathematically predictive heliocentric system. His 17th-century successors, Galileo Galilei, Johannes Kepler, and Isaac Newton, developed a modern understanding of physics that led to the gradual acceptance of the idea that Earth moves around the Sun and that the planets are governed by the same physical laws

Figure 1: *The Sun and planets of the Solar System. Pluto and the other dwarf planets are not shown. The relative sizes of objects are drawn to scale; the distances between them are not.*

Figure 2: *A photo of Earth (circled) taken by Voyager 1, 6.4 billion kilometers (4 billion miles) away. The streaks of light are diffraction spikes radiating from the Sun (off frame to the left). This photograph is known as "Pale Blue Dot".*

Figure 3: *Earth's atmosphere*

Figure 4: *Pioneer space probe's Quadrispherical Plasma Analyzer*

that govern Earth. In more recent times, this led to the investigation of geological phenomena such as mountains and craters and seasonal meteorological phenomena such as clouds, dust storms, and ice caps on the other planets.

Telescopic observations

The first exploration of the Solar System was conducted by telescope, when astronomers first began to map those objects too faint to be seen with the naked eye.

Galileo Galilei was the first to discover physical details about the individual bodies of the Solar System. He discovered that the Moon was cratered, that the Sun was marked with sunspots, and that Jupiter had four satellites in orbit around it. Christiaan Huygens followed on from Galileo's discoveries by discovering Saturn's moon Titan and the shape of the rings of Saturn. Giovanni Domenico Cassini later discovered four more moons of Saturn and the Cassini division in Saturn's rings.

Edmond Halley realised in 1705 that repeated sightings of a comet were recording the same object, returning regularly once every 75–76 years. This was the first evidence that anything other than the planets orbited the Sun. Around this time (1704), the term "Solar System" first appeared in English.

Figure 5: *A replica of Isaac Newton's telescope.*

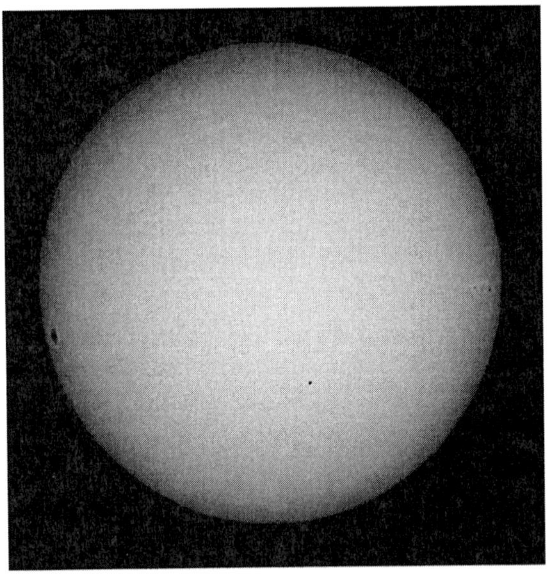

Figure 6: *The Sun photographed through a telescope with special solar filter. Sunspots and limb darkening can be clearly seen. Mercury is transiting in the lower middle of the Sun's face.*

In 1781, William Herschel was looking for binary stars in the constellation of Taurus when he observed what he thought was a new comet. Its orbit revealed that it was a new planet, Uranus, the first ever discovered.

Giuseppe Piazzi discovered Ceres in 1801, a small world between Mars and Jupiter that initially was considered a new planet. However, subsequent discoveries of thousands of other small worlds in the same region led to their eventual reclassification as asteroids.

By 1846, discrepancies in the orbit of Uranus led many to suspect a large planet must be tugging at it from farther out. Urbain Le Verrier's calculations eventually led to the discovery of Neptune. The excess perihelion precession of Mercury's orbit led Le Verrier to postulate the intra-Mercurian planet Vulcan in 1859, but that would turn out to be an irrelevant thesis.

Although it is debatable when the Solar System was truly "discovered", three 19th century observations determined its nature and place in the Universe beyond reasonable doubt. First, in 1838, Friedrich Bessel successfully measured a stellar parallax, an apparent shift in the position of a star created by Earth's motion around the Sun. This was not only the first direct, experimental proof of heliocentrism, but also revealed, for the first time, the vast distance between the Solar System and the stars. Then, in 1859, Robert Bunsen and Gustav Kirchhoff, using the newly invented spectroscope, examined the spectral signature of the Sun and discovered that it was composed of the same elements as existed on Earth, establishing for the first time a physical link between Earth and the heavens. Then, Father Angelo Secchi compared the spectral signature of the Sun with those of other stars, and found them virtually identical. The realisation that the Sun was a star led to the hypothesis that other stars could have systems of their own, though this was not to be proven for nearly 140 years.

Further apparent discrepancies in the orbits of the outer planets led Percival Lowell to conclude that yet another planet, "Planet X", must lie beyond Neptune. After his death, his Lowell Observatory conducted a search that ultimately led to Clyde Tombaugh's discovery of Pluto in 1930. Pluto was, however, found to be too small to have disrupted the orbits of the outer planets, and its discovery was therefore coincidental. Like Ceres, it was initially considered to be a planet, but after the discovery of many other similarly sized objects in its vicinity it was reclassified in 2006 as a dwarf planet by the IAU.

In 1992, the first evidence of a planetary system other than our own was discovered, orbiting the pulsar PSR B1257+12. Three years later, 51 Pegasi b, the first extrasolar planet around a Sunlike star, was discovered. As of 2017, 2687 extrasolar systems have been found.

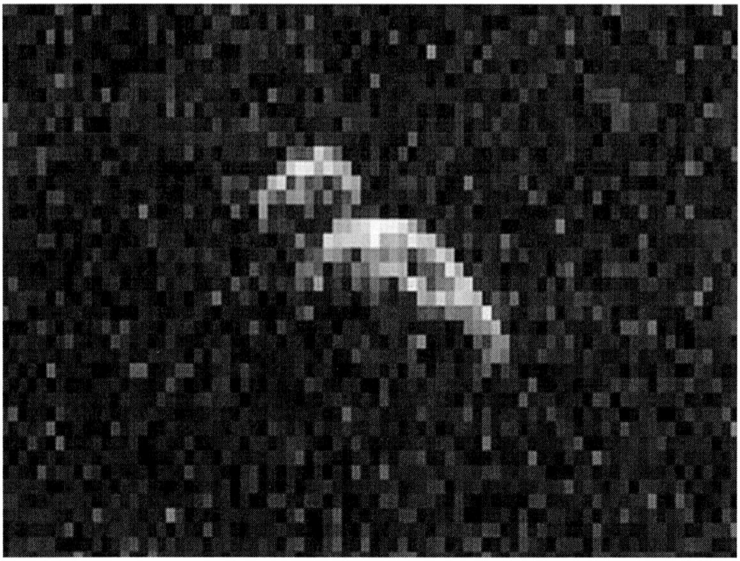

Figure 7: *Radar-imaging of 2006 DP_{14}, a small near-Earth asteroid that passed by Earth in early February 2014*

Also in 1992, astronomers David C. Jewitt of the University of Hawaii and Jane Luu of the Massachusetts Institute of Technology discovered (15760) 1992 QB1. This object proved to be the first of a new population, which became known as the Kuiper belt; an icy analogue to the asteroid belt of which such objects as Pluto and Charon were deemed a part.

Mike Brown, Chad Trujillo and David Rabinowitz announced the discovery of Eris in 2005, a scattered disc object initially thought to be larger than Pluto, which would make it the largest object discovered in orbit around the Sun since Neptune. *New Horizons*' fly-by of Pluto in July 2015 resulted in more-accurate measurements of Pluto, which is slightly larger, though less massive, than Eris.

Observations by spacecraft

Since the start of the Space Age, a great deal of exploration has been performed by robotic spacecraft missions that have been organized and executed by various space agencies.

All planets in the Solar System have now been visited to varying degrees by spacecraft launched from Earth. Through these unmanned missions, humans have been able to get close-up photographs of all the planets and, in the case of landers, perform tests of the soils and atmospheres of some.

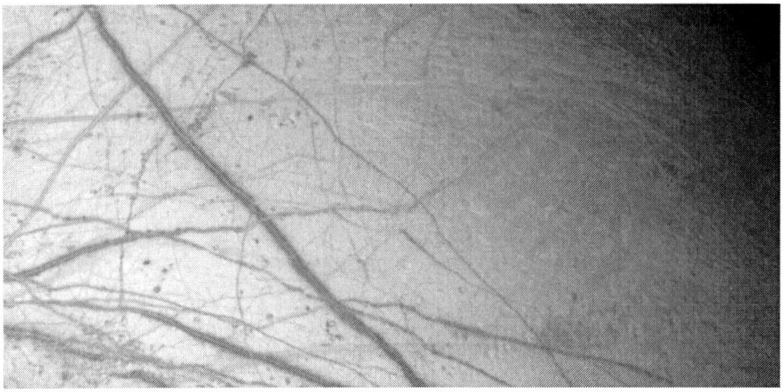

Figure 8: *Lineae on Europa by Galileo spacecraft*

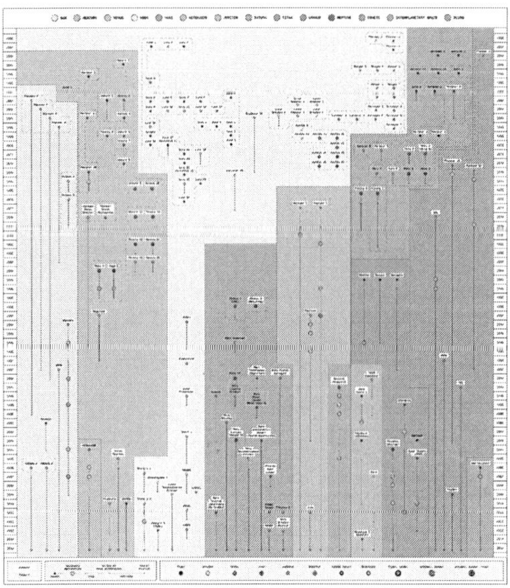

Figure 9: *Timeline of Solar System exploration.*

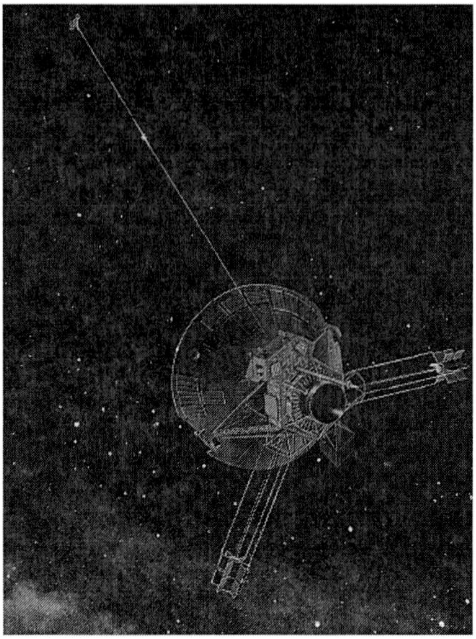

Figure 10: *Artist's conception of Pioneer 10, which passed the orbit of Pluto in 1983. The last transmission was received in January 2003, sent from approximately 82 AU away. The 44–45-year-old space probe is receding from the Sun at over 43,400 km/h (27,000 mph), so long as it has not hit anything*

The first artificial object sent into space was the Soviet satellite *Sputnik 1*, launched in 1957, which successfully orbited Earth until 4 January the following year. The American probe *Explorer 6*, launched in 1959, was the first satellite to image Earth from space.

Flybys

The first successful probe to fly by another Solar System body was *Luna 1*, which sped past the Moon in 1959. Originally meant to impact with the Moon, it instead missed its target and became the first artificial object to orbit the Sun. *Mariner 2* was the first probe to fly by another planet, Venus, in 1962. The first successful flyby of Mars was made by *Mariner 4* in 1965. *Mariner 10* first passed Mercury in 1974.

The first probe to explore the outer planets was *Pioneer 10*, which flew by Jupiter in 1973. *Pioneer 11* was the first to visit Saturn, in 1979. The *Voyager* probes performed a grand tour of the outer planets following their launch in 1977, with both probes passing Jupiter in 1979 and Saturn in 1980 – 1981.

Figure 11: *Curiosity rover self-portrait at "Rocknest" (October 31, 2012), with the rim of Gale Crater and the slopes of Aeolis Mons in the distance.*

Voyager 2 then went on to make close approaches to Uranus in 1986 and Neptune in 1989. The *Voyager* probes are now far beyond Neptune's orbit, and are on course to find and study the termination shock, heliosheath, and heliopause. According to NASA, both *Voyager* probes have encountered the termination shock at a distance of approximately 93 AU from the Sun.

The first flyby of a comet occurred in 1985, when the *International Cometary Explorer* (ICE) passed by the comet Giacobini–Zinner,[2] whereas the first flybys of asteroids were conducted by the *Galileo* space probe, which imaged both 951 Gaspra (in 1991) and 243 Ida (in 1993) on its way to Jupiter.

Launched on January 19, 2006, the *New Horizons* probe is the first man-made spacecraft to explore the Kuiper belt. This unmanned mission flew by Pluto in July 2015. Should it prove feasible, the mission will now be extended to observe a number of other Kuiper belt objects.

As of 2011, American scientists are concerned that exploration beyond the Asteroid Belt will hampered by a shortage of Plutonium-238.Wikipedia:Manual of Style/Dates and numbers#Chronological items

Orbiters, rovers and landers

In 1966, the Moon became the first Solar System body beyond Earth to be orbited by an artificial satellite (*Luna 10*), followed by Mars in 1971 (*Mariner 9*), Venus in 1975 (*Venera 9*), Jupiter in 1995 (*Galileo*), the asteroid 433 Eros in 2000 (*NEAR Shoemaker*), Saturn in 2004 (*Cassini–Huygens*), and Mercury and Vesta in 2011 (*MESSENGER* and *Dawn* respectively). *Dawn* is orbiting the asteroid–dwarf planet Ceres as of 2015.

The first probe to land on another Solar System body was the Soviet *Luna 2* probe, which impacted the Moon in 1959. Since then, increasingly distant planets have been reached, with probes landing on or impacting the surfaces of Venus in 1966 (*Venera 3*), Mars in 1971 (*Mars 3*, although a fully successful landing didn't occur until *Viking 1* in 1976), the asteroid 433 Eros in 2001 (*NEAR Shoemaker*), and Saturn's moon Titan (*Huygens*) and the comet Tempel 1 (*Deep Impact*) in 2005. The *Galileo* orbiter also dropped a probe into Jupiter's atmosphere in 1995; because Jupiter has no physical surface, it was destroyed by increasing temperature and pressure as it descended.

To date, only two worlds in the Solar System, the Moon and Mars, have been visited by mobile rovers. The first robotic rover to visit another celestial body was the Soviet *Lunokhod 1*, which landed on the Moon in 1970. The first to visit another planet was Sojourner, which travelled 500 metres across the surface of Mars in 1997. The only manned rover to visit another world was NASA's Lunar rover, which traveled with Apollos 15, 16 and 17 between 1971 and 1972.

Spacecraft exploration

Overview of some missions to the Solar System.

Examples of missions

#	Spacecraft	Launch year	Mercury	Venus	Mars	Ceres	Jupiter	Saturn	Uranus	Neptune	Pluto	End year
1	*Pioneer 10*	1972					Flyby					2003
2	*Pioneer 11*	1973					Flyby	Flyby				1995
3	*Mariner 10*	1973	Flyby	Flyby								1975
4	*Voyager 1*	1977					Flyby	Flyby				—

5	*Voyager 2*	1977					Flyby	Flyby	Flyby	Flyby		—
6	*Galileo*	1989		Flyby			Orbiter					2003
7	*Ulysses*	1990					Flyby					2009
8	*Cassini*	1997		Flyby			Flyby	Orbiter				2017
9	*Mars Odyssey*	2001			Orbiter							—
10	*MER-A / B*	2003			Rovers							2010 / —
11	*Mars Express*	2003			Orbiter							—
12	*MESSENGER*	2004	Orbiter	Flyby								2015
13	*MRO*	2005			Orbiter							—
14	*Venus Express*	2005		Orbiter								2014
15	*New Horizons*	2006					Flyby				Flyby	—
16	*Dawn*	2007				Orbiter						—
17	*Juno*	2011					Orbiter					—
18	*Curiosity (MSL)*	2011			Rover							—

See also the categories for missions to comets, asteroids, the Moon, and the Sun.

Crewed exploration

Crewed exploration of the Solar System is currently confined to Earth's immediate environs. The first human being to reach space (defined as an altitude of over 100 km) and to orbit Earth was Yuri Gagarin, a Soviet cosmonaut who was launched in *Vostok 1* on April 12, 1961. The first human to walk on the surface of another Solar System body was Neil Armstrong, who stepped onto

Figure 12: *Owen Garriott on an Earth orbit EVA, 1973*

the Moon on July 21, 1969 during the *Apollo 11* mission; five more Moon landings occurred through 1972. The United States' Space Shuttle, which debuted in 1981, is the only reusable spacecraft to successfully make multiple orbital flights. The five shuttles that have been built have flown a total of 121 missions, with two of the craft destroyed in accidents. The first orbital space station to host more than one crew was NASA's Skylab, which successfully held three crews from 1973 to 1974. The first true human settlement in space was the Soviet space station Mir, which was continuously occupied for close to ten years, from 1989 to 1999. It was decommissioned in 2001, and its successor, the International Space Station, has maintained a continuous human presence in space since then. In 2004, SpaceShipOne became the first privately funded vehicle to reach space on a suborbital flight. That year, U.S. President George W. Bush announced the Vision for Space Exploration, which called for a replacement for the aging Shuttle, a return to the Moon and, ultimately, a crewed mission to Mars.

Exploration survey

Bodies imaged up close:

Solar System

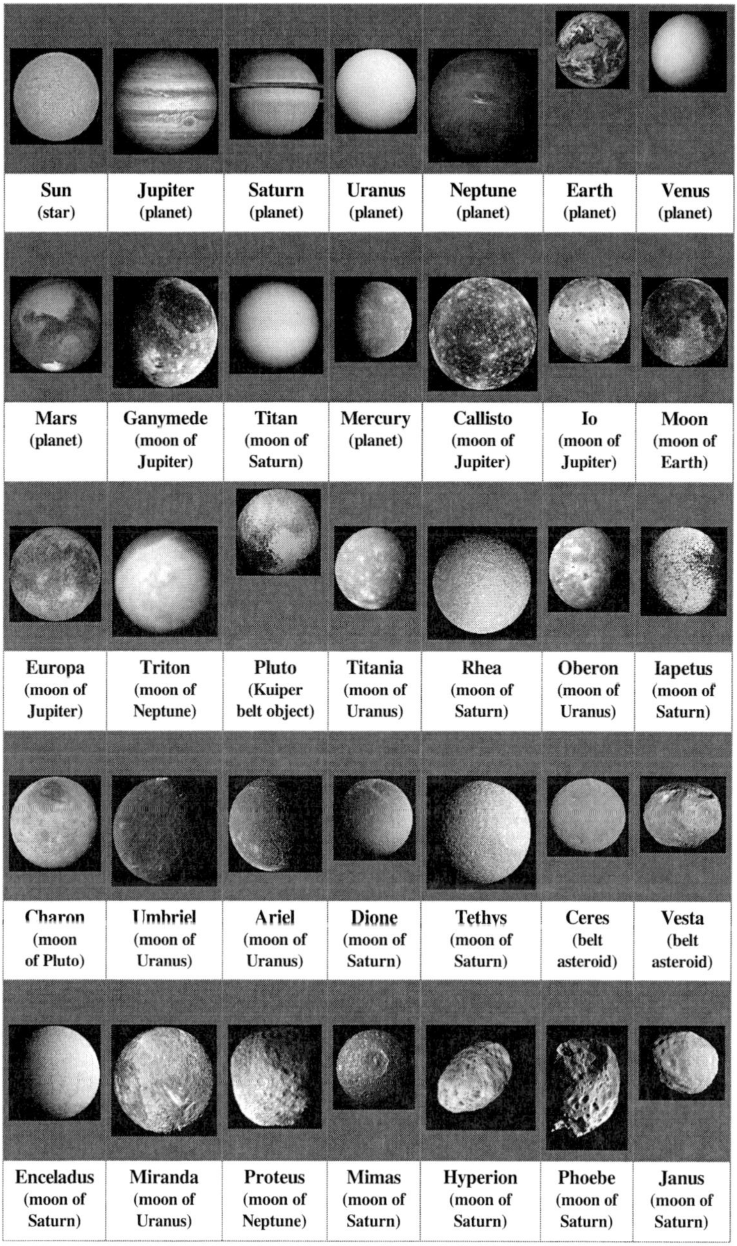

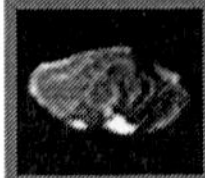		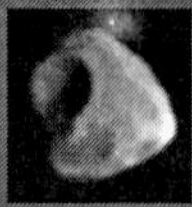		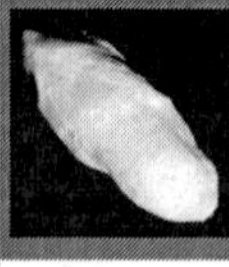	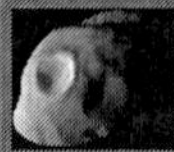	
Amalthea (moon of Jupiter)	**Epimetheus** (moon of Saturn)	**Thebe** (moon of Jupiter)	**Lutetia** (belt asteroid)	**Prometheus** (moon of Saturn)	**Pandora** (moon of Saturn)	**Mathilde** (belt asteroid)
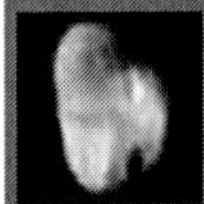	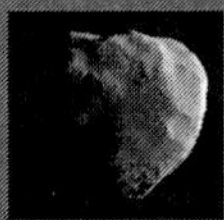	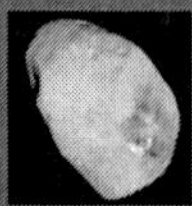	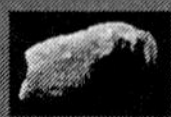	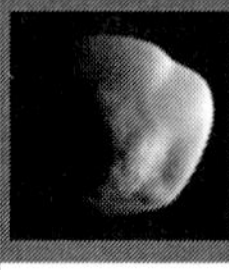	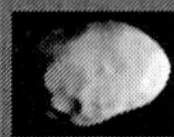	
Hydra (moon of Pluto)	**Helene** (moon of Saturn)	**Nix** (moon of Pluto)	**Ida** (belt asteroid)	**Atlas** (moon of Saturn)	**Telesto** (moon of Saturn)	**Calypso** (moon of Saturn)
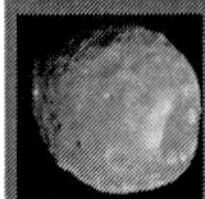	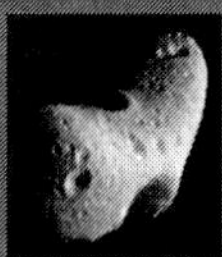		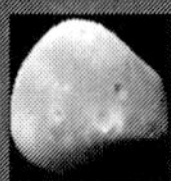			
Phobos (moon of Mars)	**Eros** (near-Earth asteroid)	**Pan** (moon of Saturn)	**Deimos** (moon of Mars)	**Gaspra** (belt asteroid)	**Tempel 1** (comet)	**Šteins** (belt asteroid)
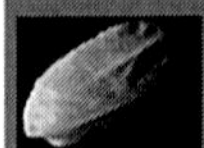	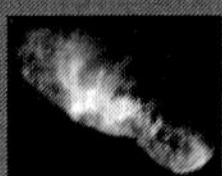		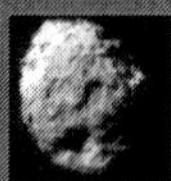			
Daphnis (moon of Saturn)	**Borrelly** (comet)	**Churyu-mov–Gerasi-menko** (comet)	**Wild 2** (comet)	**Methone** (moon of Saturn)	**Hartley 2** (comet)	**Dactyl** (moon of Ida)

Closely imaged object with only non-free images: Toutatis, Itokawa
Objects imaged only at low resolution: Halley's Comet, Styx, Kerberos, Braille, Annefrank, Nereid, and Larissa; see also the radar images at "near-Earth object"

This box:

- view
- talk
- edit[3]

Figure 13: *A Moon rock returned by Apollo 17*

Sample return

- Moon (Misc. missions including Apollo and Luna)
- Comet dust (*Stardust*)
- Solar wind (*Genesis*)
- Asteroid (*Hayabusa*)

See also: Meteorites and Cosmic dust

Formation and evolution of the Solar System

The **formation and evolution of the Solar System** began 4.6 billion years ago with the gravitational collapse of a small part of a giant molecular cloud. Most of the collapsing mass collected in the center, forming the Sun, while the rest flattened into a protoplanetary disk out of which the planets, moons, asteroids, and other small Solar System bodies formed.

This model, known as the nebular hypothesis, was first developed in the 18th century by Emanuel Swedenborg, Immanuel Kant, and Pierre-Simon Laplace. Its subsequent development has interwoven a variety of scientific disciplines including astronomy, physics, geology, and planetary science. Since the dawn of the space age in the 1950s and the discovery of extrasolar planets in the 1990s, the model has been both challenged and refined to account for new observations.

The Solar System has evolved considerably since its initial formation. Many moons have formed from circling discs of gas and dust around their parent planets, while other moons are thought to have formed independently and later been captured by their planets. Still others, such as Earth's Moon, may be the result of giant collisions. Collisions between bodies have occurred continually up to the present day and have been central to the evolution of the Solar System. The positions of the planets often shifted due to gravitational interactions. This planetary migration is now thought to have been responsible for much of the Solar System's early evolution.

In roughly 5 billion years, the Sun will cool and expand outward to many times its current diameter (becoming a red giant), before casting off its outer layers as a planetary nebula and leaving behind a stellar remnant known as a white dwarf. In the far distant future, the gravity of passing stars will gradually reduce the Sun's retinue of planets. Some planets will be destroyed, others ejected into interstellar space. Ultimately, over the course of tens of billions of years, it is likely that the Sun will be left with none of the original bodies in orbit around it.

History

Ideas concerning the origin and fate of the world date from the earliest known writings; however, for almost all of that time, there was no attempt to link such theories to the existence of a "Solar System", simply because it was not generally thought that the Solar System, in the sense we now understand it,

Figure 14: *Artist's conception of a protoplanetary disk*

existed. The first step toward a theory of Solar System formation and evolution was the general acceptance of heliocentrism, which placed the Sun at the centre of the system and the Earth in orbit around it. This concept had developed for millennia (Aristarchus of Samos had suggested it as early as 250 BC), but was not widely accepted until the end of the 17th century. The first recorded use of the term "Solar System" dates from 1704.

The current standard theory for Solar System formation, the nebular hypothesis, has fallen into and out of favour since its formulation by Emanuel Swedenborg, Immanuel Kant, and Pierre-Simon Laplace in the 18th century. The most significant criticism of the hypothesis was its apparent inability to explain the Sun's relative lack of angular momentum when compared to the planets. However, since the early 1980s studies of young stars have shown them to be surrounded by cool discs of dust and gas, exactly as the nebular hypothesis predicts, which has led to its re-acceptance.

Understanding of how the Sun is expected to continue to evolve required an understanding of the source of its power. Arthur Stanley Eddington's confirmation of Albert Einstein's theory of relativity led to his realisation that the Sun's energy comes from nuclear fusion reactions in its core, fusing hydrogen into helium. In 1935, Eddington went further and suggested that other elements also might form within stars. Fred Hoyle elaborated on this premise by arguing that evolved stars called red giants created many elements heavier than hydrogen and helium in their cores. When a red giant finally casts off its outer layers, these elements would then be recycled to form other star systems.

Formation

Nature timeline

Axis scale: billions of years.

Also see: *Human timeline* and *Life timeline*

Pre-solar nebula

The nebular hypothesis says that the Solar System formed from the gravitational collapse of a fragment of a giant molecular cloud. The cloud was about 20 parsec (65 light years) across, while the fragments were roughly 1 parsec (three and a quarter light-years) across. The further collapse of the fragments led to the formation of dense cores 0.01–0.1 pc (2,000–20,000 AU) in size.[4] One of these collapsing fragments (known as the *pre-solar nebula*) formed what became the Solar System. The composition of this region with a mass just over that of the Sun ($M_\odot$) was about the same as that of the Sun today, with hydrogen, along with helium and trace amounts of lithium produced by Big Bang nucleosynthesis, forming about 98% of its mass. The remaining 2% of the mass consisted of heavier elements that were created by nucleosynthesis in earlier generations of stars.[5] Late in the life of these stars, they ejected heavier elements into the interstellar medium.

The oldest inclusions found in meteorites, thought to trace the first solid material to form in the pre-solar nebula, are 4568.2 million years old, which is one definition of the age of the Solar System. Studies of ancient meteorites reveal traces of stable daughter nuclei of short-lived isotopes, such as iron-60, that only form in exploding, short-lived stars. This indicates that one or more supernovae occurred near the Sun while it was forming. A shock wave from a supernova may have triggered the formation of the Sun by creating relatively dense regions within the cloud, causing these regions to collapse. Because only massive, short-lived stars produce supernovae, the Sun must have formed in a large star-forming region that produced massive stars, possibly similar to the Orion Nebula. Studies of the structure of the Kuiper belt and of anomalous materials within it suggest that the Sun formed within a cluster of between 1,000 and 10,000 stars with a diameter of between 6.5 and 19.5 light years and a collective mass of 3,000 $M_\odot$. This cluster began to break apart between 135 million and 535 million years after formation. Several simulations of our young Sun interacting with close-passing stars over the first 100 million years of its life produce anomalous orbits observed in the outer Solar System, such as detached objects.

Because of the conservation of angular momentum, the nebula spun faster as it collapsed. As the material within the nebula condensed, the atoms within it began to collide with increasing frequency, converting their kinetic energy into heat. The centre, where most of the mass collected, became increasingly hotter than the surrounding disc. Over about 100,000 years, the competing forces of gravity, gas pressure, magnetic fields, and rotation caused the contracting nebula to flatten into a spinning protoplanetary disc with a diameter of about

200 AU and form a hot, dense protostar (a star in which hydrogen fusion has not yet begun) at the centre.

At this point in its evolution, the Sun is thought to have been a T Tauri star. Studies of T Tauri stars show that they are often accompanied by discs of pre-planetary matter with masses of 0.001–0.1 $M_\odot$. These discs extend to several hundred AU—the Hubble Space Telescope has observed protoplanetary discs of up to 1000 AU in diameter in star-forming regions such as the Orion Nebula—and are rather cool, reaching a surface temperature of only about 1000 kelvin at their hottest. Within 50 million years, the temperature and pressure at the core of the Sun became so great that its hydrogen began to fuse, creating an internal source of energy that countered gravitational contraction until hydrostatic equilibrium was achieved. This marked the Sun's entry into the prime phase of its life, known as the main sequence. Main-sequence stars derive energy from the fusion of hydrogen into helium in their cores. The Sun remains a main-sequence star today.

Formation of the planets

The various planets are thought to have formed from the solar nebula, the disc-shaped cloud of gas and dust left over from the Sun's formation. The currently accepted method by which the planets formed is accretion, in which the planets began as dust grains in orbit around the central protostar. Through direct contact, these grains formed into clumps up to 200 metres in diameter, which in turn collided to form larger bodies (planetesimals) of ~10 kilometres (km) in size. These gradually increased through further collisions, growing at the rate of centimetres per year over the course of the next few million years.

The inner Solar System, the region of the Solar System inside 4 AU, was too warm for volatile molecules like water and methane to condense, so the planetesimals that formed there could only form from compounds with high melting points, such as metals (like iron, nickel, and aluminium) and rocky silicates. These rocky bodies would become the terrestrial planets (Mercury, Venus, Earth, and Mars). These compounds are quite rare in the Universe, comprising only 0.6% of the mass of the nebula, so the terrestrial planets could not grow very large. The terrestrial embryos grew to about 0.05 Earth masses ($M_\oplus$) and ceased accumulating matter about 100,000 years after the formation of the Sun; subsequent collisions and mergers between these planet-sized bodies allowed terrestrial planets to grow to their present sizes (see Terrestrial planets below).

When the terrestrial planets were forming, they remained immersed in a disk of gas and dust. The gas was partially supported by pressure and so did not orbit the Sun as rapidly as the planets. The resulting drag and, more importantly, gravitational interactions with the surrounding material caused a transfer of angular momentum, and as a result the planets gradually migrated to new orbits. Models show that density and temperature variations in the disk governed this rate of migration, but the net trend was for the inner planets to migrate inward as the disk dissipated, leaving the planets in their current orbits.

The giant planets (Jupiter, Saturn, Uranus, and Neptune) formed further out, beyond the frost line, which is the point between the orbits of Mars and Jupiter where the material is cool enough for volatile icy compounds to remain solid. The ices that formed the Jovian planets were more abundant than the metals and silicates that formed the terrestrial planets, allowing the giant planets to grow massive enough to capture hydrogen and helium, the lightest and most abundant elements. Planetesimals beyond the frost line accumulated up to 4 $M_\oplus$ within about 3 million years. Today, the four giant planets comprise just under 99% of all the mass orbiting the Sun.[6] Theorists believe it is no accident that Jupiter lies just beyond the frost line. Because the frost line accumulated large amounts of water via evaporation from infalling icy material, it created a region of lower pressure that increased the speed of orbiting dust particles and halted their motion toward the Sun. In effect, the frost line acted as a barrier that caused material to accumulate rapidly at ~5 AU from the Sun. This excess material coalesced into a large embryo (or core) on the order of 10 $M_\oplus$, which began to accumulate an envelope via accretion of gas from the surrounding disc at an ever-increasing rate. Once the envelope mass became about equal to the solid core mass, growth proceeded very rapidly, reaching about 150 Earth masses $\sim 10^5$ years thereafter and finally topping out at 318 $M_\oplus$. Saturn may owe its substantially lower mass simply to having formed a few million years after Jupiter, when there was less gas available to consume.

T Tauri stars like the young Sun have far stronger stellar winds than more stable, older stars. Uranus and Neptune are thought to have formed after Jupiter and Saturn did, when the strong solar wind had blown away much of the disc material. As a result, the planets accumulated little hydrogen and helium—not more than 1 $M_\oplus$ each. Uranus and Neptune are sometimes referred to as failed cores. The main problem with formation theories for these planets is the timescale of their formation. At the current locations it would have taken millions of years for their cores to accrete. This means that Uranus and Neptune may have formed closer to the Sun—near or even between Jupiter and Saturn—and later migrated or were ejected outward (see Planetary migration below). Motion in the planetesimal era was not all inward toward the Sun; the *Stardust* sample return from Comet Wild 2 has suggested that materials from the early formation of the Solar System migrated from the warmer inner Solar System to the region of the Kuiper belt.

After between three and ten million years, the young Sun's solar wind would have cleared away all the gas and dust in the protoplanetary disc, blowing it into interstellar space, thus ending the growth of the planets.

Subsequent evolution

The planets were originally thought to have formed in or near their current orbits. However, this view underwent radical change during the late 20th and early 21st centuries. Currently, it is thought that the Solar System looked very different after its initial formation: several objects at least as massive as Mercury were present in the inner Solar System, the outer Solar System was much more compact than it is now, and the Kuiper belt was much closer to the Sun.

Terrestrial planets

At the end of the planetary formation epoch the inner Solar System was populated by 50–100 Moon- to Mars-sized planetary embryos. Further growth was possible only because these bodies collided and merged, which took less than 100 million years. These objects would have gravitationally interacted with one another, tugging at each other's orbits until they collided, growing larger until the four terrestrial planets we know today took shape. One such giant collision is thought to have formed the Moon (see Moons below), while another removed the outer envelope of the young Mercury.

One unresolved issue with this model is that it cannot explain how the initial orbits of the proto-terrestrial planets, which would have needed to be highly eccentric to collide, produced the remarkably stable and nearly circular orbits they have today. One hypothesis for this "eccentricity dumping" is that the terrestrials formed in a disc of gas still not expelled by the Sun. The "gravitational drag" of this residual gas would have eventually lowered the planets' energy, smoothing out their orbits. However, such gas, if it existed, would have prevented the terrestrial planets' orbits from becoming so eccentric in the first place. Another hypothesis is that gravitational drag occurred not between the planets and residual gas but between the planets and the remaining small bodies. As the large bodies moved through the crowd of smaller objects, the smaller objects, attracted by the larger planets' gravity, formed a region of higher density, a "gravitational wake", in the larger objects' path. As they did so, the increased gravity of the wake slowed the larger objects down into more regular orbits.

Asteroid belt

The outer edge of the terrestrial region, between 2 and 4 AU from the Sun, is called the asteroid belt. The asteroid belt initially contained more than enough matter to form 2–3 Earth-like planets, and, indeed, a large number of planetesimals formed there. As with the terrestrials, planetesimals in this region later coalesced and formed 20–30 Moon- to Mars-sized planetary embryos; however, the proximity of Jupiter meant that after this planet formed, 3 million years after the Sun, the region's history changed dramatically. Orbital resonances with Jupiter and Saturn are particularly strong in the asteroid belt, and gravitational interactions with more massive embryos scattered many planetesimals into those resonances. Jupiter's gravity increased the velocity of objects within these resonances, causing them to shatter upon collision with other bodies, rather than accrete.

As Jupiter migrated inward following its formation (see Planetary migration below), resonances would have swept across the asteroid belt, dynamically exciting the region's population and increasing their velocities relative to each other. The cumulative action of the resonances and the embryos either scattered the planetesimals away from the asteroid belt or excited their orbital inclinations and eccentricities. Some of those massive embryos too were ejected by Jupiter, while others may have migrated to the inner Solar System and played a role in the final accretion of the terrestrial planets. During this primary depletion period, the effects of the giant planets and planetary embryos left the asteroid belt with a total mass equivalent to less than 1% that of the Earth, composed mainly of small planetesimals. This is still 10–20 times more than the current mass in the main belt, which is now about 1/2,000 $M_{\oplus}$. A secondary depletion period that brought the asteroid belt down close to its present mass is thought to have followed when Jupiter and Saturn entered a temporary 2:1 orbital resonance (see below).

The inner Solar System's period of giant impacts probably played a role in the Earth acquiring its current water content ($\sim 6 \times 10^{21}$ kg) from the early asteroid belt. Water is too volatile to have been present at Earth's formation and must have been subsequently delivered from outer, colder parts of the Solar System. The water was probably delivered by planetary embryos and small planetesimals thrown out of the asteroid belt by Jupiter. A population of main-belt comets discovered in 2006 has been also suggested as a possible source for Earth's water. In contrast, comets from the Kuiper belt or farther regions delivered not more than about 6% of Earth's water. The panspermia hypothesis holds that life itself may have been deposited on Earth in this way, although this idea is not widely accepted.

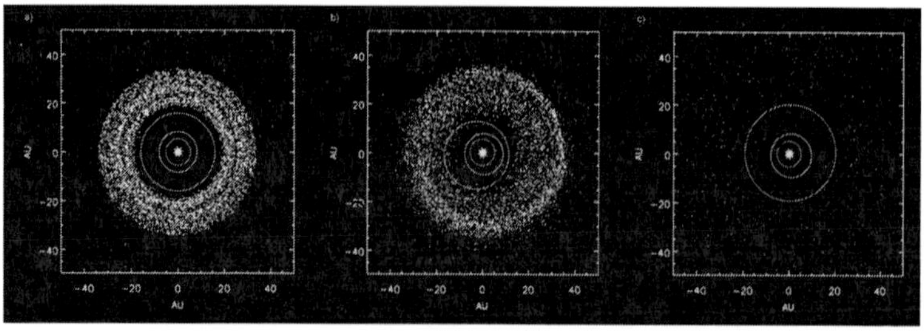

Figure 15: *Simulation showing outer planets and Kuiper belt: a) Before Jupiter/Saturn 2:1 resonance b) Scattering of Kuiper belt objects into the Solar System after the orbital shift of Neptune c) After ejection of Kuiper belt bodies by Jupiter*

Planetary migration

According to the nebular hypothesis, the outer two planets may be in the "wrong place". Uranus and Neptune (known as the "ice giants") exist in a region where the reduced density of the solar nebula and longer orbital times render their formation highly implausible. The two are instead thought to have formed in orbits near Jupiter and Saturn, where more material was available, and to have migrated outward to their current positions over hundreds of millions of years.

The migration of the outer planets is also necessary to account for the existence and properties of the Solar System's outermost regions. Beyond Neptune, the Solar System continues into the Kuiper belt, the scattered disc, and the Oort cloud, three sparse populations of small icy bodies thought to be the points of origin for most observed comets. At their distance from the Sun, accretion was too slow to allow planets to form before the solar nebula dispersed, and thus the initial disc lacked enough mass density to consolidate into a planet. The Kuiper belt lies between 30 and 55 AU from the Sun, while the farther scattered disc extends to over 100 AU, and the distant Oort cloud begins at about 50,000 AU. Originally, however, the Kuiper belt was much denser and closer to the Sun, with an outer edge at approximately 30 AU. Its inner edge would have been just beyond the orbits of Uranus and Neptune, which were in turn far closer to the Sun when they formed (most likely in the range of 15–20 AU), and in 50% of simulations ended up opposite locations, with Uranus farther from the Sun than Neptune.

According to the Nice model, after the formation of the Solar System, the orbits of all the giant planets continued to change slowly, influenced by their interaction with the large number of remaining planetesimals. After

500–600 million years (about 4 billion years ago) Jupiter and Saturn fell into a 2:1 resonance: Saturn orbited the Sun once for every two Jupiter orbits. This resonance created a gravitational push against the outer planets, possibly causing Neptune to surge past Uranus and plough into the ancient Kuiper belt. The planets scattered the majority of the small icy bodies inwards, while themselves moving outwards. These planetesimals then scattered off the next planet they encountered in a similar manner, moving the planets' orbits outwards while they moved inwards. This process continued until the planetesimals interacted with Jupiter, whose immense gravity sent them into highly elliptical orbits or even ejected them outright from the Solar System. This caused Jupiter to move slightly inward.[7] Those objects scattered by Jupiter into highly elliptical orbits formed the Oort cloud; those objects scattered to a lesser degree by the migrating Neptune formed the current Kuiper belt and scattered disc. This scenario explains the Kuiper belt's and scattered disc's present low mass. Some of the scattered objects, including Pluto, became gravitationally tied to Neptune's orbit, forcing them into mean-motion resonances. Eventually, friction within the planetesimal disc made the orbits of Uranus and Neptune circular again.

In contrast to the outer planets, the inner planets are not thought to have migrated significantly over the age of the Solar System, because their orbits have remained stable following the period of giant impacts.

Another question is why Mars came out so small compared with Earth. A study by Southwest Research Institute, San Antonio, Texas, published June 6, 2011 (called the Grand Tack Hypothesis), proposes that Jupiter had migrated inward to 1.5 AU. After Saturn formed, migrated inward, and established the 2:3 mean motion resonance with Jupiter, the study assumes that both planets migrated back to their present positions. Jupiter thus would have consumed much of the material that would have created a bigger Mars. The same simulations also reproduce the characteristics of the modern asteroid belt, with dry asteroids and water-rich objects similar to comets. However, it is unclear whether conditions in the solar nebula would have allowed Jupiter and Saturn to move back to their current positions, and according to current estimates this possibility appears unlikely. Moreover, alternative explanations for the small mass of Mars exist.

Late Heavy Bombardment and after

Life timeline

0 —
500 —
1000 —
1500 —
2000 —
2500 —
3000 —
3500 —
4000 —
4500 —

Axis scale: millions of years.

Orange labels: known ***ice ages***.

Also see: *Human timeline* and *Nature timeline*

Gravitational disruption from the outer planets' migration would have sent large numbers of asteroids into the inner Solar System, severely depleting the original belt until it reached today's extremely low mass. This event may have triggered the Late Heavy Bombardment that occurred approximately 4 billion years ago, 500–600 million years after the formation of the Solar System. This period of heavy bombardment lasted several hundred million years and is evident in the cratering still visible on geologically dead bodies of the inner Solar System such as the Moon and Mercury. The oldest known evidence for life on Earth dates to 3.8 billion years ago—almost immediately after the end of the Late Heavy Bombardment.

Impacts are thought to be a regular (if currently infrequent) part of the evolution of the Solar System. That they continue to happen is evidenced by the collision of Comet Shoemaker–Levy 9 with Jupiter in 1994, the 2009 Jupiter impact event, the Tunguska event, the Chelyabinsk meteor and the impact feature Meteor Crater in Arizona. The process of accretion, therefore, is not complete, and may still pose a threat to life on Earth.

Over the course of the Solar System's evolution, comets were ejected out of the inner Solar System by the gravity of the giant planets, and sent thousands of AU outward to form the Oort cloud, a spherical outer swarm of cometary nuclei at the farthest extent of the Sun's gravitational pull. Eventually, after about 800 million years, the gravitational disruption caused by galactic tides, passing stars and giant molecular clouds began to deplete the cloud, sending comets into the inner Solar System. The evolution of the outer Solar System also appears to have been influenced by space weathering from the solar wind, micrometeorites, and the neutral components of the interstellar medium.

The evolution of the asteroid belt after Late Heavy Bombardment was mainly governed by collisions. Objects with large mass have enough gravity to retain any material ejected by a violent collision. In the asteroid belt this usually is not the case. As a result, many larger objects have been broken apart, and sometimes newer objects have been forged from the remnants in less violent collisions. Moons around some asteroids currently can only be explained as consolidations of material flung away from the parent object without enough energy to entirely escape its gravity.

Moons

Moons have come to exist around most planets and many other Solar System bodies. These natural satellites originated by one of three possible mechanisms:

- Co-formation from a circumplanetary disc (only in the cases of the giant planets);
- Formation from impact debris (given a large enough impact at a shallow angle); and
- Capture of a passing object.

Jupiter and Saturn have several large moons, such as Io, Europa, Ganymede and Titan, which may have originated from discs around each giant planet in much the same way that the planets formed from the disc around the Sun.[8] This origin is indicated by the large sizes of the moons and their proximity to the planet. These attributes are impossible to achieve via capture, while the gaseous nature of the primaries also make formation from collision debris unlikely. The outer moons of the giant planets tend to be small and have eccentric orbits with arbitrary inclinations. These are the characteristics expected of captured bodies. Most such moons orbit in the direction opposite the rotation of their primary. The largest irregular moon is Neptune's moon Triton, which is thought to be a captured Kuiper belt object.

Moons of solid Solar System bodies have been created by both collisions and capture. Mars's two small moons, Deimos and Phobos, are thought to be captured asteroids.[9] The Earth's Moon is thought to have formed as a result of a single, large head-on collision. The impacting object probably had a mass comparable to that of Mars, and the impact probably occurred near the end of the period of giant impacts. The collision kicked into orbit some of the impactor's mantle, which then coalesced into the Moon. The impact was probably the last in the series of mergers that formed the Earth. It has been further hypothesized that the Mars-sized object may have formed at one of the stable Earth–Sun Lagrangian points (either L_4 or L_5) and drifted from its position. The moons of trans-Neptunian objects Pluto (Charon) and Orcus (Vanth) may also have formed by means of a large collision: the Pluto–Charon, Orcus–Vanth and Earth–Moon systems are unusual in the Solar System in that the satellite's mass is at least 1% that of the larger body.

Future

Astronomers estimate that the Solar System as we know it today will not change drastically until the Sun has fused almost all the hydrogen fuel in its core into helium, beginning its evolution from the main sequence of the

Hertzsprung–Russell diagram and into its red-giant phase. Even so, the Solar System will continue to evolve until then.

Long-term stability

The Solar System is chaotic over million- and billion-year timescales, with the orbits of the planets open to long-term variations. One notable example of this chaos is the Neptune–Pluto system, which lies in a 3:2 orbital resonance. Although the resonance itself will remain stable, it becomes impossible to predict the position of Pluto with any degree of accuracy more than 10–20 million years (the Lyapunov time) into the future. Another example is Earth's axial tilt, which, due to friction raised within Earth's mantle by tidal interactions with the Moon (see below), will be incomputable at some point between 1.5 and 4.5 billion years from now.

The outer planets' orbits are chaotic over longer timescales, with a Lyapunov time in the range of 2–230 million years. In all cases this means that the position of a planet along its orbit ultimately becomes impossible to predict with any certainty (so, for example, the timing of winter and summer become uncertain), but in some cases the orbits themselves may change dramatically. Such chaos manifests most strongly as changes in eccentricity, with some planets' orbits becoming significantly more—or less—elliptical.

Ultimately, the Solar System is stable in that none of the planets are likely to collide with each other or be ejected from the system in the next few billion years. Beyond this, within five billion years or so Mars's eccentricity may grow to around 0.2, such that it lies on an Earth-crossing orbit, leading to a potential collision. In the same timescale, Mercury's eccentricity may grow even further, and a close encounter with Venus could theoretically eject it from the Solar System altogether or send it on a collision course with Venus or Earth. This could happen within a billion years, according to numerical simulations in which Mercury's orbit is perturbed.

Moon–ring systems

The evolution of moon systems is driven by tidal forces. A moon will raise a tidal bulge in the object it orbits (the primary) due to the differential gravitational force across diameter of the primary. If a moon is revolving in the same direction as the planet's rotation and the planet is rotating faster than the orbital period of the moon, the bulge will constantly be pulled ahead of the moon. In this situation, angular momentum is transferred from the rotation of the primary to the revolution of the satellite. The moon gains energy and gradually spirals outward, while the primary rotates more slowly over time.

The Earth and its Moon are one example of this configuration. Today, the Moon is tidally locked to the Earth; one of its revolutions around the Earth (currently about 29 days) is equal to one of its rotations about its axis, so it always shows one face to the Earth. The Moon will continue to recede from Earth, and Earth's spin will continue to slow gradually. In about 50 billion years, if they survive the Sun's expansion, the Earth and Moon will become tidally locked to each other; each will be caught up in what is called a "spin–orbit resonance" in which the Moon will circle the Earth in about 47 days and both Moon and Earth will rotate around their axes in the same time, each only visible from one hemisphere of the other. Other examples are the Galilean moons of Jupiter (as well as many of Jupiter's smaller moons) and most of the larger moons of Saturn.

A different scenario occurs when the moon is either revolving around the primary faster than the primary rotates, or is revolving in the direction opposite the planet's rotation. In these cases, the tidal bulge lags behind the moon in its orbit. In the former case, the direction of angular momentum transfer is reversed, so the rotation of the primary speeds up while the satellite's orbit shrinks. In the latter case, the angular momentum of the rotation and revolution have opposite signs, so transfer leads to decreases in the magnitude of each (that cancel each other out).[10] In both cases, tidal deceleration causes the moon to spiral in towards the primary until it either is torn apart by tidal stresses, potentially creating a planetary ring system, or crashes into the planet's surface or atmosphere. Such a fate awaits the moons Phobos of Mars (within 30 to 50 million years), Triton of Neptune (in 3.6 billion years), Metis and Adrastea of Jupiter, and at least 16 small satellites of Uranus and Neptune. Uranus's Desdemona may even collide with one of its neighboring moons.[11]

A third possibility is where the primary and moon are tidally locked to each other. In that case, the tidal bulge stays directly under the moon, there is no transfer of angular momentum, and the orbital period will not change. Pluto and Charon are an example of this type of configuration.

Prior to the 2004 arrival of the *Cassini–Huygens* spacecraft, the rings of Saturn were widely thought to be much younger than the Solar System and were not expected to survive beyond another 300 million years. Gravitational interactions with Saturn's moons were expected to gradually sweep the rings' outer edge toward the planet, with abrasion by meteorites and Saturn's gravity eventually taking the rest, leaving Saturn unadorned. However, data from the *Cassini* mission led scientists to revise that early view. Observations revealed 10 km-wide icy clumps of material that repeatedly break apart and reform, keeping the rings fresh. Saturn's rings are far more massive than the rings of the other giant planets. This large mass is thought to have preserved Saturn's

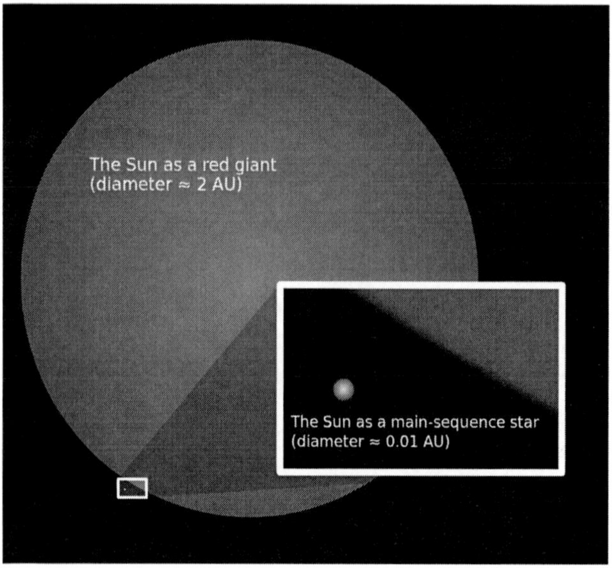

Figure 16: *Relative size of the Sun as it is now (inset) compared to its estimated future size as a red giant*

rings since it first formed 4.5 billion years ago, and is likely to preserve them for billions of years to come.

The Sun and planetary environments

In the long term, the greatest changes in the Solar System will come from changes in the Sun itself as it ages. As the Sun burns through its supply of hydrogen fuel, it gets hotter and burns the remaining fuel even faster. As a result, the Sun is growing brighter at a rate of ten percent every 1.1 billion years. In one billion years' time, as the Sun's radiation output increases, its circumstellar habitable zone will move outwards, making the Earth's surface too hot for liquid water to exist there naturally. At this point, all life on land will become extinct. Evaporation of water, a potent greenhouse gas, from the oceans' surface could accelerate temperature increase, potentially ending all life on Earth even sooner. During this time, it is possible that as Mars's surface temperature gradually rises, carbon dioxide and water currently frozen under the surface regolith will release into the atmosphere, creating a greenhouse effect that will heat the planet until it achieves conditions parallel to Earth today, providing a potential future abode for life. By 3.5 billion years from now, Earth's surface conditions will be similar to those of Venus today.

Around 5.4 billion years from now, the core of the Sun will become hot enough to trigger hydrogen fusion in its surrounding shell. This will cause the outer layers of the star to expand greatly, and the star will enter a phase of its life in which it is called a red giant.[12] Within 7.5 billion years, the Sun will have expanded to a radius of 1.2 AU—256 times its current size. At the tip of the red giant branch, as a result of the vastly increased surface area, the Sun's surface will be much cooler (about 2600 K) than now and its luminosity much higher—up to 2,700 current solar luminosities. For part of its red giant life, the Sun will have a strong stellar wind that will carry away around 33% of its mass.[13] During these times, it is possible that Saturn's moon Titan could achieve surface temperatures necessary to support life.

As the Sun expands, it will swallow the planets Mercury and Venus. Earth's fate is less clear; although the Sun will envelop Earth's current orbit, the star's loss of mass (and thus weaker gravity) will cause the planets' orbits to move farther out. If it were only for this, Venus and Earth would probably escape incineration, but a 2008 study suggests that Earth will likely be swallowed up as a result of tidal interactions with the Sun's weakly bound outer envelope.

Gradually, the hydrogen burning in the shell around the solar core will increase the mass of the core until it reaches about 45% of the present solar mass. At this point the density and temperature will become so high that the fusion of helium into carbon will begin, leading to a helium flash; the Sun will shrink from around 250 to 11 times its present (main-sequence) radius. Consequently, its luminosity will decrease from around 3,000 to 54 times its current level, and its surface temperature will increase to about 4770 K. The Sun will become a horizontal giant, burning helium in its core in a stable fashion much like it burns hydrogen today. The helium-fusing stage will last only 100 million years. Eventually, it will have to again resort to the reserves of hydrogen and helium in its outer layers and will expand a second time, turning into what is known as an asymptotic giant. Here the luminosity of the Sun will increase again, reaching about 2,090 present luminosities, and it will cool to about 3500 K. This phase lasts about 30 million years, after which, over the course of a further 100,000 years, the Sun's remaining outer layers will fall away, ejecting a vast stream of matter into space and forming a halo known (misleadingly) as a planetary nebula. The ejected material will contain the helium and carbon produced by the Sun's nuclear reactions, continuing the enrichment of the interstellar medium with heavy elements for future generations of stars.

This is a relatively peaceful event, nothing akin to a supernova, which the Sun is too small to undergo as part of its evolution. Any observer present to witness this occurrence would see a massive increase in the speed of the solar wind, but not enough to destroy a planet completely. However, the star's loss of mass could send the orbits of the surviving planets into chaos, causing some to collide, others to be ejected from the Solar System, and still others to be torn apart by tidal interactions. Afterwards, all that will remain of the Sun is a white dwarf, an extraordinarily dense object, 54% its original mass but only the size of the Earth. Initially, this white dwarf may be 100 times as luminous as the Sun is now. It will consist entirely of degenerate carbon and oxygen, but will never reach temperatures hot enough to fuse these elements. Thus the white dwarf Sun will gradually cool, growing dimmer and dimmer.

As the Sun dies, its gravitational pull on the orbiting bodies such as planets, comets and asteroids will weaken due to its mass loss. All remaining planets' orbits will expand; if Venus, Earth, and Mars still exist, their orbits will lie roughly at 1.4 AU (210,000,000 km), 1.9 AU (280,000,000 km), and 2.8 AU (420,000,000 km). They and the other remaining planets will become dark, frigid hulks, completely devoid of any form of life. They will continue to orbit their star, their speed slowed due to their increased distance from the Sun and the Sun's reduced gravity. Two billion years later, when the Sun has cooled to the 6000–8000K range, the carbon and oxygen in the Sun's core will freeze, with over 90% of its remaining mass assuming a crystalline structure. Eventually, after billions more years, the Sun will finally cease to shine altogether, becoming a black dwarf.

Galactic interaction

The Solar System travels alone through the Milky Way in a circular orbit approximately 30,000 light years from the Galactic Centre. Its speed is about 220 km/s. The period required for the Solar System to complete one revolution around the Galactic Centre, the galactic year, is in the range of 220–250 million years. Since its formation, the Solar System has completed at least 20 such revolutions.

Various scientists have speculated that the Solar System's path through the galaxy is a factor in the periodicity of mass extinctions observed in the Earth's fossil record. One hypothesis supposes that vertical oscillations made by the Sun as it orbits the Galactic Centre cause it to regularly pass through the galactic plane. When the Sun's orbit takes it outside the galactic disc, the influence of the galactic tide is weaker; as it re-enters the galactic disc, as it does every 20–25 million years, it comes under the influence of the far stronger "disc tides", which, according to mathematical models, increase the flux of Oort cloud comets into the Solar System by a factor of 4, leading to a massive increase in the likelihood of a devastating impact.

However, others argue that the Sun is currently close to the galactic plane, and yet the last great extinction event was 15 million years ago. Therefore, the Sun's vertical position cannot alone explain such periodic extinctions, and that extinctions instead occur when the Sun passes through the galaxy's spiral arms. Spiral arms are home not only to larger numbers of molecular clouds, whose gravity may distort the Oort cloud, but also to higher concentrations of bright blue giants, which live for relatively short periods and then explode violently as supernovae.

Galactic collision and planetary disruption

Although the vast majority of galaxies in the Universe are moving away from the Milky Way, the Andromeda Galaxy, the largest member of the Local Group of galaxies, is heading toward it at about 120 km/s. In 4 billion years, Andromeda and the Milky Way will collide, causing both to deform as tidal forces distort their outer arms into vast tidal tails. If this initial disruption occurs, astronomers calculate a 12% chance that the Solar System will be pulled outward into the Milky Way's tidal tail and a 3% chance that it will become gravitationally bound to Andromeda and thus a part of that galaxy. After a further series of glancing blows, during which the likelihood of the Solar System's ejection rises to 30%, the galaxies' supermassive black holes will merge. Eventually, in roughly 6 billion years, the Milky Way and Andromeda will complete their merger into a giant elliptical galaxy. During the merger, if there is enough gas, the increased gravity will force the gas to the centre of the forming elliptical galaxy. This may lead to a short period of intensive star formation called a starburst. In addition, the infalling gas will feed the newly formed black hole, transforming it into an active galactic nucleus. The force of these interactions will likely push the Solar System into the new galaxy's outer halo, leaving it relatively unscathed by the radiation from these collisions.

It is a common misconception that this collision will disrupt the orbits of the planets in the Solar System. Although it is true that the gravity of passing stars can detach planets into interstellar space, distances between stars are so great that the likelihood of the Milky Way–Andromeda collision causing such disruption to any individual star system is negligible. Although the Solar System as a whole could be affected by these events, the Sun and planets are not expected to be disturbed.

However, over time, the cumulative probability of a chance encounter with a star increases, and disruption of the planets becomes all but inevitable. Assuming that the Big Crunch or Big Rip scenarios for the end of the Universe do not occur, calculations suggest that the gravity of passing stars will have completely stripped the dead Sun of its remaining planets within 1 quadrillion (10^{15}) years. This point marks the end of the Solar System. Although the Sun and planets may survive, the Solar System, in any meaningful sense, will cease to exist.

Chronology

The time frame of the Solar System's formation has been determined using radiometric dating. Scientists estimate that the Solar System is 4.6 billion years old. The oldest known mineral grains on Earth are approximately 4.4 billion years old. Rocks this old are rare, as Earth's surface is constantly being reshaped by erosion, volcanism, and plate tectonics. To estimate the age of the Solar System, scientists use meteorites, which were formed during the early condensation of the solar nebula. Almost all meteorites (see the Canyon Diablo meteorite) are found to have an age of 4.6 billion years, suggesting that the Solar System must be at least this old.

Studies of discs around other stars have also done much to establish a time frame for Solar System formation. Stars between one and three million years old have discs rich in gas, whereas discs around stars more than 10 million years old have little to no gas, suggesting that giant planets within them have ceased forming.

Timeline of Solar System evolution

A graphical timeline is available at
Graphical timeline of Earth and Sun

Note: All dates and times in this chronology are approximate and should be taken as an order of magnitude indicator only.

Chronology of the formation and evolution of the Solar System

Phase	Time since formation of the Sun	Time from present (approximate)	Event
Pre-Solar System	Billions of years before the formation of the Solar System	Over 4.6 billion years ago (bya)	Previous generations of stars live and die, injecting heavy elements into the interstellar medium out of which the Solar System formed.
	$\sim$ 50 million years before formation of the Solar System	4.6 bya	If the Solar System formed in an Orion nebula-like star-forming region, the most massive stars are formed, live their lives, die, and explode in supernova. One particular supernova, called the *primal supernova*, possibly triggers the formation of the Solar System.

Formation of Sun	0–100,000 years	4.6 bya	Pre-solar nebula forms and begins to collapse. Sun begins to form.
	100,000 – 50 million years	4.6 bya	Sun is a T Tauri protostar.
	100,000 – 10 million years	4.6 bya	By 10 million years, gas in the protoplanetary disc has been blown away, and outer planet formation is likely complete.
	10 million – 100 million years	4.5–4.6 bya	Terrestrial planets and the Moon form. Giant impacts occur. Water delivered to Earth.
Main sequence	50 million years	4.5 bya	Sun becomes a main-sequence star.
	200 million years	4.4 bya	Oldest known rocks on the Earth formed.
	500 million – 600 million years	4.0–4.1 bya	Resonance in Jupiter and Saturn's orbits moves Neptune out into the Kuiper belt. Late Heavy Bombardment occurs in the inner Solar System.
	800 million years	3.8 bya	Oldest known life on Earth. Oort cloud reaches maximum mass.
	4.6 billion years	**Today**	Sun remains a main-sequence star, continually growing warmer and brighter by ~10% every 1 billion years.
	6 billion years	1.4 billion years in the future	Sun's habitable zone moves outside of the Earth's orbit, possibly shifting onto Mars's orbit.
	7 billion years	2.4 billion years in the future	The Milky Way and Andromeda Galaxy begin to collide. Slight chance the Solar System could be captured by Andromeda before the two galaxies fuse completely.
Post-main sequence	10 billion – 12 billion years	5–7 billion years in the future	Sun starts burning hydrogen in a shell surrounding its core, ending its main sequence life. Sun begins to ascend the red giant branch of the Hertzsprung–Russell diagram, growing dramatically more luminous (by a factor of up to 2,700), larger (by a factor of up to 250 in radius), and cooler (down to 2600 K): Sun is now a red giant. Mercury and possibly Venus and Earth are swallowed. Saturn's moon Titan may become habitable.
	~ 12 billion years	~ 7 billion years in the future	Sun passes through helium-burning horizontal-branch and asymptotic-giant-branch phases, losing a total of ~30% of its mass in all post-main-sequence phases. The asymptotic-giant-branch phase ends with the ejection of a planetary nebula, leaving the core of the Sun behind as a white dwarf.
Remnant Sun	~ 1 quadrillion years (10^{15} years)	~ 1 quadrillion years in the future	Sun cools to 5 K. Gravity of passing stars detaches planets from orbits. Solar System ceases to exist.

Bibliography

- Duncan, Martin J.; Lissauer, Jack J. (1997). "Orbital Stability of the Uranian Satellite System". *Icarus*. **125** (1): 1–12. Bibcode: 1997Icar..125....1D[14]. doi: 10.1006/icar.1996.5568[15].
- Zeilik, Michael A.; Gregory, Stephen A. (1998). *Introductory Astronomy & Astrophysics* (4th ed.). Saunders College Publishing. ISBN 0-03-006228-4.

External links

- 7M animation[16] from skyandtelescope.com[17] showing the early evolution of the outer Solar System.
- QuickTime animation of the future collision between the Milky Way and Andromeda[18]
- How the Sun Will Die: And What Happens to Earth[19] (Video at Space.com)

<indicator name="featured-star"> </indicator>

Sun

<indicator name="pp-default"> </indicator> <indicator name="featured-star"> </indicator>

The Sun

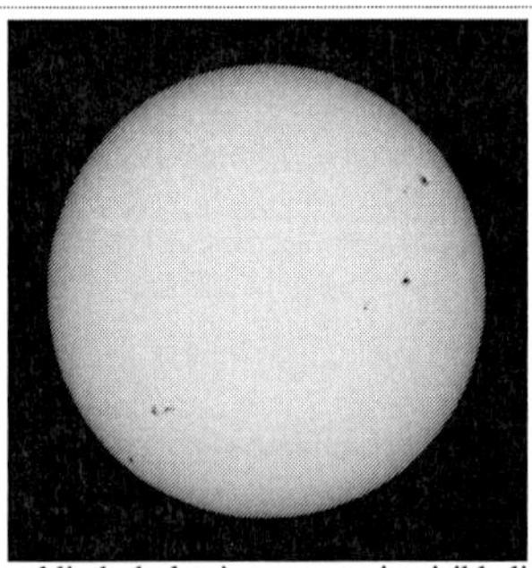

Sun with sunspots and limb darkening as seen in visible light with solar filter.

Observation data	
Mean distance from Earth	1 au ≈ 1.496×10^8 km 8 min 19 s at light speed

Visual brightness (*V*)	–26.74
Absolute magnitude	4.83
Spectral classification	G2V
Metallicity	$Z = 0.0122$
Angular size	31.6–32.7 minutes of arc
Adjectives	Solar
Orbital characteristics	
Mean distance from Milky Way core	$\approx 2.7 \times 10^{17}$ km 27,200 light-years
Galactic period	$(2.25–2.50) \times 10^8$ yr
Velocity	≈ 220 km/s (orbit around the center of the Milky Way) ≈ 20 km/s (relative to average velocity of other stars in stellar neighborhood) ≈ 370 km/s (relative to the cosmic microwave background)
Physical characteristics	
Equatorial radius	695,700 km, 696,392 km 109 $R_\oplus$
Equatorial circumference	4.379×10^6 km 109 × Earth
Flattening	9×10^{-6}
Surface area	6.09×10^{12} km^2 12,000 × Earth
Volume	1.41×10^{18} km^3 1,300,000 × Earth
Mass	$(1.98855 \pm 0.00025) \times 10^{30}$ kg 333,000 $M_\oplus$
Average density	1.408 g/cm^3 0.255 × Earth
Center density (modeled)	162.2 g/cm^3 12.4 × Earth
Equatorial surface gravity	274.0 m/s^2 27.94 *g* 27,542.29 *cgs* 28 × Earth
Moment of inertia factor	0.070 (estimate)
Escape velocity (from the surface)	617.7 km/s 55 × Earth

Temperature	Center (modeled): 1.57×10^7 K Photosphere (effective): 5,772 K Corona: $\approx 5 \times 10^6$ K
Luminosity (L_{sol})	3.828×10^{26} W $\approx 3.75 \times 10^{28}$ lm ≈ 98 lm/W efficacy
Mean radiance (I_{sol})	2.009×10^7 W·m^{-2}·sr^{-1}
Age	$\approx$ 4.6 billion years
	Rotation characteristics
Obliquity	7.25° (to the ecliptic) 67.23° (to the galactic plane)
Right ascension of North pole	286.13° 19 h 4 min 30 s
Declination of North pole	+63.87° 63° 52' North
Sidereal rotation period (at equator)	25.05 d
(at 16° latitude)	25.38 d 25 d 9 h 7 min 12 s
(at poles)	34.4 d
Rotation velocity (at equator)	7.189×10^3 km/h
	Photospheric composition (by mass)
Hydrogen	73.46%[20]
Helium	24.85%
Oxygen	0.77%
Carbon	0.29%
Iron	0.16%
Neon	0.12%
Nitrogen	0.09%
Silicon	0.07%
Magnesium	0.05%
Sulfur	0.04%

The **Sun** is the star at the center of the Solar System. It is a nearly perfect sphere of hot plasma, with internal convective motion that generates a magnetic field via a dynamo process. It is by far the most important source of energy for life on Earth. Its diameter is about 1.39 million kilometers, i.e. 109 times that of Earth, and its mass is about 330,000 times that of Earth, accounting for about

99.86% of the total mass of the Solar System. About three quarters of the Sun's mass consists of hydrogen (~73%); the rest is mostly helium (~25%), with much smaller quantities of heavier elements, including oxygen, carbon, neon, and iron.

The Sun is a G-type main-sequence star (G2V) based on its spectral class. As such, it is informally referred to as a yellow dwarf. It formed approximately 4.6 billion[21] years ago from the gravitational collapse of matter within a region of a large molecular cloud. Most of this matter gathered in the center, whereas the rest flattened into an orbiting disk that became the Solar System. The central mass became so hot and dense that it eventually initiated nuclear fusion in its core. It is thought that almost all stars form by this process.

The Sun is roughly middle-aged; it has not changed dramatically for more than four billion years, and will remain fairly stable for more than another five billion years. After hydrogen fusion in its core has diminished to the point at which it is no longer in hydrostatic equilibrium, the core of the Sun will experience a marked increase in density and temperature while its outer layers expand to eventually become a red giant. It is calculated that the Sun will become sufficiently large to engulf the current orbits of Mercury and Venus, and render Earth uninhabitable.

The enormous effect of the Sun on Earth has been recognized since prehistoric times, and the Sun has been regarded by some cultures as a deity. The synodic rotation of Earth and its orbit around the Sun are the basis of solar calendars, one of which is the predominant calendar in use today.

Name and etymology

The English proper name *Sun* developed from Old English *sunne* and may be related to *south*. Cognates to English *sun* appear in other Germanic languages, including Old Frisian *sunne*, *sonne*, Old Saxon *sunna*, Middle Dutch *sonne*, modern Dutch *zon*, Old High German *sunna*, modern German *Sonne*, Old Norse *sunna*, and Gothic *sunnō*. All Germanic terms for the Sun stem from Proto-Germanic **sunnōn*.

The English weekday name *Sunday* stems from Old English (*Sunnandæg*; "Sun's day", from before 700) and is ultimately a result of a Germanic interpretation of Latin *dies solis*, itself a translation of the Greek ἡμέρα ἡλίου (*hēméra hēlíou*). The Latin name for the Sun, *Sol*, is not common in general English language use; the adjectival form is the related word *solar*. The term *sol* is also used by planetary astronomers to refer to the duration of a solar day on another planet, such as Mars. A mean Earth solar day is approximately 24 hours, whereas a mean Martian 'sol' is 24 hours, 39 minutes, and 35.244 seconds.

Religious aspects

Solar deities play a major role in many world religions and mythologies. The ancient Sumerians believed that the sun was Utu, the god of justice and twin brother of Inanna, the Queen of Heaven, who was identified as the planet Venus. Later, Utu was identified with the East Semitic god Shamash. Utu was regarded as a helper-deity, who aided those in distress, and, in iconography, he is usually portrayed with a long beard and clutching a saw, which represented his role as the dispenser of justice.

From at least the 4th Dynasty of Ancient Egypt, the Sun was worshipped as the god Ra, portrayed as a falcon-headed divinity surmounted by the solar disk, and surrounded by a serpent. In the New Empire period, the Sun became identified with the dung beetle, whose spherical ball of dung was identified with the Sun. In the form of the Sun disc Aten, the Sun had a brief resurgence during the Amarna Period when it again became the preeminent, if not only, divinity for the Pharaoh Akhenaton.

In Proto-Indo-European religion, the sun was personified as the goddess **Seh$_2$ul*. Derivatives of this goddess in Indo-European languages include the Old Norse *Sól*, Sanskrit *Surya*, Gaulish *Sulis*, Lithuanian *Saulė*, and Slavic *Solntse*. In ancient Greek religion, the sun deity was the male god Helios, but traces of an earlier female solar deity are preserved in Helen of Troy. In later times, Helios was syncretized with Apollo.

In ancient Roman culture, Sunday was the day of the Sun god. It was adopted as the Sabbath day by Christians who did not have a Jewish background. The symbol of light was a pagan device adopted by Christians, and perhaps the most important one that did not come from Jewish traditions. In paganism, the Sun was a source of life, giving warmth and illumination to mankind. It was the center of a popular cult among Romans, who would stand at dawn to catch the first rays of sunshine as they prayed. The celebration of the winter solstice (which influenced Christmas) was part of the Roman cult of the unconquered Sun (Sol Invictus). Christian churches were built with an orientation so that the congregation faced toward the sunrise in the East.

Tonatiuh, the Aztec god of the sun, was usually depicted holding arrows and a shield and was closely associated with the practice of human sacrifice. The sun goddess Amaterasu is an important deity in the Shinto religion.

Characteristics

The Sun is a G-type main-sequence star that comprises about 99.86% of the mass of the Solar System. The Sun has an absolute magnitude of +4.83, estimated to be brighter than about 85% of the stars in the Milky Way, most of which are red dwarfs. The Sun is a Population I, or heavy-element-rich, star. The formation of the Sun may have been triggered by shockwaves from one or more nearby supernovae. This is suggested by a high abundance of heavy elements in the Solar System, such as gold and uranium, relative to the abundances of these elements in so-called Population II, heavy-element-poor, stars. The heavy elements could most plausibly have been produced by endothermic nuclear reactions during a supernova, or by transmutation through neutron absorption within a massive second-generation star.

The Sun is by far the brightest object in the Earth's sky, with an apparent magnitude of −26.74. This is about 13 billion times brighter than the next brightest star, Sirius, which has an apparent magnitude of −1.46. The mean distance of the Sun's center to Earth's center is approximately 1 astronomical unit (about 150,000,000 km; 93,000,000 mi), though the distance varies as Earth moves from perihelion in January to aphelion in July. At this average distance, light travels from the Sun's horizon to Earth's horizon in about 8 minutes and 19 seconds, while light from the closest points of the Sun and Earth takes about two seconds less. The energy of this sunlight supports almost all life[22] on Earth by photosynthesis, and drives Earth's climate and weather.

The Sun does not have a definite boundary, but its density decreases exponentially with increasing height above the photosphere. For the purpose of measurement, however, the Sun's radius is considered to be the distance from its center to the edge of the photosphere, the apparent visible surface of the Sun. By this measure, the Sun is a near-perfect sphere with an oblateness estimated at about 9 millionths, which means that its polar diameter differs from its equatorial diameter by only 10 kilometres (6.2 mi). The tidal effect of the planets is weak and does not significantly affect the shape of the Sun. The Sun rotates faster at its equator than at its poles. This differential rotation is caused by convective motion due to heat transport and the Coriolis force due to the Sun's rotation. In a frame of reference defined by the stars, the rotational period is approximately 25.6 days at the equator and 33.5 days at the poles. Viewed from Earth as it orbits the Sun, the *apparent rotational period* of the Sun at its equator is about 28 days.

Sunlight

The solar constant is the amount of power that the Sun deposits per unit area that is directly exposed to sunlight. The solar constant is equal to approximately 1,368 W/m^2 (watts per square meter) at a distance of one astronomical unit (AU) from the Sun (that is, on or near Earth). Sunlight on the surface of Earth is attenuated by Earth's atmosphere, so that less power arrives at the surface (closer to 1,000 W/m^2) in clear conditions when the Sun is near the zenith.

Sunlight at the top of Earth's atmosphere is composed (by total energy) of about 50% infrared light, 40% visible light, and 10% ultraviolet light.[23] The atmosphere in particular filters out over 70% of solar ultraviolet, especially at the shorter wavelengths. Solar ultraviolet radiation ionizes Earth's dayside upper atmosphere, creating the electrically conducting ionosphere.

The Sun's color is white, with a CIE color-space index near (0.3, 0.3), when viewed from space or when the Sun is high in the sky. When measuring all the photons emitted, the Sun is actually emitting more photons in the green portion of the spectrum than any other. When the Sun is low in the sky, atmospheric scattering renders the Sun yellow, red, orange, or magenta. Despite its typical whiteness, most people mentally picture the Sun as yellow; the reasons for this are the subject of debate. The Sun is a G2V star, with *G2* indicating its surface temperature of approximately 5,778 K (5,505 °C, 9,941 °F), and *V* that it, like most stars, is a main-sequence star. The average luminance of the Sun is about 1.88 giga candela per square metre, but as viewed through Earth's atmosphere, this is lowered to about 1.44 Gcd/m^2.[24] However, the luminance is not constant across the disk of the Sun (limb darkening).

Composition

The Sun is composed primarily of the chemical elements hydrogen and helium; they account for 74.9% and 23.8% of the mass of the Sun in the photosphere, respectively. All heavier elements, called *metals* in astronomy, account for less than 2% of the mass, with oxygen (roughly 1% of the Sun's mass), carbon (0.3%), neon (0.2%), and iron (0.2%) being the most abundant.

The Sun inherited its chemical composition from the interstellar medium out of which it formed. The hydrogen and helium in the Sun were produced by Big Bang nucleosynthesis, and the heavier elements were produced by stellar nucleosynthesis in generations of stars that completed their stellar evolution and returned their material to the interstellar medium before the formation of the Sun. The chemical composition of the photosphere is normally considered representative of the composition of the primordial Solar System. However, since the Sun formed, some of the helium and heavy elements have gravitationally settled from the photosphere. Therefore, in today's photosphere the helium fraction is reduced, and the metallicity is only 84% of what it was in the protostellar phase (before nuclear fusion in the core started). The protostellar Sun's composition is believed to have been 71.1% hydrogen, 27.4% helium, and 1.5% heavier elements.

Today, nuclear fusion in the Sun's core has modified the composition by converting hydrogen into helium, so the innermost portion of the Sun is now roughly 60% helium, with the abundance of heavier elements unchanged. Because heat is transferred from the Sun's core by radiation rather than by convection (see Radiative zone below), none of the fusion products from the core have risen to the photosphere.

The reactive core zone of "hydrogen burning", where hydrogen is converted into helium, is starting to surround an inner core of "helium ash". This development will continue and will eventually cause the Sun to leave the main sequence, to become a red giant.[25]

The solar heavy-element abundances described above are typically measured both using spectroscopy of the Sun's photosphere and by measuring abundances in meteorites that have never been heated to melting temperatures. These meteorites are thought to retain the composition of the protostellar Sun and are thus not affected by settling of heavy elements. The two methods generally agree well.

Singly ionized iron-group elements

In the 1970s, much research focused on the abundances of iron-group elements in the Sun.[26] Although significant research was done, until 1978 it was difficult to determine the abundances of some iron-group elements (e.g. cobalt and manganese) via spectrography because of their hyperfine structures.

The first largely complete set of oscillator strengths of singly ionized iron-group elements were made available in the 1960s,[27] and these were subsequently improved.[28] In 1978, the abundances of singly ionized elements of the iron group were derived.

Isotopic composition

Various authors have considered the existence of a gradient in the isotopic compositions of solar and planetary noble gases,[29] e.g. correlations between isotopic compositions of neon and xenon in the Sun and on the planets.[30]

Prior to 1983, it was thought that the whole Sun has the same composition as the solar atmosphere. In 1983, it was claimed that it was fractionation in the Sun itself that caused the isotopic-composition relationship between the planetary and solar-wind-implanted noble gases.

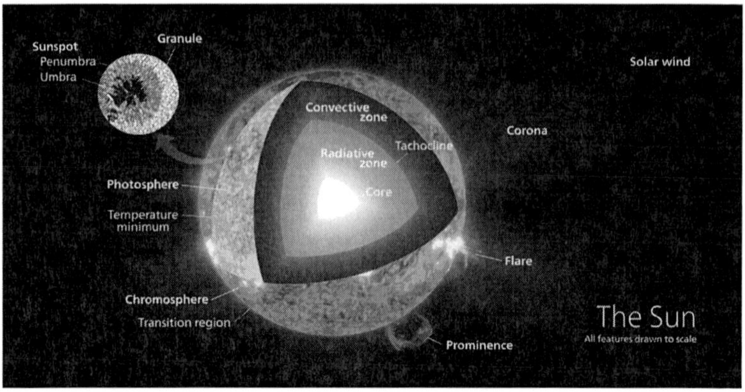

Figure 17: *The structure of the Sun*

Structure and energy production

Core

The core of the Sun extends from the center to about 20–25% of the solar radius. It has a density of up to 150 g/cm^3 (about 150 times the density of water) and a temperature of close to 15.7 million kelvins (K). By contrast, the Sun's surface temperature is approximately 5,800 K. Recent analysis of SOHO mission data favors a faster rotation rate in the core than in the radiative zone above. Through most of the Sun's life, energy has been produced by nuclear fusion in the core region through a series of steps called the p–p (proton–proton) chain; this process converts hydrogen into helium. Only 0.8% of the energy generated in the Sun comes from the CNO cycle, though this proportion is expected to increase as the Sun becomes older.

The core is the only region in the Sun that produces an appreciable amount of thermal energy through fusion; 99% of the power is generated within 24% of the Sun's radius, and by 30% of the radius, fusion has stopped nearly entirely. The remainder of the Sun is heated by this energy as it is transferred outwards through many successive layers, finally to the solar photosphere where it escapes into space as sunlight or the kinetic energy of particles.

The proton–proton chain occurs around 9.2×10^{37} times each second in the core, converting about 3.7×10^{38} protons into alpha particles (helium nuclei) every second (out of a total of $\sim8.9\times10^{56}$ free protons in the Sun), or about 6.2×10^{11} kg/s. Fusing four free protons (hydrogen nuclei) into a single alpha particle (helium nucleus) releases around 0.7% of the fused mass as energy, so the Sun releases energy at the mass–energy conversion rate of 4.26 million metric tons per second (which requires 600 metric megatons of hydrogen), for 384.6 yottawatts (3.846×10^{26} W), or 9.192×10^{10} megatons of TNT per second. Theoretical models of the Sun's interior indicate a power density of approximately 276.5 W/m^3, a value that more nearly approximates that of reptile metabolism or a compost pile than of a thermonuclear bomb.

The fusion rate in the core is in a self-correcting equilibrium: a slightly higher rate of fusion would cause the core to heat up more and expand slightly against the weight of the outer layers, reducing the density and hence the fusion rate and correcting the perturbation; and a slightly lower rate would cause the core to cool and shrink slightly, increasing the density and increasing the fusion rate and again reverting it to its present rate.

Radiative zone

From the core out to about 0.7 solar radii, thermal radiation is the primary means of energy transfer. The temperature drops from approximately 7 million to 2 million kelvins with increasing distance from the core. This temperature gradient is less than the value of the adiabatic lapse rate and hence cannot drive convection, which explains why the transfer of energy through this zone is by radiation instead of thermal convection. Ions of hydrogen and helium emit photons, which travel only a brief distance before being reabsorbed by other ions. The density drops a hundredfold (from 20 g/cm^3 to 0.2 g/cm^3) from 0.25 solar radii to the 0.7 radii, the top of the radiative zone.

Tachocline

The radiative zone and the convective zone are separated by a transition layer, the tachocline. This is a region where the sharp regime change between the uniform rotation of the radiative zone and the differential rotation of the convection zone results in a large shear between the two—a condition where successive horizontal layers slide past one another. Presently, it is hypothesized (see Solar dynamo) that a magnetic dynamo within this layer generates the Sun's magnetic field.

Convective zone

The Sun's convection zone extends from 0.7 solar radii (200,000 km) to near the surface. In this layer, the solar plasma is not dense enough or hot enough to transfer the heat energy of the interior outward via radiation. Instead, the density of the plasma is low enough to allow convective currents to develop and move the Sun's energy outward towards its surface. Material heated at the tachocline picks up heat and expands, thereby reducing its density and allowing it to rise. As a result, an orderly motion of the mass develops into

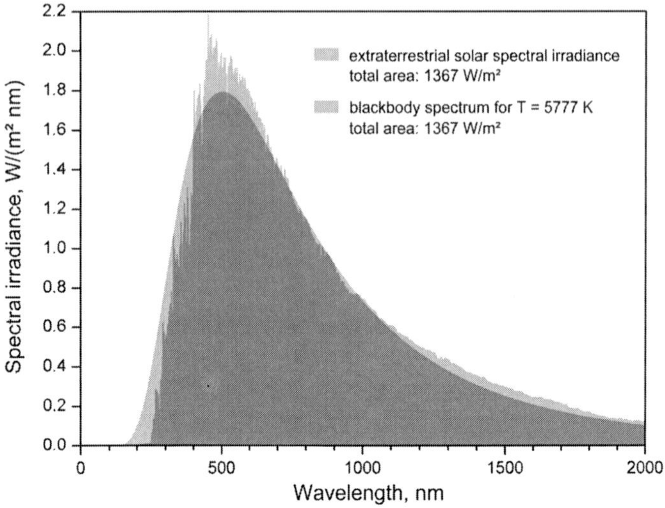

Figure 18: *The effective temperature, or black body temperature, of the Sun (5,777 K) is the temperature a black body of the same size must have to yield the same total emissive power.*

thermal cells that carry the majority of the heat outward to the Sun's photosphere above. Once the material diffusively and radiatively cools just beneath the photospheric surface, its density increases, and it sinks to the base of the convection zone, where it again picks up heat from the top of the radiative zone and the convective cycle continues. At the photosphere, the temperature has dropped to 5,700 K and the density to only 0.2 g/m^3 (about 1/6,000 the density of air at sea level).

The thermal columns of the convection zone form an imprint on the surface of the Sun giving it a granular appearance called the solar granulation at the smallest scale and supergranulation at larger scales. Turbulent convection in this outer part of the solar interior sustains "small-scale" dynamo action over the near-surface volume of the Sun. The Sun's thermal columns are Bénard cells and take the shape of hexagonal prisms.

Photosphere

The visible surface of the Sun, the photosphere, is the layer below which the Sun becomes opaque to visible light. Above the photosphere visible sunlight is free to propagate into space, and almost all of its energy escapes the Sun entirely. The change in opacity is due to the decreasing amount of H^- ions, which absorb visible light easily. Conversely, the visible light we see is produced as electrons react with hydrogen atoms to produce H^- ions. The photosphere is tens to hundreds of kilometers thick, and is slightly less opaque than air on Earth. Because the upper part of the photosphere is cooler than the lower part, an image of the Sun appears brighter in the center than on the edge or *limb* of the solar disk, in a phenomenon known as limb darkening. The spectrum of sunlight has approximately the spectrum of a black-body radiating at about 6,000 K, interspersed with atomic absorption lines from the tenuous layers above the photosphere. The photosphere has a particle density of $\sim 10^{23}$ m^{-3} (about 0.37% of the particle number per volume of Earth's atmosphere at sea level). The photosphere is not fully ionized—the extent of ionization is about 3%, leaving almost all of the hydrogen in atomic form.

During early studies of the optical spectrum of the photosphere, some absorption lines were found that did not correspond to any chemical elements then known on Earth. In 1868, Norman Lockyer hypothesized that these absorption lines were caused by a new element that he dubbed *helium*, after the Greek Sun god Helios. Twenty-five years later, helium was isolated on Earth.

Atmosphere

During a total solar eclipse, when the disk of the Sun is covered by that of the Moon, parts of the Sun's surrounding atmosphere can be seen. It is composed of four distinct parts: the chromosphere, the transition region, the corona and the heliosphere.

The coolest layer of the Sun is a temperature minimum region extending to about 500 km above the photosphere, and has a temperature of about 4,100 K. This part of the Sun is cool enough to allow the existence of simple molecules such as carbon monoxide and water, which can be detected via their absorption spectra.

The chromosphere, transition region, and corona are much hotter than the surface of the Sun. The reason is not well understood, but evidence suggests that Alfvén waves may have enough energy to heat the corona.

Above the temperature minimum layer is a layer about 2,000 km thick, dominated by a spectrum of emission and absorption lines. It is called the *chromosphere* from the Greek root *chroma*, meaning color, because the chromosphere is visible as a colored flash at the beginning and end of total solar eclipses. The temperature of the chromosphere increases gradually with altitude, ranging up to around 20,000 K near the top. In the upper part of the chromosphere helium becomes partially ionized.

Figure 19: *During a total solar eclipse, the solar corona can be seen with the naked eye, during the brief period of totality.*

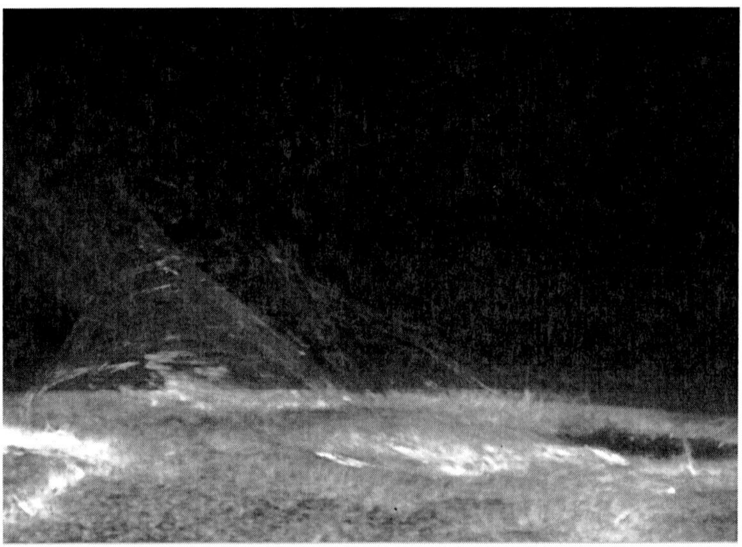

Figure 20: *Taken by Hinode's Solar Optical Telescope on 12 January 2007, this image of the Sun reveals the filamentary nature of the plasma connecting regions of different magnetic polarity.*

Above the chromosphere, in a thin (about 200 km) transition region, the temperature rises rapidly from around 20,000 K in the upper chromosphere to coronal temperatures closer to 1,000,000 K. The temperature increase is facilitated by the full ionization of helium in the transition region, which significantly reduces radiative cooling of the plasma. The transition region does not occur at a well-defined altitude. Rather, it forms a kind of nimbus around chromospheric features such as spicules and filaments, and is in constant, chaotic motion. The transition region is not easily visible from Earth's surface, but is readily observable from space by instruments sensitive to the extreme ultraviolet portion of the spectrum.

The corona is the next layer of the Sun. The low corona, near the surface of the Sun, has a particle density around 10^{15} m^{-3} to 10^{16} m^{-3}. The average temperature of the corona and solar wind is about 1,000,000–2,000,000 K; however, in the hottest regions it is 8,000,000–20,000,000 K. Although no complete theory yet exists to account for the temperature of the corona, at least some of its heat is known to be from magnetic reconnection. The corona is the extended atmosphere of the Sun, which has a volume much larger than the volume enclosed by the Sun's photosphere. A flow of plasma outward from the Sun into interplanetary space is the solar wind.

The heliosphere, the tenuous outermost atmosphere of the Sun, is filled with the solar wind plasma. This outermost layer of the Sun is defined to begin at the distance where the flow of the solar wind becomes *superalfvénic*—that is, where the flow becomes faster than the speed of Alfvén waves, at approximately 20 solar radii (0.1 AU). Turbulence and dynamic forces in the heliosphere cannot affect the shape of the solar corona within, because the information can only travel at the speed of Alfvén waves. The solar wind travels outward continuously through the heliosphere, forming the solar magnetic field into a spiral shape, until it impacts the heliopause more than 50 AU from the Sun. In December 2004, the Voyager 1 probe passed through a shock front that is thought to be part of the heliopause. In late 2012 Voyager 1 recorded a marked increase in cosmic ray collisions and a sharp drop in lower energy particles from the solar wind, which suggested that the probe had passed through the heliopause and entered the interstellar medium.

Photons and neutrinos

High-energy gamma-ray photons initially released with fusion reactions in the core are almost immediately absorbed by the solar plasma of the radiative zone, usually after traveling only a few millimeters. Re-emission happens in a random direction and usually at a slightly lower energy. With this sequence of emissions and absorptions, it takes a long time for radiation to reach the Sun's surface. Estimates of the photon travel time range between 10,000 and

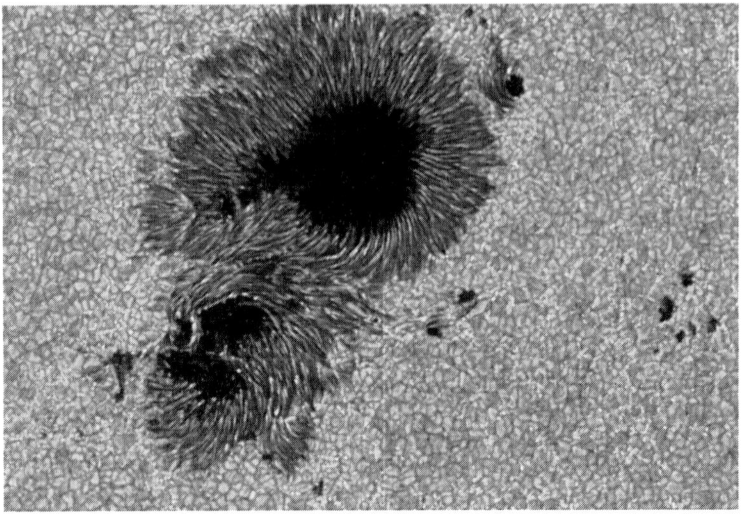

Figure 21: *Visible light photograph of sunspot, 13 December 2006*

170,000 years. In contrast, it takes only 2.3 seconds for the neutrinos, which account for about 2% of the total energy production of the Sun, to reach the surface. Because energy transport in the Sun is a process that involves photons in thermodynamic equilibrium with matter, the time scale of energy transport in the Sun is longer, on the order of 30,000,000 years. This is the time it would take the Sun to return to a stable state, if the rate of energy generation in its core were suddenly changed.

Neutrinos are also released by the fusion reactions in the core, but, unlike photons, they rarely interact with matter, so almost all are able to escape the Sun immediately. For many years measurements of the number of neutrinos produced in the Sun were lower than theories predicted by a factor of 3. This discrepancy was resolved in 2001 through the discovery of the effects of neutrino oscillation: the Sun emits the number of neutrinos predicted by the theory, but neutrino detectors were missing $^2/_3$ of them because the neutrinos had changed flavor by the time they were detected.

Figure 22: *In this false-color ultraviolet image, the Sun shows a C3-class solar flare (white area on upper left), a solar tsunami (wave-like structure, upper right) and multiple filaments of plasma following a magnetic field, rising from the stellar surface.*

Magnetism and activity

Magnetic field

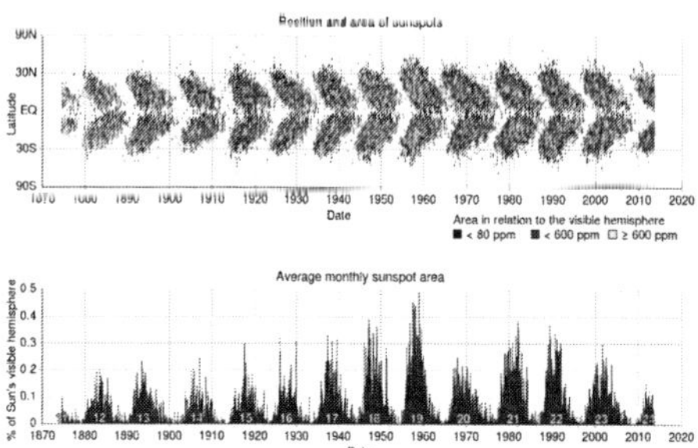

Butterfly diagram showing paired sunspot pattern. Graph is of sunspot area.

The Sun has a magnetic field that varies across the surface of the Sun. Its polar field is 1–2 gauss (0.0001–0.0002 T), whereas the field is typically 3,000 gauss (0.3 T) in features on the Sun called sunspots and 10–100 gauss (0.001–0.01 T) in solar prominences.

Figure 23: *The heliospheric current sheet extends to the outer reaches of the Solar System, and results from the influence of the Sun's rotating magnetic field on the plasma in the interplanetary medium.*

The magnetic field also varies in time and location. The quasi-periodic 11-year solar cycle is the most prominent variation in which the number and size of sunspots waxes and wanes.

Sunspots are visible as dark patches on the Sun's photosphere, and correspond to concentrations of magnetic field where the convective transport of heat is inhibited from the solar interior to the surface. As a result, sunspots are slightly cooler than the surrounding photosphere, and, so, they appear dark. At a typical solar minimum, few sunspots are visible, and occasionally none can be seen at all. Those that do appear are at high solar latitudes. As the solar cycle progresses towards its maximum, sunspots tend form closer to the solar equator, a phenomenon known as Spörer's law. The largest sunspots can be tens of thousands of kilometers across.

An 11-year sunspot cycle is half of a 22-year Babcock–Leighton dynamo cycle, which corresponds to an oscillatory exchange of energy between toroidal and poloidal solar magnetic fields. At solar-cycle maximum, the external poloidal dipolar magnetic field is near its dynamo-cycle minimum strength, but an internal toroidal quadrupolar field, generated through differential rotation within the tachocline, is near its maximum strength. At this point in the dynamo cycle, buoyant upwelling within the convective zone forces emergence of toroidal magnetic field through the photosphere, giving rise to pairs of sunspots, roughly aligned east–west and having footprints with opposite magnetic polarities. The magnetic polarity of sunspot pairs alternates every solar cycle, a phenomenon known as the Hale cycle.

During the solar cycle's declining phase, energy shifts from the internal toroidal magnetic field to the external poloidal field, and sunspots diminish in number and size. At solar-cycle minimum, the toroidal field is, correspondingly, at minimum strength, sunspots are relatively rare, and the poloidal field is at its maximum strength. With the rise of the next 11-year sunspot cycle, differential rotation shifts magnetic energy back from the poloidal to the toroidal field, but with a polarity that is opposite to the previous cycle. The process carries on continuously, and in an idealized, simplified scenario, each 11-year sunspot cycle corresponds to a change, then, in the overall polarity of the Sun's large-scale magnetic field.

The solar magnetic field extends well beyond the Sun itself. The electrically conducting solar wind plasma carries the Sun's magnetic field into space, forming what is called the interplanetary magnetic field. In an approximation known as ideal magnetohydrodynamics, plasma particles only move along the magnetic field lines. As a result, the outward-flowing solar wind stretches the interplanetary magnetic field outward, forcing it into a roughly radial structure. For a simple dipolar solar magnetic field, with opposite hemispherical polarities on either side of the solar magnetic equator, a thin current sheet is formed in the solar wind. At great distances, the rotation of the Sun twists the dipolar magnetic field and corresponding current sheet into an Archimedean spiral structure called the Parker spiral. The interplanetary magnetic field is much stronger than the dipole component of the solar magnetic field. The Sun's dipole magnetic field of 50–400 μT (at the photosphere) reduces with the inverse-cube of the distance to about 0.1 nT at the distance of Earth. However, according to spacecraft observations the interplanetary field at Earth's location is around 5 nT, about a hundred times greater. The difference is due to magnetic fields generated by electrical currents in the plasma surrounding the Sun.

Variation in activity

The Sun's magnetic field leads to many effects that are collectively called solar activity. Solar flares and coronal-mass ejections tend to occur at sunspot groups. Slowly changing high-speed streams of solar wind are emitted from coronal holes at the photospheric surface. Both coronal-mass ejections and high-speed streams of solar wind carry plasma and interplanetary magnetic field outward into the Solar System. The effects of solar activity on Earth

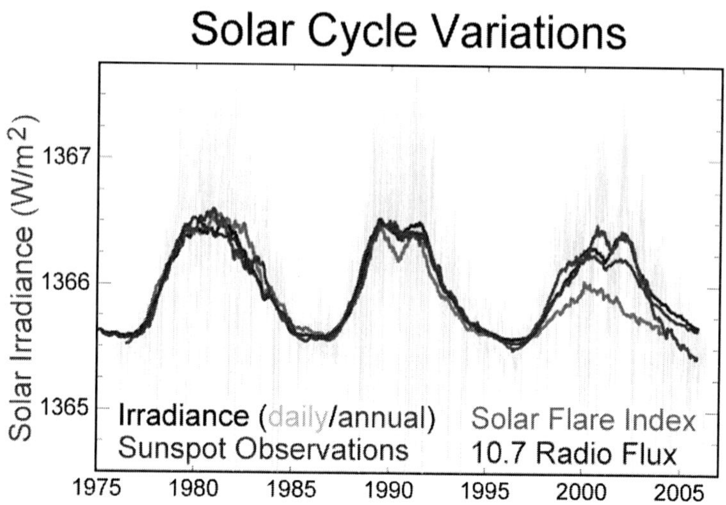

Figure 24: *Measurements of solar cycle variation during the last 30 years*

include auroras at moderate to high latitudes and the disruption of radio communications and electric power. Solar activity is thought to have played a large role in the formation and evolution of the Solar System.

With solar-cycle modulation of sunspot number comes a corresponding modulation of space weather conditions, including those surrounding Earth where technological systems can be affected.

Long-term change

Long-term secular change in sunspot number is thought, by some scientists, to be correlated with long-term change in solar irradiance, which, in turn, might influence Earth's long-term climate. For example, in the 17th century, the solar cycle appeared to have stopped entirely for several decades; few sunspots were observed during a period known as the Maunder minimum. This coincided in time with the era of the Little Ice Age, when Europe experienced unusually cold temperatures. Earlier extended minima have been discovered through analysis of tree rings and appear to have coincided with lower-than-average global temperatures.

A recent theory claims that there are magnetic instabilities in the core of the Sun that cause fluctuations with periods of either 41,000 or 100,000 years. These could provide a better explanation of the ice ages than the Milankovitch cycles.

Life phases

The Sun today is roughly halfway through the most stable part of its life. It has not changed dramatically for over four billion years, and will remain fairly stable for more than five billion more. However, after hydrogen fusion in its core has stopped, the Sun will undergo severe changes, both internally and externally.

Formation

The Sun formed about 4.6 billion years ago from the collapse of part of a giant molecular cloud that consisted mostly of hydrogen and helium and that probably gave birth to many other stars. This age is estimated using computer models of stellar evolution and through nucleocosmochronology. The result is consistent with the radiometric date of the oldest Solar System material, at 4.567 billion years ago. Studies of ancient meteorites reveal traces of stable daughter nuclei of short-lived isotopes, such as iron-60, that form only in exploding, short-lived stars. This indicates that one or more supernovae must have occurred near the location where the Sun formed. A shock wave from a nearby supernova would have triggered the formation of the Sun by compressing the matter within the molecular cloud and causing certain regions to collapse under their own gravity. As one fragment of the cloud collapsed it also began to rotate because of conservation of angular momentum and heat up with the increasing pressure. Much of the mass became concentrated in the center, whereas the rest flattened out into a disk that would become the planets and other Solar System bodies. Gravity and pressure within the core of the cloud generated a lot of heat as it accreted more matter from the surrounding disk, eventually triggering nuclear fusion. Thus, the Sun was born.

Main sequence

The Sun is about halfway through its main-sequence stage, during which nuclear fusion reactions in its core fuse hydrogen into helium. Each second, more than four million tonnes of matter are converted into energy within the Sun's core, producing neutrinos and solar radiation. At this rate, the Sun has so far converted around 100 times the mass of Earth into energy, about 0.03% of the total mass of the Sun. The Sun will spend a total of approximately 10 billion years as a main-sequence star. The Sun is gradually becoming hotter during its time on the main sequence, because the helium atoms in the core occupy less volume than the hydrogen atoms that were fused. The core is therefore shrinking, allowing the outer layers of the Sun to move closer to the centre and experience a stronger gravitational force, according to the inverse-square law. This stronger force increases the pressure on the core, which is resisted

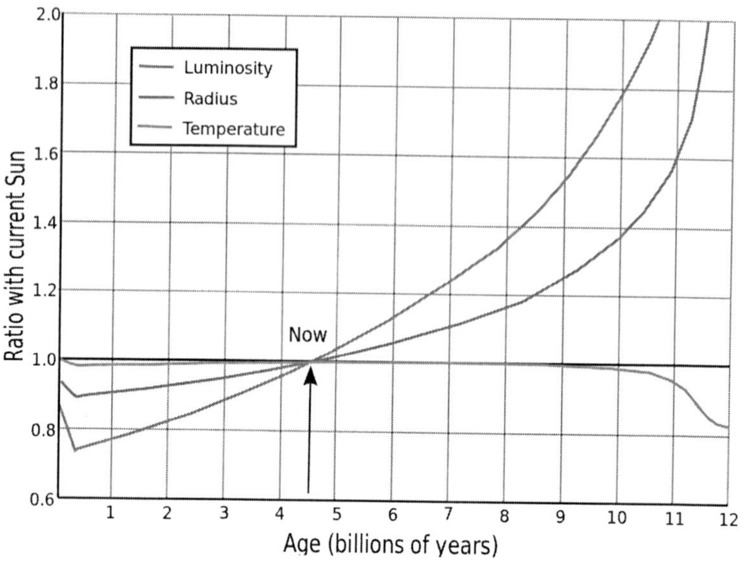

Figure 25: *Evolution of the Sun's luminosity, radius and effective temperature compared to the present Sun. After Ribas (2010)*

by a gradual increase in the rate at which fusion occurs. This process speeds up as the core gradually becomes denser. It is estimated that the Sun has become 30% brighter in the last 4.5 billion years. At present, it is increasing in brightness by about 1% every 100 million years.

After core hydrogen exhaustion

The Sun does not have enough mass to explode as a supernova. Instead it will exit the main sequence in approximately 5 billion years and start to turn into a red giant. As a red giant, the Sun will grow so large that it will engulf Mercury, Venus, and probably Earth.

Even before it becomes a red giant, the luminosity of the Sun will have nearly doubled, and Earth will receive as much sunlight as Venus receives today. Once the core hydrogen is exhausted in 5.4 billion years, the Sun will expand into a subgiant phase and slowly double in size over about half a billion years. It will then expand more rapidly over about half a billion years until it is over two hundred times larger than today and a couple of thousand times more luminous. This then starts the red-giant-branch phase where the Sun will spend around a billion years and lose around a third of its mass.

After the red-giant branch the Sun has approximately 120 million years of active life left, but much happens. First, the core, full of degenerate helium

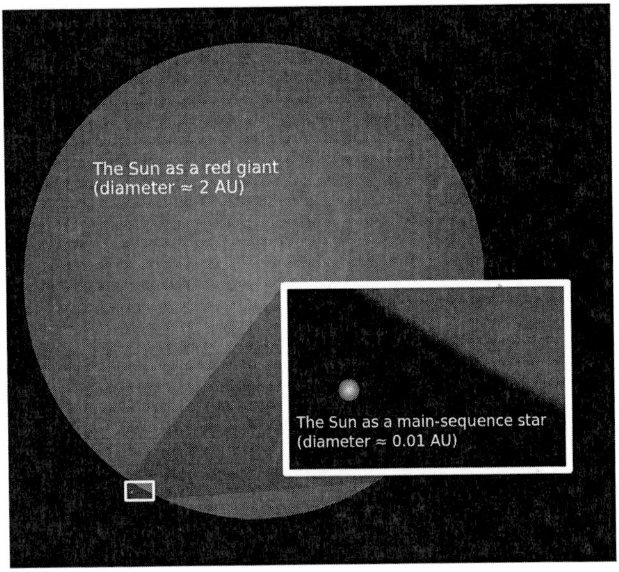

Figure 26: *The size of the current Sun (now in the main sequence) compared to its estimated size during its red-giant phase in the future*

ignites violently in the helium flash, where it is estimated that 6% of the core, itself 40% of the Sun's mass, will be converted into carbon within a matter of minutes through the triple-alpha process. The Sun then shrinks to around 10 times its current size and 50 times the luminosity, with a temperature a little lower than today. It will then have reached the red clump or horizontal branch, but a star of the Sun's mass does not evolve blueward along the horizontal branch. Instead, it just becomes moderately larger and more luminous over about 100 million years as it continues to burn helium in the core.

When the helium is exhausted, the Sun will repeat the expansion it followed when the hydrogen in the core was exhausted, except that this time it all happens faster, and the Sun becomes larger and more luminous. This is the asymptotic-giant-branch phase, and the Sun is alternately burning hydrogen in a shell or helium in a deeper shell. After about 20 million years on the early asymptotic giant branch, the Sun becomes increasingly unstable, with rapid mass loss and thermal pulses that increase the size and luminosity for a few hundred years every 100,000 years or so. The thermal pulses become larger each time, with the later pulses pushing the luminosity to as much as 5,000 times the current level and the radius to over 1 AU. According to a 2008 model, Earth's orbit is shrinking due to tidal forces (and, eventually, drag from the lower chromosphere), so that it will be engulfed by the Sun near the tip of

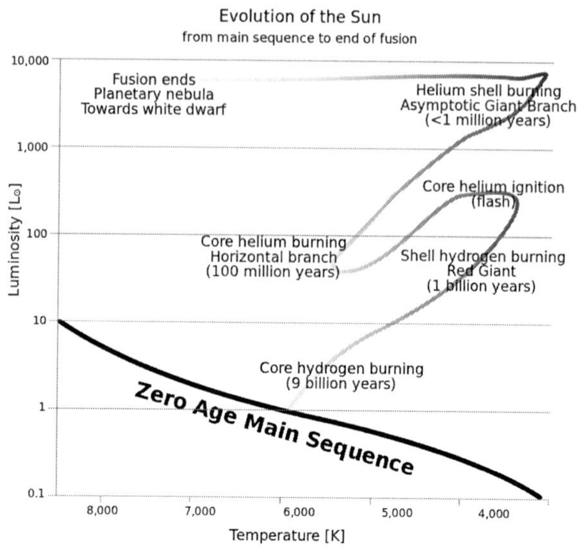

Figure 27: *Evolution of a Sun-like star. The track of a one solar mass star on the Hertzsprung–Russell diagram is shown from the main sequence to the post-asymptotic-giant-branch stage.*

the red giant branch phase, 1 and 3.8 million years after Mercury and Venus have respectively suffered the same fate. Models vary depending on the rate and timing of mass loss. Models that have higher mass loss on the red-giant branch produce smaller, less luminous stars at the tip of the asymptotic giant branch, perhaps only 2,000 times the luminosity and less than 200 times the radius. For the Sun, four thermal pulses are predicted before it completely loses its outer envelope and starts to make a planetary nebula. By the end of that phase – lasting approximately 500,000 years – the Sun will only have about half of its current mass.

The post-asymptotic-giant-branch evolution is even faster. The luminosity stays approximately constant as the temperature increases, with the ejected half of the Sun's mass becoming ionised into a planetary nebula as the exposed core reaches 30,000 K. The final naked core, a white dwarf, will have a temperature of over 100,000 K, and contain an estimated 54.05% of the Sun's present day mass. The planetary nebula will disperse in about 10,000 years, but the white dwarf will survive for trillions of years before fading to a hypothetical black dwarf.

Motion and location

Orbit in Milky Way

The Sun lies close to the inner rim of the Milky Way's Orion Arm, in the Local Interstellar Cloud or the Gould Belt, at a distance of 7.5–8.5 kpc (25,000–28,000 light-years) from the Galactic Center.[31] The Sun is contained within the Local Bubble, a space of rarefied hot gas, possibly produced by the supernova remnant Geminga. The distance between the local arm and the next arm out, the Perseus Arm, is about 6,500 light-years. The Sun, and thus the Solar System, is found in what scientists call the galactic habitable zone. The *Apex of the Sun's Way*, or the solar apex, is the direction that the Sun travels relative to other nearby stars. This motion is towards a point in the constellation Hercules, near the star Vega. Of the 50 nearest stellar systems within 17 light-years from Earth (the closest being the red dwarf Proxima Centauri at approximately 4.2 light-years), the Sun ranks fourth in mass.

The Sun orbits the center of the Milky Way, and it is presently moving in the direction of constellation of Cygnus. The Sun's orbit around the Milky Way is roughly elliptical with orbital perturbations due to the non-uniform mass distribution in Milky Way, such as that in and between the galactic spiral arms. In addition, the Sun oscillates up and down relative to the galactic plane approximately 2.7 times per orbit. It has been argued that the Sun's passage through the higher density spiral arms often coincides with mass extinctions on Earth, perhaps due to increased impact events. It takes the Solar System about 225–250 million years to complete one orbit through the Milky Way (a *galactic year*), so it is thought to have completed 20–25 orbits during the lifetime of the Sun. The orbital speed of the Solar System about the center of the Milky Way is approximately 251 km/s (156 mi/s). At this speed, it takes around 1,190 years for the Solar System to travel a distance of 1 light-year, or 7 days to travel 1 AU.

The Milky Way is moving with respect to the cosmic microwave background radiation (CMB) in the direction of the constellation Hydra with a speed of 550 km/s, and the Sun's resultant velocity with respect to the CMB is about 370 km/s in the direction of Crater or Leo.

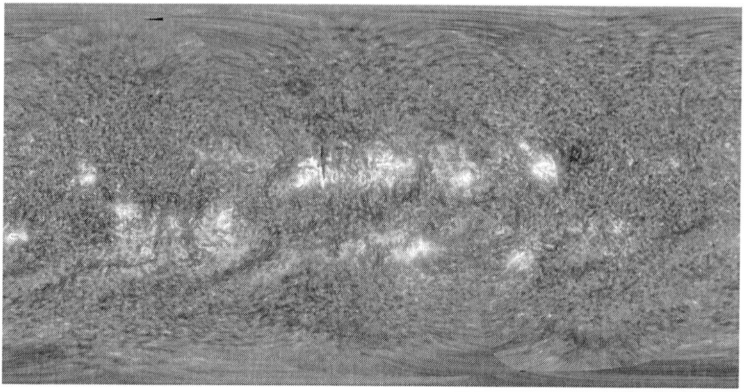

Figure 28: *Map of the full Sun by STEREO and SDO spacecraft*

Theoretical problems

Coronal heating problem

The temperature of the photosphere is approximately 6,000 K, whereas the temperature of the corona reaches 1,000,000–2,000,000 K. The high temperature of the corona shows that it is heated by something other than direct heat conduction from the photosphere.

It is thought that the energy necessary to heat the corona is provided by turbulent motion in the convection zone below the photosphere, and two main mechanisms have been proposed to explain coronal heating. The first is wave heating, in which sound, gravitational or magnetohydrodynamic waves are produced by turbulence in the convection zone. These waves travel upward and dissipate in the corona, depositing their energy in the ambient matter in the form of heat. The other is magnetic heating, in which magnetic energy is continuously built up by photospheric motion and released through magnetic reconnection in the form of large solar flares and myriad similar but smaller events—nanoflares.

Currently, it is unclear whether waves are an efficient heating mechanism. All waves except Alfvén waves have been found to dissipate or refract before reaching the corona. In addition, Alfvén waves do not easily dissipate in the corona. Current research focus has therefore shifted towards flare heating mechanisms.

Faint young Sun problem

Theoretical models of the Sun's development suggest that 3.8 to 2.5 billion years ago, during the Archean eon, the Sun was only about 75% as bright as it is today. Such a weak star would not have been able to sustain liquid water on Earth's surface, and thus life should not have been able to develop. However, the geological record demonstrates that Earth has remained at a fairly constant temperature throughout its history, and that the young Earth was somewhat warmer than it is today. One theory among scientists is that the atmosphere of the young Earth contained much larger quantities of greenhouse gases (such as carbon dioxide, methane) than are present today, which trapped enough heat to compensate for the smaller amount of solar energy reaching it.

However, examination of Archaean sediments appears inconsistent with the hypothesis of high greenhouse concentrations. Instead, the moderate temperature range may be explained by a lower surface albedo brought about by less continental area and the "lack of biologically induced cloud condensation nuclei". This would have led to increased absorption of solar energy, thereby compensating for the lower solar output.

History of observation

The enormous effect of the Sun on Earth has been recognized since prehistoric times, and the Sun has been regarded by some cultures as a deity.

Early understanding

The Sun has been an object of veneration in many cultures throughout human history. Humanity's most fundamental understanding of the Sun is as the luminous disk in the sky, whose presence above the horizon creates day and whose absence causes night. In many prehistoric and ancient cultures, the Sun was thought to be a solar deity or other supernatural entity. Worship of the Sun was central to civilizations such as the ancient Egyptians, the Inca of South America and the Aztecs of what is now Mexico. In religions such as Hinduism, the Sun is still considered a god. Many ancient monuments were constructed with solar phenomena in mind; for example, stone megaliths accurately mark the summer or winter solstice (some of the most prominent megaliths are located in Nabta Playa, Egypt; Mnajdra, Malta and at Stonehenge, England); Newgrange, a prehistoric human-built mount in Ireland, was designed to detect the winter solstice; the pyramid of El Castillo at Chichén Itzá in Mexico is designed to cast shadows in the shape of serpents climbing the pyramid at the vernal and autumnal equinoxes.

Figure 29: *The Trundholm sun chariot pulled by a horse is a sculpture believed to be illustrating an important part of Nordic Bronze Age mythology. The sculpture is probably from around 1350 BC. It is displayed at the National Museum of Denmark.*

The Egyptians portrayed the god Ra as being carried across the sky in a solar barque, accompanied by lesser gods, and to the Greeks, he was Helios, carried by a chariot drawn by fiery horses. From the reign of Elagabalus in the late Roman Empire the Sun's birthday was a holiday celebrated as Sol Invictus (literally "Unconquered Sun") soon after the winter solstice, which may have been an antecedent to Christmas. Regarding the fixed stars, the Sun appears from Earth to revolve once a year along the ecliptic through the zodiac, and so Greek astronomers categorized it as one of the seven planets (Greek *planetes*, "wanderer"); the naming of the days of the weeks after the seven planets dates to the Roman era.

Development of scientific understanding

In the early first millennium BC, Babylonian astronomers observed that the Sun's motion along the ecliptic is not uniform, though they did not know why; it is today known that this is due to the movement of Earth in an elliptic orbit around the Sun, with Earth moving faster when it is nearer to the Sun at perihelion and moving slower when it is farther away at aphelion.

One of the first people to offer a scientific or philosophical explanation for the Sun was the Greek philosopher Anaxagoras. He reasoned that it was not the chariot of Helios, but instead a giant flaming ball of metal even larger than the land of the Peloponnesus and that the Moon reflected the light of the Sun. For teaching this heresy, he was imprisoned by the authorities and sentenced to death, though he was later released through the intervention of Pericles. Eratosthenes estimated the distance between Earth and the Sun in the 3rd century BC as "of stadia myriads 400 and 80000", the translation of which is ambiguous, implying either 4,080,000 stadia (755,000 km) or 804,000,000 stadia (148 to 153 million kilometers or 0.99 to 1.02 AU); the latter value is correct to within a few percent. In the 1st century AD, Ptolemy estimated the distance as 1,210 times the radius of Earth, approximately 7.71 million kilometers (0.0515 AU).

The theory that the Sun is the center around which the planets orbit was first proposed by the ancient Greek Aristarchus of Samos in the 3rd century BC, and later adopted by Seleucus of Seleucia (see Heliocentrism). This view was developed in a more detailed mathematical model of a heliocentric system in the 16th century by Nicolaus Copernicus.

Observations of sunspots were recorded during the Han Dynasty (206 BC–AD 220) by Chinese astronomers, who maintained records of these observations for centuries. Averroes also provided a description of sunspots in the 12th century. The invention of the telescope in the early 17th century permitted detailed observations of sunspots by Thomas Harriot, Galileo Galilei and other astronomers. Galileo posited that sunspots were on the surface of the Sun rather than small objects passing between Earth and the Sun.

Arabic astronomical contributions include Albatenius' discovery that the direction of the Sun's apogee (the place in the Sun's orbit against the fixed stars where it seems to be moving slowest) is changing.[32] (In modern heliocentric terms, this is caused by a gradual motion of the aphelion of the *Earth's* orbit). Ibn Yunus observed more than 10,000 entries for the Sun's position for many years using a large astrolabe.[33]

From an observation of a transit of Venus in 1032, the Persian astronomer and polymath Avicenna concluded that Venus is closer to Earth than the Sun. In 1672 Giovanni Cassini and Jean Richer determined the distance to Mars and were thereby able to calculate the distance to the Sun.

In 1666, Isaac Newton observed the Sun's light using a prism, and showed that it is made up of light of many colors. In 1800, William Herschel discovered infrared radiation beyond the red part of the solar spectrum. The 19th century saw advancement in spectroscopic studies of the Sun; Joseph von Fraunhofer recorded more than 600 absorption lines in the spectrum, the strongest

Figure 30: *Sol, the Sun, from a 1550 edition of Guido Bonatti's Liber astronomiae.*

of which are still often referred to as Fraunhofer lines. In the early years of the modern scientific era, the source of the Sun's energy was a significant puzzle. Lord Kelvin suggested that the Sun is a gradually cooling liquid body that is radiating an internal store of heat. Kelvin and Hermann von Helmholtz then proposed a gravitational contraction mechanism to explain the energy output, but the resulting age estimate was only 20 million years, well short of the time span of at least 300 million years suggested by some geological discoveries of that time. In 1890 Joseph Lockyer, who discovered helium in the solar spectrum, proposed a meteoritic hypothesis for the formation and evolution of the Sun.

Not until 1904 was a documented solution offered. Ernest Rutherford suggested that the Sun's output could be maintained by an internal source of heat, and suggested radioactive decay as the source. However, it would be Albert Einstein who would provide the essential clue to the source of the Sun's energy output with his mass-energy equivalence relation $E = mc^2$. In 1920, Sir Arthur Eddington proposed that the pressures and temperatures at the core of the Sun could produce a nuclear fusion reaction that merged hydrogen (protons) into helium nuclei, resulting in a production of energy from the net change in mass. The preponderance of hydrogen in the Sun was confirmed in 1925 by Cecilia Payne using the ionization theory developed by Meghnad Saha, an Indian physicist. The theoretical concept of fusion was developed in the 1930s

Figure 31: *The Sun giving out a large geomagnetic storm on 1:29 pm, EST, 13 March 2012*

by the astrophysicists Subrahmanyan Chandrasekhar and Hans Bethe. Hans Bethe calculated the details of the two main energy-producing nuclear reactions that power the Sun. In 1957, Margaret Burbidge, Geoffrey Burbidge, William Fowler and Fred Hoyle showed that most of the elements in the universe have been synthesized by nuclear reactions inside stars, some like the Sun.

Solar space missions

The first satellites designed to observe the Sun were NASA's Pioneers 5, 6, 7, 8 and 9, which were launched between 1959 and 1968. These probes orbited the Sun at a distance similar to that of Earth, and made the first detailed measurements of the solar wind and the solar magnetic field. Pioneer 9 operated for a particularly long time, transmitting data until May 1983.

In the 1970s, two Helios spacecraft and the Skylab Apollo Telescope Mount provided scientists with significant new data on solar wind and the solar corona. The Helios 1 and 2 probes were U.S.–German collaborations that studied the solar wind from an orbit carrying the spacecraft inside Mercury's orbit at perihelion. The Skylab space station, launched by NASA in 1973, included a solar observatory module called the Apollo Telescope Mount that

Figure 32: *A lunar transit of the Sun captured during calibration of STEREO B's ultraviolet imaging cameras*

was operated by astronauts resident on the station. Skylab made the first time-resolved observations of the solar transition region and of ultraviolet emissions from the solar corona. Discoveries included the first observations of coronal mass ejections, then called "coronal transients", and of coronal holes, now known to be intimately associated with the solar wind.

In 1980, the Solar Maximum Mission was launched by NASA. This spacecraft was designed to observe gamma rays, X-rays and UV radiation from solar flares during a time of high solar activity and solar luminosity. Just a few months after launch, however, an electronics failure caused the probe to go into standby mode, and it spent the next three years in this inactive state. In 1984 Space Shuttle *Challenger* mission STS-41C retrieved the satellite and repaired its electronics before re-releasing it into orbit. The Solar Maximum Mission subsequently acquired thousands of images of the solar corona before re-entering Earth's atmosphere in June 1989.

Launched in 1991, Japan's Yohkoh (*Sunbeam*) satellite observed solar flares at X-ray wavelengths. Mission data allowed scientists to identify several different types of flares, and demonstrated that the corona away from regions of peak activity was much more dynamic and active than had previously been supposed. Yohkoh observed an entire solar cycle but went into standby mode

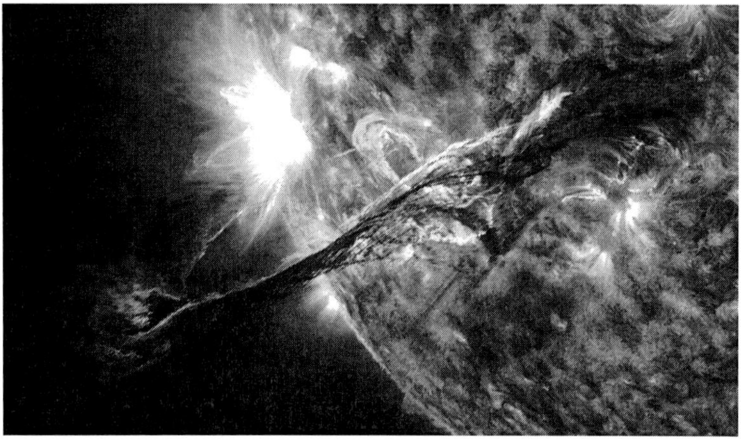

Figure 33: *A solar prominence erupts in August 2012, as captured by SDO*

when an annular eclipse in 2001 caused it to lose its lock on the Sun. It was destroyed by atmospheric re-entry in 2005.

One of the most important solar missions to date has been the Solar and Heliospheric Observatory, jointly built by the European Space Agency and NASA and launched on 2 December 1995. Originally intended to serve a two-year mission, a mission extension through 2012 was approved in October 2009. It has proven so useful that a follow-on mission, the Solar Dynamics Observatory (SDO), was launched in February 2010. Situated at the Lagrangian point between Earth and the Sun (at which the gravitational pull from both is equal), SOHO has provided a constant view of the Sun at many wavelengths since its launch. Besides its direct solar observation, SOHO has enabled the discovery of a large number of comets, mostly tiny sungrazing comets that incinerate as they pass the Sun.

All these satellites have observed the Sun from the plane of the ecliptic, and so have only observed its equatorial regions in detail. The Ulysses probe was launched in 1990 to study the Sun's polar regions. It first travelled to Jupiter, to "slingshot" into an orbit that would take it far above the plane of the ecliptic. Once Ulysses was in its scheduled orbit, it began observing the solar wind and magnetic field strength at high solar latitudes, finding that the solar wind from high latitudes was moving at about 750 km/s, which was slower than expected, and that there were large magnetic waves emerging from high latitudes that scattered galactic cosmic rays.

Figure 34: *During certain atmospheric conditions, the Sun becomes clearly visible to the naked eye, and can be observed without stress to the eyes. Click on this photo to see the full cycle of a sunset, as observed from the high plains of the Mojave Desert.*

Elemental abundances in the photosphere are well known from spectroscopic studies, but the composition of the interior of the Sun is more poorly understood. A solar wind sample return mission, Genesis, was designed to allow astronomers to directly measure the composition of solar material.

The Solar Terrestrial Relations Observatory (STEREO) mission was launched in October 2006. Two identical spacecraft were launched into orbits that cause them to (respectively) pull further ahead of and fall gradually behind Earth. This enables stereoscopic imaging of the Sun and solar phenomena, such as coronal mass ejections.

The Indian Space Research Organisation has scheduled the launch of a 100 kg satellite named Aditya for 2017–18. Its main instrument will be a coronagraph for studying the dynamics of the Solar corona.

Observation and effects

The brightness of the Sun can cause pain from looking at it with the naked eye; however, doing so for brief periods is not hazardous for normal non-dilated eyes. Looking directly at the Sun causes phosphene visual artifacts and temporary partial blindness. It also delivers about 4 milliwatts of sunlight to the retina, slightly heating it and potentially causing damage in eyes that cannot

Figure 35: *The Sun, as seen from low Earth orbit overlooking the International Space Station. This sunlight is not filtered by the lower atmosphere, which blocks much of the solar spectrum*

respond properly to the brightness. UV exposure gradually yellows the lens of the eye over a period of years, and is thought to contribute to the formation of cataracts, but this depends on general exposure to solar UV, and not whether one looks directly at the Sun.[34] Long-duration viewing of the direct Sun with the naked eye can begin to cause UV-induced, sunburn-like lesions on the retina after about 100 seconds, particularly under conditions where the UV light from the Sun is intense and well focused; conditions are worsened by young eyes or new lens implants (which admit more UV than aging natural eyes), Sun angles near the zenith, and observing locations at high altitude.

Viewing the Sun through light-concentrating optics such as binoculars may result in permanent damage to the retina without an appropriate filter that blocks UV and substantially dims the sunlight. When using an attenuating filter to view the Sun, the viewer is cautioned to use a filter specifically designed for that use. Some improvised filters that pass UV or IR rays, can actually harm the eye at high brightness levels. Herschel wedges, also called Solar Diagonals, are effective and inexpensive for small telescopes. The sunlight that is destined for the eyepiece is reflected from an unsilvered surface of a piece of glass. Only a very small fraction of the incident light is reflected. The rest passes through the glass and leaves the instrument. If the glass breaks because

Figure 36: *Halo with sun dogs*

of the heat, no light at all is reflected, making the device fail-safe. Simple filters made of darkened glass allow the full intensity of sunlight to pass through if they break, endangering the observer's eyesight. Unfiltered binoculars can deliver hundreds of times as much energy as using the naked eye, possibly causing immediate damage. It is claimed that even brief glances at the midday Sun through an unfiltered telescope can cause permanent damage.

Partial solar eclipses are hazardous to view because the eye's pupil is not adapted to the unusually high visual contrast: the pupil dilates according to the total amount of light in the field of view, *not* by the brightest object in the field. During partial eclipses most sunlight is blocked by the Moon passing in front of the Sun, but the uncovered parts of the photosphere have the same surface brightness as during a normal day. In the overall gloom, the pupil expands from $\sim$2 mm to $\sim$6 mm, and each retinal cell exposed to the solar image receives up to ten times more light than it would looking at the non-eclipsed Sun. This can damage or kill those cells, resulting in small permanent blind spots for the viewer. The hazard is insidious for inexperienced observers and for children, because there is no perception of pain: it is not immediately obvious that one's vision is being destroyed.

During sunrise and sunset, sunlight is attenuated because of Rayleigh scattering and Mie scattering from a particularly long passage through Earth's atmosphere, and the Sun is sometimes faint enough to be viewed comfortably with

Figure 37: *A sunrise*

the naked eye or safely with optics (provided there is no risk of bright sunlight suddenly appearing through a break between clouds). Hazy conditions, atmospheric dust, and high humidity contribute to this atmospheric attenuation.

An optical phenomenon, known as a green flash, can sometimes be seen shortly after sunset or before sunrise. The flash is caused by light from the Sun just below the horizon being bent (usually through a temperature inversion) towards the observer. Light of shorter wavelengths (violet, blue, green) is bent more than that of longer wavelengths (yellow, orange, red) but the violet and blue light is scattered more, leaving light that is perceived as green.

Ultraviolet light from the Sun has antiseptic properties and can be used to sanitize tools and water. It also causes sunburn, and has other biological effects such as the production of vitamin D and sun tanning. Ultraviolet light is strongly attenuated by Earth's ozone layer, so that the amount of UV varies greatly with latitude and has been partially responsible for many biological adaptations, including variations in human skin color in different regions of the globe.

Planetary system

The Sun has eight known planets. This includes four terrestrial planets (Mercury, Venus, Earth, and Mars), two gas giants (Jupiter and Saturn), and two ice giants (Uranus and Neptune). The Solar System also has at least five dwarf planets, an asteroid belt, numerous comets, and a large number of icy bodies which lie beyond the orbit of Neptune.

Notes

Further reading

- Cohen, Richard (2010). *Chasing the Sun: the Epic Story of the Star that Gives us Life*. Simon & Schuster. ISBN 1-4000-6875-4.
- Thompson, M. J. (2004). "Solar interior: Helioseismology and the Sun's interior". *Astronomy & Geophysics*. **45** (4): 21–25.
- Solar Activity[35] Scholarpedia Hugh Hudson 3(3):3967. doi: 10.4249/scholarpedia.3967[36]

External links

- Nasa SOHO (Solar & Heliospheric Observatory) satellite[37]
- National Solar Observatory[38]
- Astronomy Cast: The Sun[39]
- A collection of spectacular images of the Sun from various institutions[40] (*The Boston Globe*)
- Satellite observations of solar luminosity[41]
- Sun|Trek, an educational website about the Sun[42]
- The Swedish 1-meter Solar Telescope, SST[43]
- An animated explanation of the structure of the Sun[44] (University of Glamorgan)
- Animation – The Future of the Sun[45]
- Solar Conveyor Belt Speeds Up[46] – NASA – images, link to report on Science
- NASA 5-year timelapse video of the Sun[47]
- Sun in Ultra High Definition[48] NASA 11 January 2015

Interplanetary medium

The **interplanetary medium** is the material which fills the Solar System, and through which all the larger Solar System bodies, such as planets, dwarf planets, asteroids, and comets, move.

Composition and physical characteristics

The interplanetary medium includes interplanetary dust, cosmic rays and hot plasma from the solar wind. The temperature of the interplanetary medium varies. For dust particles within the asteroid belt, typical temperatures range from 200 K (–73 °C) at 2.2 AU down to 165 K (–108 °C) at 3.2 AU. The density of the interplanetary medium is very low, about 5 particles per cubic centimeter in the vicinity of the EarthWikipedia:Citation needed; it decreases with increasing distance from the Sun, in inverse proportion to the square of the distance. It is variable, and may be affected by magnetic fields and events such as coronal mass ejections. It may rise to as high as 100 particles/cm^3.

Since the interplanetary medium is a plasma, it has the characteristics of a plasma, rather than a simple gas; for example, it carries with it the Sun's magnetic field, is highly electrically conductive (resulting in the heliospheric current sheet), forms plasma double layers where it comes into contact with a planetary magnetosphere or at the heliopause, and exhibits filamentation (such as in aurorae).

The plasma in the interplanetary medium is also responsible for the strength of the Sun's magnetic field at the orbit of the Earth being over 100 times greater than originally anticipated. If space were a vacuum, then the Sun's 10^{-4} tesla magnetic dipole field would reduce with the cube of the distance to about 10^{-11} tesla. But satellite observations show that it is about 100 times greater at around 10^{-9} tesla. Magnetohydrodynamic (MHD) theory predicts that the motion of a conducting fluid (e.g., the interplanetary medium) in a magnetic field, induces electric currents which in turn generate magnetic fields, and in this respect it behaves like an MHD dynamo.

Extent of the interplanetary medium

The outer edge of the Solar System is the boundary between the flow of the solar wind and the interstellar medium. This boundary is known as the heliopause and is believed to be a fairly sharp transition of the order of 110 to 160 astronomical units from the Sun. The interplanetary medium thus fills the roughly spherical volume contained within the heliopause.

Figure 38: *The heliospheric current sheet results from the influence of the Sun's rotating magnetic field on the plasma in the interplanetary medium*[49]

Interaction with planets

How the interplanetary medium interacts with planets depends on whether they have magnetic fields or not. Bodies such as the Moon have no magnetic field and the solar wind can impact directly on their surface. Over billions of years, the lunar regolith has acted as a collector for solar wind particles, and so studies of rocks from the Moon's surface can be valuable in studies of the solar wind.

High energy particles from the solar wind impacting on the Moon's surface also cause it to emit faintly at X-ray wavelengths.

Planets with their own magnetic field, such as the Earth and Jupiter, are surrounded by a magnetosphere within which their magnetic field is dominant over the Sun's. This disrupts the flow of the solar wind, which is channelled around the magnetosphere. Material from the solar wind can 'leak' into the magnetosphere, causing aurorae and also populating the Van Allen Belts with ionised material.

Observable phenomena of the interplanetary medium

The interplanetary medium is responsible for several effects which can be seen from Earth. The Zodiacal light is a broad band of faint light sometimes seen after sunset and before sunrise, stretched along the ecliptic and brightest near the horizon. It is caused by sunlight scattering off dust particles in the interplanetary medium between the Earth and the Sun.

A similar effect is the Gegenschein, which is seen directly opposite to the Sun's position in the sky. It is much fainter than the Zodiacal light, and is caused by sunlight reflecting off dust particles outside the Earth's orbit.

History

The term "interplanetary" appears to have been first used in print in 1691 by the scientist Robert Boyle: "The air is different from the æther (or vacuum) in the... interplanetary spaces" Boyle *Hist. Air*.

The notion that space is considered to be a vacuum filled with an "aether", or just a cold, dark vacuum continued up until the 1950s (see below).

In 1898, American astronomer Charles Augustus Young wrote: "Interplanetary space is a vacuum, far more perfect than anything we can produce by artificial means..." (*The Elements of Astronomy*, Charles Augustus Young, 1898).

And Akasofu recounted that: "The view that interplanetary space is a vacuum into which the Sun intermittently emitted corpuscular streams was changed radically by Ludwig Biermann (1951, 1953) who proposed on the basis of comet tails, that the Sun continuously blows its atmosphere out in all directions at supersonic speed" (Syun-Ichi Akasofu, *Exploring the Secrets of the Aurora*, 2002)

Tufts University Professor of astronomy, Kenneth R. Lang, writing in 2000 noted, "Half a century ago, most people visualized our planet as a solitary sphere traveling in a cold, dark vacuum of space around the Sun".[50]

External links

- Bill Arnett's *The Nine Planets* page about the interplanetary medium[51]

Solar wind

The **solar wind** is a stream of charged particles released from the upper atmosphere of the Sun, called the corona. This plasma consists of mostly electrons, protons and alpha particles with thermal energies between 1.5 and 10 keV. Embedded within the solar-wind plasma is the interplanetary magnetic field. The solar wind varies in density, temperature and speed over time and over solar latitude and longitude. Its particles can escape the Sun's gravity because of their high energy resulting from the high temperature of the corona, which in turn is a result of the coronal magnetic field.

At a distance of more than a few solar radii from the Sun, the solar wind is supersonic and reaches speeds of 250 to 750 kilometers per second.[52] The flow of the solar wind is no longer supersonic at the termination shock. The Voyager 2 spacecraft crossed the shock more than five times between 30 August and 10 December 2007. Voyager 2 crossed the shock about a billion kilometers closer to the Sun than the 13.5 billion kilometer distance where Voyager 1 came upon the termination shock. The spacecraft moved outward through the termination shock into the heliosheath and onward toward the interstellar medium. Other related phenomena include the aurora (northern and southern lights), the plasma tails of comets that always point away from the Sun, and geomagnetic storms that can change the direction of magnetic field lines.

History

The existence of particles flowing outward from the Sun to the Earth was first suggested by British astronomer Richard C. Carrington. In 1859, Carrington and Richard Hodgson independently made the first observation of what would later be called a solar flare. This is a sudden, localised increase in brightness on the solar disc, which is now known to often occur in conjunction with an episodic ejection of material and magnetic flux from the Sun's atmosphere, known as a coronal mass ejection. On the following day, a geomagnetic storm was observed, and Carrington suspected that there might be a connection, which is now attributed to the arrival of the coronal mass ejection in near-Earth space and its subsequent interaction with the Earth's magnetosphere. George FitzGerald later suggested that matter was being regularly accelerated away from the Sun and was reaching the Earth after several days.

In 1910 British astrophysicist Arthur Eddington essentially suggested the existence of the solar wind, without naming it, in a footnote to an article on Comet Morehouse. The idea never fully caught on even though Eddington had also made a similar suggestion at a Royal Institution address the previous year. In

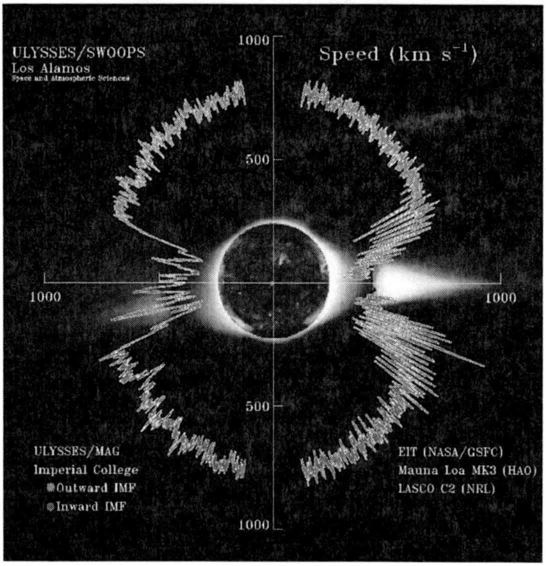

Figure 39: *Ulysses observations of solar wind speed as a function of helio latitude during solar minimum. Slow wind (~400 km/s) is confined to the equatorial regions, while fast wind (~750 km/s) is seen over the poles. Red/blue colours show inward/outward polarities of the heliospheric magnetic field.*

the latter case, he postulated that the ejected material consisted of electrons while in his study of Comet Morehouse he supposed them to be ions.

The first person to suggest that the ejected material consisted of both ions and electrons was Kristian Birkeland. His geomagnetic surveys showed that auroral activity was nearly uninterrupted. As these displays and other geomagnetic activity were being produced by particles from the Sun, he concluded that the Earth was being continually bombarded by "rays of electric corpuscles emitted by the Sun". In 1916, Birkeland proposed that, "From a physical point of view it is most probable that solar rays are neither exclusively negative nor positive rays, but of both kinds". In other words, the solar wind consists of both negative electrons and positive ions.[53] Three years later in 1919, Frederick Lindemann also suggested that particles of both polarities, protons as well as electrons, come from the Sun.[54]

Around the 1930s, scientists had determined that the temperature of the solar corona must be a million degrees Celsius because of the way it stood out into space (as seen during total eclipses). Later spectroscopic work confirmed this extraordinary temperature. In the mid-1950s Sydney Chapman calculated the

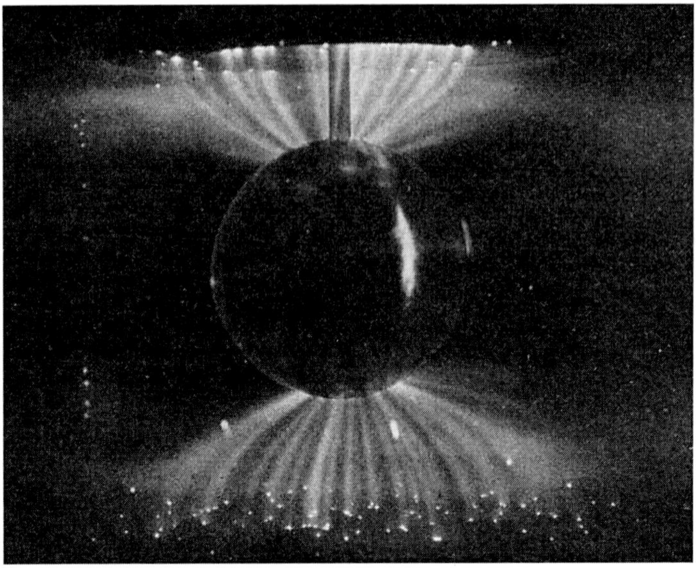

Figure 40: *Laboratory simulation of the magnetosphere's influence on the Solar Wind; these auroral-like Birkeland currents were created in a terrella, a magnetised anode globe in an evacuated chamber.*

properties of a gas at such a temperature and determined it was such a superb conductor of heat that it must extend way out into space, beyond the orbit of Earth. Also in the 1950s, Ludwig Biermann became interested in the fact that no matter whether a comet is headed towards or away from the Sun, its tail always points away from the Sun. Biermann postulated that this happens because the Sun emits a steady stream of particles that pushes the comet's tail away. Wilfried Schröder claimed that Paul Ahnert was the first to relate solar wind to comet tail direction based on observations of the comet Whipple-Fedke (1942g).

Eugene Parker realised that the heat flowing from the Sun in Chapman's model and the comet tail blowing away from the Sun in Biermann's hypothesis had to be the result of the same phenomenon, which he termed the "solar wind". Parker showed in 1958 that even though the Sun's corona is strongly attracted by solar gravity, it is such a good heat conductor that it is still very hot at large distances. Since gravity weakens as distance from the Sun increases, the outer coronal atmosphere escapes supersonically into interstellar space. Furthermore, Parker was the first person to notice that the weakening effect of the gravity has the same effect on hydrodynamic flow as a de Laval nozzle: it incites a transition from subsonic to supersonic flow.

Opposition to Parker's hypothesis on the solar wind was strong. The paper he submitted to the Astrophysical Journal in 1958 was rejected by two reviewers. It was saved by the editor Subrahmanyan Chandrasekhar (who later received the 1983 Nobel Prize in physics).

In January 1959, the Soviet satellite Luna 1 first directly observed the solar wind and measured its strength.[55,56] They were detected by hemispherical ion traps. The discovery, made by Konstantin Gringauz, was verified by Luna 2, Luna 3 and by the more distant measurements of Venera 1. Three years later its measurement was performed by Neugebauer and collaborators using the Mariner 2 spacecraft.

In the late 1990s the Ultraviolet Coronal Spectrometer (UVCS) instrument on board the SOHO spacecraft observed the acceleration region of the fast solar wind emanating from the poles of the Sun and found that the wind accelerates much faster than can be accounted for by thermodynamic expansion alone. Parker's model predicted that the wind should make the transition to supersonic flow at an altitude of about 4 solar radii from the photosphere (surface); but the transition (or "sonic point") now appears to be much lower, perhaps only 1 solar radius above the photosphere, suggesting that some additional mechanism accelerates the solar wind away from the Sun. The acceleration of the fast wind is still not understood and cannot be fully explained by Parker's theory. The gravitational and electromagnetic explanation for this acceleration is, however, detailed in an earlier paper by 1970 Nobel laureate for Physics, Hannes Alfvén.

The first numerical simulation of the solar wind in the solar corona including closed and open field lines was performed by Pneuman and Kopp in 1971. The magnetohydrodynamics equations in steady state were solved iteratively starting with an initial dipolar configuration.

In 1990, the Ulysses probe was launched to study the solar wind from high solar latitudes. All prior observations had been made at or near the Solar System's ecliptic plane.

Acceleration

While early models of the solar wind relied primarily on thermal energy to accelerate the material, by the 1960s it was clear that thermal acceleration alone cannot account for the high speed of solar wind. An additional unknown acceleration mechanism is required and likely relates to magnetic fields in the solar atmosphere.

The Sun's corona, or extended outer layer, is a region of plasma that is heated to over a million kelvin. As a result of thermal collisions, the particles within the inner corona have a range and distribution of speeds described by a Maxwellian distribution. The mean velocity of these particles is about 145 km/s, which is well below the solar escape velocity of 618 km/s. However, a few of the particles achieve energies sufficient to reach the terminal velocity of 400 km/s, which allows them to feed the solar wind. At the same temperature, electrons, due to their much smaller mass, reach escape velocity and build up an electric field that further accelerates ions away from the Sun.

The total number of particles carried away from the Sun by the solar wind is about 1.3×10^{36} per second. Thus, the total mass loss each year is about $(2–3)\times10^{-14}$ solar masses, or about one billion kilograms per second. This is equivalent to losing a mass equal to the Earth every 150 million years. However, only about 0.01% of the Sun's total mass has been lost through the solar wind. Other stars have much stronger stellar winds that result in significantly higher mass loss rates.

Properties and structure

Fast and slow solar wind

The solar wind is observed to exist in two fundamental states, termed the slow solar wind and the fast solar wind, though their differences extend well beyond their speeds. In near-Earth space, the slow solar wind is observed to have a velocity of 300–500 km/s, a temperature of $1.4–1.6\times10^{6}$ K and a composition that is a close match to the corona. By contrast, the fast solar wind has a typical velocity of 750 km/s, a temperature of 8×10^{5} K and it nearly matches the composition of the Sun's photosphere. The slow solar wind is twice as dense and more variable in nature than the fast solar wind.

The slow solar wind appears to originate from a region around the Sun's equatorial belt that is known as the "streamer belt", where coronal streamers are produced by magnetic flux open to the heliosphere draping over closed magnetic loops. The exact coronal structures involved in slow solar wind formation and the method by which the material is released is still under debate. Observations of the Sun between 1996 and 2001 showed that emission of the slow solar wind occurred at latitudes up to 30–35° during the solar minimum (the period of lowest solar activity), then expanded toward the poles as the solar cycle approached maximum. At solar maximum, the poles were also emitting a slow solar wind.

The fast solar wind originates from coronal holes,[57] which are funnel-like regions of open field lines in the Sun's magnetic field. Such open lines are particularly prevalent around the Sun's magnetic poles. The plasma source is small magnetic fields created by convection cells in the solar atmosphere. These fields confine the plasma and transport it into the narrow necks of the coronal funnels, which are located only 20,000 kilometers above the photosphere. The plasma is released into the funnel when these magnetic field lines reconnect.

Pressure

The wind exerts a pressure at 1 AU typically in the range of 1–6 nPa ($1\text{–}6\times10^{-9}$ N/m^2), although it can readily vary outside that range.

The dynamic pressure is a function of wind speed and density. The formula is

$$P = 1.6726\times10^{-6} * n * V^2$$

where pressure P is in nPa (nanopascals), n is the density in particles/cm^3 and V is the speed in km/s of the solar wind.

Coronal mass ejection

Both the fast and slow solar wind can be interrupted by large, fast-moving bursts of plasma called interplanetary coronal mass ejections, or ICMEs. ICMEs are the interplanetary manifestation of solar coronal mass ejections, which are caused by release of magnetic energy at the Sun. CMEs are often called "solar storms" or "space storms" in the popular media. They are sometimes, but not always, associated with solar flares, which are another manifestation of magnetic energy release at the Sun. ICMEs cause shock waves in the thin plasma of the heliosphere, launching electromagnetic waves and accelerating particles (mostly protons and electrons) to form showers of ionizing radiation that precede the CME.

When a CME impacts the Earth's magnetosphere, it temporarily deforms the Earth's magnetic field, changing the direction of compass needles and inducing large electrical ground currents in Earth itself; this is called a geomagnetic storm and it is a global phenomenon. CME impacts can induce magnetic reconnection in Earth's magnetotail (the midnight side of the magnetosphere); this launches protons and electrons downward toward Earth's atmosphere, where they form the aurora.

ICMEs are not the only cause of space weather. Different patches on the Sun are known to give rise to slightly different speeds and densities of wind depending on local conditions. In isolation, each of these different wind streams would form a spiral with a slightly different angle, with fast-moving streams moving out more directly and slow-moving streams wrapping more around the Sun. Fast moving streams tend to overtake slower streams that originate westward of them on the Sun, forming turbulent co-rotating interaction regions that give rise to wave motions and accelerated particles, and that affect Earth's magnetosphere in the same way as, but more gently than, CMEs.

Figure 41: *The heliospheric current sheet results from the influence of the Sun's rotating magnetic field on the plasma in the solar wind.*

Solar System effects

Over the Sun's lifetime, the interaction of its surface layers with the escaping solar wind has significantly decreased its surface rotation rate. The wind is considered responsible for comets' tails, along with the Sun's radiation. The solar wind contributes to fluctuations in celestial radio waves observed on the Earth, through an effect called interplanetary scintillation.

Magnetospheres

Where the solar wind intersects with a planet that has a well-developed magnetic field (such as Earth, Jupiter and Saturn), the particles are deflected by the Lorentz force. This region, known as the magnetosphere, causes the particles to travel around the planet rather than bombarding the atmosphere or surface. The magnetosphere is roughly shaped like a hemisphere on the side facing the Sun, then is drawn out in a long wake on the opposite side. The boundary of this region is called the magnetopause, and some of the particles are able to penetrate the magnetosphere through this region by partial reconnection of the magnetic field lines.

The solar wind is responsible for the overall shape of Earth's magnetosphere. Fluctuations in its speed, density, direction, and entrained magnetic field

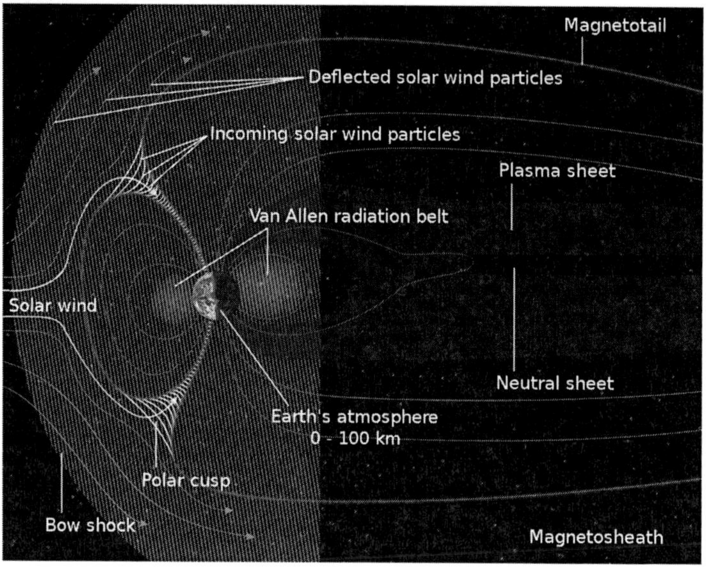

Figure 42: *Schematic of Earth's magnetosphere. The solar wind flows from left to right.*

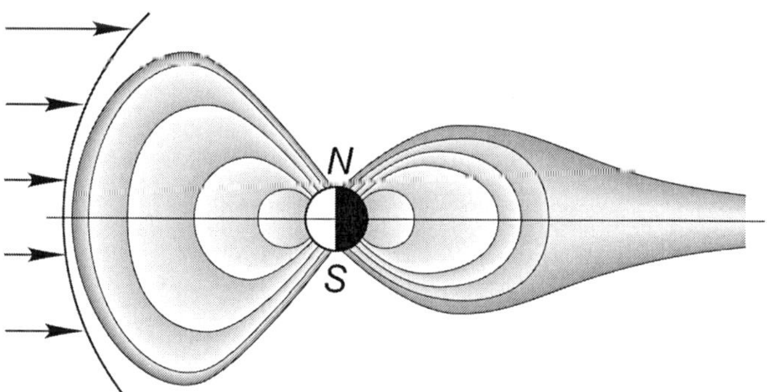

Figure 43: *Noon meridian section of magnetosphere.*

strongly affect Earth's local space environment. For example, the levels of ionizing radiation and radio interference can vary by factors of hundreds to thousands; and the shape and location of the magnetopause and bow shock wave upstream of it can change by several Earth radii, exposing geosynchronous satellites to the direct solar wind. These phenomena are collectively called space weather.

From the European Space Agency's Cluster mission, a new study has taken place that proposes that it is easier for the solar wind to infiltrate the magnetosphere than previously believed. A group of scientists directly observed the existence of certain waves in the solar wind that were not expected. A recent study shows that these waves enable incoming charged particles of solar wind to breach the magnetopause. This suggests that the magnetic bubble forms more as a filter than a continuous barrier. This latest discovery occurred through the distinctive arrangement of the four identical Cluster spacecraft, which fly in a controlled configuration through near-Earth space. As they sweep from the magnetosphere into interplanetary space and back again, the fleet provides exceptional three-dimensional insights on the phenomena that connect the sun to Earth.

The research characterized variances in formation of the interplanetary magnetic field (IMF) largely influenced by Kelvin-Helmholtz waves (which occur at the interface of two fluids) as a result of differences in thickness and numerous other characteristics of the boundary layer. Experts believe that this was the first occasion that the appearance of Kelvin-Helmholtz waves at the magnetopause had been displayed at high latitude dawnward orientation of the IMF. These waves are being seen in unforeseen places under solar wind conditions that were formerly believed to be undesired for their generation. These discoveries show how Earth's magnetosphere can be penetrated by solar particles under specific IMF circumstances. The findings are also relevant to studies of magnetospheric progressions around other planetary bodies. This study suggests that Kelvin-Helmholtz waves can be a somewhat common, and possibly constant, instrument for the entrance of solar wind into terrestrial magnetospheres under various IMF orientations.[58]

Atmospheres

The solar wind affects other incoming cosmic rays interacting with planetary atmospheres. Moreover, planets with a weak or non-existent magnetosphere are subject to atmospheric stripping by the solar wind.

Venus, the nearest and most similar planet to Earth, has 100 times denser atmosphere, with little or no geo-magnetic field. Space probes discovered a comet-like tail that extends to Earth's orbit.

Earth itself is largely protected from the solar wind by its magnetic field, which deflects most of the charged particles; however some of the charged particles are trapped in the Van Allen radiation belt. A smaller number of particles from the solar wind manage to travel, as though on an electromagnetic energy transmission line, to the Earth's upper atmosphere and ionosphere in the auroral zones. The only time the solar wind is observable on the Earth is when it is strong enough to produce phenomena such as the aurora and geomagnetic storms. Bright auroras strongly heat the ionosphere, causing its plasma to expand into the magnetosphere, increasing the size of the plasma geosphere and injecting atmospheric matter into the solar wind. Geomagnetic storms result when the pressure of plasmas contained inside the magnetosphere is sufficiently large to inflate and thereby distort the geomagnetic field.

Although Mars is larger than Mercury and four times farther from the Sun, it is thought that the solar wind has stripped away up to a third of its original atmosphere, leaving a layer 1/100th as dense as the Earth's. It is believed the mechanism for this atmospheric stripping is gas caught in bubbles of magnetic field, which are ripped off by solar winds. In 2015 the NASA Mars Atmosphere and Volatile Evolution (MAVEN) mission measured the rate of atmospheric stripping caused by the magnetic field carried by the solar wind as it flows past Mars, which generates an electric field, much as a turbine on Earth can be used to generate electricity. This electric field accelerates electrically charged gas atoms, called ions, in Mars' upper atmosphere and shoots them into space. The MAVEN mission measured the rate of atmospheric stripping at about 100 grams ($\sim$1/4 lb) per second.

Moons and planetary surfaces

Mercury, the nearest planet to the Sun, bears the full brunt of the solar wind, and since its atmosphere is vestigial and transient, its surface is bathed in radiation.

Mercury has an intrinsic magnetic field, so under normal solar wind conditions, the solar wind cannot penetrate its magnetosphere and particles only reach the surface in the cusp regions. During coronal mass ejections, however, the magnetopause may get pressed into the surface of the planet, and under these conditions, the solar wind may interact freely with the planetary surface.

The Earth's Moon has no atmosphere or intrinsic magnetic field, and consequently its surface is bombarded with the full solar wind. The Project Apollo missions deployed passive aluminum collectors in an attempt to sample the solar wind, and lunar soil returned for study confirmed that the lunar regolith is enriched in atomic nuclei deposited from the solar wind. These elements may prove useful resources for lunar colonies.

Figure 44: *Apollo's SWC experiment*

Outer limits

The solar wind "blows a bubble" in the interstellar medium (the rarefied hydrogen and helium gas that permeates the galaxy). The point where the solar wind's strength is no longer great enough to push back the interstellar medium is known as the heliopause and is often considered to be the outer border of the Solar System. The distance to the heliopause is not precisely known and probably depends on the current velocity of the solar wind and the local density of the interstellar medium, but it is far outside Pluto's orbit. Scientists hope to gain perspective on the heliopause from data acquired through the Interstellar Boundary Explorer (IBEX) mission, launched in October 2008.

Notable events

- From May 10 to May 12, 1999, NASA's Advanced Composition Explorer (ACE) and WIND spacecraft observed a 98% decrease of solar wind density. This allowed energetic electrons from the Sun to flow to Earth in narrow beams known as "strahl", which caused a highly unusual "polar rain" event, in which a visible aurora appeared over the North Pole. In addition, Earth's magnetosphere increased to between 5 and 6 times its normal size.

- On 13 December 2010, Voyager 1 determined that the velocity of the solar wind, at its location 10.8 billion miles from Earth had slowed to zero. "We have gotten to the point where the wind from the Sun, which until now has always had an outward motion, is no longer moving outward; it is only moving sideways so that it can end up going down the tail of the heliosphere, which is a comet-shaped-like object," said Voyager project scientist Edward Stone.

References

Fox, Karen C. (2012) "NASA Study Using Cluster Reveals New Insights Into Solar Wind" NASA.

S.Cuperman and N. Metzler, Role of fluctuations in the interplanetary magnetic field on the heat conduction in the Solar Wind.J.Geophys. Res. 78 (16), 3167-3168, 1973.

S. Cuperman and N. Metzler. Astrophys. J., 182 (3), 961-975, 1973.

S. Cuperman and N. Metzler, Solution of 3-fluid model equations with anomalous transport coefficients for thequiet Solar Wind. Astrophys.J., 196 (1) 205-219, 1975

S. Cuperman, N. Metzler and M. Spygelglass, Confirmation of known numerical solutions for the quiet Solar Wind equations. Astrophys. J., 198 (3), 755-759, 1975.

S.Cuperman and N. Metzler, Relative magnitude of streaming velocities of alpha particles and protons at 1AU. Astrophys. and Space Sci. 45 (2) 411-417,1976.

N. Metzler. A multi-fluid model for stellar winds.Proceedings of the L.D.de Feiter Memorial Symposium on the Study of Traveling Interplanetary Phenomena. AFGL-TR-77-0309, Air Force Systems Command, USAF, 1978.

N. Metzler and M. Dryer, A self-consistent solution of the three-fluid model of the Solar Wind. Astrophys. J., 222 (2), 689-695, 1978.

S. Cuperman and N. Metzler, Comments on Acceleration of Solar Wind He++3 effects of Resonant and nonresonant interactions with transverse waves. J. Geophys. Res. 84 (NA5), 2139-2140 (1979)

N. Metzler, S. Cuperman, M. Dryer and P. Rosenau, A time-dependent two-fluid model with thermal conduction for Solar Wind. Astrophys. J., 231 (3) 960-976, 1979.

External links

Wikimedia Commons has media related to ***Solar wind***.

Wikisourcehas original text related to this article:
Solar-wind interaction with planetary ionospheres

- Real-time plots of solar wind activity from the[59] Advanced Composition Explorer
- Current solar wind pressure[60] NOAA Space Weather Prediction Center
- Solar wind[61] MSFC solar wind site
- Sun|trek website[62] An educational resource for teachers and students about the Sun and its effect on the Earth
- Cluster shows how solar wind is heated[63]
- Could solar wind power Earth?[64] October 4, 2010 by Miranda Marquit
- Solar Or Wind Power? Why Not Both?[65] By Eric Bland, Wed September 29, 2010
- The Solar Wind Power Satellite as an alternative to a traditional Dyson Sphere and its implications for remote detection[66] International Journal of Astrobiology (2010), 9: 89-99

Terrestrial planet

A **terrestrial planet**, **telluric planet**, or **rocky planet** is a planet that is composed primarily of silicate rocks or metals. Within the Solar System, the terrestrial planets are the inner planets closest to the Sun, i.e. Mercury, Venus, Earth, and Mars. The terms "terrestrial planet" and "telluric planet" are derived from Latin words for Earth (*Terra* and *Tellus*), as these planets are, in terms of structure, "Earth-like".

Terrestrial planets have a solid planetary surface, making them substantially different from the larger giant planets, which are composed mostly of some combination of hydrogen, helium, and water existing in various physical states.

Figure 45: *The terrestrial planets of the Solar System: Mercury, Venus, Earth, and Mars, sized to scale*

Structure

All terrestrial planets in the Solar System have the same basic type of structure, such as a central metallic core, mostly iron, with a surrounding silicate mantle. The Moon is similar, but has a much smaller iron core. Io and Europa are also satellites that have internal structures similar to that of terrestrial planets. Terrestrial planets can have canyons, craters, mountains, volcanoes, and other surface structures, depending on the presence of water and tectonic activity. Terrestrial planets have secondary atmospheres, generated through volcanism or comet impacts, in contrast to the giant planets, whose atmospheres are primary, captured directly from the original solar nebula.

Solar System's terrestrial planets

The Solar System has four terrestrial planets: Mercury, Venus, Earth, and Mars. Only one terrestrial planet, Earth, is known to have an active hydrosphere.

During the formation of the Solar System, there were probably many more terrestrial planetesimals, but most merged with or were ejected by the four terrestrial planets.

Dwarf planets, such as Ceres, Pluto and Eris, and small Solar System bodies are similar to terrestrial planets in the fact that they do have a solid surface, but are, on average, composed of more icy materials (Ceres, Pluto and Eris have densities of 2.17, 1.87 and 2.52 $g{\cdot}cm^{-3}$, respectively, and Haumea's density is similar to Pallas's 2.8 $g{\cdot}cm^{-3}$). The Earth's Moon has a density of 3.4 $g{\cdot}cm^{-3}$ and Jupiter's satellites, Io, 3.528 and Europa, 3.013 $g{\cdot}cm^{-3}$; other satellites typically have densities less than 2 $g{\cdot}cm^{-3}$.Wikipedia:Citation needed

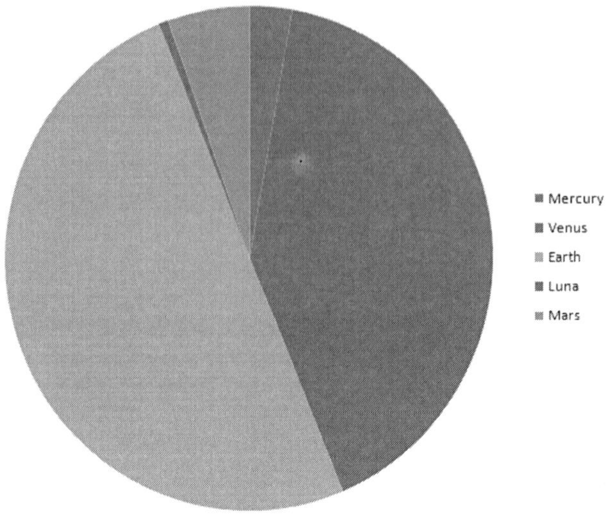

Figure 46: *Relative masses of the terrestrial planets of the Solar System, including the Moon (designated here as "Luna")*

Density trends

The uncompressed density of a terrestrial planet is the average density its materials would have at zero pressure. A greater uncompressed density indicates greater metal content. Uncompressed density differs from the true average density because compression within planet cores increases their density; the average density depends on planet size as well as composition.

Densities of the terrestrial planets

Object	Density (g·cm^{-3})		Semi-major axis (AU)
	Mean	Uncompressed	
Mercury	5.4	5.3	0.39
Venus	5.2	4.4	0.72
Earth	5.5	4.4	1.0
Mars	3.9	3.8	1.5

The uncompressed density of terrestrial planets trends towards lower values as the distance from the Sun increases. The rocky minor planet Vesta orbiting outside of Mars is less dense than Mars still, at 3.4 g·cm^{-3}.

It is unknown whether extrasolar terrestrial planets in general will show to follow this trend.

Extrasolar terrestrial planets

Most of the planets discovered outside the Solar System are giant planets, because they are more easily detectable.[67,68,69] But since 2005, hundreds of potentially terrestrial extrasolar planets have been found, with several being confirmed as terrestrial. Most of these are super-Earths, i.e. planets with masses between Earth's and Neptune's; super-Earths may be gas planets or terrestrial, depending on their mass and other parameters.

During the early 1990s, the first extrasolar planets were discovered orbiting the pulsar PSR B1257+12, with masses of 0.02, 4.3, and 3.9 times that of Earth's, by pulsar timing.

When 51 Pegasi b, the first planet found around a star still undergoing fusion, was discovered, many astronomers assumed it to be a gigantic terrestrial,Wikipedia:Citation needed because it was assumed no gas giant could exist as close to its star (0.052 AU) as 51 Pegasi b did. It was later found to be a gas giant.

In 2005, the first planets around main-sequence stars that may be terrestrial were found: Gliese 876 d, has a mass 7 to 9 times that of Earth and an orbital period of just two Earth days. It orbits the red dwarf Gliese 876, 15 light years from Earth. OGLE-2005-BLG-390Lb, about 5.5 times the mass of Earth, orbits a star about 21,000 light years away in the constellation Scorpius. From 2007 to 2010, three (possibly four) potential terrestrial planets were found orbiting within the Gliese 581 planetary system. The smallest, Gliese 581e, is only about 1.9 Earth mass, but orbits very close to the star. An ideal terrestrial planet would be 2 Earth masses with a 25-day orbital period around a red dwarf. Two others, Gliese 581c and Gliese 581d, as well as a disputed planet, Gliese 581g, are more-massive super-Earths orbiting in or close to the habitable zone of the star, so they could potentially be habitable, with Earth-like temperatures.

Another possibly terrestrial planet, HD 85512 b, was discovered in 2011; it has at least 3.6 times the mass of Earth. The radius and composition of all these planets are unknown.

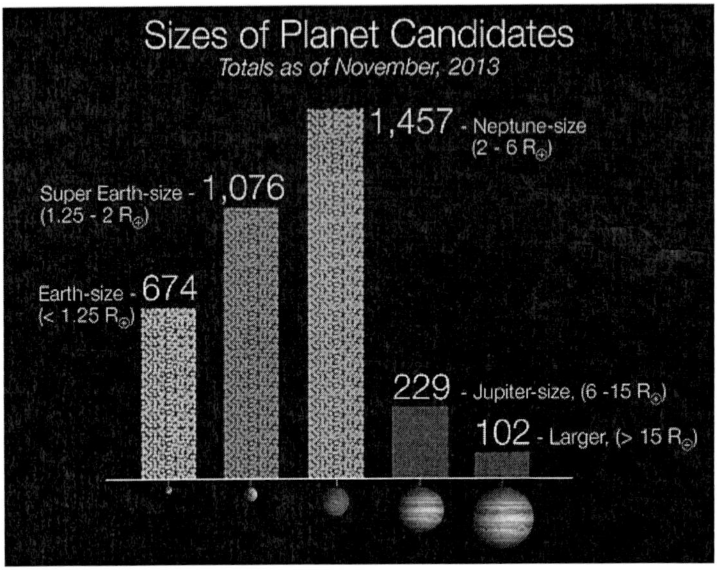

Figure 47: *Sizes of Kepler planet candidates based on 2,740 candidates orbiting 2,036 stars as of November 4, 2013 (NASA).*

The first confirmed terrestrial exoplanet, Kepler-10b, was found in 2011 by the Kepler Mission, specifically designed to discover Earth-size planets around other stars using the transit method.[70]

In the same year, the Kepler Space Observatory Mission team released a list of 1235 extrasolar planet candidates, including six that are "Earth-size" or "super-Earth-size" (i.e. they have a radius less than 2 Earth radii)[71] and in the habitable zone of their star. Since then, Kepler has discovered hundreds of planets ranging from Moon-sized to super-Earths, with many more candidates in this size range (see image).

List of terrestrial exoplanets

The following exoplanets have a density of at least 5 g/cm^3 and a mass below Neptune's and are thus very likely terrestrial:

Kepler-10b, Kepler-20b, Kepler-36b, Kepler-48d, Kepler 68c, Kepler-78b, Kepler-89b, Kepler-93b, Kepler-97b, Kepler-99b, Kepler-100b, Kepler-101c, Kepler-102b, Kepler-102d, Kepler-113b, Kepler-131b, Kepler-131c, Kepler-138c, Kepler-406b, Kepler-406c, Kepler-409b.

The Neptune-mass planet Kepler-10c also has a density >5 g/cm^3 and is thus very likely terrestrial.

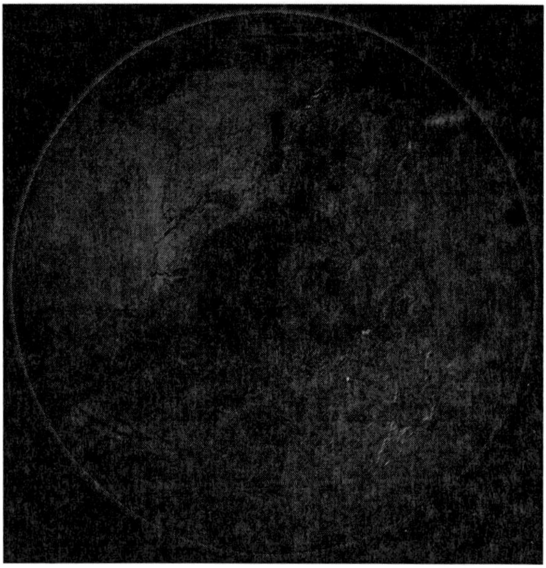

Figure 48: *Artist's impression of a carbon planet*

Frequency

In 2013, astronomers reported, based on *Kepler* space mission data, that there could be as many as 40 billion Earth- and super-Earth-sized planets orbiting in the habitable zones of Sun-like stars and red dwarfs within the Milky Way. 11 billion of these estimated planets may be orbiting Sun-like stars. The nearest such planet may be 12 light-years away, according to the scientists. However, this does not give estimates for the number of extrasolar terrestrial planets, because there are planets as small as Earth that have been shown to be gas planets (see KOI-314c).[72]

Types

Several possible classifications for terrestrial planets have been proposed:

Silicate planet

The standard type of terrestrial planet seen in the Solar System, made primarily of silicon-based rocky mantle with a metallic (iron) core.

Iron planet

A theoretical type of terrestrial planet that consists almost entirely of iron and therefore has a greater density and a smaller radius than other terrestrial planets of comparable mass. Mercury in the Solar System has a metallic core equal to 60–70% of its planetary mass. Iron planets are thought to form in the high-temperature regions close to a star, like Mercury, and if the protoplanetary disk is rich in iron.

Coreless planet

A theoretical type of terrestrial planet that consists of silicate rock but has no metallic core, i.e. the opposite of an iron planet. Although the Solar System contains no coreless planets, chondrite asteroids and meteorites are common in the Solar System. Coreless planets are thought to form farther from the star where volatile oxidizing material is more common.

Carbon planet (also called ”diamond planet”)

A theoretical class of planets, composed of a metal core surrounded by primarily carbon-based minerals. They may be considered a type of terrestrial planet if the metal content dominates. The Solar System contains no carbon planets, but does have carbonaceous asteroids.

References

Mercury (planet)

<indicator name=”pp-default”> </indicator> <indicator name=”featured-star”> </indicator>

Mercury

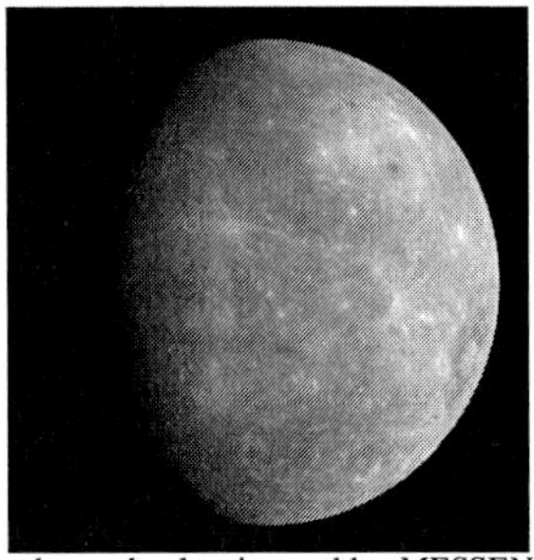

Mercury in enhanced color, imaged by *MESSENGER* (2008)

Designations	
Pronunciation	/ˈmɜːrkjəri/ 🔊 Media:en-us-Mercury.ogg
Adjectives	Mercurian, mercurial
Orbital characteristics	
Epoch J2000	
Aphelion	• 0.466 697 AU • 69,816,900 km
Perihelion	• 0.307 499 AU • 46,001,200 km
Semi-major axis	• 0.387 098 AU • 57,909,050 km
Eccentricity	0.205 630
Orbital period	• 87.969 1 d • 0.240 846 yr • 0.5 Mercury solar day
Synodic period	115.88 d
Average orbital speed	47.362 km/s
Mean anomaly	174.796°
Inclination	• 7.005° to ecliptic • 3.38° to Sun's equator • 6.34° to invariable plane
Longitude of ascend-ing node	48.331°
Argument of perihelion	29.124°
Satellites	None
Physical characteristics	

Mean radius	• 2,439.7±1.0 km • 0.3829 Earths
Flattening	0
Surface area	• 7.48×10^7 km^2 • 0.147 Earths
Volume	• 6.083×10^{10} km^3 • 0.056 Earths
Mass	• 3.3011×10^{23} kg • 0.055 Earths
Mean density	5.427 g/cm^3
Surface gravity	• 3.7 m/s^2 • 0.38 g
Moment of inertia factor	0.346±0.014
Escape velocity	4.25 km/s
Sidereal rotation period	• 58.646 d • 1407.5 h
Equatorial rotation velocity	10.892 km/h (3.026 m/s)
Axial tilt	2.04′ ± 0.08′ (to orbit) (0.034°)
North pole right ascension	• $18^h 44^m 2^s$ • 281.01°
North pole declination	61.45°
Albedo	• 0.068 (Bond) • 0.142 (geom.)

Surface temp.	**min**	**mean**	**max**
0°N, 0°W	100 K	340 K	700 K
85°N, 0°W	80 K	200 K	380 K

Apparent magnitude	–2.6 to 5.7
Angular diameter	4.5–13”
	Atmosphere
Surface pressure	trace (≲ 0.5 nPa)
Composition by volume	• 42% molecular oxygen • 29.0% sodium • 22.0% hydrogen • 6.0% helium • 0.5% potassium • Trace amounts of argon, nitrogen, carbon dioxide, water vapor, xenon, krypton, and neon

Mercury is the smallest and innermost planet in the Solar System. Its orbital period around the Sun of 88 days is the shortest of all the planets in the Solar System. It is named after the Roman deity Mercury, the messenger to the gods.

Like Venus, Mercury orbits the Sun within Earth's orbit as an *inferior planet*, so it can only be seen visually in the morning or the evening sky, and never exceeds 28° away from the Sun. Also, like Venus and the Moon, the planet displays the complete range of phases as it moves around its orbit relative to Earth. Seen from Earth, this cycle of phases reoccurs approximately every 116 days, the so-called synodic period. Although Mercury can appear as a bright star-like object when viewed from Earth, its proximity to the Sun often makes it more difficult to see than Venus.

Mercury is gravitationally locked with the Sun in a 3:2 spin-orbit resonance,[73] and rotates in a way that is unique in the Solar System. As seen relative to the fixed stars, it rotates on its axis exactly three times for every two revolutions it makes around the Sun. As seen from the Sun, in a frame of reference that rotates with the orbital motion, it appears to rotate only once every two Mercurian years. An observer on Mercury would therefore see only one day every two years.

Mercury's axis has the smallest tilt of any of the Solar System's planets (about $^1/_{30}$ degree), and its orbital eccentricity is the largest of all known planets in the Solar System. At aphelion, Mercury is about 1.5 times as far from the Sun as it is at perihelion. Mercury's surface appears heavily cratered and is similar in appearance to the Moon's, indicating that it has been geologically inactive for billions of years. Having almost no atmosphere to retain heat, it has surface temperatures that vary diurnally more than on any other planet in the Solar System, ranging from 100 K (–173 °C; –280 °F) at night to 700 K (427 °C; 800 °F) during the day across the equatorial regions. The polar regions are constantly below 180 K (–93 °C; –136 °F). The planet has no known natural satellites.

Two spacecraft have visited Mercury: *Mariner 10* flew by in 1974 and 1975; and *MESSENGER*, launched in 2004, orbited Mercury over 4,000 times in four years before exhausting its fuel and crashing into the planet's surface on April 30, 2015.

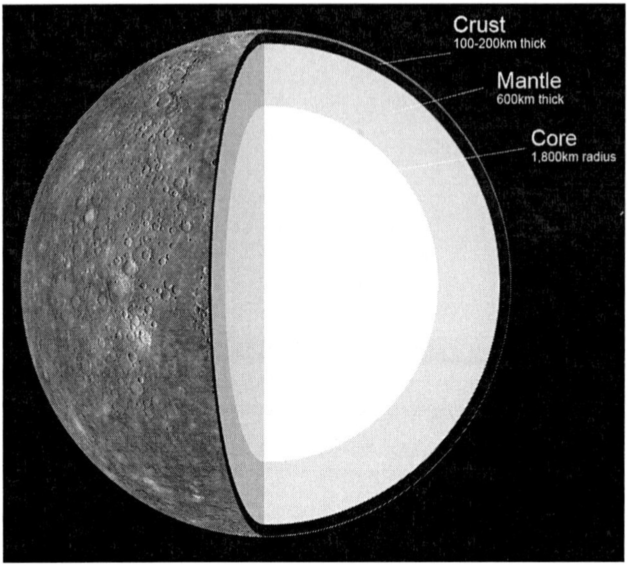

Figure 49:
Internal structure of Mercury:

1. *Crust: 100–300 km thick*
2. *Mantle: 600 km thick*
3. *Core: 1,800 km radius*

Physical characteristics

Internal structure

Mercury is one of four terrestrial planets in the Solar System, and is a rocky body like Earth. It is the smallest planet in the Solar System, with an equatorial radius of 2,439.7 kilometres (1,516.0 mi). Mercury is also smaller—albeit more massive—than the largest natural satellites in the Solar System, Ganymede and Titan. Mercury consists of approximately 70% metallic and 30% silicate material. Mercury's density is the second highest in the Solar System at 5.427 g/cm^3, only slightly less than Earth's density of 5.515 g/cm^3. If the effect of gravitational compression were to be factored out from both planets, the materials of which Mercury is made would be denser than those of Earth, with an uncompressed density of 5.3 g/cm^3 versus Earth's 4.4 g/cm^3.

Mercury's density can be used to infer details of its inner structure. Although Earth's high density results appreciably from gravitational compression, particularly at the core, Mercury is much smaller and its inner regions are not as

Figure 50: *Gravity anomalies on Mercury—mass concentrations (red) suggest subsurface structure and evolution*

compressed. Therefore, for it to have such a high density, its core must be large and rich in iron.

Geologists estimate that Mercury's core occupies about 55% of its volume; for Earth this proportion is 17%. Research published in 2007 suggests that Mercury has a molten core. Surrounding the core is a 500–700 km mantle consisting of silicates.[74] Based on data from the Mariner 10 mission and Earth-based observation, Mercury's crust is estimated to be 35 km thick. One distinctive feature of Mercury's surface is the presence of numerous narrow ridges, extending up to several hundred kilometers in length. It is thought that these were formed as Mercury's core and mantle cooled and contracted at a time when the crust had already solidified.

Mercury's core has a higher iron content than that of any other major planet in the Solar System, and several theories have been proposed to explain this. The most widely accepted theory is that Mercury originally had a metal–silicate ratio similar to common chondrite meteorites, thought to be typical of the Solar System's rocky matter, and a mass approximately 2.25 times its current mass. Early in the Solar System's history, Mercury may have been struck by a planetesimal of approximately 1/6 that mass and several thousand kilometers across. The impact would have stripped away much of the original crust and mantle, leaving the core behind as a relatively major component. A similar process, known as the giant impact hypothesis, has been proposed to explain the formation of the Moon.

Alternatively, Mercury may have formed from the solar nebula before the Sun's energy output had stabilized. It would initially have had twice its present mass, but as the protosun contracted, temperatures near Mercury could have been between 2,500 and 3,500 K and possibly even as high as 10,000 K. Much of Mercury's surface rock could have been vaporized at such temperatures, forming an atmosphere of "rock vapor" that could have been carried away by the solar wind.

A third hypothesis proposes that the solar nebula caused drag on the particles from which Mercury was accreting, which meant that lighter particles were lost from the accreting material and not gathered by Mercury. Each hypothesis predicts a different surface composition, and there are two space missions set to make observations. *MESSENGER*, which ended in 2015, found higher-than-expected potassium and sulfur levels on the surface, suggesting that the giant impact hypothesis and vaporization of the crust and mantle did not occur because potassium and sulfur would have been driven off by the extreme heat of these events. *BepiColombo*, which will arrive at Mercury in 2025, will make observations to test these hypotheses. The findings so far would seem to favor the third hypothesis; however, further analysis of the data is needed.

Surface geology

Topography

Map of Mercury's northern hemisphere by the MLA instrument on *MESSENGER* lowest (purple) to 10 km (6.2 mi) highest (red).

Mercury's surface is similar in appearance to that of the Moon, showing extensive mare-like plains and heavy cratering, indicating that it has been geologically inactive for billions of years. Because knowledge of Mercury's geology had been based only on the 1975 Mariner 10 flyby and terrestrial observations,

Figure 51: *Mercury's surface*

it is the least understood of the terrestrial planets. As data from *MESSENGER* orbiter are processed, this knowledge will increase. For example, an unusual crater with radiating troughs has been discovered that scientists called "the spider". It was later named Apollodorus.

Albedo features are areas of markedly different reflectivity, as seen by telescopic observation. Mercury has dorsa (also called "wrinkle-ridges"), Moon like highlands, montes (mountains), planitiae (plains), rupes (escarpments), and valles (valleys).

Names for features on Mercury come from a variety of sources. Names coming from people are limited to the deceased. Craters are named for artists, musicians, painters, and authors who have made outstanding or fundamental contributions to their field. Ridges, or dorsa, are named for scientists who have contributed to the study of Mercury. Depressions or fossae are named for works of architecture. Montes are named for the word "hot" in a variety of languages. Plains or planitiae are named for Mercury in various languages. Escarpments or rupēs are named for ships of scientific expeditions. Valleys or valles are named for radio telescope facilities.

Mercury was heavily bombarded by comets and asteroids during and shortly following its formation 4.6 billion years ago, as well as during a possibly separate subsequent episode called the Late Heavy Bombardment that ended 3.8

Figure 52: *MASCS spectrum scan of Mercury's surface by MESSENGER*

billion years ago. During this period of intense crater formation, Mercury received impacts over its entire surface, facilitated by the lack of any atmosphere to slow impactors down. During this time Mercury was volcanically active; basins such as the Caloris Basin were filled by magma, producing smooth plains similar to the maria found on the Moon.

Data from the October 2008 flyby of *MESSENGER* gave researchers a greater appreciation for the jumbled nature of Mercury's surface. Mercury's surface is more heterogeneous than either Mars's or the Moon's, both of which contain significant stretches of similar geology, such as maria and plateaus.

Impact basins and craters

Craters on Mercury range in diameter from small bowl-shaped cavities to multi-ringed impact basins hundreds of kilometers across. They appear in all states of degradation, from relatively fresh rayed craters to highly degraded crater remnants. Mercurian craters differ subtly from lunar craters in that the area blanketed by their ejecta is much smaller, a consequence of Mercury's stronger surface gravity. According to IAU rules, each new crater must be named after an artist that was famous for more than fifty years, and dead for more than three years, before the date the crater is named.

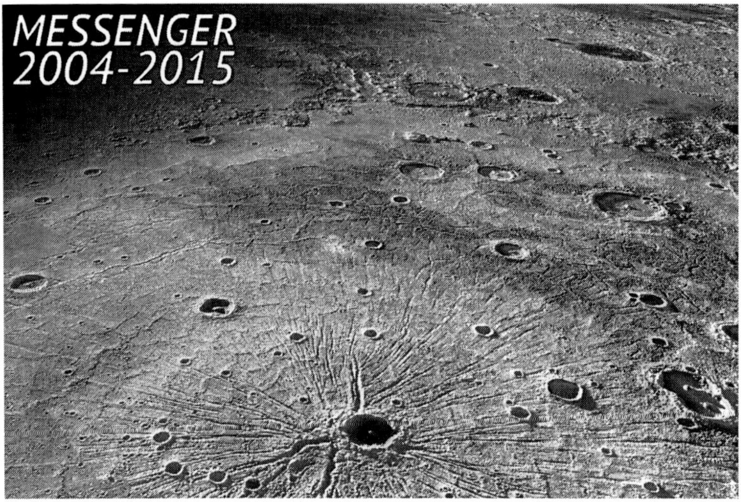

Figure 53: *Perspective view of Caloris Basin – high (red); low (blue).*

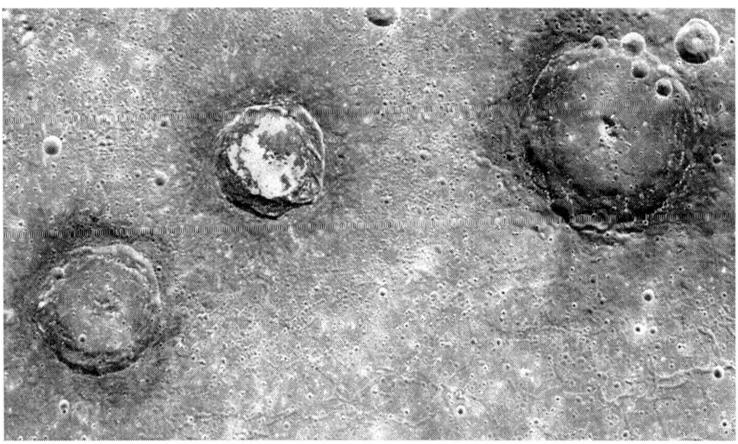

Figure 54: *Enhanced-color image of Munch, Sander and Poe craters amid volcanic plains (orange) near Caloris Basin*

Figure 55:
Interior of Abedin crater

The largest known crater is Caloris Basin, with a diameter of 1,550 km. The impact that created the Caloris Basin was so powerful that it caused lava eruptions and left a concentric ring over 2 km tall surrounding the impact crater. At the antipode of the Caloris Basin is a large region of unusual, hilly terrain known as the "Weird Terrain". One hypothesis for its origin is that shock waves generated during the Caloris impact traveled around Mercury, converging at the basin's antipode (180 degrees away). The resulting high stresses fractured the surface. Alternatively, it has been suggested that this terrain formed as a result of the convergence of ejecta at this basin's antipode.

Overall, about 15 impact basins have been identified on the imaged part of Mercury. A notable basin is the 400 km wide, multi-ring Tolstoj Basin that has an ejecta blanket extending up to 500 km from its rim and a floor that has been filled by smooth plains materials. Beethoven Basin has a similar-sized ejecta blanket and a 625 km diameter rim. Like the Moon, the surface of Mercury has likely incurred the effects of space weathering processes, including Solar wind and micrometeorite impacts.

Plains

There are two geologically distinct plains regions on Mercury. Gently rolling, hilly plains in the regions between craters are Mercury's oldest visible surfaces, predating the heavily cratered terrain. These inter-crater plains appear to have obliterated many earlier craters, and show a general paucity of smaller craters below about 30 km in diameter.

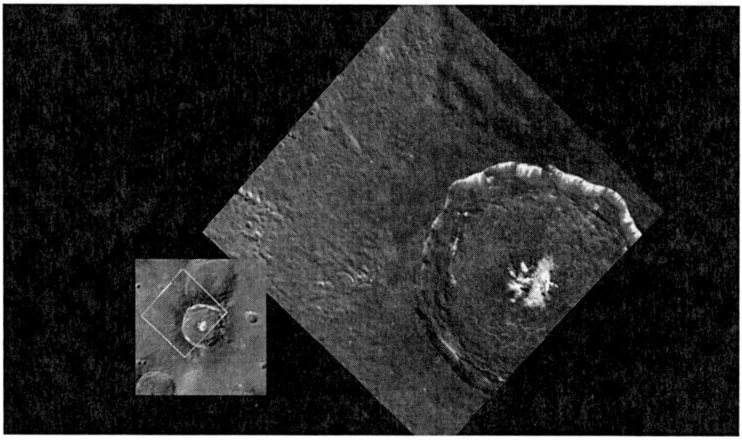

Figure 56: *Degas crater*

The so-called "Weird Terrain" formed at the point antipodal to the Caloris Basin impact

Smooth plains are widespread flat areas that fill depressions of various sizes and bear a strong resemblance to the lunar maria. Notably, they fill a wide ring surrounding the Caloris Basin. Unlike lunar maria, the smooth plains of Mercury have the same albedo as the older inter-crater plains. Despite a lack of unequivocally volcanic characteristics, the localisation and rounded, lobate shape of these plains strongly support volcanic origins. All the smooth plains of Mercury formed significantly later than the Caloris basin, as evidenced by appreciably smaller crater densities than on the Caloris ejecta blanket. The floor of the Caloris Basin is filled by a geologically distinct flat plain, broken up by ridges and fractures in a roughly polygonal pattern. It is not clear whether they are volcanic lavas induced by the impact, or a large sheet of impact melt.

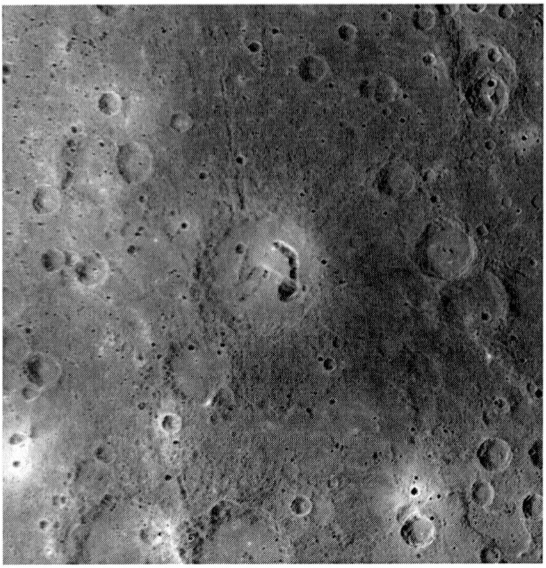

Figure 57: *Picasso crater — the large arc-shaped pit located on the eastern side of its floor are postulated to have formed when subsurface magma subsided or drained, causing the surface to collapse into the resulting void.*

Compressional features

One unusual feature of Mercury's surface is the numerous compression folds, or rupes, that crisscross the plains. As Mercury's interior cooled, it contracted and its surface began to deform, creating wrinkle ridges and lobate scarps associated with thrust faults. The scarps can reach lengths of 1000 km and heights of 3 km. These compressional features can be seen on top of other features, such as craters and smooth plains, indicating they are more recent. Mapping of the features has suggested a total shrinkage of Mercury's radius in the range of ~1 to 7 km. Small-scale thrust fault scarps have been found, tens of meters in height and with lengths in the range of a few km, that appear to be less than 50 million years old, indicating that compression of the interior and consequent surface geological activity continue to the present.

The Lunar Reconnaissance Orbiter discovered that similar small thrust faults exist on the Moon.

Volcanology

Images obtained by *MESSENGER* have revealed evidence for pyroclastic flows on Mercury from low-profile shield volcanoes. *MESSENGER* data has helped

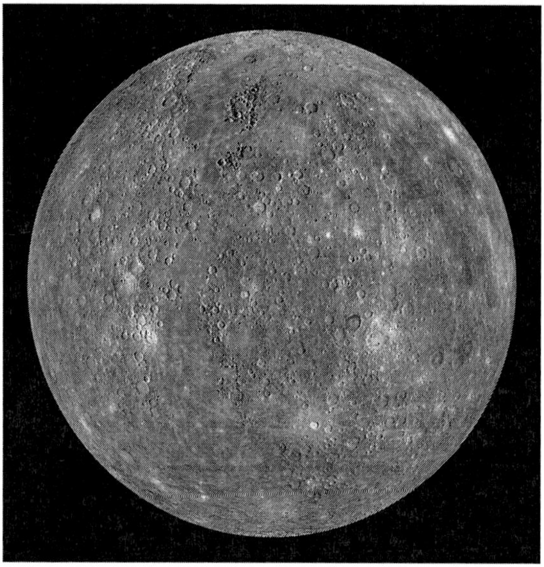

Figure 58: *Composite image of Mercury taken by MESSENGER*

identify 51 pyroclastic deposits on the surface, where 90% of them are found within impact craters. A study of the degradation state of the impact craters that host pyroclastic deposits suggests that pyroclastic activity occurred on Mercury over a prolonged interval.

A "rimless depression" inside the southwest rim of the Caloris Basin consists of at least nine overlapping volcanic vents, each individually up to 8 km in diameter. It is thus a "compound volcano". The vent floors are at a least 1 km below their brinks and they bear a closer resemblance to volcanic craters sculpted by explosive eruptions or modified by collapse into void spaces created by magma withdrawal back down into a conduit. The scientists could not quantify the age of the volcanic complex system, but reported that it could be of the order of a billion years.

Surface conditions and exosphere

The surface temperature of Mercury ranges from 100 K to 700 K at the most extreme places: 0°N, 0°W, or 180°W. It never rises above 180 K at the poles, due to the absence of an atmosphere and a steep temperature gradient between the equator and the poles. The subsolar point reaches about 700 K during perihelion (0°W or 180°W), but only 550 K at aphelion (90° or 270°W). On the dark side of the planet, temperatures average 110 K. The intensity of sunlight

Figure 59: *Radar image of Mercury's north pole*

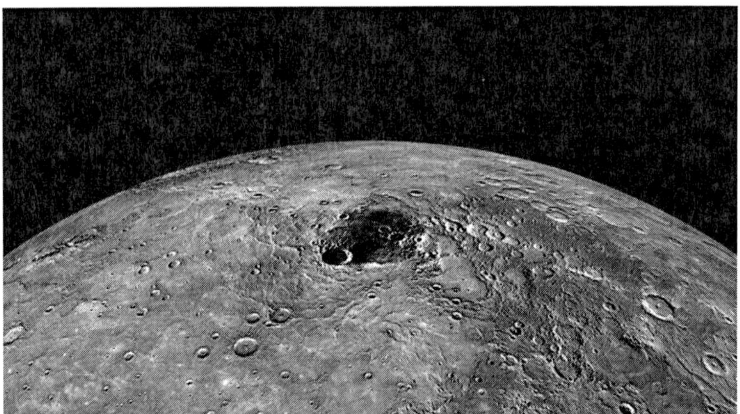

Figure 60: *Composite of the north pole of Mercury, where NASA confirmed the discovery of a large volume of water ice, in permanently dark craters that exist there.*

on Mercury's surface ranges between 4.59 and 10.61 times the solar constant (1,370 $W{\cdot}m^{-2}$).

Although the daylight temperature at the surface of Mercury is generally extremely high, observations strongly suggest that ice (frozen water) exists on Mercury. The floors of deep craters at the poles are never exposed to direct sunlight, and temperatures there remain below 102 K; far lower than the global average. Water ice strongly reflects radar, and observations by the 70-meter Goldstone Solar System Radar and the VLA in the early 1990s revealed that there are patches of high radar reflection near the poles. Although ice was not the only possible cause of these reflective regions, astronomers think it was the most likely.

The icy regions are estimated to contain about 10^{14}–10^{15} kg of ice, and may be covered by a layer of regolith that inhibits sublimation. By comparison, the Antarctic ice sheet on Earth has a mass of about 4×10^{18} kg, and Mars's south polar cap contains about 10^{16} kg of water. The origin of the ice on Mercury is not yet known, but the two most likely sources are from outgassing of water from the planet's interior or deposition by impacts of comets.

Mercury is too small and hot for its gravity to retain any significant atmosphere over long periods of time; it does have a tenuous surface-bounded exosphere containing hydrogen, helium, oxygen, sodium, calcium, potassium and others at a surface pressure of less than approximately 0.5 nPa (0.005 picobars). This exosphere is not stable—atoms are continuously lost and replenished from a variety of sources. Hydrogen atoms and helium atoms probably come from the solar wind, diffusing into Mercury's magnetosphere before later escaping back into space. Radioactive decay of elements within Mercury's crust is another source of helium, as well as sodium and potassium. *MESSENGER* found high proportions of calcium, helium, hydroxide, magnesium, oxygen, potassium, silicon and sodium. Water vapor is present, released by a combination of processes such as: comets striking its surface, sputtering creating water out of hydrogen from the solar wind and oxygen from rock, and sublimation from reservoirs of water ice in the permanently shadowed polar craters. The detection of high amounts of water-related ions like O^+, OH^-, and H_2O^+ was a surprise. Because of the quantities of these ions that were detected in Mercury's space environment, scientists surmise that these molecules were blasted from the surface or exosphere by the solar wind.

Sodium, potassium and calcium were discovered in the atmosphere during the 1980–1990s, and are thought to result primarily from the vaporization of surface rock struck by micrometeorite impacts including presently from Comet Encke. In 2008, magnesium was discovered by *MESSENGER*. Studies indicate that, at times, sodium emissions are localized at points that correspond

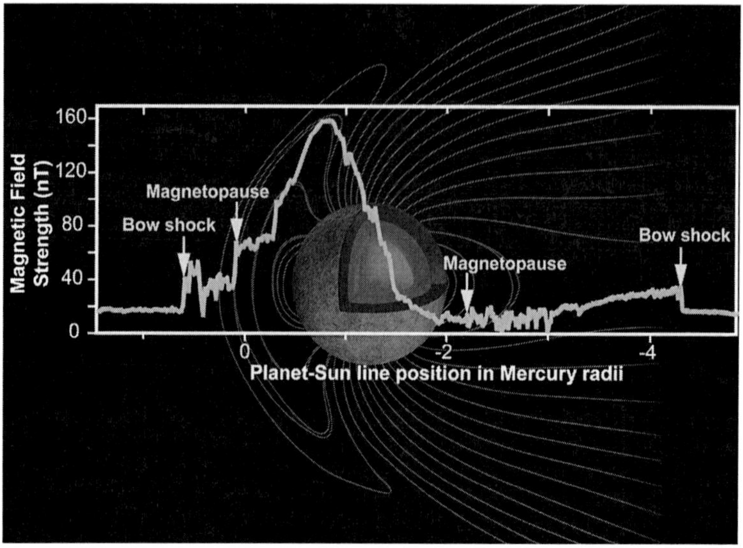

Figure 61: *Graph showing relative strength of Mercury's magnetic field*

to the planet's magnetic poles. This would indicate an interaction between the magnetosphere and the planet's surface.

On November 29, 2012, NASA confirmed that images from *MESSENGER* had detected that craters at the north pole contained water ice. *MESSENGER*'s principal investigator Sean Solomon is quoted in *The New York Times* estimating the volume of the ice to be large enough to "encase Washington, D.C., in a frozen block two and a half miles deep".[75]

Magnetic field and magnetosphere

Despite its small size and slow 59-day-long rotation, Mercury has a significant, and apparently global, magnetic field. According to measurements taken by Mariner 10, it is about 1.1% the strength of Earth's. The magnetic-field strength at Mercury's equator is about 300 nT. Like that of Earth, Mercury's magnetic field is dipolar. Unlike Earth's, Mercury's poles are nearly aligned with the planet's spin axis. Measurements from both the Mariner 10 and *MESSENGER* space probes have indicated that the strength and shape of the magnetic field are stable.

It is likely that this magnetic field is generated by a dynamo effect, in a manner similar to the magnetic field of Earth. This dynamo effect would result from the circulation of the planet's iron-rich liquid core. Particularly strong tidal effects caused by the planet's high orbital eccentricity would serve to keep the core in the liquid state necessary for this dynamo effect.

Mercury's magnetic field is strong enough to deflect the solar wind around the planet, creating a magnetosphere. The planet's magnetosphere, though small enough to fit within Earth, is strong enough to trap solar wind plasma. This contributes to the space weathering of the planet's surface. Observations taken by the Mariner 10 spacecraft detected this low energy plasma in the magnetosphere of the planet's nightside. Bursts of energetic particles in the planet's magnetotail indicate a dynamic quality to the planet's magnetosphere.

During its second flyby of the planet on October 6, 2008, *MESSENGER* discovered that Mercury's magnetic field can be extremely "leaky". The spacecraft encountered magnetic "tornadoes" – twisted bundles of magnetic fields connecting the planetary magnetic field to interplanetary space – that were up to 800 km wide or a third of the radius of the planet. These twisted magnetic flux tubes, technically known as flux transfer events, form open windows in the planet's magnetic shield through which the solar wind may enter and directly impact Mercury's surface via magnetic reconnection This also occurs in Earth's magnetic field. The *MESSENGER* observations showed the reconnection rate is ten times higher at Mercury, but its proximity to the Sun only accounts for about a third of the reconnection rate observed by *MESSENGER*.

Orbit, rotation, and longitude

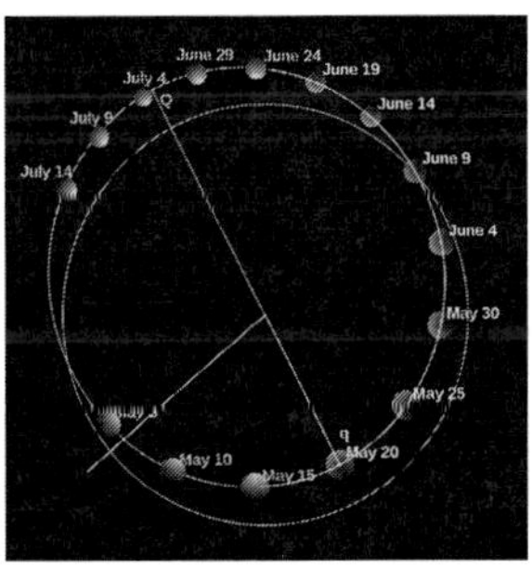

Orbit of Mercury (yellow). Dates refer to 2006.

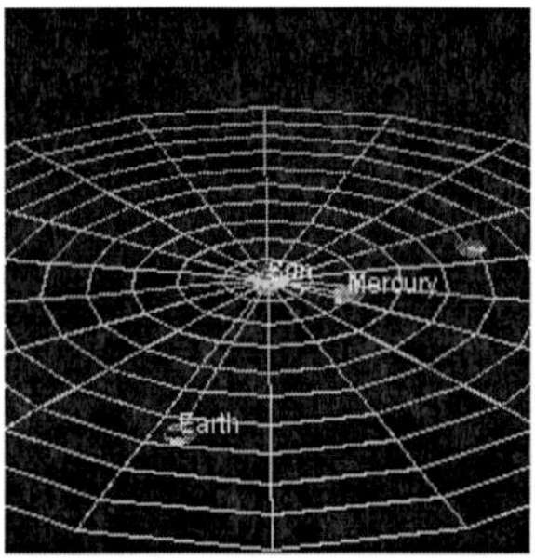

Animation of Mercury's and Earth's revolution around the Sun

Mercury has the most eccentric orbit of all the planets; its eccentricity is 0.21 with its distance from the Sun ranging from 46,000,000 to 70,000,000 km (29,000,000 to 43,000,000 mi). It takes 87.969 Earth days to complete an orbit. The diagram on the right illustrates the effects of the eccentricity, showing Mercury's orbit overlaid with a circular orbit having the same semi-major axis. Mercury's higher velocity when it is near perihelion is clear from the greater distance it covers in each 5-day interval. In the diagram the varying distance of Mercury to the Sun is represented by the size of the planet, which is inversely proportional to Mercury's distance from the Sun. This varying distance to the Sun leads to Mercury's surface being flexed by tidal bulges raised by the Sun that are about 17 times stronger than the Moon's on Earth. Combined with a 3:2 spin–orbit resonance of the planet's rotation around its axis, it also results in complex variations of the surface temperature. The resonance makes a single solar day on Mercury last exactly two Mercury years, or about 176 Earth days.

Mercury's orbit is inclined by 7 degrees to the plane of Earth's orbit (the ecliptic), as shown in the diagram on the right. As a result, transits of Mercury across the face of the Sun can only occur when the planet is crossing the plane of the ecliptic at the time it lies between Earth and the Sun. This occurs about every seven years on average.

Mercury's axial tilt is almost zero, with the best measured value as low as 0.027 degrees. This is significantly smaller than that of Jupiter, which has the second smallest axial tilt of all planets at 3.1 degrees. This means that to an observer at Mercury's poles, the center of the Sun never rises more than 2.1 arcminutes above the horizon.

At certain points on Mercury's surface, an observer would be able to see the Sun rise about halfway, then reverse and set before rising again, all within the same Mercurian day. This is because approximately four Earth days before perihelion, Mercury's angular orbital velocity equals its angular rotational velocity so that the Sun's apparent motion ceases; closer to perihelion, Mercury's angular orbital velocity then exceeds the angular rotational velocity. Thus, to a hypothetical observer on Mercury, the Sun appears to move in a retrograde direction. Four Earth days after perihelion, the Sun's normal apparent motion resumes. A similar effect would have occurred if Mercury had been in synchronous rotation: the alternating gain and loss of rotation over revolution would have caused a libration of 23.65° in longitude.[76]

For the same reason, there are two points on Mercury's equator, 180 degrees apart in longitude, at either of which, around perihelion in alternate Mercurian years (once a Mercurian day), the Sun passes overhead, then reverses its apparent motion and passes overhead again, then reverses a second time and passes overhead a third time, taking a total of about 16 Earth-days for this entire process. In the other alternate Mercurian years, the same thing happens at the other of these two points. The amplitude of the retrograde motion is small, so the overall effect is that, for two or three weeks, the Sun is almost stationary overhead, and is at its most brilliant because Mercury is at perihelion, its closest to the Sun. This prolonged exposure to the Sun at its brightest makes these two points the hottest places on Mercury. Conversely, there are two other points on the equator, 90 degrees of longitude apart from the first ones, where the Sun passes overhead only when the planet is at aphelion in alternate years, when the apparent motion of the Sun in Mercury's sky is relatively rapid. These points, which are the ones on the equator where the apparent retrograde motion of the Sun happens when it is crossing the horizon as described in the preceding paragraph, receive much less solar heat than the first ones described above.

Mercury attains inferior conjunction (nearest approach to Earth) every 116 Earth days on average, but this interval can range from 105 days to 129 days due to the planet's eccentric orbit. Mercury can come as near as 82.2 gigametres (0.549 astronomical units; 51.1 million miles) to Earth, and that is slowly declining: The next approach to within 82.1 Gm (51.0 million miles) is in 2679, and to within 82.0 Gm (51.0 million miles) in 4487, but it will not be closer to Earth than 80 Gm (50 million miles) until 28,622.[77] Its period of retrograde motion as seen from Earth can vary from 8 to 15 days on either side of inferior conjunction. This large range arises from the planet's high orbital eccentricity.

Longitude convention

The longitude convention for Mercury puts the zero of longitude at one of the two hottest points on the surface, as described above. However, when this area was first visited, by Mariner 10, this zero meridian was in darkness, so it was impossible to select a feature on the surface to define the exact position of the meridian. Therefore, a small crater further west was chosen, called Hun Kal, which provides the exact reference point for measuring longitude.

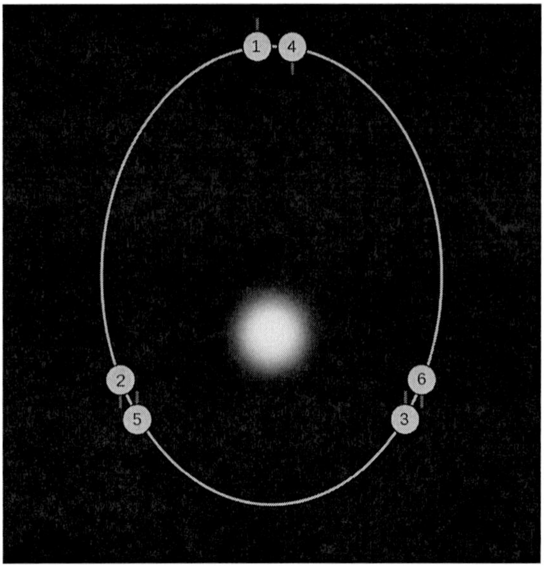

Figure 62: *After one orbit, Mercury has rotated 1.5 times, so after two complete orbits the same hemisphere is again illuminated.*

The center of Hun Kal defines the 20° West meridian. A 1970 International Astronomical Union resolution suggests that longitudes be measured positively in the westerly direction on Mercury. The two hottest places on the equator are therefore at longitudes 0°W and 180°W, and the coolest points on the equator are at longitudes 90°W and 270°W. However, the *MESSENGER* project uses an east-positive convention.

Spin–orbit resonance

For many years it was thought that Mercury was synchronously tidally locked with the Sun, rotating once for each orbit and always keeping the same face directed towards the Sun, in the same way that the same side of the Moon always faces Earth. Radar observations in 1965 proved that the planet has a 3:2 spin–orbit resonance, rotating three times for every two revolutions around the Sun. The eccentricity of Mercury's orbit makes this resonance stable—at perihelion, when the solar tide is strongest, the Sun is nearly still in Mercury's sky.

The original reason astronomers thought it was synchronously locked was that, whenever Mercury was best placed for observation, it was always nearly at the same point in its 3:2 resonance, hence showing the same face. This is because, coincidentally, Mercury's rotation period is almost exactly half of its synodic period with respect to Earth. Due to Mercury's 3:2 spin–orbit resonance, a solar day (the length between two meridian transits of the Sun) lasts about 176 Earth days. A sidereal day (the period of rotation) lasts about 58.7 Earth days.

Simulations indicate that the orbital eccentricity of Mercury varies chaotically from nearly zero (circular) to more than 0.45 over millions of years due to perturbations from the other planets. This was thought to explain Mercury's 3:2 spin–orbit resonance (rather than the more usual 1:1), because this state is more likely to arise during a period of high eccentricity. However, accurate modeling based on a realistic model of tidal response has demonstrated that Mercury was captured into the 3:2 spin–orbit state at a very early stage of its history, within 20 (more likely, 10) million years after its formation.

Numerical simulations show that a future secular orbital resonant perihelion interaction with Jupiter may cause the eccentricity of Mercury's orbit to increase to the point where there is a 1% chance that the planet may collide with Venus within the next five billion years.

Advance of perihelion

In 1859, the French mathematician and astronomer Urbain Le Verrier reported that the slow precession of Mercury's orbit around the Sun could not be completely explained by Newtonian mechanics and perturbations by the known planets. He suggested, among possible explanations, that another planet (or perhaps instead a series of smaller 'corpuscules') might exist in an orbit even closer to the Sun than that of Mercury, to account for this perturbation.[78] (Other explanations considered included a slight oblateness of the Sun.) The success of the search for Neptune based on its perturbations of the orbit of Uranus led astronomers to place faith in this possible explanation, and the hypothetical planet was named Vulcan, but no such planet was ever found.

The perihelion precession of Mercury is 5,600 arcseconds (1.5556°) per century relative to Earth, or 574.10±0.65 arcseconds per century relative to the inertial ICRF. Newtonian mechanics, taking into account all the effects from the other planets, predicts a precession of 5,557 arcseconds (1.5436°) per century. In the early 20th century, Albert Einstein's general theory of relativity provided the explanation for the observed precession. The effect is small: just 42.98 arcseconds per century for Mercury; it therefore requires a little over twelve million orbits for a full excess turn. Similar, but much smaller, effects exist for other Solar System bodies: 8.62 arcseconds per century for Venus, 3.84 for Earth, 1.35 for Mars, and 10.05 for 1566 Icarus.

Albert Einstein's formula for the perihelion shift per revolution is $\epsilon = 24\pi^3 \frac{a^2}{T^2c^2(1-e^2)}$, where e is the orbital eccentricity, a the semi-major

Figure 63: *Image mosaic by Mariner 10, 1974*

axis, and T the orbital period. Filling in the values gives a result of 0.1035 arcseconds per revolution or 0.4297 arcseconds per Earth year, i.e., 42.97 arcseconds per century.

Observation

Mercury's apparent magnitude varies between –2.6 (brighter than the brightest star Sirius) and about +5.7 (approximating the theoretical limit of naked-eye visibility). The extremes occur when Mercury is close to the Sun in the sky. Observation of Mercury is complicated by its proximity to the Sun, as it is lost in the Sun's glare for much of the time. Mercury can be observed for only a brief period during either morning or evening twilight.

Mercury can, like several other planets and the brightest stars, be seen during a total solar eclipse.

Like the Moon and Venus, Mercury exhibits phases as seen from Earth. It is "new" at inferior conjunction and "full" at superior conjunction. The planet is rendered invisible from Earth on both of these occasions because of its being obscured by the Sun, except its new phase during a transit.

Mercury is technically brightest as seen from Earth when it is at a full phase. Although Mercury is farthest from Earth when it is full, the greater illuminated

Figure 64: *False-color map showing the maximum temperatures of the north polar region*

area that is visible and the opposition brightness surge more than compensates for the distance. The opposite is true for Venus, which appears brightest when it is a crescent, because it is much closer to Earth than when gibbous.

Nonetheless, the brightest (full phase) appearance of Mercury is an essentially impossible time for practical observation, because of the extreme proximity of the Sun. Mercury is best observed at the first and last quarter, although they are phases of lesser brightness. The first and last quarter phases occur at greatest elongation east and west of the Sun, respectively. At both of these times Mercury's separation from the Sun ranges anywhere from 17.9° at perihelion to 27.8° at aphelion.[79,80] At greatest *western* elongation, Mercury rises at its earliest before sunrise, and at greatest *eastern* elongation, it sets at its latest after sunset.

Mercury can be easily seen from the tropics and subtropics more than from higher latitudes. Viewed from low latitudes and at the right times of year, the ecliptic intersects the horizon at a steep angle. Mercury is 10° above the horizon when the planet appears directly above the Sun (i.e. its orbit appears vertical) and is at maximum elongation from the Sun (28°) and also when the Sun is 18° below the horizon, so the sky is just completely dark.[81] This angle is the maximum altitude at which Mercury is visible in a completely dark sky.

Figure 65: *False-color image of Carnegie Rupes, a tectonic landform—high terrain (red); low (blue).*

At middle latitudes, Mercury is more often and easily visible from the Southern Hemisphere than from the Northern. This is because Mercury's maximum western elongation occurs only during early autumn in the Southern Hemisphere, whereas its greatest eastern elongation happens only during late winter in the Southern Hemisphere. In both of these cases, the angle at which the planet's orbit intersects the horizon is maximized, allowing it to rise several hours before sunrise in the former instance and not set until several hours after sundown in the latter from southern mid-latitudes, such as Argentina and South Africa.

An alternate method for viewing Mercury involves observing the planet during daylight hours when conditions are clear, ideally when it is at its greatest elongation. This allows the planet to be found easily, even when using telescopes with 8 cm (3.1 in) apertures. Care must be taken to ensure the instrument isn't pointed directly towards the Sun because of the risk for eye damage. This method bypasses the limitation of twilight observing when the ecliptic is located at a low elevation (e.g. on autumn evenings).

Ground-based telescope observations of Mercury reveal only an illuminated partial disk with limited detail. The first of two spacecraft to visit the planet was Mariner 10, which mapped about 45% of its surface from 1974 to 1975. The second is the *MESSENGER* spacecraft, which after three Mercury flybys

Figure 66: *Mercury, from Liber astronomiae, 1550*

between 2008 and 2009, attained orbit around Mercury on March 17, 2011, to study and map the rest of the planet.

The Hubble Space Telescope cannot observe Mercury at all, due to safety procedures that prevent its pointing too close to the Sun.

Because the shift of 0.15 revolutions in a year makes up a seven-year cycle ($0.15 \times 7 \approx 1.0$), in the seventh year Mercury follows almost exactly (earlier by 7 days) the sequence of phenomena it showed seven years before.

Observation history

Ancient astronomers

The earliest known recorded observations of Mercury are from the Mul.Apin tablets. These observations were most likely made by an Assyrian astronomer around the 14th century BC. The cuneiform name used to designate Mercury on the Mul.Apin tablets is transcribed as Udu.Idim.Gu\u$_4$.Ud ("the jumping planet"). Babylonian records of Mercury date back to the 1st millennium BC. The Babylonians called the planet Nabu after the messenger to the gods in their mythology.

The ancient Greeks knew the planet as Στίλβων (*Stilbon*), meaning "the gleaming", Ἑρμάων (*Hermaon*) and Ἑρμής (*Hermes*),[82] a planetary name that is retained in modern Greek (Ερμής: *Ermis*).[83] The Romans named the planet after the swift-footed Roman messenger god, Mercury (Latin *Mercurius*), which they equated with the Greek Hermes, because it moves across

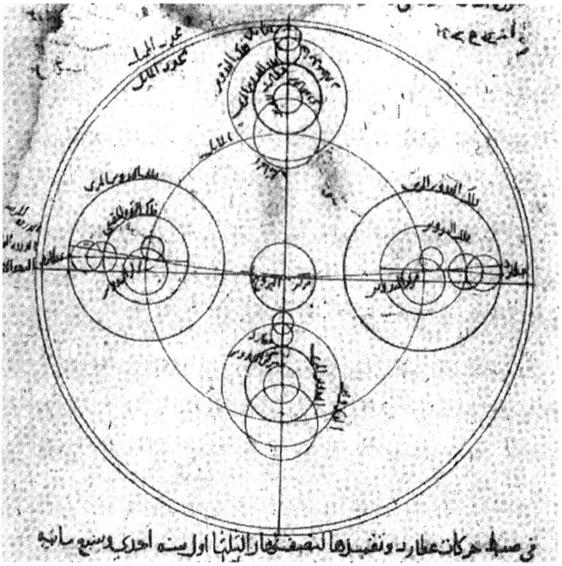

Figure 67: *Ibn al-Shatir's model for the appearances of Mercury, showing the multiplication of epicycles using the Tusi couple, thus eliminating the Ptolemaic eccentrics and equant.*

the sky faster than any other planet. The astronomical symbol for Mercury is a stylized version of Hermes' caduceus.

The Roman-Egyptian astronomer Ptolemy wrote about the possibility of planetary transits across the face of the Sun in his work *Planetary Hypotheses*. He suggested that no transits had been observed either because planets such as Mercury were too small to see, or because the transits were too infrequent.

In ancient China, Mercury was known as "the Hour Star" (*Chen-xing* 辰星). It was associated with the direction north and the phase of water in the Five Phases system of metaphysics. Modern Chinese, Korean, Japanese and Vietnamese cultures refer to the planet literally as the "water star" (水星), based on the Five elements. Hindu mythology used the name Budha for Mercury, and this god was thought to preside over Wednesday. The god Odin (or Woden) of Germanic paganism was associated with the planet Mercury and Wednesday. The Maya may have represented Mercury as an owl (or possibly four owls; two for the morning aspect and two for the evening) that served as a messenger to the underworld.

In medieval Islamic astronomy, the Andalusian astronomer Abū Ishāq Ibrāhīm al-Zarqālī in the 11th century described the deferent of Mercury's geocentric

Figure 68: *Transit of Mercury. Mercury is visible as a black dot below and to the left of center. The dark area above the center of the solar disk is a sunspot.*

orbit as being oval, like an egg or a pignon, although this insight did not influence his astronomical theory or his astronomical calculations.[84] In the 12th century, Ibn Bajjah observed "two planets as black spots on the face of the Sun", which was later suggested as the transit of Mercury and/or Venus by the Maragha astronomer Qotb al-Din Shirazi in the 13th century. (Note that most such medieval reports of transits were later taken as observations of sunspots.)

In India, the Kerala school astronomer Nilakantha Somayaji in the 15th century developed a partially heliocentric planetary model in which Mercury orbits the Sun, which in turn orbits Earth, similar to the Tychonic system later proposed by Tycho Brahe in the late 16th century.

Ground-based telescopic research

The first telescopic observations of Mercury were made by Galileo in the early 17th century. Although he observed phases when he looked at Venus, his telescope was not powerful enough to see the phases of Mercury. In 1631, Pierre Gassendi made the first telescopic observations of the transit of a planet across the Sun when he saw a transit of Mercury predicted by Johannes Kepler. In 1639, Giovanni Zupi used a telescope to discover that the planet had

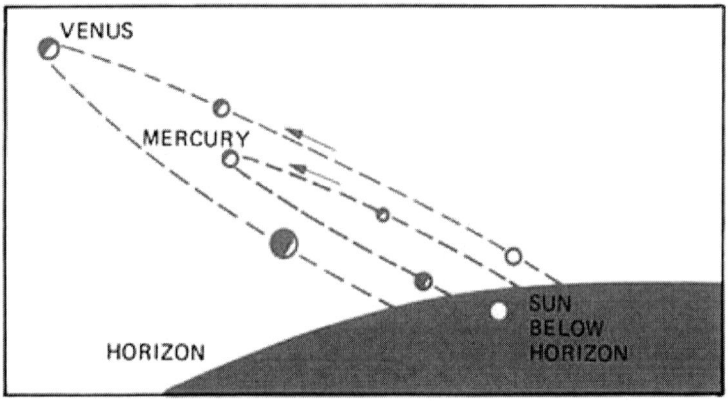

Figure 69: *Elongation is the angle between the Sun and the planet, with Earth as the reference point. Mercury appears close to the Sun.*

orbital phases similar to Venus and the Moon. The observation demonstrated conclusively that Mercury orbited around the Sun.

A rare event in astronomy is the passage of one planet in front of another (occultation), as seen from Earth. Mercury and Venus occult each other every few centuries, and the event of May 28, 1737 is the only one historically observed, having been seen by John Bevis at the Royal Greenwich Observatory. The next occultation of Mercury by Venus will be on December 3, 2133.

The difficulties inherent in observing Mercury mean that it has been far less studied than the other planets. In 1800, Johann Schröter made observations of surface features, claiming to have observed 20-kilometre-high (12 mi) mountains. Friedrich Bessel used Schröter's drawings to erroneously estimate the rotation period as 24 hours and an axial tilt of 70°. In the 1880s, Giovanni Schiaparelli mapped the planet more accurately, and suggested that Mercury's rotational period was 88 days, the same as its orbital period due to tidal locking. This phenomenon is known as synchronous rotation. The effort to map the surface of Mercury was continued by Eugenios Antoniadi, who published a book in 1934 that included both maps and his own observations. Many of the planet's surface features, particularly the albedo features, take their names from Antoniadi's map.

In June 1962, Soviet scientists at the Institute of Radio-engineering and Electronics of the USSR Academy of Sciences, led by Vladimir Kotelnikov, became the first to bounce a radar signal off Mercury and receive it, starting radar observations of the planet. Three years later, radar observations by Americans

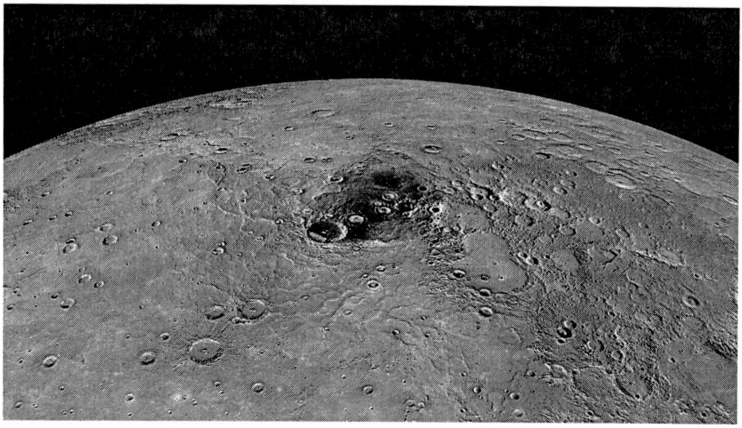

Figure 70: *Water ice (yellow) at Mercury's north polar region*

Gordon Pettengill and R. Dyce, using the 300-meter Arecibo Observatory radio telescope in Puerto Rico, showed conclusively that the planet's rotational period was about 59 days.[85] The theory that Mercury's rotation was synchronous had become widely held, and it was a surprise to astronomers when these radio observations were announced. If Mercury were tidally locked, its dark face would be extremely cold, but measurements of radio emission revealed that it was much hotter than expected. Astronomers were reluctant to drop the synchronous rotation theory and proposed alternative mechanisms such as powerful heat-distributing winds to explain the observations.

Italian astronomer Giuseppe Colombo noted that the rotation value was about two-thirds of Mercury's orbital period, and proposed that the planet's orbital and rotational periods were locked into a 3:2 rather than a 1:1 resonance. Data from Mariner 10 subsequently confirmed this view. This means that Schiaparelli's and Antoniadi's maps were not "wrong". Instead, the astronomers saw the same features during every *second* orbit and recorded them, but disregarded those seen in the meantime, when Mercury's other face was toward the Sun, because the orbital geometry meant that these observations were made under poor viewing conditions.

Ground-based optical observations did not shed much further light on Mercury, but radio astronomers using interferometry at microwave wavelengths, a technique that enables removal of the solar radiation, were able to discern physical and chemical characteristics of the subsurface layers to a depth of several meters.[86] Not until the first space probe flew past Mercury did many of its most fundamental morphological properties become known. Moreover,

Figure 71: *MESSENGER being prepared for launch*

recent technological advances have led to improved ground-based observations. In 2000, high-resolution lucky imaging observations were conducted by the Mount Wilson Observatory 1.5 meter Hale telescope. They provided the first views that resolved surface features on the parts of Mercury that were not imaged in the Mariner 10 mission. Most of the planet has been mapped by the Arecibo radar telescope, with 5 km (3.1 mi) resolution, including polar deposits in shadowed craters of what may be water ice.

Research with space probes

Reaching Mercury from Earth poses significant technical challenges, because it orbits so much closer to the Sun than Earth. A Mercury-bound spacecraft launched from Earth must travel over 91 million kilometres (57 million miles) into the Sun's gravitational potential well. Mercury has an orbital speed of 48 km/s (30 mi/s), whereas Earth's orbital speed is 30 km/s (19 mi/s). Therefore, the spacecraft must make a large change in velocity (delta-v) to enter a Hohmann transfer orbit that passes near Mercury, as compared to the delta-v required for other planetary missions.

The potential energy liberated by moving down the Sun's potential well becomes kinetic energy; requiring another large delta-v change to do anything other than rapidly pass by Mercury. To land safely or enter a stable orbit the spacecraft would rely entirely on rocket motors. Aerobraking is ruled out because Mercury has a negligible atmosphere. A trip to Mercury requires more

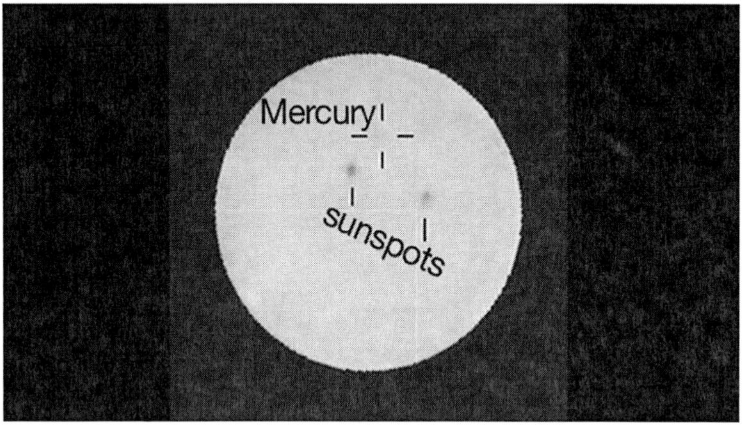

Figure 72: *Mercury transiting the Sun as viewed by the Mars rover Curiosity (June 3, 2014).*

rocket fuel than that required to escape the Solar System completely. As a result, only two space probes have visited it so far. A proposed alternative approach would use a solar sail to attain a Mercury-synchronous orbit around the Sun.

Mariner 10

Mariner 10, the first probe to visit Mercury

The first spacecraft to visit Mercury was NASA's Mariner 10 (1974–1975). The spacecraft used the gravity of Venus to adjust its orbital velocity so that it could approach Mercury, making it both the first spacecraft to use this gravitational "slingshot" effect and the first NASA mission to visit multiple planets. Mariner 10 provided the first close-up images of Mercury's surface, which immediately showed its heavily cratered nature, and revealed many other types of geological features, such as the giant scarps that were later ascribed to the effect of the planet shrinking slightly as its iron core cools. Unfortunately, the same face of the planet was lit at each of Mariner 10's close approaches. This

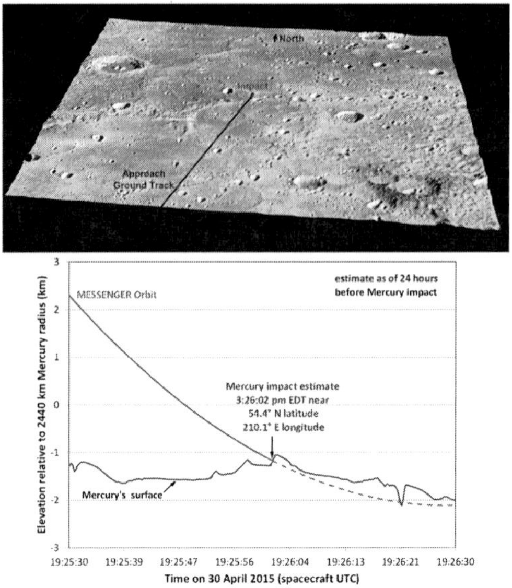

Figure 73: *Estimated details of the impact of MESSENGER on 30 April 2015*

made close observation of both sides of the planet impossible, and resulted in the mapping of less than 45% of the planet's surface.

The spacecraft made three close approaches to Mercury, the closest of which took it to within 327 km (203 mi) of the surface. At the first close approach, instruments detected a magnetic field, to the great surprise of planetary geologists—Mercury's rotation was expected to be much too slow to generate a significant dynamo effect. The second close approach was primarily used for imaging, but at the third approach, extensive magnetic data were obtained. The data revealed that the planet's magnetic field is much like Earth's, which deflects the solar wind around the planet. The origin of Mercury's magnetic field is still the subject of several competing theories.

On March 24, 1975, just eight days after its final close approach, Mariner 10 ran out of fuel. Because its orbit could no longer be accurately controlled, mission controllers instructed the probe to shut down. Mariner 10 is thought to be still orbiting the Sun, passing close to Mercury every few months.

MESSENGER

A second NASA mission to Mercury, named *MESSENGER* (MErcury Surface, Space ENvironment, GEochemistry, and Ranging), was launched on 3

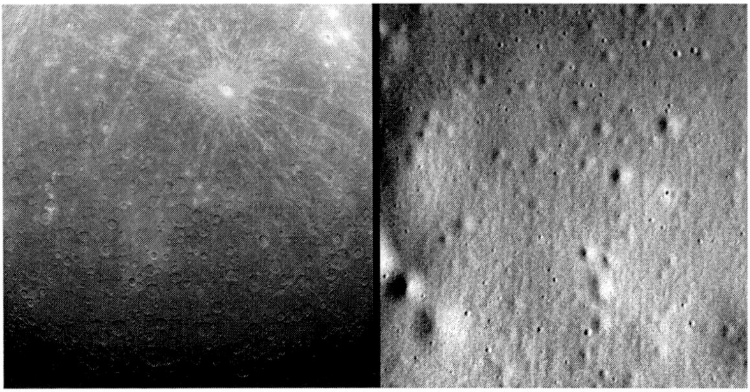

Figure 74: *First (29 March 2011) and last (30 April 2015) images of Mercury by MESSENGER*

August 2004. It made a fly-by of Earth in August 2005, and of Venus in October 2006 and June 2007 to place it onto the correct trajectory to reach an orbit around Mercury. A first fly-by of Mercury occurred on January 14, 2008, a second on October 6, 2008, and a third on September 29, 2009. Most of the hemisphere not imaged by *Mariner 10* was mapped during these fly-bys. The probe successfully entered an elliptical orbit around the planet on March 18, 2011. The first orbital image of Mercury was obtained on March 29, 2011. The probe finished a one-year mapping mission, and then entered a one-year extended mission into 2013. In addition to continued observations and mapping of Mercury, *MESSENGER* observed the 2012 solar maximum.[87]

The mission was designed to clear up six key issues: Mercury's high density, its geological history, the nature of its magnetic field, the structure of its core, whether it has ice at its poles, and where its tenuous atmosphere comes from. To this end, the probe carried imaging devices that gathered much-higher resolution images of much more of Mercury than Mariner 10, assorted spectrometers to determine abundances of elements in the crust, and magnetometers and devices to measure velocities of charged particles. Measurements of changes in the probe's orbital velocity were expected to be used to infer details of the planet's interior structure. *MESSENGER*'s final maneuver was on April 24, 2015, and it crashed into Mercury's surface on April 30, 2015. The spacecraft's impact with Mercury occurred near 3:26 PM EDT on April 30, 2015, leaving a crater estimated to be 16 m (52 ft) in diameter.

BepiColombo

The European Space Agency is planning a joint mission with Japan called BepiColombo, which will orbit Mercury with two probes: one to map the planet and the other to study its magnetosphere. Once launched in 2018, BepiColombo is expected to reach Mercury in 2025. It will release a magnetometer probe into an elliptical orbit, then chemical rockets will fire to deposit the mapper probe into a circular orbit. Both probes will operate for one terrestrial year. The mapper probe will carry an array of spectrometers similar to those on *MESSENGER*, and will study the planet at many different wavelengths including infrared, ultraviolet, X-ray and gamma ray.

Comparison

Size comparison with other Solar System objects

Mercury, Earth

Mercury, Venus, Earth, Mars

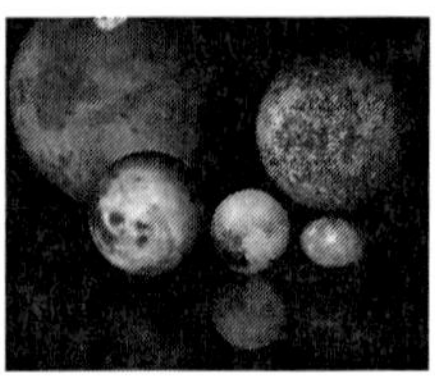

Back row: Mars, Mercury
Front: Moon, Pluto, Haumea

External links

<indicator name="spoken-icon"> </indicator>

- Mariner 10 Atlas of Mercury – NASA[88]
- Mercury nomenclature[89] and map with feature names[90] from the USGS planetary nomenclature page
- *MESSENGER* Mission web site[91]
- Mercury QuickMap from *MESSENGER* web site[92]
- SolarViews.com – Mercury[93]
- Astronomy Cast: Mercury[94]
- Geody Mercury[95] World's search engine that supports NASA World Wind, Celestia, and other applications.
- A Day On Mercury[96] flash animation
- Mercury articles in Planetary Science Research Discoveries[97]
- 'BepiColombo', ESA's Mercury Mission[98] 5 June 2013
- Bauer, Amanda; Merrifield, Michael (2009). "Mercury"[99]. *Sixty Symbols*. Brady Haran for the University of Nottingham.
- NASA Astronomy Picture of the Day: *MESSENGER* Satellite's False Color Mercury Globe Spin (12 June 2013)[100]
- Rotating relief model[101] of Mercury by Seán Doran (starts with Mozart just above center and Caloris Planitia above it)

Venus

<indicator name=”pp-default”> </indicator>

Venus

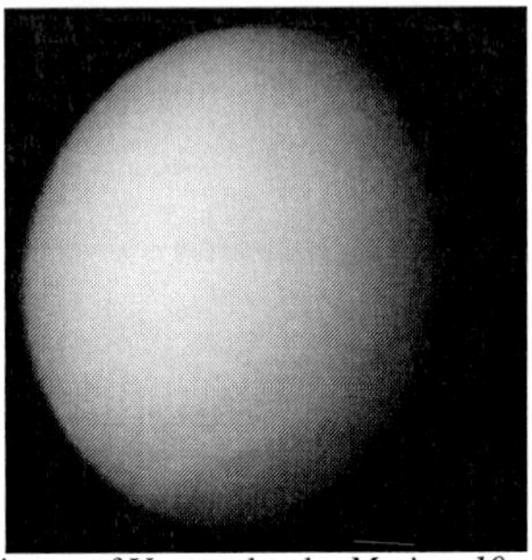 A real-colour image of Venus taken by *Mariner 10* processed from two filters. The surface is obscured by thick sulfuric acid clouds.	
Designations	
Pronunciation	/ˈviːnəs/ Media:en-us-Venus.ogg
Adjectives	Venusian or (rarely) Cytherean, Venerean
Orbital characteristics	
Epoch J2000	
Aphelion	• 0.728213 AU • 108,939,000 km
Perihelion	• 0.718440 AU • 107,477,000 km
Semi-major axis	• 0.723332 AU • 108,208,000 km
Eccentricity	0.006772
Orbital period	• 224.701 d • 0.615198 yr • 1.92 Venus solar day
Synodic period	583.92 days
Average orbital speed	35.02 km/s
Mean anomaly	50.115°
Inclination	• 3.39458° to ecliptic • 3.86° to Sun’s equator • 2.19° to invariable plane

Longitude of ascend-ing node	76.680°
Argument of perihelion	54.884°
Satellites	None
Physical characteristics	
Mean radius	• 6,051.8±1.0 km • 0.9499 Earths
Flattening	0
Surface area	• 4.6023×10^8 km^2 • 0.902 Earths
Volume	• 9.2843×10^{11} km^3 • 0.866 Earths
Mass	• 4.8675×10^{24} kg • 0.815 Earths
Mean density	5.243 g/cm^3
Surface gravity	• 8.87 m/s^2 • 0.904 *g*
Escape velocity	10.36 km/s (6.44 mi/s)
Sidereal rotation period	–243.025 d (retrograde)
Equatorial rota-tion velocity	6.52 km/h (1.81 m/s)
Axial tilt	2.64° (for retrograde rotation) 177.36° (to orbit)[102]
North pole right ascension	• $18^h\ 11^m\ 2^s$ • 272.76°
North pole declination	67.16°
Albedo	• 0.689 (geometric) • 0.77 (Bond) • 0.65 (visual geometric)

Surface temp.	min	mean	max
Kelvin		737 K	
Celsius		462 °C	

Apparent magnitude	–4.9 (crescent) to –3.8 (full)
Angular diameter	9.7”–66.0”
Atmosphere	

Surface pressure	92 bar (9.2 MPa)
Composition by volume	• 96.5% carbon dioxide • 3.5% nitrogen • 0.015% sulfur dioxide • 0.007% argon • 0.002% water vapour • 0.0017% carbon monoxide • 0.0012% helium • 0.0007% neon • trace carbonyl sulfide • trace hydrogen chloride • trace hydrogen fluoride

Venus is the second planet from the Sun, orbiting it every 224.7 Earth days. It has the longest rotation period (243 days) of any planet in the Solar System and rotates in the opposite direction to most other planets. It has no natural satellites. It is named after the Roman goddess of love and beauty. It is the second-brightest natural object in the night sky after the Moon, reaching an apparent magnitude of –4.6 – bright enough to cast shadows at night and, rarely, visible to the naked eye in broad daylight. Orbiting within Earth's orbit, Venus is an inferior planet and never appears to venture far from the Sun; its maximum angular distance from the Sun (elongation) is 47.8°.

Venus is a terrestrial planet and is sometimes called Earth's "sister planet" because of their similar size, mass, proximity to the Sun, and bulk composition. It is radically different from Earth in other respects. It has the densest atmosphere of the four terrestrial planets, consisting of more than 96% carbon dioxide. The atmospheric pressure at the planet's surface is 92 times that of Earth, or roughly the pressure found 900 m (3,000 ft) underwater on Earth. Venus is by far the hottest planet in the Solar System, with a mean surface temperature of 735 K (462 °C; 863 °F), even though Mercury is closer to the Sun. Venus is shrouded by an opaque layer of highly reflective clouds of sulfuric acid, preventing its surface from being seen from space in visible light. It may have had water oceans in the past, but these would have vaporized as the temperature rose due to a runaway greenhouse effect. The water has probably photodissociated, and the free hydrogen has been swept into interplanetary space by the solar wind because of the lack of a planetary magnetic field. Venus's surface is a dry desertscape interspersed with slab-like rocks and is periodically resurfaced by volcanism.

As one of the brightest objects in the sky, Venus has been a major fixture in human culture for as long as records have existed. It has been made sacred to gods of many cultures, and has been a prime inspiration for writers and poets as the "morning star" and "evening star". Venus was the first planet to have its motions plotted across the sky, as early as the second millennium BC.

Figure 75: *Size comparison with Earth*

As the closest planet to Earth, Venus has been a prime target for early interplanetary exploration. It was the first planet beyond Earth visited by a spacecraft (*Mariner 2* in 1962), and the first to be successfully landed on (by *Venera 7* in 1970). Venus's thick clouds render observation of its surface impossible in visible light, and the first detailed maps did not emerge until the arrival of the Magellan orbiter in 1991. Plans have been proposed for rovers or more complex missions, but they are hindered by Venus's hostile surface conditions.

Physical characteristics

Venus is one of the four terrestrial planets in the Solar System, meaning that it is a rocky body like Earth. It is similar to Earth in size and mass, and is often described as Earth's "sister" or "twin". The diameter of Venus is 12,103.6 km (7,520.8 mi)—only 638.4 km (396.7 mi) less than Earth's—and its mass is 81.5% of Earth's. Conditions on the Venusian surface differ radically from those on Earth because its dense atmosphere is 96.5% carbon dioxide, with most of the remaining 3.5% being nitrogen.

Geography

The Venusian surface was a subject of speculation until some of its secrets were revealed by planetary science in the 20th century. Venera landers in 1975 and 1982 returned images of a surface covered in sediment and relatively angular rocks. The surface was mapped in detail by *Magellan* in 1990–91. The ground shows evidence of extensive volcanism, and the sulfur in the atmosphere may indicate that there have been some recent eruptions.

About 80% of the Venusian surface is covered by smooth, volcanic plains, consisting of 70% plains with wrinkle ridges and 10% smooth or lobate plains. Two highland "continents" make up the rest of its surface area, one lying in the planet's northern hemisphere and the other just south of the equator. The northern continent is called Ishtar Terra after Ishtar, the Babylonian goddess of love, and is about the size of Australia. Maxwell Montes, the highest mountain on Venus, lies on Ishtar Terra. Its peak is 11 km (7 mi) above the Venusian average surface elevation. The southern continent is called Aphrodite Terra, after the Greek goddess of love, and is the larger of the two highland regions at roughly the size of South America. A network of fractures and faults covers much of this area.

The absence of evidence of lava flow accompanying any of the visible calderas remains an enigma. The planet has few impact craters, demonstrating that the surface is relatively young, approximately 300–600 million years old. Venus has some unique surface features in addition to the impact craters, mountains, and valleys commonly found on rocky planets. Among these are flat-topped volcanic features called "farra", which look somewhat like pancakes and range in size from 20 to 50 km (12 to 31 mi) across, and from 100 to 1,000 m (330 to 3,280 ft) high; radial, star-like fracture systems called "novae"; features with both radial and concentric fractures resembling spider webs, known as "arachnoids"; and "coronae", circular rings of fractures sometimes surrounded by a depression. These features are volcanic in origin.

Most Venusian surface features are named after historical and mythological women. Exceptions are Maxwell Montes, named after James Clerk Maxwell, and highland regions Alpha Regio, Beta Regio, and Ovda Regio. The latter three features were named before the current system was adopted by the International Astronomical Union, the body which oversees planetary nomenclature.

The longitudes of physical features on Venus are expressed relative to its prime meridian. The original prime meridian passed through the radar-bright spot at the centre of the oval feature Eve, located south of Alpha Regio. After the Venera missions were completed, the prime meridian was redefined to pass through the central peak in the crater Ariadne.

Surface geology

Much of the Venusian surface appears to have been shaped by volcanic activity. Venus has several times as many volcanoes as Earth, and it has 167 large volcanoes that are over 100 km (62 mi) across. The only volcanic complex of this size on Earth is the Big Island of Hawaii.[:154] This is not because Venus is more volcanically active than Earth, but because its crust is older. Earth's

Figure 76: *False-colour image of Maat Mons with a vertical exaggeration of 22.5*

oceanic crust is continually recycled by subduction at the boundaries of tectonic plates, and has an average age of about 100 million years, whereas the Venusian surface is estimated to be 300–600 million years old.

Several lines of evidence point to ongoing volcanic activity on Venus. During the Soviet Venera program, the *Venera 9* orbiter obtained spectroscopic evidence of lightning on Venus, and the *Venera 12* descent probe obtained additional evidence of lightning and thunder. The European Space Agency's *Venus Express* in 2007 detected whistler waves further confirming the occurrence of lightning on Venus. One possibility is that ash from a volcanic eruption was generating the lightning. Another piece of evidence comes from measurements of sulfur dioxide concentrations in the atmosphere, which dropped by a factor of 10 between 1978 and 1986, jumped in 2006, and again declined 10-fold. This may mean that levels had been boosted several times by large volcanic eruptions.

In 2008 and 2009, the first direct evidence for ongoing volcanism was observed by *Venus Express*, in the form of four transient localized infrared hot spots within the rift zone Ganis Chasma,[103] near the shield volcano Maat Mons. Three of the spots were observed in more than one successive orbit. These spots are thought to represent lava freshly released by volcanic eruptions. The actual temperatures are not known, because the size of the hot

Figure 77: *Impact craters on the surface of Venus (false-colour image reconstructed from radar data)*

spots could not be measured, but are likely to have been in the 800–1,100 K (527–827 °C; 980–1,520 °F) range, relative to a normal temperature of 740 K (467 °C; 872 °F).

Almost a thousand impact craters on Venus are evenly distributed across its surface. On other cratered bodies, such as Earth and the Moon, craters show a range of states of degradation. On the Moon, degradation is caused by subsequent impacts, whereas on Earth it is caused by wind and rain erosion. On Venus, about 85% of the craters are in pristine condition. The number of craters, together with their well-preserved condition, indicates the planet underwent a global resurfacing event about 300–600 million years ago, followed by a decay in volcanism. Whereas Earth's crust is in continuous motion, Venus is thought to be unable to sustain such a process. Without plate tectonics to dissipate heat from its mantle, Venus instead undergoes a cyclical process in which mantle temperatures rise until they reach a critical level that weakens the crust. Then, over a period of about 100 million years, subduction occurs on an enormous scale, completely recycling the crust.

Venusian craters range from 3 to 280 km (2 to 174 mi) in diameter. No craters are smaller than 3 km, because of the effects of the dense atmosphere on incoming objects. Objects with less than a certain kinetic energy are slowed

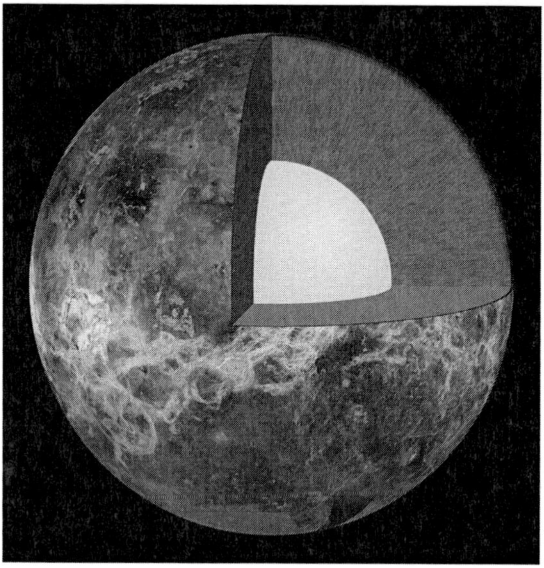

Figure 78: *The internal structure of Venus – the crust (outer layer), the mantle (middle layer) and the core (yellow inner layer)*

down so much by the atmosphere that they do not create an impact crater. Incoming projectiles less than 50 m (160 ft) in diameter will fragment and burn up in the atmosphere before reaching the ground.

Internal structure

Without seismic data or knowledge of its moment of inertia, little direct information is available about the internal structure and geochemistry of Venus. The similarity in size and density between Venus and Earth suggests they share a similar internal structure: a core, mantle, and crust. Like that of Earth, the Venusian core is at least partially liquid because the two planets have been cooling at about the same rate. The slightly smaller size of Venus means pressures are 24% lower in its deep interior than Earth's. The principal difference between the two planets is the lack of evidence for plate tectonics on Venus, possibly because its crust is too strong to subduct without water to make it less viscous. This results in reduced heat loss from the planet, preventing it from cooling and providing a likely explanation for its lack of an internally generated magnetic field. Instead, Venus may lose its internal heat in periodic major resurfacing events.

Atmosphere and climate

Cloud structure in the Venusian atmosphere in 1979, revealed by observations in the ultraviolet band by Pioneer Venus Orbiter

Global radar view of Venus (without the clouds) from *Magellan* between 1990 and 1994

Venus has an extremely dense atmosphere composed of 96.5% carbon dioxide, 3.5% nitrogen, and traces of other gases, most notably sulfur dioxide. The mass of its atmosphere is 93 times that of Earth's, whereas the pressure at its surface is about 92 times that at Earth's—a pressure equivalent to that at a depth of nearly 1 kilometre (0.62 mi) under Earth's oceans. The density at the surface is 65 kg/m^3, 6.5% that of water or 50 times as dense as Earth's atmosphere at 293 K (20 °C; 68 °F) at sea level. The CO_2-rich atmosphere generates the strongest greenhouse effect in the Solar System, creating surface temperatures of at least 735 K (462 °C; 864 °F). This makes Venus's surface hotter than Mercury's, which has a minimum surface temperature of 53 K (–220 °C; –364 °F) and maximum surface temperature of 693 K (420 °C; 788 °F), even though Venus is nearly twice Mercury's distance from the Sun and thus receives only 25% of Mercury's solar irradiance. This temperature is higher than that used for sterilization. The surface of Venus is often said to resemble traditional accounts of Hell.

Studies have suggested that billions of years ago Venus's atmosphere was much more like Earth's than it is now, and that there may have been substantial quantities of liquid water on the surface, but after a period of 600 million to several billion years, a runaway greenhouse effect was caused by the evaporation of that original water, which generated a critical level of greenhouse gases in its atmosphere. Although the surface conditions on Venus are no longer hospitable to any Earthlike life that may have formed before this event, there is speculation on the possibility that life exists in the upper cloud layers of Venus, 50 km (31 mi) up from the surface, where the temperature ranges between 303 and 353 K (30 and 80 °C; 86 and 176 °F) but the environment is acidic.

Thermal inertia and the transfer of heat by winds in the lower atmosphere mean that the temperature of Venus's surface does not vary significantly between the night and day sides, despite Venus's extremely slow rotation. Winds at the surface are slow, moving at a few kilometres per hour, but because of the high density of the atmosphere at the surface, they exert a significant amount of force against obstructions, and transport dust and small stones across the surface. This alone would make it difficult for a human to walk through, even if the heat, pressure, and lack of oxygen were not a problem.

Above the dense CO_2 layer are thick clouds consisting mainly of sulfuric acid droplets. The clouds also contain sulfur aerosol, about 1% ferric chloride and some water. Other possible constituents of the cloud particles are ferric sulfate, aluminium chloride and phosphoric anhydride. Clouds at different levels have different compositions and particle size distributions. These clouds reflect and scatter about 90% of the sunlight that falls on them back into space, and prevent visual observation of Venus's surface. The permanent cloud cover means that although Venus is closer than Earth to the Sun, it receives less sunlight on the ground. Strong 300 km/h (185 mph) winds at the cloud tops go around Venus about every four to five Earth days. Winds on Venus move at up to 60 times the speed of its rotation, whereas Earth's fastest winds are only 10–20% rotation speed.

The surface of Venus is effectively isothermal; it retains a constant temperature not only between day and night sides but between the equator and the poles. Venus's minute axial tilt—less than 3°, compared to 23° on Earth—also minimises seasonal temperature variation. The only appreciable variation in temperature occurs with altitude. The highest point on Venus, Maxwell Montes, is therefore the coolest point on Venus, with a temperature of about 655 K (380 °C; 715 °F) and an atmospheric pressure of about 4.5 MPa (45 bar). In 1995, the *Magellan* spacecraft imaged a highly reflective substance at the tops of the highest mountain peaks that bore a strong resemblance to terrestrial snow. This substance likely formed from a similar process to snow, albeit at a far higher temperature. Too volatile to condense on the surface, it rose in gaseous form to higher elevations, where it is cooler and could precipitate. The identity of this substance is not known with certainty, but speculation has ranged from elemental tellurium to lead sulfide (galena).

The clouds of Venus may be capable of producing lightning. The existence of lightning in the atmosphere of Venus has been controversial since the first suspected bursts were detected by the Soviet Venera probes. In 2006–07, *Venus Express* clearly detected whistler mode waves, the signatures of lightning. Their intermittent appearance indicates a pattern associated with weather activity. According to these measurements, the lightning rate is at least half of that on Earth. In 2007, *Venus Express* discovered that a huge double atmospheric vortex exists at the south pole.

Venus Express also discovered, in 2011, that an ozone layer exists high in the atmosphere of Venus. On 29 January 2013, ESA scientists reported that the ionosphere of Venus streams outwards in a manner similar to "the ion tail seen streaming from a comet under similar conditions."

In December 2015 and to a lesser extent in April and May 2016, researchers working on Japan's *Akatsuki* mission observed bow shapes in the atmosphere of Venus. This was considered direct evidence of the existence of perhaps the largest stationary gravity waves in the solar system.

Atmospheric composition

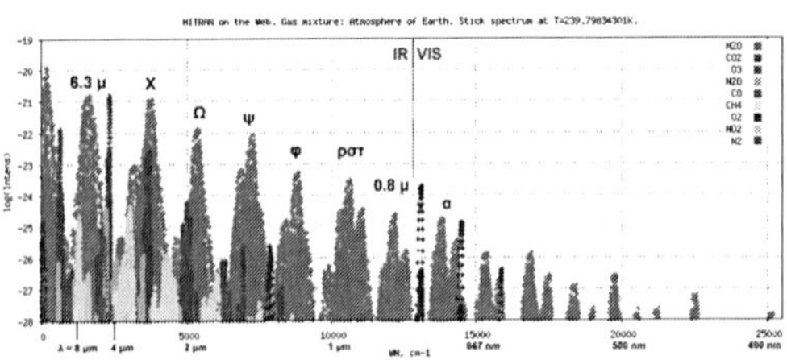

Absorption spectrum of a simple gas mixture corresponding to Earth's atmosphere

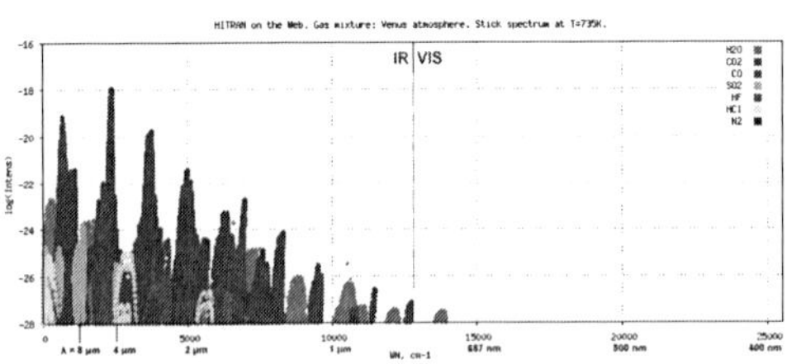

The composition of the atmosphere of Venus based on HITRAN data created using HITRAN on the Web system.

Green colour – water vapour, red – carbon dioxide, WN – wavenumber (other colours have different meanings, lower wavelengths on the right, higher on the left).

Magnetic field and core

In 1967, *Venera 4* found Venus's magnetic field to be much weaker than that of Earth. This magnetic field is induced by an interaction between the ionosphere and the solar wind, rather than by an internal dynamo as in the Earth's core. Venus's small induced magnetosphere provides negligible protection to the atmosphere against cosmic radiation.

The lack of an intrinsic magnetic field at Venus was surprising, given that it is similar to Earth in size, and was expected also to contain a dynamo at its core. A dynamo requires three things: a conducting liquid, rotation, and convection. The core is thought to be electrically conductive and, although its rotation is often thought to be too slow, simulations show it is adequate to produce a dynamo. This implies that the dynamo is missing because of a lack of convection in Venus's core. On Earth, convection occurs in the liquid outer layer of the core because the bottom of the liquid layer is much hotter than the top. On Venus, a global resurfacing event may have shut down plate tectonics and led to a reduced heat flux through the crust. This caused the mantle temperature to increase, thereby reducing the heat flux out of the core. As a result, no internal geodynamo is available to drive a magnetic field. Instead, the heat from the core is being used to reheat the crust.

One possibility is that Venus has no solid inner core, or that its core is not cooling, so that the entire liquid part of the core is at approximately the same temperature. Another possibility is that its core has already completely solidified. The state of the core is highly dependent on the concentration of sulfur, which is unknown at present.

The weak magnetosphere around Venus means that the solar wind is interacting directly with its outer atmosphere. Here, ions of hydrogen and oxygen are being created by the dissociation of neutral molecules from ultraviolet radiation. The solar wind then supplies energy that gives some of these ions sufficient velocity to escape Venus's gravity field. This erosion process results in a steady loss of low-mass hydrogen, helium, and oxygen ions, whereas higher-mass molecules, such as carbon dioxide, are more likely to be retained. Atmospheric erosion by the solar wind probably led to the loss of most of Venus's water during the first billion years after it formed. The erosion has increased the ratio of higher-mass deuterium to lower-mass hydrogen in the atmosphere 100 times compared to the rest of the solar system.

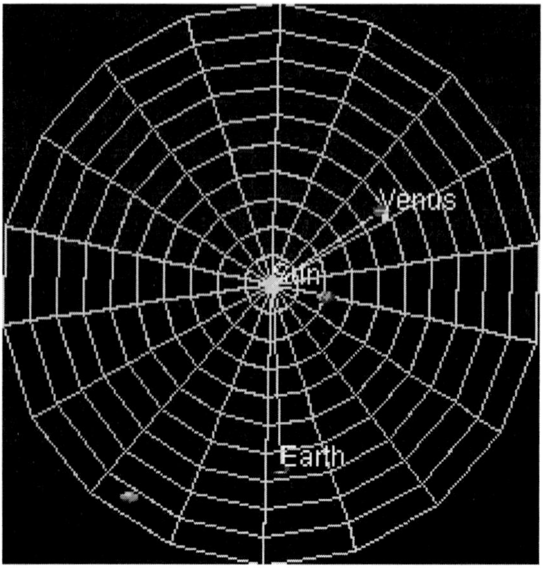

Figure 79: *Venus orbits the Sun at an average distance of about 108 million kilometres (about 0.7 AU) and completes an orbit every 224.7 days. Venus is the second planet from the Sun and orbits the Sun approximately 1.6 times (yellow trail) in Earth's 365 days (blue trail)*

Orbit and rotation

Venus orbits the Sun at an average distance of about 0.72 AU (108 million km; 67 million mi), and completes an orbit every 224.7 days. Although all planetary orbits are elliptical, Venus's orbit is the closest to circular, with an eccentricity of less than 0.01. When Venus lies between Earth and the Sun in inferior conjunction, it makes the closest approach to Earth of any planet at an average distance of 41 million km (25 million mi). The planet reaches inferior conjunction every 584 days, on average. Because of the decreasing eccentricity of Earth's orbit, the minimum distances will become greater over tens of thousands of years. From the year 1 to 5383, there are 526 approaches less than 40 million km; then there are none for about 60,158 years.[104]

All the planets in the Solar System orbit the Sun in an anti-clockwise direction as viewed from above Earth's north pole. Most planets also rotate on their axes in an anti-clockwise direction, but Venus rotates clockwise in retrograde rotation once every 243 Earth days—the slowest rotation of any planet. Because its rotation is so slow, Venus is very close to spherical. A Venusian sidereal day thus lasts longer than a Venusian year (243 versus 224.7 Earth days). Venus's equator rotates at 6.52 km/h (4.05 mph), whereas Earth's rotates at 1,669.8 km/h (1,037.6 mph). Venus's rotation has slowed down in the 16 years between the *Magellan* spacecraft and *Venus Express* visits; each Venusian sidereal day has increased by 6.5 minutes in that time span. Because of the retrograde rotation, the length of a solar day on Venus is significantly shorter than the sidereal day, at 116.75 Earth days (making the Venusian solar day shorter than Mercury's 176 Earth days). One Venusian year is about 1.92 Venusian solar days. To an observer on the surface of Venus, the Sun would rise in the west and set in the east, although Venus's opaque clouds prevent observing the Sun from the planet's surface.

Venus may have formed from the solar nebula with a different rotation period and obliquity, reaching its current state because of chaotic spin changes caused by planetary perturbations and tidal effects on its dense atmosphere, a change that would have occurred over the course of billions of years. The rotation period of Venus may represent an equilibrium state between tidal locking to the Sun's gravitation, which tends to slow rotation, and an atmospheric tide created by solar heating of the thick Venusian atmosphere. The 584-day average interval between successive close approaches to Earth is almost exactly equal to 5 Venusian solar days, but the hypothesis of a spin–orbit resonance with Earth has been discounted.

Venus has no natural satellites. It has several trojan asteroids: the quasi-satellite 2002 VE68 and two other temporary trojans, 2001 CK_{32} and 2012 XE133. In the 17th century, Giovanni Cassini reported a moon orbiting Venus, which was named Neith and numerous sightings were reported over the following 200 years, but most were determined to be stars in the vicinity. Alex Alemi's and David Stevenson's 2006 study of models of the early Solar System at the California Institute of Technology shows Venus likely had at least one moon created by a huge impact event billions of years ago. About 10 million years later, according to the study, another impact reversed the planet's spin direction and caused the Venusian moon gradually to spiral inward until it collided with Venus. If later impacts created moons, these were removed in the same way. An alternative explanation for the lack of satellites is the effect of strong solar tides, which can destabilize large satellites orbiting the inner terrestrial planets.

Observation

To the naked eye, Venus appears as a white point of light brighter than any other planet or star (apart from the Sun). Its brightest apparent magnitude, –4.9,[105] occurs during crescent phase, only 36 days before or after inferior conjunction. Venus will be brightest on 30 April 2017, then grow dimmer for nearly a year. Venus fades to about magnitude –3 when it is backlit by the Sun.

Figure 80: *Venus is always brighter than all other planets or stars as seen from Earth. The second brightest object on the image is Jupiter.*

The planet is bright enough to be seen in a clear midday sky and is more easily visible when the Sun is low on the horizon or setting. As an inferior planet, it always lies within about 47° of the Sun.

Venus "overtakes" Earth every 584 days as it orbits the Sun. As it does so, it changes from the "Evening Star", visible after sunset, to the "Morning Star", visible before sunrise. Although Mercury, the other inferior planet, reaches a maximum elongation of only 28° and is often difficult to discern in twilight, Venus is hard to miss when it is at its brightest. Its greater maximum elongation means it is visible in dark skies long after sunset. As the brightest point-like object in the sky, Venus is a commonly misreported "unidentified flying object".

Phases

As it orbits the Sun, Venus displays phases like those of the Moon in a telescopic view. The planet appears as a small and "full" disc when it is on the opposite side of the Sun (at superior conjunction). Venus shows a larger disc and "quarter phase" at its maximum elongations from the Sun, and appears its brightest in the night sky. The planet presents a much larger thin "crescent" in telescopic views as it passes along the near side between Earth and the Sun. Venus displays its largest size and "new phase" when it is between Earth and the Sun (at inferior conjunction). Its atmosphere is visible through telescopes by the halo of sunlight refracted around it.

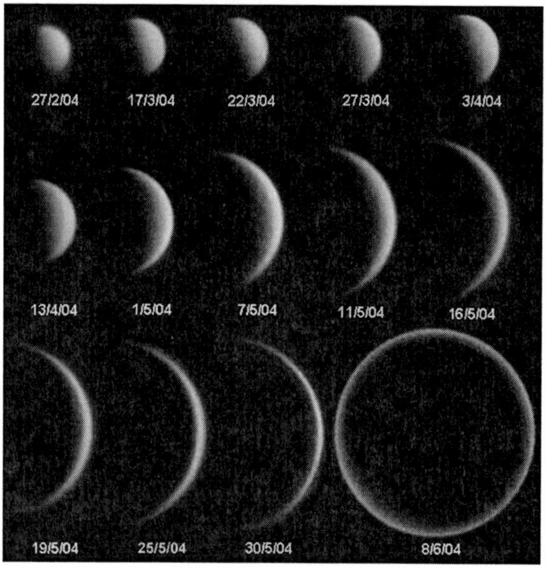

Figure 81: *The phases of Venus and evolution of its apparent diameter*

Transits

The Venusian orbit is slightly inclined relative to Earth's orbit; thus, when the planet passes between Earth and the Sun, it usually does not cross the face of the Sun. Transits of Venus occur when the planet's inferior conjunction coincides with its presence in the plane of Earth's orbit. Transits of Venus occur in cycles of 243 years with the current pattern of transits being pairs of transits separated by eight years, at intervals of about 105.5 years or 121.5 years—a pattern first discovered in 1639 by the English astronomer Jeremiah Horrocks.

The latest pair was June 8, 2004 and June 5–6, 2012. The transit could be watched live from many online outlets or observed locally with the right equipment and conditions.

The preceding pair of transits occurred in December 1874 and December 1882; the following pair will occur in December 2117 and December 2125. The oldest film known is the 1874 *Passage de Venus*, showing the 1874 Venus transit of the sun. Historically, transits of Venus were important, because they allowed astronomers to determine the size of the astronomical unit, and hence the size of the Solar System as shown by Horrocks in 1639. Captain Cook's exploration of the east coast of Australia came after he had sailed to Tahiti in 1768 to observe a transit of Venus.

Figure 82: *2004 transit of Venus*

Pentagram of Venus

The pentagram of Venus is the path that Venus makes as observed from Earth. Successive inferior conjunctions of Venus repeat very near a 13:8 orbital resonance (Earth orbits 8 times for every 13 orbits of Venus), shifting 144° upon sequential inferior conjunctions. The resonance 13:8 ratio is approximate. 8/13 is approximately 0.615385 while Venus orbits the Sun in 0.615187 years.

Daylight apparitions

Naked eye observations of Venus during daylight hours exist in several anecdotes and records. Astronomer Edmund Halley calculated its maximum naked eye brightness in 1716, when many Londoners were alarmed by its appearance in the daytime. French emperor Napoleon Bonaparte once witnessed a daytime apparition of the planet while at a reception in Luxembourg. Another historical daytime observation of the planet took place during the inauguration of the American president Abraham Lincoln in Washington, D.C., on 4 March 1865. Although naked eye visibility of Venus' phases is disputed, records exist of observations of its crescent.

Figure 83: *The pentagram of Venus. Earth is positioned at the centre of the diagram, and the curve represents the direction and distance of Venus as a function of time.*

Ashen light

A long-standing mystery of Venus observations is the so-called ashen light—an apparent weak illumination of its dark side, seen when the planet is in the crescent phase. The first claimed observation of ashen light was made in 1643, but the existence of the illumination has never been reliably confirmed. Observers have speculated it may result from electrical activity in the Venusian atmosphere, but it could be illusory, resulting from the physiological effect of observing a bright, crescent-shaped object.

Studies

Early studies

Venus was known to ancient civilizations both as the "morning star" and as the "evening star", names that reflect the early assumption that these were two separate objects. The ancient Sumerians, who recognized Venus as a single object, believed that it was their goddess Inanna. Inanna's movements in several of her myths, including *Inanna and Shukaletuda* and *Inanna's Descent into the Underworld* appear to parallel the motion of the planet Venus. The

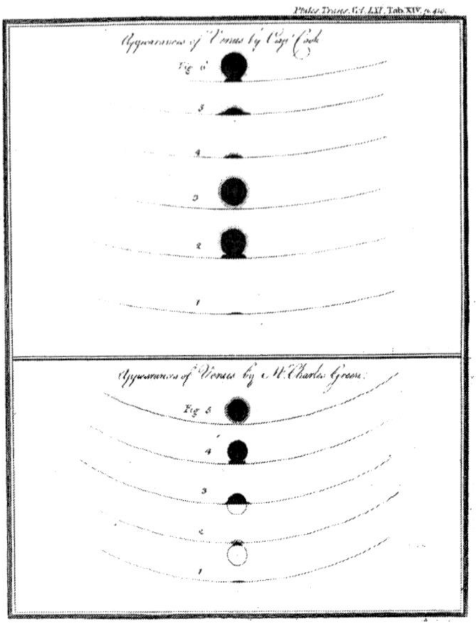

Figure 84: *The "black drop effect" as recorded during the 1769 transit*

Venus tablet of Ammisaduqa, believed to have been compiled around the mid-seventeenth century BCE, shows the Babylonians understood the two were a single object, referred to in the tablet as the "bright queen of the sky", and could support this view with detailed observations.

The ancient Greeks thought that Venus was two separate stars: Phosphorus and Hesperus. Pliny the Elder credited the realization that they were a single object to Pythagoras in the sixth century BCE, while Diogenes Laertius argued that Parmenides was probably responsible. The ancient Chinese referred to the morning Venus as "the Great White" (*Tai-bai* 太白) or "the Opener (Starter) of Brightness" (*Qi-ming* 啓明), and the evening Venus as "the Excellent West One" (*Chang-geng* 長庚). The Romans designated the morning aspect of Venus as Lucifer, literally "Light-Bringer", and the evening aspect as Vesper, both literal translations of the respective Greek names.

In the second century, in his astronomical treatise *Almagest*, Ptolemy theorized that both Mercury and Venus are located between the Sun and the Earth. The 11th century Persian astronomer Avicenna claimed to have observed the transit of Venus, which later astronomers took as confirmation of Ptolemy's theory. In the 12th century, the Andalusian astronomer Ibn Bajjah observed "two planets as black spots on the face of the Sun", which were later identified as the transits of Venus and Mercury by the Maragha astronomer Qotb al-Din Shirazi in the

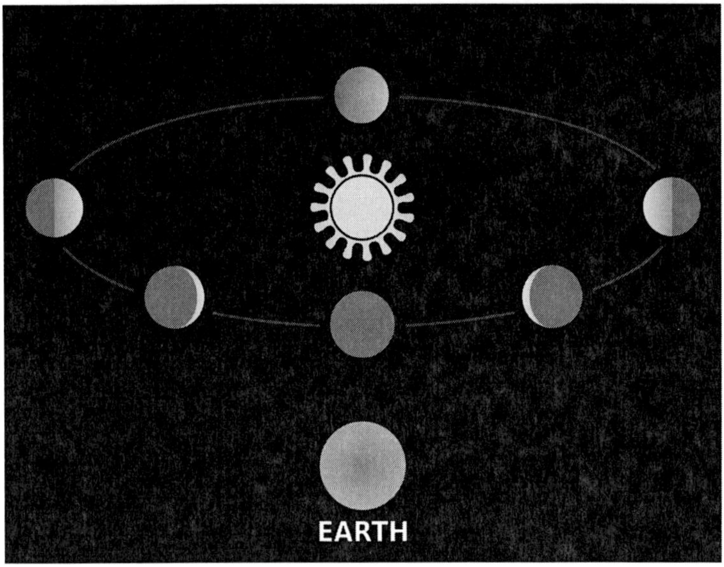

Figure 85: *Galileo's discovery that Venus showed phases (although remaining near the Sun in Earth's sky) proved that it orbits the Sun and not Earth*

13th century.[106] Avicenna did not record the date of his observation. There was a transit of Venus within his lifetime, on 24 May 1032, although it is questionable whether it would have been visible from his location.</ref>

When the Italian physicist Galileo Galilei first observed the planet in the early 17th century, he found it showed phases like the Moon, varying from crescent to gibbous to full and vice versa. When Venus is furthest from the Sun in the sky, it shows a half-lit phase, and when it is closest to the Sun in the sky, it shows as a crescent or full phase. This could be possible only if Venus orbited the Sun, and this was among the first observations to clearly contradict the Ptolemaic geocentric model that the Solar System was concentric and centred on Earth.

The 1639 transit of Venus was accurately predicted by Jeremiah Horrocks and observed by him and his friend, William Crabtree, at each of their respective homes, on 4 December 1639 (24 November under the Julian calendar in use at that time).

The atmosphere of Venus was discovered in 1761 by Russian polymath Mikhail Lomonosov. Venus's atmosphere was observed in 1790 by German astronomer Johann Schröter. Schröter found when the planet was a thin crescent, the cusps extended through more than 180°. He correctly surmised this

Figure 86: *Modern telescopic view of Venus from Earth's surface*

was due to scattering of sunlight in a dense atmosphere. Later, American astronomer Chester Smith Lyman observed a complete ring around the dark side of the planet when it was at inferior conjunction, providing further evidence for an atmosphere. The atmosphere complicated efforts to determine a rotation period for the planet, and observers such as Italian-born astronomer Giovanni Cassini and Schröter incorrectly estimated periods of about 24 h from the motions of markings on the planet's apparent surface.

Ground-based research

Little more was discovered about Venus until the 20th century. Its almost featureless disc gave no hint what its surface might be like, and it was only with the development of spectroscopic, radar and ultraviolet observations that more of its secrets were revealed. The first ultraviolet observations were carried out in the 1920s, when Frank E. Ross found that ultraviolet photographs revealed considerable detail that was absent in visible and infrared radiation. He suggested this was due to a dense, yellow lower atmosphere with high cirrus clouds above it.

Spectroscopic observations in the 1900s gave the first clues about the Venusian rotation. Vesto Slipher tried to measure the Doppler shift of light from Venus, but found he could not detect any rotation. He surmised the planet must have

Figure 87: *Artist's impression of Mariner 2, launched in 1962, a skeletal, bottle-shaped spacecraft with a large radio dish on top*

a much longer rotation period than had previously been thought. Later work in the 1950s showed the rotation was retrograde. Radar observations of Venus were first carried out in the 1960s, and provided the first measurements of the rotation period, which were close to the modern value.

Radar observations in the 1970s revealed details of the Venusian surface for the first time. Pulses of radio waves were beamed at the planet using the 300 m (980 ft) radio telescope at Arecibo Observatory, and the echoes revealed two highly reflective regions, designated the Alpha and Beta regions. The observations also revealed a bright region attributed to mountains, which was called Maxwell Montes. These three features are now the only ones on Venus that do not have female names.

Exploration

The first robotic space probe mission to Venus, and the first to any planet, began with the Soviet Venera program in 1961. The United States' exploration of Venus had its first success with the Mariner 2 mission on 14 December 1962, becoming the world's first successful interplanetary mission, passing 34,833 km (21,644 mi) above the surface of Venus, and gathering data on the planet's atmosphere.

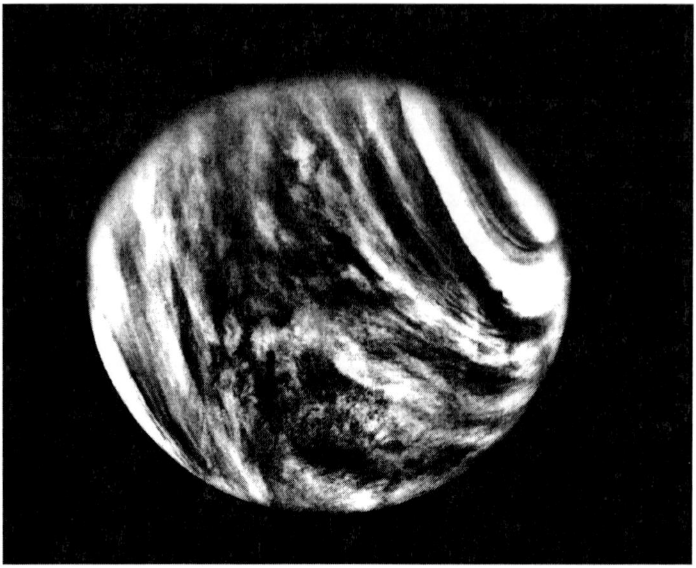

Figure 88: *Global view of Venus in ultraviolet light done by Mariner 10.*

On 18 October 1967, the Soviet *Venera 4* successfully entered the atmosphere and deployed science experiments. *Venera 4* showed the surface temperature was hotter than *Mariner 2* had calculated, at almost 500 °C, determined that the atmosphere is 95% carbon dioxide (CO
2), and discovered that Venus's atmosphere was considerably denser than *Venera 4*'s designers had anticipated. The joint *Venera 4–Mariner 5* data were analysed by a combined Soviet–American science team in a series of colloquia over the following year, in an early example of space cooperation.

In 1974, *Mariner 10* swung by Venus on its way to Mercury and took ultraviolet photographs of the clouds, revealing the extraordinarily high wind speeds in the Venusian atmosphere.

In 1975, the Soviet *Venera 9* and *10* landers transmitted the first images from the surface of Venus, which were in black and white. In 1982 the first colour images of the surface were obtained with the Soviet *Venera 13* and *14* landers.

NASA obtained additional data in 1978 with the Pioneer Venus project that consisted of two separate missions: Pioneer Venus Orbiter and Pioneer Venus Multiprobe. The successful Soviet Venera program came to a close in October 1983, when *Venera 15* and *16* were placed in orbit to conduct detailed mapping of 25% of Venus's terrain (from the north pole to 30°N latitude)

Several other Venus flybys took place in the 1980s and 1990s that increased the understanding of Venus, including *Vega 1* (1985), *Vega 2* (1985), *Galileo* (1990), *Magellan* (1994), *Cassini–Huygens* (1998), and *MESSENGER* (2006). Then, *Venus Express* by the European Space Agency (ESA) entered orbit around Venus in April 2006. Equipped with seven scientific instruments, *Venus Express* provided unprecedented long-term observation of Venus's atmosphere. ESA concluded that mission in December 2014.

As of 2016, Japan's *Akatsuki* is in a highly elliptical orbit around Venus since 7 December 2015, and there are several probing proposals under study by Roscosmos, NASA, and India's ISRO.

In 2016, NASA announced that it was planning a rover, the Automaton Rover for Extreme Environments, designed to survive for an extended time in Venus' environmental conditions. It would be controlled by a mechanical computer and driven by wind power.

In culture

Venus is a primary feature of the night sky, and so has been of remarkable importance in mythology, astrology and fiction throughout history and in different cultures. Classical poets such as Homer, Sappho, Ovid and Virgil spoke of the star and its light. Romantic poets such as William Blake, Robert Frost, Alfred Lord Tennyson and William Wordsworth wrote odes to it. With the invention of the telescope, the idea that Venus was a physical world and possible destination began to take form.

The impenetrable Venusian cloud cover gave science fiction writers free rein to speculate on conditions at its surface; all the more so when early observations showed that not only was it similar in size to Earth, it possessed a substantial atmosphere. Closer to the Sun than Earth, the planet was frequently depicted as warmer, but still habitable by humans. The genre reached its peak between the 1930s and 1950s, at a time when science had revealed some aspects of Venus, but not yet the harsh reality of its surface conditions. Findings from the first missions to Venus showed the reality to be quite different, and brought this particular genre to an end. As scientific knowledge of Venus advanced, so science fiction authors tried to keep pace, particularly by conjecturing human attempts to terraform Venus.

Symbol

♀ The astronomical symbol for Venus is the same as that used in biology for the female sex: a circle with a small cross beneath. The Venus symbol also represents femininity, and in Western alchemy stood for the metal copper. Polished copper has been used for mirrors from antiquity, and the symbol for Venus has sometimes been understood to stand for the mirror of the goddess.

Habitability

The speculation of the existence of life on Venus decreased significantly since the early 1960s, when spacecraft began studying Venus and it became clear that the conditions on Venus are extreme compared to those on Earth.

The fact that Venus is located closer to the Sun than Earth, raising temperatures on the surface to nearly 735 K (462 °C; 863 °F), the atmospheric pressure is ninety times that of Earth, and the extreme impact of the greenhouse effect, make water-based life as we know it unlikely. A few scientists have speculated that thermoacidophilic extremophile microorganisms might exist in the lower-temperature, acidic upper layers of the Venusian atmosphere.[107] The atmospheric pressure and temperature fifty kilometres above the surface are similar to those at Earth's surface. This has led to proposals to use aerostats (lighter-than-air balloons) for initial exploration and ultimately for permanent "floating cities" in the Venusian atmosphere. Among the many engineering challenges are the dangerous amounts of sulfuric acid at these heights.

External links

Wikimedia Commons has media related to ***Venus***.

Wikiquote has quotations related to: ***Venus***

- Venus profile[108] at NASA's Solar System Exploration site
- Missions to Venus[109] and Image catalog[110] at the National Space Science Data Center
- Soviet Exploration of Venus[111] and Image catalog[112] at Mentallandscape.com
- Venus page[113] at *The Nine Planets*
- Transits of Venus[114] at NASA.gov
- Geody Venus[115], a search engine for surface features

Cartographic resources

- Map-a-Planet: Venus[116] by the U.S. Geological Survey
- Gazetteer of Planetary Nomenclature: Venus[117] by the International Astronomical Union
- Venus crater database[118] by the Lunar and Planetary Institute
- Map of Venus[119] by Eötvös Loránd University

<indicator name="featured-star"> </indicator>

Earth

<indicator name="pp-default"> </indicator> <indicator name="featured-star"> </indicator>

Earth

The Blue Marble photograph of Earth, taken during the Apollo 17 lunar mission in 1972

Orbital characteristics	
Epoch J2000	
Aphelion	152100000 km (94500000 mi; 1.017 AU)
Perihelion	147095000 km (91401000 mi; 0.98327 AU)
Semi-major axis	149598023 km (92955902 mi; 1.00000102 AU)
Eccentricity	0.0167086
Orbital period	365.256363004 d (1.00001742096 yr)
Average orbital speed	29.78 km/s (107200 km/h; 66600 mph)
Mean anomaly	358.617°

Inclination	• 7.155° to the Sun's equator; • 1.57869° to invariable plane; • 0.00005° to J2000 ecliptic
Longitude of ascending node	–11.26064° to J2000 ecliptic
Argument of perihelion	114.20783°
Satellites	• 1 natural satellite: the Moon • 5 quasi-satellites • >1 400 operational artificial satellites • >16 000 space debris
Physical characteristics	
Mean radius	6371.0 km (3958.8 mi)
Equatorial radius	6378.1 km (3963.2 mi)
Polar radius	6356.8 km (3949.9 mi)
Flattening	0.0033528 1/298.257222101 (ETRS89)
Circumference	• 40075.017 km equatorial (24901.461 mi)[120] • 40007.86 km meridional (24859.73 mi)[121]
Surface area	• 510072000 km^2 (196940000 sq mi) • 148940000 km^2 land (57510000 sq mi; 29.2%) • 361132000 km^2 water (139434000 sq mi; 70.8%)
Volume	1.08321×10^{12} km^3 (2.59876×10^{11} cu mi)
Mass	5.97237×10^{24} kg (1.31668×10^{25} lb) (3.0×10^{-6} $M_{\odot}$)
Mean density	5.514 g/cm^3 (0.1992 lb/cu in)
Surface gravity	9.807 m/s^2 (1 *g*; 32.18 ft/s^2)
Moment of inertia factor	0.3307
Escape velocity	11.186 km/s (40270 km/h; 25020 mph)
Sidereal rotation period	0.99726968 d (23h 56m 4.100s)
Equatorial rotation velocity	0.4651 km/s (1674.4 km/h; 1040.4 mph)
Axial tilt	23.4392811°
Albedo	• 0.367 geometric • 0.306 Bond

Surface temp.	min	mean	max
Kelvin	184 K	288 K	330 K
Celsius	–89.2 °C	15 °C	56.7 °C
Fahrenheit	–128.5 °F	59 °F	134 °F

Atmosphere

Surface pressure	101.325 kPa (at MSL)
Composition by volume	• 78.08% nitrogen (N_2; dry air) • 20.95% oxygen (O_2) • 0.930% argon • 0.0402% carbon dioxide • ∼ 1% water vapor (climate-variable)

Earth is the third planet from the Sun and the only object in the Universe known to harbor life. According to radiometric dating and other sources of evidence, Earth formed over 4 billion years ago. Earth's gravity interacts with other objects in space, especially the Sun and the Moon, Earth's only natural satellite. Earth revolves around the Sun in 365.26 days, a period known as an Earth year. During this time, Earth rotates about its axis about 366.26 times.

Earth's axis of rotation is tilted, producing seasonal variations on the planet's surface. The gravitational interaction between the Earth and Moon causes ocean tides, stabilizes the Earth's orientation on its axis, and gradually slows its rotation. Earth is the densest planet in the Solar System and the largest of the four terrestrial planets.

Earth's lithosphere is divided into several rigid tectonic plates that migrate across the surface over periods of many millions of years. About 71% of Earth's surface is covered with water, mostly by oceans. The remaining 29% is land consisting of continents and islands that together have many lakes, rivers and other sources of water that contribute to the hydrosphere. The majority of Earth's polar regions are covered in ice, including the Antarctic ice sheet and the sea ice of the Arctic ice pack. Earth's interior remains active with a solid iron inner core, a liquid outer core that generates the Earth's magnetic field, and a convecting mantle that drives plate tectonics.

Within the first billion years of Earth's history, life appeared in the oceans and began to affect the Earth's atmosphere and surface, leading to the proliferation of aerobic and anaerobic organisms. Some geological evidence indicates that life may have arisen as much as 4.1 billion years ago. Since then, the combination of Earth's distance from the Sun, physical properties, and geological history have allowed life to evolve and thrive.[122] In the history of the Earth, biodiversity has gone through long periods of expansion, occasionally punctuated by mass extinction events. Over 99% of all species that ever lived on Earth are extinct. Estimates of the number of species on Earth today vary widely; most species have not been described. Over 7.4 billion humans live on Earth and depend on its biosphere and natural resources for their survival. Humans have developed diverse societies and cultures; politically, the world has about 200 sovereign states.

Name and etymology

The modern English word *Earth* developed from a wide variety of Middle English forms,[123] which derived from an Old English noun most often spelled *eorðe*.[124] It has cognates in every Germanic language, and their proto-Germanic root has been reconstructed as **erþō*. In its earliest appearances, *eorðe* was already being used to translate the many senses of Latin *terra* and Greek γῆ (*gē*): the ground,[125]</ref> its soil,[126]</ref> dry land,[127]
"And God called the dry land **Earth**; and the gathering together of the waters called he Seas."[128]</ref> the human world,[129]</ref> the surface of the world (including the sea),[130]
"Here first with mighty power the Everlasting Lord, the Helm of all created things, Almighty King, made **earth** and heaven, raised up the sky and founded the spacious land."[131]</ref> and the globe itself.[132]</ref> As with Terra and Gaia, Earth was a personified goddess in Germanic paganism: the Angles were listed by Tacitus as among the devotees of Nerthus,[133] and later Norse mythology included Jörð, a giantess often given as the mother of Thor.[134]

Originally, *earth* was written in lowercase, and from early Middle English, its definite sense as "the globe" was expressed as *the earth*. By early Modern English, many nouns were capitalized, and *the earth* became (and often remained) *the Earth*, particularly when referenced along with other heavenly bodies. More recently, the name is sometimes simply given as *Earth*, by analogy with the names of the other planets. House styles now vary: Oxford spelling recognizes the lowercase form as the most common, with the capitalized form an acceptable variant. Another convention capitalizes "Earth" when appearing as a name (e.g. "Earth's atmosphere") but writes it in lowercase when preceded by *the* (e.g. "the atmosphere of the earth"). It almost always appears in lowercase in colloquial expressions such as "what on earth are you doing?"[135]

Chronology

Formation

The oldest material found in the Solar System is dated to 4.5672±0.0006 billion years ago (Bya). By 4.54±0.04 Bya the primordial Earth had formed. The formation and evolution of Solar System bodies occurred along with the Sun. In theory, a solar nebula partitions a volume out of a molecular cloud by gravitational collapse, which begins to spin and flatten into a circumstellar disk, and then the planets grow out of that disk along with the Sun. A nebula contains gas, ice grains, and dust (including primordial nuclides). According to nebular

Figure 89: *Artist's impression of the early Solar System's planetary disk*

theory, planetesimals formed by accretion, with the primordial Earth taking 10–20 million years (Mys) to form.

A subject of on-going research is the formation of the Moon, some 4.53 Bya. A working hypothesis is that it was formed by accretion from material loosed from Earth after a Mars-sized object, named Theia, impacted Earth. In this scenario, the mass of Theia was approximately 10% of that of Earth, it impacted Earth with a glancing blow, and some of its mass merged with Earth. Between approximately 4.1 and 3.8 Bya, numerous asteroid impacts during the Late Heavy Bombardment caused significant changes to the greater surface environment of the Moon, and by inference, to that of Earth.

Geological history

Earth's atmosphere and oceans were formed by volcanic activity and outgassing that included water vapor. The origin of the world's oceans was condensation augmented by water and ice delivered by asteroids, protoplanets, and comets. In this model, atmospheric "greenhouse gases" kept the oceans from freezing when the newly forming Sun had only 70% of its current luminosity. By 3.5 Bya, Earth's magnetic field was established, which helped prevent the atmosphere from being stripped away by the solar wind.

A crust formed when the molten outer layer of Earth cooled to form a solid. The two models that explain land mass propose either a steady growth to the present-day forms or, more likely, a rapid growth early in Earth history followed by a long-term steady continental area. Continents formed by plate tectonics, a process ultimately driven by the continuous loss of heat from Earth's

Figure 90: *Hoodoos at the Bryce Canyon National Park, Utah*

interior. On time scales lasting hundreds of millions of years, the supercontinents have assembled and broken apart. Roughly 750 million years ago (Mya), one of the earliest known supercontinents, Rodinia, began to break apart. The continents later recombined to form Pannotia, 600–540 Mya, then finally Pangaea, which also broke apart 180 Mya.

The present pattern of ice ages began about 40 Mya and then intensified during the Pleistocene about 3 Mya. High-latitude regions have since undergone repeated cycles of glaciation and thaw, repeating about every 40,000–100000 years. The last continental glaciation ended 10,000 years ago.

Origin of life and evolution

Life timeline

0 —
500 —
1000 —
1500 —
2000 —
2500 —
3000 —
3500 —
4000 —
4500 —

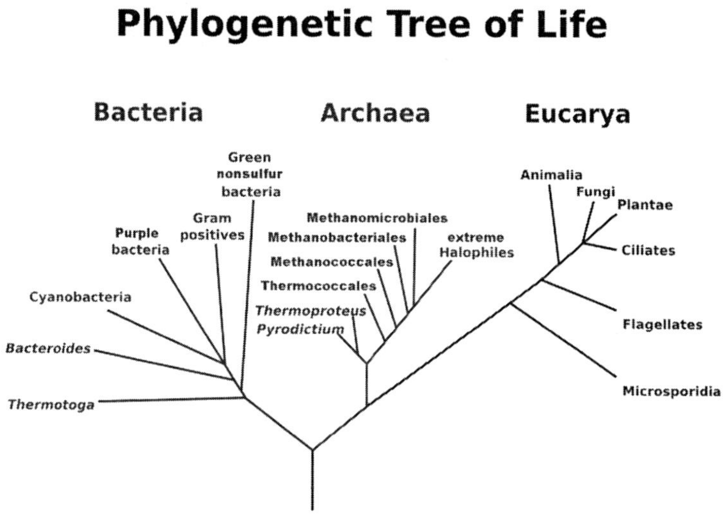

Figure 91: *Speculative phylogenetic tree of life on Earth based on rRNA analysis*

Axis scale: millions of years.

Orange labels: known *ice ages*.

Also see: *Human timeline* and *Nature timeline*

Chemical reactions led to the first self-replicating molecules about four billion years ago. A half billion years later, the last common ancestor of all life arose. The evolution of photosynthesis allowed the Sun's energy to be harvested directly by life forms. The resultant molecular oxygen (O_2) accumulated in the atmosphere and due to interaction with ultraviolet solar radiation, formed a protective ozone layer (O_3) in the upper atmosphere. The incorporation of smaller cells within larger ones resulted in the development of complex cells called eukaryotes. True multicellular organisms formed as cells within colonies became increasingly specialized. Aided by the absorption of harmful ultraviolet radiation by the ozone layer, life colonized Earth's surface. Among the earliest fossil evidence for life is microbial mat fossils found in 3.48 billion-year-old sandstone in Western Australia,[136,137] biogenic graphite found in 3.7 billion-year-old metasedimentary rocks in Western Greenland, remains of biotic material found in 4.1 billion-year-old rocks in Western Australia.

During the Neoproterozoic, 750 to 580 Mya, much of Earth might have been covered in ice. This hypothesis has been termed "Snowball Earth", and it is of particular interest because it preceded the Cambrian explosion, when multicellular life forms significantly increased in complexity. Following the Cambrian explosion, 535 Mya, there have been five major mass extinctions. The most recent such event was 66 Mya, when an asteroid impact triggered the extinction of the non-avian dinosaurs and other large reptiles, but spared some small animals such as mammals, which then resembled shrews. Over the past 66 Mys, mammalian life has diversified, and several million years ago an African ape-like animal such as *Orrorin tugenensis* gained the ability to stand upright. This facilitated tool use and encouraged communication that provided the nutrition and stimulation needed for a larger brain, which allowed the evolution of humans. The development of agriculture, and then civilization, led to humans having an influence on Earth and the nature and quantity of other life forms that continues today.

Future

Earth's expected long-term future is closely tied to that of the Sun. Over the next 1.1 Bys, solar luminosity will increase by 10%, and over the next 3.5 Bys by 40%. The Earth's increasing surface temperature will accelerate the inorganic CO_2 cycle, reducing its concentration to levels lethally low for plants (10 ppm for C4 photosynthesis) in approximately 500–900 Mys. The lack of vegetation will result in the loss of oxygen in the atmosphere, and animal life will become extinct. After another billion years all surface water will have disappeared and the mean global temperature will reach 70 °C (158 °F). From that point, the Earth is expected to be habitable for another 500 Ma, possibly up to 2.3 Ga if nitrogen is removed from the atmosphere. Even if the Sun were eternal and stable, 27% of the water in the modern oceans will descend to the mantle in one billion years, due to reduced steam venting from mid-ocean ridges.

The Sun will evolve to become a red giant in about 5 Bys. Models predict that the Sun will expand to roughly 1 AU (150 million km; 93 million mi), which is about 250 times its present radius. Earth's fate is less clear. As a red giant, the Sun will lose roughly 30% of its mass, so, without tidal effects, Earth will move to an orbit 1.7 AU (250 million km; 160 million mi) from the Sun when the star reaches its maximum radius. Most, if not all, remaining life will be destroyed by the Sun's increased luminosity (peaking at about 5,000 times its present level). A 2008 simulation indicates that Earth's orbit will eventually decay due to tidal effects and drag, causing it to enter the Sun's atmosphere and be vaporized.

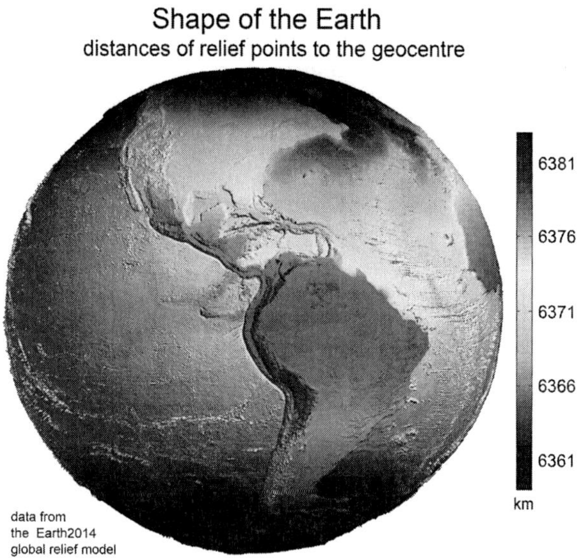

Figure 92: *Shape of planet Earth. Shown are distances between surface relief and the geocentre. The South American Andes summits are visible as elevated areas. Data from the Earth2014 global relief model.*

Physical characteristics

Shape

The shape of Earth is approximately oblate spheroidal. Due to rotation, the Earth is flattened along the geographic axis and bulging around the equator. The diameter of the Earth at the equator is 43 kilometres (27 mi) larger than the pole-to-pole diameter. Thus the point on the surface farthest from Earth's center of mass is the summit of the equatorial Chimborazo volcano in Ecuador. The average diameter of the reference spheroid is 12,742 kilometres (7,918 mi). Local topography deviates from this idealized spheroid, although on a global scale these deviations are small compared to Earth's radius: The maximum deviation of only 0.17% is at the Mariana Trench (10,911 metres (35,797 ft) below local sea level), whereas Mount Everest (8,848 metres (29,029 ft) above local sea level) represents a deviation of 0.14%.[138]</ ref>

In geodesy, the exact shape that Earth's oceans would adopt in the absence of land and perturbations such as tides and winds is called the geoid. More precisely, the geoid is the surface of gravitational equipotential at mean sea level.

Chemical composition

Chemical composition of the crust

Compound	Formula	Composition	
		Continental	Oceanic
silica	SiO_2	60.2%	48.6%
alumina	Al_2O_3	15.2%	16.5%
lime	CaO	5.5%	12.3%
magnesia	MgO	3.1%	6.8%
iron(II) oxide	FeO	3.8%	6.2%
sodium oxide	Na_2O	3.0%	2.6%
potassium oxide	K_2O	2.8%	0.4%
iron(III) oxide	Fe_2O_3	2.5%	2.3%
water	H_2O	1.4%	1.1%
carbon dioxide	CO_2	1.2%	1.4%
titanium dioxide	TiO_2	0.7%	1.4%
phosphorus pentoxide	P_2O_5	0.2%	0.3%
Total		**99.6%**	**99.9%**

Earth's mass is approximately 5.97×10^{24} kg (5,970 Yg). It is composed mostly of iron (32.1%), oxygen (30.1%), silicon (15.1%), magnesium (13.9%), sulfur (2.9%), nickel (1.8%), calcium (1.5%), and aluminium (1.4%), with the remaining 1.2% consisting of trace amounts of other elements. Due to mass segregation, the core region is estimated to be primarily composed of iron (88.8%), with smaller amounts of nickel (5.8%), sulfur (4.5%), and less than 1% trace elements.

The most common rock constituents of the crust are nearly all oxides: chlorine, sulfur, and fluorine are the important exceptions to this and their total amount in any rock is usually much less than 1%. Over 99% of the crust is composed of 11 oxides, principally silica, alumina, iron oxides, lime, magnesia, potash, and soda.

Internal structure

Earth's interior, like that of the other terrestrial planets, is divided into layers by their chemical or physical (rheological) properties. The outer layer is a chemically distinct silicate solid crust, which is underlain by a highly viscous solid mantle. The crust is separated from the mantle by the Mohorovičić discontinuity. The thickness of the crust varies from about 6 km (kilometers) under the oceans to 30–50 km for the continents. The crust and the cold, rigid, top of the upper mantle are collectively known as the lithosphere, and it is of the lithosphere that the tectonic plates are composed. Beneath the lithosphere is the asthenosphere, a relatively low-viscosity layer on which the lithosphere rides. Important changes in crystal structure within the mantle occur at 410 and 660 km below the surface, spanning a transition zone that separates the upper and lower mantle. Beneath the mantle, an extremely low viscosity liquid outer core lies above a solid inner core. The Earth's inner core might rotate at a slightly higher angular velocity than the remainder of the planet, advancing by 0.1–0.5° per year. The radius of the inner core is about one fifth of that of Earth.

Geologic layers of Earth

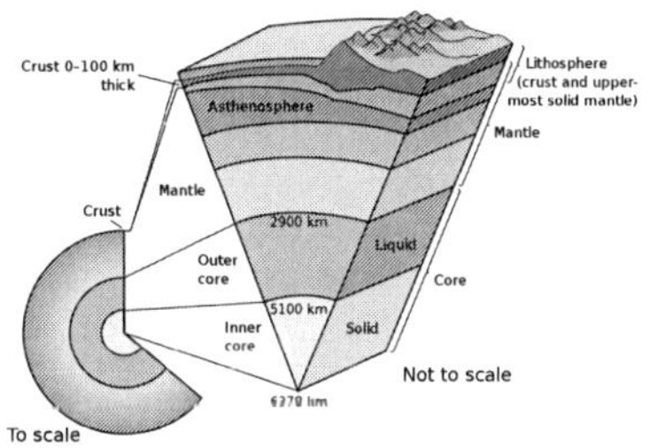

Earth cutaway from core to exosphere. Not to scale.

Depth km	Component layer	Density g/cm^3
0–60	Litho sphere[139]	—
0–35	Crust[140]	2.2–2.9
35–60	Upper mantle	3.4–4.4
35–2890	Mantle	3.4–5.6
100–700	Asthenosphere	—
2890–5100	Outer core	9.9–12.2
5100–6378	Inner core	12.8–13.1

Heat

Earth's internal heat comes from a combination of residual heat from planetary accretion (about 20%) and heat produced through radioactive decay (80%). The major heat-producing isotopes within Earth are potassium-40, uranium-238, and thorium-232. At the center, the temperature may be up to 6,000 °C (10,830 °F), and the pressure could reach 360 GPa. Because much of the heat is provided by radioactive decay, scientists postulate that early in Earth's history, before isotopes with short half-lives were depleted, Earth's heat production was much higher. At approximately 3 Ga, twice the present-day heat would have been produced, increasing the rates of mantle convection and plate tectonics, and allowing the production of uncommon igneous rocks such as komatiites that are rarely formed today.

Present-day major heat-producing isotopes

Isotope	Heat release W/kg isotope	Half-life years	Mean mantle concentration kg isotope/kg mantle	Heat release W/kg mantle
^{238}U	94.6×10^{-6}	4.47×10^{9}	30.8×10^{-9}	2.91×10^{-12}
^{235}U	569×10^{-6}	0.704×10^{9}	0.22×10^{-9}	0.125×10^{-12}
^{232}Th	26.4×10^{-6}	14.0×10^{9}	124×10^{-9}	3.27×10^{-12}
^{40}K	29.2×10^{-6}	1.25×10^{9}	36.9×10^{-9}	1.08×10^{-12}

The mean heat loss from Earth is 87 mW m^{-2}, for a global heat loss of 4.42×10^{13} W. A portion of the core's thermal energy is transported toward the crust by mantle plumes, a form of convection consisting of upwellings of higher-temperature rock. These plumes can produce hotspots and flood basalts. More of the heat in Earth is lost through plate tectonics, by mantle upwelling associated with mid-ocean ridges. The final major mode of heat loss is through conduction through the lithosphere, the majority of which occurs under the oceans because the crust there is much thinner than that of the continents.

Tectonic plates

Earth's major plates

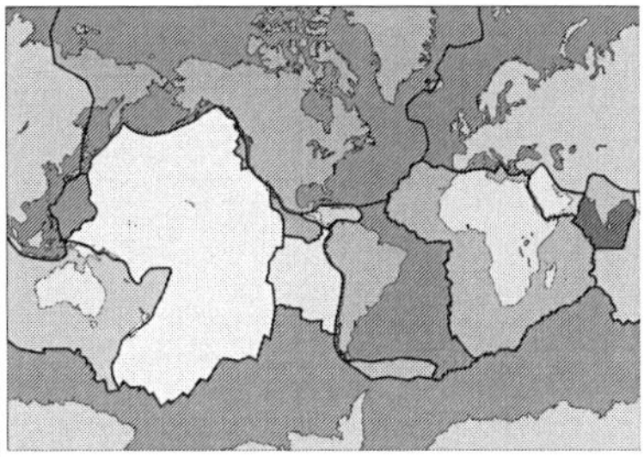

Plate name	Area 10^6 km^2
Pacific Plate	103.3
African Plate	78.0
North American Plate	75.9
Eurasian Plate	67.8
Antarctic Plate	60.9
Indo-Australian Plate	47.2
South American Plate	43.6

The mechanically rigid outer layer of Earth, the lithosphere, is divided into pieces called tectonic plates. These plates are rigid segments that move in relation to one another at one of three types of plate boundaries: convergent boundaries, at which two plates come together, divergent boundaries, at which two plates are pulled apart, and transform boundaries, in which two plates slide past one another laterally. Earthquakes, volcanic activity, mountain-building, and oceanic trench formation can occur along these plate boundaries. The tectonic plates ride on top of the asthenosphere, the solid but less-viscous part of the upper mantle that can flow and move along with the plates.

As the tectonic plates migrate, oceanic crust is subducted under the leading edges of the plates at convergent boundaries. At the same time, the upwelling of mantle material at divergent boundaries creates mid-ocean ridges. The combination of these processes recycles the oceanic crust back into the mantle. Due to this recycling, most of the ocean floor is less than 100 Ma old in age. The oldest oceanic crust is located in the Western Pacific and has an estimated age of 200 Ma. By comparison, the oldest dated continental crust is 4030 Ma.

The seven major plates are the Pacific, North American, Eurasian, African, Antarctic, Indo-Australian, and South American. Other notable plates include

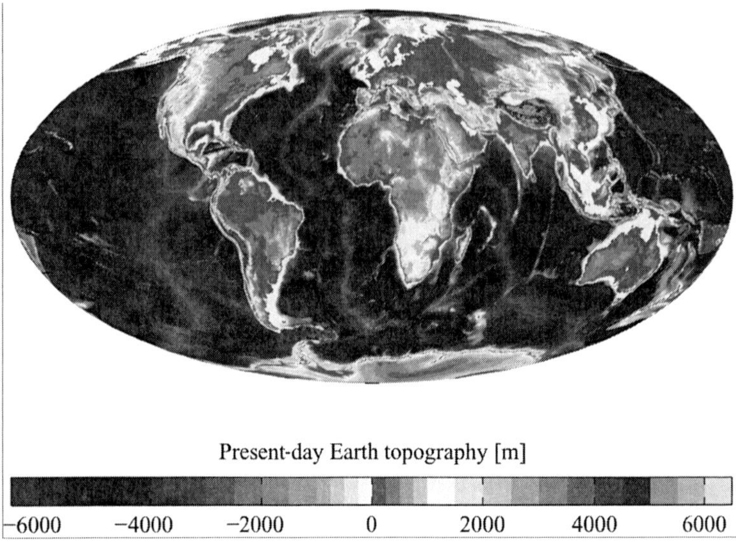

Figure 93: *Present-day Earth altimetry and bathymetry. Data from the National Geophysical Data Center.*

the Arabian Plate, the Caribbean Plate, the Nazca Plate off the west coast of South America and the Scotia Plate in the southern Atlantic Ocean. The Australian Plate fused with the Indian Plate between 50 and 55 Mya. The fastest-moving plates are the oceanic plates, with the Cocos Plate advancing at a rate of 75 mm/year and the Pacific Plate moving 52–69 mm/year. At the other extreme, the slowest-moving plate is the Eurasian Plate, progressing at a typical rate of 21 mm/year.

Surface

The total surface area of the Earth is about 510 million km^2 (197 million sq mi). Of this, 70.8%, or 361.13 million km^2 (139.43 million sq mi), is below sea level and covered by ocean water. Below the ocean's surface are much of the continental shelf, mountains, volcanoes, oceanic trenches, submarine canyons, oceanic plateaus, abyssal plains, and a globe-spanning mid-ocean ridge system. The remaining 29.2% (148.94 million km^2, or 57.51 million sq mi) not covered by water has terrain that varies greatly from place to place and consists of mountains, deserts, plains, plateaus, and other landforms. Tectonics and erosion, volcanic eruptions, flooding, weathering, glaciation, the growth of coral reefs, and meteorite impacts are among the processes that constantly reshape the Earth's surface over geological time.

The continental crust consists of lower density material such as the igneous rocks granite and andesite. Less common is basalt, a denser volcanic rock that is the primary constituent of the ocean floors. Sedimentary rock is formed from the accumulation of sediment that becomes buried and compacted together. Nearly 75% of the continental surfaces are covered by sedimentary rocks, although they form about 5% of the crust. The third form of rock material found on Earth is metamorphic rock, which is created from the transformation of pre-existing rock types through high pressures, high temperatures, or both. The most abundant silicate minerals on Earth's surface include quartz, feldspars, amphibole, mica, pyroxene and olivine. Common carbonate minerals include calcite (found in limestone) and dolomite.

The elevation of the land surface varies from the low point of –418 m at the Dead Sea, to a maximum altitude of 8,848 m at the top of Mount Everest. The mean height of land above sea level is about 797 metres (2,615 ft)[141].

The pedosphere is the outermost layer of Earth's continental surface and is composed of soil and subject to soil formation processes. The total arable land is 10.9% of the land surface, with 1.3% being permanent cropland. Close to 40% of Earth's land surface is used for cropland and pasture, or an estimated 1.3×10^7 km^2 of cropland and 3.4×10^7 km^2 of pastureland.

Hydrosphere

The abundance of water on Earth's surface is a unique feature that distinguishes the "Blue Planet" from other planets in the Solar System. Earth's hydrosphere consists chiefly of the oceans, but technically includes all water surfaces in the world, including inland seas, lakes, rivers, and underground waters down to a depth of 2,000 m. The deepest underwater location is Challenger Deep of the Mariana Trench in the Pacific Ocean with a depth of 10,911.4 m.

The mass of the oceans is approximately 1.35×10^{18} metric tons or about 1/4400 of Earth's total mass. The oceans cover an area of 3.618×10^8 km^2 with a mean depth of 3682 m, resulting in an estimated volume of 1.332×10^9 km^3. If all of Earth's crustal surface were at the same elevation as a smooth sphere, the depth of the resulting world ocean would be 2.7 to 2.8 km.

About 97.5% of the water is saline; the remaining 2.5% is fresh water. Most fresh water, about 68.7%, is present as ice in ice caps and glaciers.

The average salinity of Earth's oceans is about 35 grams of salt per kilogram of sea water (3.5% salt). Most of this salt was released from volcanic activity or extracted from cool igneous rocks. The oceans are also a reservoir of dissolved atmospheric gases, which are essential for the survival of many aquatic life forms. Sea water has an important influence on the world's climate, with the oceans acting as a large heat reservoir. Shifts in the oceanic temperature

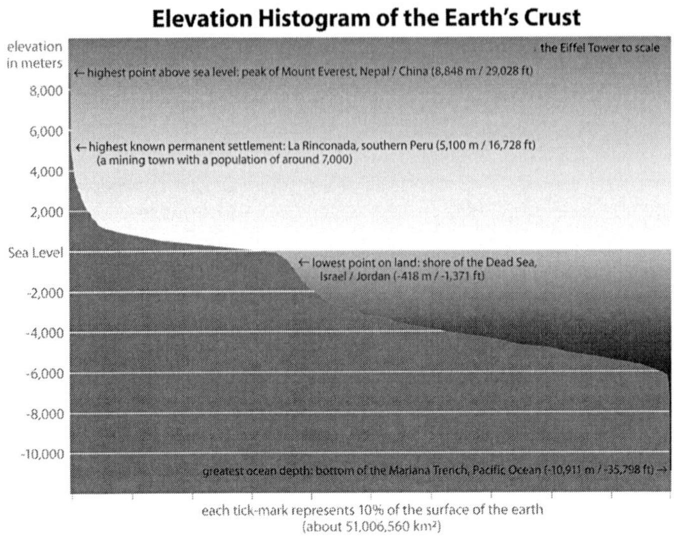

Figure 94: *Elevation histogram of Earth's surface*

distribution can cause significant weather shifts, such as the El Niño-Southern Oscillation.

Atmosphere

The atmospheric pressure on Earth's surface averages 101.325 kPa, with a scale height of about 8.5 km. It has a composition of 78% nitrogen and 21% oxygen, with trace amounts of water vapor, carbon dioxide, and other gaseous molecules. The height of the troposphere varies with latitude, ranging between 8 km at the poles to 17 km at the equator, with some variation resulting from weather and seasonal factors.

Earth's biosphere has significantly altered its atmosphere. Oxygenic photosynthesis evolved 2.7 Gya, forming the primarily nitrogen–oxygen atmosphere of today. This change enabled the proliferation of aerobic organisms and, indirectly, the formation of the ozone layer due to the subsequent conversion of atmospheric O_2 into O_3. The ozone layer blocks ultraviolet solar radiation, permitting life on land. Other atmospheric functions important to life include transporting water vapor, providing useful gases, causing small meteors to burn up before they strike the surface, and moderating temperature. This last phenomenon is known as the greenhouse effect: trace molecules within the atmosphere serve to capture thermal energy emitted from the ground, thereby

Figure 95: *Satellite image of Earth cloud cover using NASA's Moderate-Resolution Imaging Spectroradiometer*

raising the average temperature. Water vapor, carbon dioxide, methane, nitrous oxide, and ozone are the primary greenhouse gases in the atmosphere. Without this heat-retention effect, the average surface temperature would be −18 °C, in contrast to the current +15 °C, and life on Earth probably would not exist in its current form. In May 2017, glints of light, seen as twinkling from an orbiting satellite a million miles away, were found to be reflected light from ice crystals in the atmosphere.

Weather and climate

Hurricane Felix seen from low Earth orbit, September 2007

Lenticular cloud with a pressure ridge near Mount Discovery, Antarctica, November 2013

Massive clouds above the Mojave Desert, February 2016

Earth's atmosphere has no definite boundary, slowly becoming thinner and fading into outer space. Three-quarters of the atmosphere's mass is contained within the first 11 km (6.8 mi) of the surface. This lowest layer is called the troposphere. Energy from the Sun heats this layer, and the surface below, causing expansion of the air. This lower-density air then rises and is replaced by cooler, higher-density air. The result is atmospheric circulation that drives the weather and climate through redistribution of thermal energy.

The primary atmospheric circulation bands consist of the trade winds in the equatorial region below 30° latitude and the westerlies in the mid-latitudes between 30° and 60°. Ocean currents are also important factors in determining climate, particularly the thermohaline circulation that distributes thermal energy from the equatorial oceans to the polar regions.

Water vapor generated through surface evaporation is transported by circulatory patterns in the atmosphere. When atmospheric conditions permit an uplift of warm, humid air, this water condenses and falls to the surface as precipitation. Most of the water is then transported to lower elevations by river systems and usually returned to the oceans or deposited into lakes. This water cycle is a vital mechanism for supporting life on land and is a primary factor in the erosion of surface features over geological periods. Precipitation patterns vary widely, ranging from several meters of water per year to less than a millimeter. Atmospheric circulation, topographic features, and temperature differences determine the average precipitation that falls in each region.

The amount of solar energy reaching Earth's surface decreases with increasing latitude. At higher latitudes, the sunlight reaches the surface at lower angles, and it must pass through thicker columns of the atmosphere. As a result, the mean annual air temperature at sea level decreases by about 0.4 °C (0.7 °F) per degree of latitude from the equator. Earth's surface can be subdivided into specific latitudinal belts of approximately homogeneous climate. Ranging from the equator to the polar regions, these are the tropical (or equatorial), subtropical, temperate and polar climates.

This latitudinal rule has several anomalies:

- Proximity to oceans moderates the climate. For example, the Scandinavian Peninsula has more moderate climate than similarly northern latitudes of northern Canada.
- The wind enables this moderating effect. The windward side of a land mass experiences more moderation than the leeward side. In the Northern Hemisphere, the prevailing wind is west-to-east, and western coasts tend to be milder than eastern coasts. This is seen in Eastern North America and Western Europe, where rough continental climates appear on the east coast on parallels with mild climates on the other side of the ocean. In the Southern Hemisphere, the prevailing wind is east-to-west, and the eastern coasts are milder.
- The distance from the Earth to the Sun varies. The Earth is closest to the Sun (at perihelion) in January, which is summer in the Southern Hemisphere. It is furthest away (at aphelion) in July, which is summer in the Northern Hemisphere, and only 93.55% of the solar radiation from the Sun falls on a given square area of land than at perihelion. Despite this, there are larger land masses in the Northern Hemisphere, which are easier to heat than the seas. Consequently, summers are 2.3 °C (4 °F) warmer in the Northern Hemisphere than in the Southern Hemisphere under similar conditions.
- The climate is colder at high altitudes than at sea level because of the decreased air density.

The commonly used Köppen climate classification system has five broad groups (humid tropics, arid, humid middle latitudes, continental and cold polar), which are further divided into more specific subtypes. The Köppen system rates regions of terrain based on observed temperature and precipitation.

The highest air temperature ever measured on Earth was 56.7 °C (134.1 °F) in Furnace Creek, California, in Death Valley, in 1913. The lowest air temperature ever directly measured on Earth was –89.2 °C (–128.6 °F) at Vostok Station in 1983, but satellites have used remote sensing to measure temperatures as low as –94.7 °C (–138.5 °F) in East Antarctica. These temperature records are only measurements made with modern instruments from the 20th century onwards and likely do not reflect the full range of temperature on Earth.

Upper atmosphere

Above the troposphere, the atmosphere is usually divided into the stratosphere, mesosphere, and thermosphere. Each layer has a different lapse rate, defining the rate of change in temperature with height. Beyond these, the exosphere thins out into the magnetosphere, where the geomagnetic fields interact with the solar wind. Within the stratosphere is the ozone layer, a component that partially shields the surface from ultraviolet light and thus is important for life

Figure 96: *This view from orbit shows the full Moon partially obscured by Earth's atmosphere.*

on Earth. The Kármán line, defined as 100 km above Earth's surface, is a working definition for the boundary between the atmosphere and outer space.

Thermal energy causes some of the molecules at the outer edge of the atmosphere to increase their velocity to the point where they can escape from Earth's gravity. This causes a slow but steady loss of the atmosphere into space. Because unfixed hydrogen has a low molecular mass, it can achieve escape velocity more readily, and it leaks into outer space at a greater rate than other gases. The leakage of hydrogen into space contributes to the shifting of Earth's atmosphere and surface from an initially reducing state to its current oxidizing one. Photosynthesis provided a source of free oxygen, but the loss of reducing agents such as hydrogen is thought to have been a necessary precondition for the widespread accumulation of oxygen in the atmosphere. Hence the ability of hydrogen to escape from the atmosphere may have influenced the nature of life that developed on Earth. In the current, oxygen-rich atmosphere most hydrogen is converted into water before it has an opportunity to escape. Instead, most of the hydrogen loss comes from the destruction of methane in the upper atmosphere.

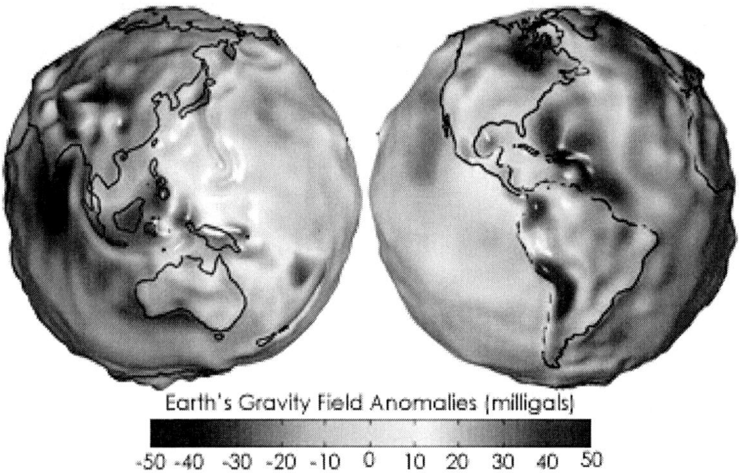

Figure 97: *Earth's gravity measured by NASA's GRACE mission, showing deviations from the theoretical gravity. Red shows where gravity is stronger than the smooth, standard value, and blue shows where it is weaker.*

Gravitational field

The gravity of Earth is the acceleration that is imparted to objects due to the distribution of mass within the Earth. Near the Earth's surface, gravitational acceleration is approximately 9.8 m/s^2 (32 ft/s^2). Local differences in topography, geology, and deeper tectonic structure cause local and broad, regional differences in the Earth's gravitational field, known as gravitational anomalies.

Magnetic field

The main part of Earth's magnetic field is generated in the core, the site of a dynamo process that converts the kinetic energy of thermally and compositionally driven convection into electrical and magnetic field energy. The field extends outwards from the core, through the mantle, and up to Earth's surface, where it is, approximately, a dipole. The poles of the dipole are located close to Earth's geographic poles. At the equator of the magnetic field, the magnetic-field strength at the surface is 3.05×10^{-5} T, with global magnetic dipole moment of 7.91×10^{15} T m^3. The convection movements in the core are chaotic; the magnetic poles drift and periodically change alignment. This causes secular variation of the main field and field reversals at irregular intervals averaging a few times every million years. The most recent reversal occurred approximately 700,000 years ago.

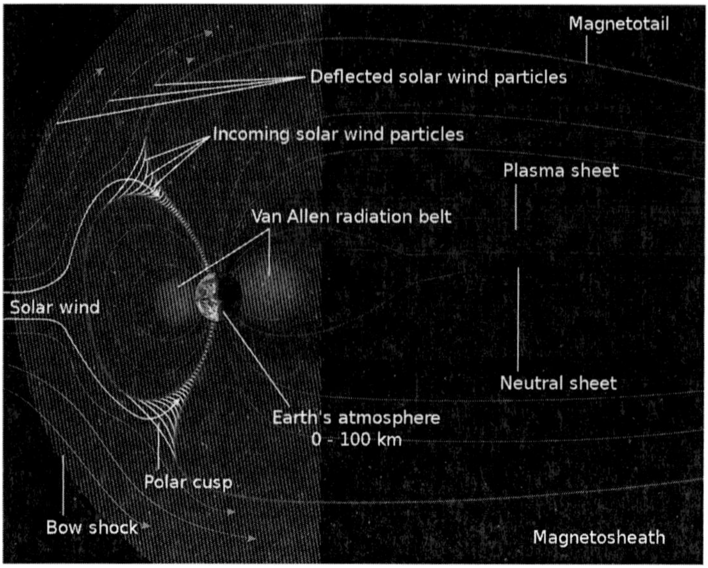

Figure 98: *Schematic of Earth's magnetosphere. The solar wind flows from left to right*

Magnetosphere

The extent of Earth's magnetic field in space defines the magnetosphere. Ions and electrons of the solar wind are deflected by the magnetosphere; solar wind pressure compresses the dayside of the magnetosphere, to about 10 Earth radii, and extends the nightside magnetosphere into a long tail. Because the velocity of the solar wind is greater than the speed at which waves propagate through the solar wind, a supersonic bowshock precedes the dayside magnetosphere within the solar wind. Charged particles are contained within the magnetosphere; the plasmasphere is defined by low-energy particles that essentially follow magnetic field lines as Earth rotates; the ring current is defined by medium-energy particles that drift relative to the geomagnetic field, but with paths that are still dominated by the magnetic field, and the Van Allen radiation belt are formed by high-energy particles whose motion is essentially random, but otherwise contained by the magnetosphere.

During magnetic storms and substorms, charged particles can be deflected from the outer magnetosphere and especially the magnetotail, directed along field lines into Earth's ionosphere, where atmospheric atoms can be excited and ionized, causing the aurora.

Figure 99: *Earth's rotation imaged by DSCOVR EPIC on 29 May 2016, a few weeks before the solstice.*

Orbit and rotation

Rotation

Earth's rotation period relative to the Sun—its mean solar day—is 86,400 seconds of mean solar time (86,400.0025 SI seconds). Because Earth's solar day is now slightly longer than it was during the 19th century due to tidal deceleration, each day varies between 0 and 2 SI ms longer.

Earth's rotation period relative to the fixed stars, called its *stellar day* by the International Earth Rotation and Reference Systems Service (IERS), is 86,164.0989 seconds of mean solar time (UT1), or 23^h 56^m 4.0989^s. Earth's rotation period relative to the precessing or moving mean vernal equinox, misnamed its *sidereal day*, is 86,164.0905 seconds of mean solar time (UT1) (23^h 56^m 4.0905^s). Thus the sidereal day is shorter than the stellar day by about 8.4 ms. The length of the mean solar day in SI seconds is available from the IERS for the periods 1623–2005 and 1962–2005.

Apart from meteors within the atmosphere and low-orbiting satellites, the main apparent motion of celestial bodies in Earth's sky is to the west at a rate of 15°/h = 15'/min. For bodies near the celestial equator, this is equivalent to an apparent diameter of the Sun or the Moon every two minutes; from Earth's

Figure 100: *The Pale Blue Dot photo taken in 1990 by the Voyager 1 spacecraft showing Earth (center right) from nearly 6.4 billion km (4 billion mi) away*

surface, the apparent sizes of the Sun and the Moon are approximately the same.

Orbit

Earth orbits the Sun at an average distance of about 150 million km (93 million mi) every 365.2564 mean solar days, or one sidereal year. This gives an apparent movement of the Sun eastward with respect to the stars at a rate of about 1°/day, which is one apparent Sun or Moon diameter every 12 hours. Due to this motion, on average it takes 24 hours—a solar day—for Earth to complete a full rotation about its axis so that the Sun returns to the meridian. The orbital speed of Earth averages about 29.78 km/s (107,200 km/h; 66,600 mph), which is fast enough to travel a distance equal to Earth's diameter, about 12,742 km (7,918 mi), in seven minutes, and the distance to the Moon, 384,000 km (239,000 mi), in about 3.5 hours.

The Moon and Earth orbit a common barycenter every 27.32 days relative to the background stars. When combined with the Earth–Moon system's common orbit around the Sun, the period of the synodic month, from new moon to new moon, is 29.53 days. Viewed from the celestial north pole, the motion of Earth, the Moon, and their axial rotations are all counterclockwise. Viewed

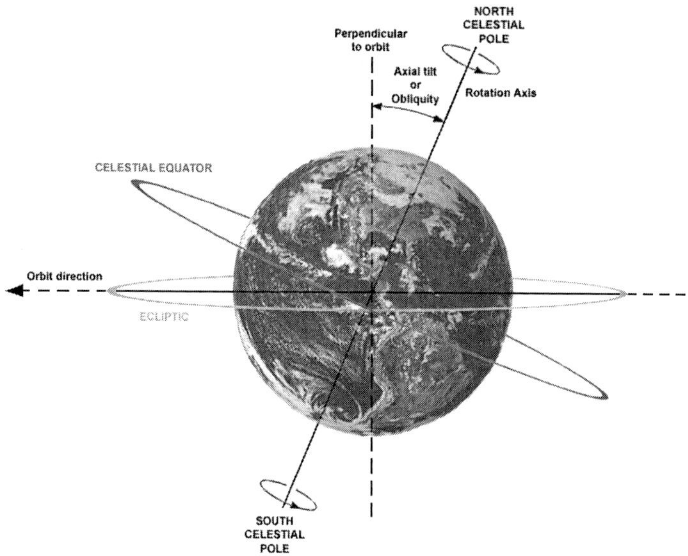

Figure 101: *Earth's axial tilt (or obliquity) and its relation to the rotation axis and plane of orbit*

from a vantage point above the north poles of both the Sun and Earth, Earth orbits in a counterclockwise direction about the Sun. The orbital and axial planes are not precisely aligned: Earth's axis is tilted some 23.44 degrees from the perpendicular to the Earth–Sun plane (the ecliptic), and the Earth–Moon plane is tilted up to ±5.1 degrees against the Earth–Sun plane. Without this tilt, there would be an eclipse every two weeks, alternating between lunar eclipses and solar eclipses.

The Hill sphere, or the sphere of gravitational influence, of the Earth is about 1.5 million kilometres (930,000 mi) in radius. This is the maximum distance at which the Earth's gravitational influence is stronger than the more distant Sun and planets. Objects must orbit the Earth within this radius, or they can become unbound by the gravitational perturbation of the Sun.

Earth, along with the Solar System, is situated in the Milky Way and orbits about 28,000 light-years from its center. It is about 20 light-years above the galactic plane in the Orion Arm.

Axial tilt and seasons

The axial tilt of the Earth is approximately 23.439281° with the axis of its orbit plane, always pointing towards the Celestial Poles. Due to Earth's axial tilt, the amount of sunlight reaching any given point on the surface varies over the course of the year. This causes the seasonal change in climate, with summer in the Northern Hemisphere occurring when the Tropic of Cancer is facing the Sun, and winter taking place when the Tropic of Capricorn in the Southern Hemisphere faces the Sun. During the summer, the day lasts longer, and the Sun climbs higher in the sky. In winter, the climate becomes cooler and the days shorter. In northern temperate latitudes, the Sun rises north of true east during the summer solstice, and sets north of true west, reversing in the winter. The Sun rises south of true east in the summer for the southern temperate zone and sets south of true west.

Above the Arctic Circle, an extreme case is reached where there is no daylight at all for part of the year, up to six months at the North Pole itself, a polar night. In the Southern Hemisphere, the situation is exactly reversed, with the South Pole oriented opposite the direction of the North Pole. Six months later, this pole will experience a midnight sun, a day of 24 hours, again reversing with the South Pole.

By astronomical convention, the four seasons can be determined by the solstices—the points in the orbit of maximum axial tilt toward or away from the Sun—and the equinoxes, when the direction of the tilt and the direction to the Sun are perpendicular. In the Northern Hemisphere, winter solstice currently occurs around 21 December; summer solstice is near 21 June, spring equinox is around 20 March and autumnal equinox is about 22 or 23 September. In the Southern Hemisphere, the situation is reversed, with the summer and winter solstices exchanged and the spring and autumnal equinox dates swapped.

The angle of Earth's axial tilt is relatively stable over long periods of time. Its axial tilt does undergo nutation; a slight, irregular motion with a main period of 18.6 years. The orientation (rather than the angle) of Earth's axis also changes over time, precessing around in a complete circle over each 25,800 year cycle; this precession is the reason for the difference between a sidereal year and a tropical year. Both of these motions are caused by the varying attraction of the Sun and the Moon on Earth's equatorial bulge. The poles also migrate a few meters across Earth's surface. This polar motion has multiple, cyclical components, which collectively are termed quasiperiodic motion. In addition to an annual component to this motion, there is a 14-month cycle called the Chandler wobble. Earth's rotational velocity also varies in a phenomenon known as length-of-day variation.

In modern times, Earth's perihelion occurs around 3 January, and its aphelion around 4 July. These dates change over time due to precession and other orbital factors, which follow cyclical patterns known as Milankovitch cycles. The changing Earth–Sun distance causes an increase of about 6.9% in solar energy

Figure 102: *The Rocky Mountains in Canada overlook Moraine Lake.*

reaching Earth at perihelion relative to aphelion. Because the Southern Hemisphere is tilted toward the Sun at about the same time that Earth reaches the closest approach to the Sun, the Southern Hemisphere receives slightly more energy from the Sun than does the northern over the course of a year. This effect is much less significant than the total energy change due to the axial tilt, and most of the excess energy is absorbed by the higher proportion of water in the Southern Hemisphere.

A study from 2016 suggested that Planet Nine tilted all solar system planets, including Earth's by about 6 degrees.

Habitability

A planet that can sustain life is termed habitable, even if life did not originate there. Earth provides liquid water—an environment where complex organic molecules can assemble and interact, and sufficient energy to sustain metabolism. The distance of Earth from the Sun, as well as its orbital eccentricity, rate of rotation, axial tilt, geological history, sustaining atmosphere, and magnetic field all contribute to the current climatic conditions at the surface.

Biosphere

A planet's life forms inhabit ecosystems, whose total is sometimes said to form a "biosphere". Earth's biosphere is thought to have begun evolving about 3.5 Gya. The biosphere is divided into a number of biomes, inhabited by broadly similar plants and animals. On land, biomes are separated primarily by differences in latitude, height above sea level and humidity. Terrestrial biomes lying within the Arctic or Antarctic Circles, at high altitudes or in extremely arid areas are relatively barren of plant and animal life; species diversity reaches a peak in humid lowlands at equatorial latitudes.

In July 2016, scientists reported identifying a set of 355 genes from the Last Universal Common Ancestor (LUCA) of all organisms living on Earth.

Natural resources and land use

Estimated human land use, 2000

Land use	Mha
Cropland	1,510–1,611
Pastures	2,500–3,410
Natural forests	3,143–3,871
Planted forests	126–215
Urban areas	66–351
Unused, productive land	356–445

Earth has resources that have been exploited by humans. Those termed non-renewable resources, such as fossil fuels, only renew over geological timescales.

Large deposits of fossil fuels are obtained from Earth's crust, consisting of coal, petroleum, and natural gas. These deposits are used by humans both for energy production and as feedstock for chemical production. Mineral ore bodies have also been formed within the crust through a process of ore genesis, resulting from actions of magmatism, erosion, and plate tectonics. These bodies form concentrated sources for many metals and other useful elements.

Earth's biosphere produces many useful biological products for humans, including food, wood, pharmaceuticals, oxygen, and the recycling of many organic wastes. The land-based ecosystem depends upon topsoil and fresh water, and the oceanic ecosystem depends upon dissolved nutrients washed down from the land. In 1980, 5,053 Mha (50.53 million km^2) of Earth's land surface consisted of forest and woodlands, 6,788 Mha (67.88 million km^2) was grasslands and pasture, and 1,501 Mha (15.01 million km^2) was cultivated as

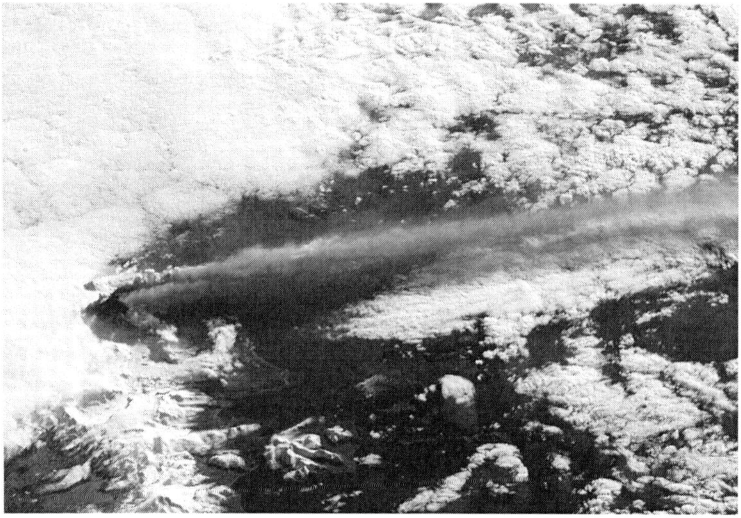

Figure 103: *A volcano injecting hot ash into the atmosphere*

croplands. The estimated amount of irrigated land in 1993 was 2,481,250 square kilometres (958,020 sq mi). Humans also live on the land by using building materials to construct shelters.

Natural and environmental hazards

Large areas of Earth's surface are subject to extreme weather such as tropical cyclones, hurricanes, or typhoons that dominate life in those areas. From 1980 to 2000, these events caused an average of 11,800 human deaths per year. Many places are subject to earthquakes, landslides, tsunamis, volcanic eruptions, tornadoes, sinkholes, blizzards, floods, droughts, wildfires, and other calamities and disasters.

Many localized areas are subject to human-made pollution of the air and water, acid rain and toxic substances, loss of vegetation (overgrazing, deforestation, desertification), loss of wildlife, species extinction, soil degradation, soil depletion and erosion.

There is a scientific consensus linking human activities to global warming due to industrial carbon dioxide emissions. This is predicted to produce changes such as the melting of glaciers and ice sheets, more extreme temperature ranges, significant changes in weather and a global rise in average sea levels.

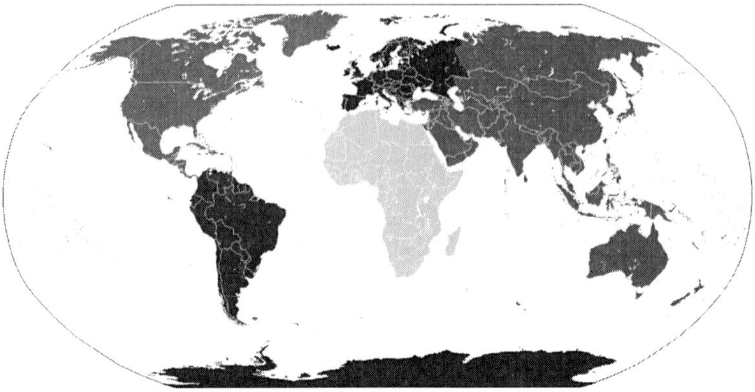

Figure 104:
The seven continents of Earth[142]

Human geography

Cartography, the study and practice of map-making, and geography, the study of the lands, features, inhabitants and phenomena on Earth, have historically been the disciplines devoted to depicting Earth. Surveying, the determination of locations and distances, and to a lesser extent navigation, the determination of position and direction, have developed alongside cartography and geography, providing and suitably quantifying the requisite information.

Earth's human population reached approximately seven billion on 31 October 2011. Projections indicate that the world's human population will reach 9.2 billion in 2050. Most of the growth is expected to take place in developing nations. Human population density varies widely around the world, but a majority live in Asia. By 2020, 60% of the world's population is expected to be living in urban, rather than rural, areas.

It is estimated that one-eighth of Earth's surface is suitable for humans to live on – three-quarters of Earth's surface is covered by oceans, leaving one-quarter as land. Half of that land area is desert (14%), high mountains (27%), or other unsuitable terrains. The northernmost permanent settlement in the world is Alert, on Ellesmere Island in Nunavut, Canada. (82°28′N) The southernmost is the Amundsen–Scott South Pole Station, in Antarctica, almost exactly at the South Pole. (90°S)

Independent sovereign nations claim the planet's entire land surface, except for some parts of Antarctica, a few land parcels along the Danube river's western bank, and the unclaimed area of Bir Tawil between Egypt and Sudan. As of

2015[143], there are 193 sovereign states that are member states of the United Nations, plus two observer states and 72 dependent territories and states with limited recognition. Earth has never had a sovereign government with authority over the entire globe, although some nation-states have striven for world domination and failed.

The United Nations is a worldwide intergovernmental organization that was created with the goal of intervening in the disputes between nations, thereby avoiding armed conflict. The U.N. serves primarily as a forum for international diplomacy and international law. When the consensus of the membership permits, it provides a mechanism for armed intervention.

The first human to orbit Earth was Yuri Gagarin on 12 April 1961. In total, about 487 people have visited outer space and reached orbit as of 30 July 2010[143], and, of these, twelve have walked on the Moon. Normally, the only humans in space are those on the International Space Station. The station's crew, made up of six people, is usually replaced every six months. The farthest that humans have traveled from Earth is 400,171 km, achieved during the Apollo 13 mission in 1970.

Moon

Characteristics

Diameter	3,474.8 km
Mass	7.349×10^{22} kg
Semi-major axis	384,400 km
Orbital period	27 d 7 h 43.7 m

The Moon is a relatively large, terrestrial, planet-like natural satellite, with a diameter about one-quarter of Earth's. It is the largest moon in the Solar System relative to the size of its planet, although Charon is larger relative to the

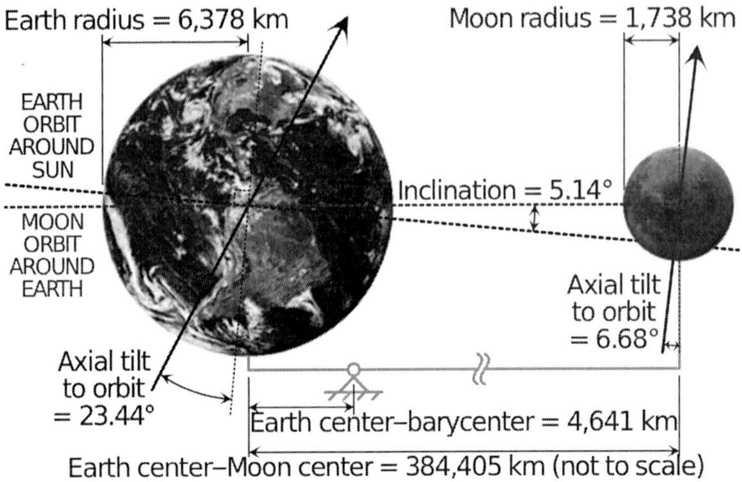

Figure 105: *Details of the Earth–Moon system, showing the radius of each object and the Earth–Moon barycenter. The Moon's axis is located by Cassini's third law.*

dwarf planet Pluto. The natural satellites of other planets are also referred to as "moons", after Earth's.

The gravitational attraction between Earth and the Moon causes tides on Earth. The same effect on the Moon has led to its tidal locking: its rotation period is the same as the time it takes to orbit Earth. As a result, it always presents the same face to the planet. As the Moon orbits Earth, different parts of its face are illuminated by the Sun, leading to the lunar phases; the dark part of the face is separated from the light part by the solar terminator.

Due to their tidal interaction, the Moon recedes from Earth at the rate of approximately 38 mm/yr. Over millions of years, these tiny modifications—and the lengthening of Earth's day by about 23 µs/yr—add up to significant changes. During the Devonian period, for example, (approximately 410 Mya) there were 400 days in a year, with each day lasting 21.8 hours.

The Moon may have dramatically affected the development of life by moderating the planet's climate. Paleontological evidence and computer simulations show that Earth's axial tilt is stabilized by tidal interactions with the Moon. Some theorists think that without this stabilization against the torques applied by the Sun and planets to Earth's equatorial bulge, the rotational axis might be chaotically unstable, exhibiting chaotic changes over millions of years, as appears to be the case for Mars.

Viewed from Earth, the Moon is just far enough away to have almost the same apparent-sized disk as the Sun. The angular size (or solid angle) of these two bodies match because, although the Sun's diameter is about 400 times as large as the Moon's, it is also 400 times more distant. This allows total and annular solar eclipses to occur on Earth.

The most widely accepted theory of the Moon's origin, the giant-impact hypothesis, states that it formed from the collision of a Mars-size protoplanet called Theia with the early Earth. This hypothesis explains (among other things) the Moon's relative lack of iron and volatile elements and the fact that its composition is nearly identical to that of Earth's crust.

Asteroids and artificial satellites

Earth has at least five co-orbital asteroids, including 3753 Cruithne and 2002 AA29. A trojan asteroid companion, 2010 TK7, is librating around the leading Lagrange triangular point, L4, in the Earth's orbit around the Sun.

The tiny near-Earth asteroid 2006 RH120 makes close approaches to the Earth–Moon system roughly every twenty years. During these approaches, it can orbit Earth for brief periods of time.

As of June 2016[143], there were 1,419 operational, human-made satellites orbiting Earth. There are also inoperative satellites, including Vanguard 1, the oldest satellite currently in orbit, and over 16,000 pieces of tracked space debris. Earth's largest artificial satellite is the International Space Station.

Cultural and historical viewpoint

The standard astronomical symbol of Earth consists of a cross circumscribed by a circle, ⊕, representing the four corners of the world.

Human cultures have developed many views of the planet. Earth is sometimes personified as a deity. In many cultures it is a mother goddess that is also the primary fertility deity, and by the mid-20th century, the Gaia Principle compared Earth's environments and life as a single self-regulating organism leading to broad stabilization of the conditions of habitability.[144] Creation myths in many religions involve the creation of Earth by a supernatural deity or deities.

Scientific investigation has resulted in several culturally transformative shifts in people's view of the planet. In the West, belief in a flat Earth was displaced by the idea of spherical Earth, credited to Pythagoras in the 6th century BC. Earth was further believed to be the center of the universe until the 16th century when scientists first theorized that it was a moving object, comparable to the other planets in the Solar System. Due to the efforts of influential Christian scholars and clerics such as James Ussher, who sought to determine the age of Earth through analysis of genealogies in Scripture, Westerners before the 19th

Figure 106: *Earthrise, taken by astronauts on board Apollo 8*

century generally believed Earth to be a few thousand years old at most. It was only during the 19th century that geologists realized Earth's age was at least many millions of years.

Lord Kelvin used thermodynamics to estimate the age of Earth to be between 20 million and 400 million years in 1864, sparking a vigorous debate on the subject; it was only when radioactivity and radioactive dating were discovered in the late 19th and early 20th centuries that a reliable mechanism for determining Earth's age was established, proving the planet to be billions of years old. The perception of Earth shifted again in the 20th century when humans first viewed it from orbit, and especially with photographs of Earth returned by the Apollo program.

References

Further reading

- Comins, Neil F. (2001). *Discovering the Essential Universe*[145] (2nd ed.). New York: W. H. Freeman. Bibcode: 2003deu..book.....C[146]. ISBN 0-7167-5804-0. OCLC 52082611[147].

External links

<indicator name="spoken-icon"> </indicator>

- *National Geographic* encyclopedic entry about Earth[148]
- Earth – Profile[149] – Solar System Exploration[150] – NASA
- Earth – Climate Changes Cause Shape to Change[151] – NASA
- United States Geological Survey[152] – USGS
- Earth – Astronaut Photography Gateway[153] – NASA
- Earth Observatory[154] – NASA
- Earth – Audio (29:28) – Cain/Gay – Astronomy Cast (2007)[155]
- Earth – Videos – International Space Station:
  - Video (01:02)[156] – Earth (time-lapse)
  - Video (00:27)[157] – Earth and Auroras (time-lapse)

Mars

<indicator name="pp-default"> </indicator>

Mars

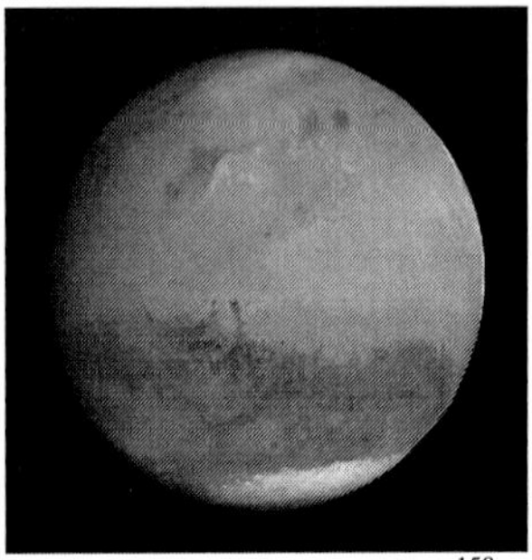

Mars in natural colour in 2007[158]

Designations	
Pronunciation	UK English: /mɑːz/ US English: /ˈmɑːrz/ Media:en-us-Mars.ogg
Adjectives	Martian
Orbital characteristics	
Epoch J2000	
Aphelion	1.6660 AU 249.2 Gm
Perihelion	1.3814 AU 206.7 Gm
Semi-major axis	1.523679 AU 227.9392 Gm
Eccentricity	0.0934
Orbital period	1.8808 Julian years 686.971 d 668.5991 sols
Synodic period	779.96 days 2.135 Julian years
Average orbital speed	24.077 km/s
Mean anomaly	19.373°
Inclination	1.850° to ecliptic 5.65° to Sun's equator 1.67° to invariable plane
Longitude of ascending node	49.558°

Argument of perihelion	286.502°
Satellites	2
	Physical characteristics
Mean radius	3,389.5±0.2 km
Equatorial radius	3,396.2±0.1 km 0.533 Earths
Polar radius	3,376.2±0.1 km 0.531 Earths
Flattening	0.00589±0.00015
Surface area	144,798,500 km^2 0.284 Earths
Volume	1.6318×10^{11} km^3 0.151 Earths
Mass	6.4171×10^{23} kg 0.107 Earths
Mean density	3.9335±0.0004 g/cm^3
Surface gravity	3.711 m/s^2 0.376 *g*
Moment of inertia factor	0.3662±0.0017
Escape velocity	5.027 km/s
Sidereal rotation period	1.025957 d $24^h\ 37^m\ 22^s$
Equatorial rotation velocity	868.22 km/h (241.17 m/s)
Axial tilt	25.19° to its orbital plane
North pole right ascension	$21^h\ 10^m\ 44^s$ 317.68143°
North pole declination	52.88650°
Albedo	0.170 (geometric) 0.25 (Bond)

Surface temp.	**min**	**mean**	**max**
Kelvin	130 K	210 K	308 K
Celsius	−143 °C	−63 °C	35 °C
Fahrenheit	−226 °F	−82 °F	95 °F

Apparent magnitude	+1.6 to −3.0
Angular diameter	3.5–25.1”
	Atmosphere
Surface pressure	0.636 (0.4–0.87) kPa 0.00628 atm

Composition by volume	• 95.97% carbon dioxide • 1.93% argon • 1.89% nitrogen • 0.146% oxygen • 0.0557% carbon monoxide

Mars is the fourth planet from the Sun and the second-smallest planet in the Solar System after Mercury. In English, Mars carries a name of the Roman god of war, and is often referred to as the "**Red Planet**" because the reddish iron oxide prevalent on its surface gives it a reddish appearance that is distinctive among the astronomical bodies visible to the naked eye. Mars is a terrestrial planet with a thin atmosphere, having surface features reminiscent both of the impact craters of the Moon and the valleys, deserts, and polar ice caps of Earth.

The rotational period and seasonal cycles of Mars are likewise similar to those of Earth, as is the tilt that produces the seasons. Mars is the site of Olympus Mons, the largest volcano and second-highest known mountain in the Solar System, and of Valles Marineris, one of the largest canyons in the Solar System. The smooth Borealis basin in the northern hemisphere covers 40% of the planet and may be a giant impact feature. Mars has two moons, Phobos and Deimos, which are small and irregularly shaped. These may be captured asteroids, similar to 5261 Eureka, a Mars trojan.

There are ongoing investigations assessing the past habitability potential of Mars, as well as the possibility of extant life. Future astrobiology missions are planned, including the Mars 2020 and ExoMars rovers. Liquid water cannot exist on the surface of Mars due to low atmospheric pressure, which is less than 1% of the Earth's, except at the lowest elevations for short periods. The two polar ice caps appear to be made largely of water. The volume of water ice in the south polar ice cap, if melted, would be sufficient to cover the entire planetary surface to a depth of 11 meters (36 ft). In November 2016, NASA reported finding a large amount of underground ice in the Utopia Planitia region of Mars. The volume of water detected has been estimated to be equivalent to the volume of water in Lake Superior.

Mars can easily be seen from Earth with the naked eye, as can its reddish coloring. Its apparent magnitude reaches −2.91, which is surpassed only by Jupiter, Venus, the Moon, and the Sun. Optical ground-based telescopes are typically limited to resolving features about 300 kilometers (190 mi) across when Earth and Mars are closest because of Earth's atmosphere.

Physical characteristics

Mars is approximately half the diameter of Earth with a surface area only slightly less than the total area of Earth's dry land. Mars is less dense than Earth, having about 15% of Earth's volume and 11% of Earth's mass, resulting in about 38% of Earth's surface gravity. The red-orange appearance of the Martian surface is caused by iron(III) oxide, or rust. It can look like butterscotch; other common surface colors include golden, brown, tan, and greenish, depending on the minerals present.[159]

Comparison: Earth and Mars

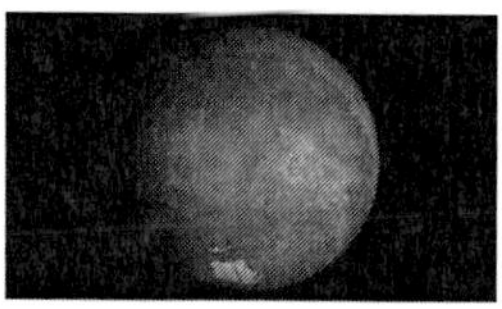

Animation (00:40) showing major features of Mars

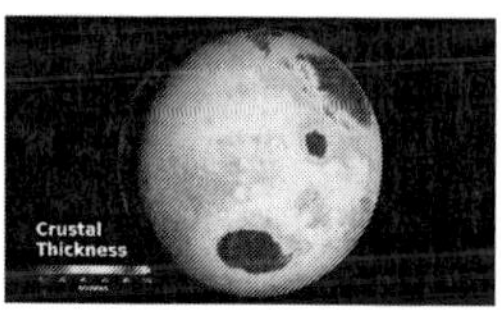

Video (01:28) showing how three NASA orbiters mapped the gravity field of Mars

Internal structure

Like Earth, Mars has differentiated into a dense metallic core overlaid by less dense materials. Current models of its interior imply a core with a radius of about 1,794 ± 65 kilometers (1,115 ± 40 mi), consisting primarily of iron and nickel with about 16–17% sulfur. This iron(II) sulfide core is thought to be twice as rich in lighter elements as Earth's. The core is surrounded by a silicate mantle that formed many of the tectonic and volcanic features on the planet, but it appears to be dormant. Besides silicon and oxygen, the most abundant elements in the Martian crust are iron, magnesium, aluminum, calcium, and potassium. The average thickness of the planet's crust is about 50 km (31 mi), with a maximum thickness of 125 km (78 mi). Earth's crust averages 40 km (25 mi).

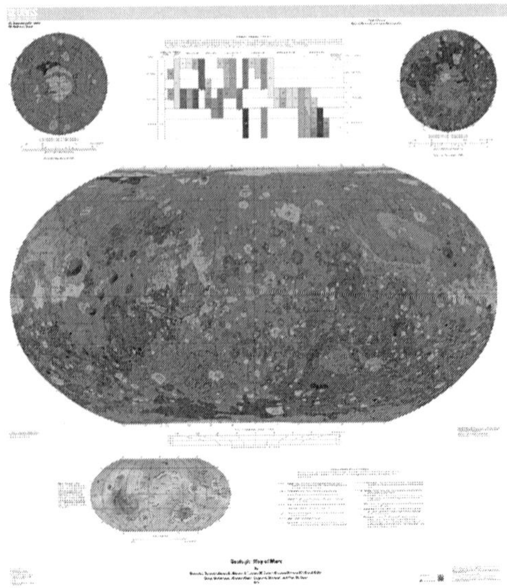

Figure 107: *Geologic map of Mars (USGS, 2014)*

Surface geology

Mars is a terrestrial planet that consists of minerals containing silicon and oxygen, metals, and other elements that typically make up rock. The surface of Mars is primarily composed of tholeiitic basalt, although parts are more silica-rich than typical basalt and may be similar to andesitic rocks on Earth or silica glass. Regions of low albedo suggest concentrations of plagioclase feldspar, with northern low albedo regions displaying higher than normal concentrations of sheet silicates and high-silicon glass. Parts of the southern highlands include detectable amounts of high-calcium pyroxenes. Localized concentrations of hematite and olivine have been found. Much of the surface is deeply covered by finely grained iron(III) oxide dust.

Although Mars has no evidence of a structured global magnetic field, observations show that parts of the planet's crust have been magnetized, suggesting that alternating polarity reversals of its dipole field have occurred in the past. This paleomagnetism of magnetically susceptible minerals is similar to the alternating bands found on Earth's ocean floors. One theory, published in 1999 and re-examined in October 2005 (with the help of the *Mars Global Surveyor*), is that these bands suggest plate tectonic activity on Mars four billion years ago, before the planetary dynamo ceased to function and the planet's magnetic field faded.

Figure 108: *Artist's impression of how Mars may have looked four billion years ago*

It is thought that, during the Solar System's formation, Mars was created as the result of a stochastic process of run-away accretion of material from the protoplanetary disk that orbited the Sun. Mars has many distinctive chemical features caused by its position in the Solar System. Elements with comparatively low boiling points, such as chlorine, phosphorus, and sulphur, are much more common on Mars than Earth; these elements were probably pushed outward by the young Sun's energetic solar wind.

After the formation of the planets, all were subjected to the so-called "Late Heavy Bombardment". About 60% of the surface of Mars shows a record of impacts from that era, whereas much of the remaining surface is probably underlain by immense impact basins caused by those events. There is evidence of an enormous impact basin in the northern hemisphere of Mars, spanning 10,600 by 8,500 km (6,600 by 5,300 mi), or roughly four times the size of the Moon's South Pole – Aitken basin, the largest impact basin yet discovered. This theory suggests that Mars was struck by a Pluto-sized body about four billion years ago. The event, thought to be the cause of the Martian hemispheric dichotomy, created the smooth Borealis basin that covers 40% of the planet.

The geological history of Mars can be split into many periods, but the following are the three primary periods:

- **Noachian period** (named after Noachis Terra): Formation of the oldest extant surfaces of Mars, 4.5 to 3.5 billion years ago. Noachian age surfaces are scarred by many large impact craters. The Tharsis bulge,

Figure 109: *Exposure of silica-rich dust uncovered by the Spirit rover*

a volcanic upland, is thought to have formed during this period, with extensive flooding by liquid water late in the period.

- **Hesperian period** (named after Hesperia Planum): 3.5 to between 3.3 and 2.9 billion years ago. The Hesperian period is marked by the formation of extensive lava plains.
- **Amazonian period** (named after Amazonis Planitia): between 3.3 and 2.9 billion years ago to the present. Amazonian regions have few meteorite impact craters, but are otherwise quite varied. Olympus Mons formed during this period, with lava flows elsewhere on Mars.

Geological activity is still taking place on Mars. The Athabasca Valles is home to sheet-like lava flows created about 200 Mya. Water flows in the grabens called the Cerberus Fossae occurred less than 20 Mya, indicating equally recent volcanic intrusions. On February 19, 2008, images from the *Mars Reconnaissance Orbiter* showed evidence of an avalanche from a 700-metre-high (2,300 ft) cliff.

Soil

The *Phoenix* lander returned data showing Martian soil to be slightly alkaline and containing elements such as magnesium, sodium, potassium and chlorine. These nutrients are found in soils on Earth, and they are necessary for growth of plants. Experiments performed by the lander showed that the Martian soil has a basic pH of 7.7, and contains 0.6% of the salt perchlorate.

Streaks are common across Mars and new ones appear frequently on steep slopes of craters, troughs, and valleys. The streaks are dark at first and get lighter with age. The streaks can start in a tiny area, then spread out for hundreds of metres. They have been seen to follow the edges of boulders and other obstacles in their path. The commonly accepted theories include that they are dark underlying layers of soil revealed after avalanches of bright dust or dust devils. Several other explanations have been put forward, including those that involve water or even the growth of organisms.

Hydrology

Liquid water cannot exist on the surface of Mars due to low atmospheric pressure, which is less than 1% that of Earth's, except at the lowest elevations for short periods. The two polar ice caps appear to be made largely of water. The volume of water ice in the south polar ice cap, if melted, would be sufficient to cover the entire planetary surface to a depth of 11 meters (36 ft). A permafrost mantle stretches from the pole to latitudes of about 60°. Large quantities of water ice are thought to be trapped within the thick cryosphere of Mars. Radar data from *Mars Express* and the *Mars Reconnaissance Orbiter* show large quantities of water ice at both poles (July 2005) and at middle latitudes (November 2008). The Phoenix lander directly sampled water ice in shallow Martian soil on July 31, 2008.

Landforms visible on Mars strongly suggest that liquid water has existed on the planet's surface. Huge linear swathes of scoured ground, known as outflow channels, cut across the surface in about 25 places. These are thought to be a record of erosion caused by the catastrophic release of water from subsurface aquifers, though some of these structures have been hypothesized to result from the action of glaciers or lava. One of the larger examples, Ma'adim Vallis is 700 km (430 mi) long, much greater than the Grand Canyon, with a width of 20 km (12 mi) and a depth of 2 km (1.2 mi) in places. It is thought to have been carved by flowing water early in Mars's history. The youngest of these channels are thought to have formed as recently as only a few million years ago. Elsewhere, particularly on the oldest areas of the Martian surface, finer-scale, dendritic networks of valleys are spread across significant proportions of the landscape. Features of these valleys and their distribution strongly imply that they were carved by runoff resulting from precipitation in early Mars history. Subsurface water flow and groundwater sapping may play important subsidiary roles in some networks, but precipitation was probably the root cause of the incision in almost all cases.

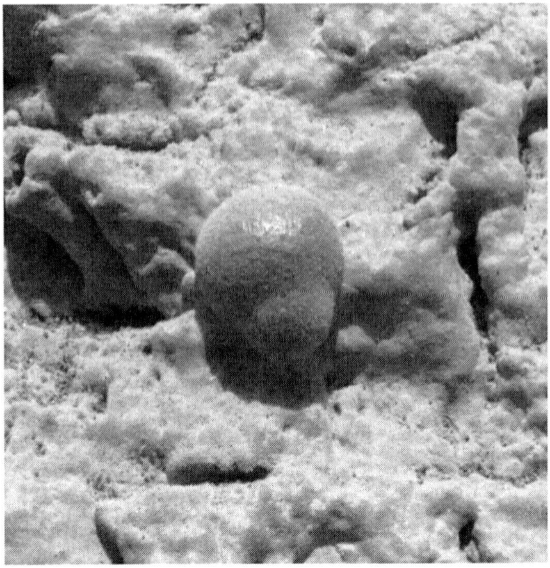

Figure 110: *Photomicrograph by Opportunity showing a gray hematite concretion, nicknamed "blueberries", indicative of the past existence of liquid water*

Along crater and canyon walls, there are thousands of features that appear similar to terrestrial gullies. The gullies tend to be in the highlands of the southern hemisphere and to face the Equator; all are poleward of 30° latitude. A number of authors have suggested that their formation process involves liquid water, probably from melting ice, although others have argued for formation mechanisms involving carbon dioxide frost or the movement of dry dust. No partially degraded gullies have formed by weathering and no superimposed impact craters have been observed, indicating that these are young features, possibly still active. Other geological features, such as deltas and alluvial fans preserved in craters, are further evidence for warmer, wetter conditions at an interval or intervals in earlier Mars history. Such conditions necessarily require the widespread presence of crater lakes across a large proportion of the surface, for which there is independent mineralogical, sedimentological and geomorphological evidence.

Further evidence that liquid water once existed on the surface of Mars comes from the detection of specific minerals such as hematite and goethite, both of which sometimes form in the presence of water. In 2004, *Opportunity* detected the mineral jarosite. This forms only in the presence of acidic water, which demonstrates that water once existed on Mars. More recent evidence for liquid water comes from the finding of the mineral gypsum on the surface

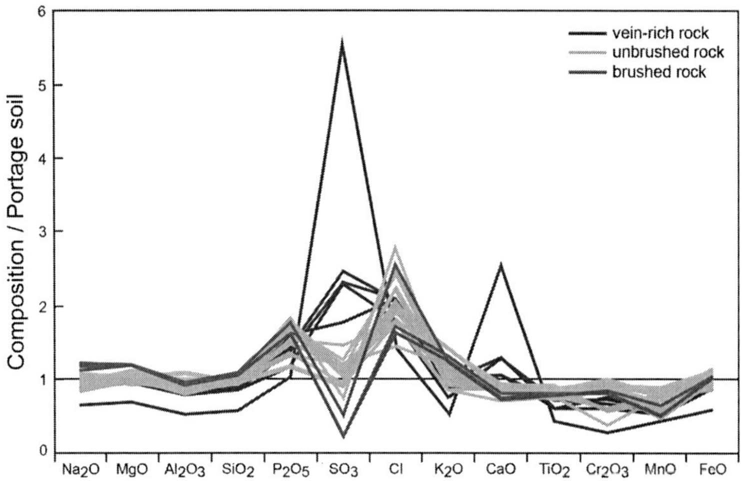

Figure 111: *Composition of "Yellowknife Bay" rocks. Rock veins are higher in calcium and sulfur than "portage" soil (Curiosity, APXS, 2013).*

by NASA's Mars rover Opportunity in December 2011. It is believed that the amount of water in the upper mantle of Mars, represented by hydroxyl ions contained within the minerals of Mars's geology, is equal to or greater than that of Earth at 50–300 parts per million of water, which is enough to cover the entire planet to a depth of 200–1,000 m (660–3,280 ft).

On March 18, 2013, NASA reported evidence from instruments on the *Curiosity* rover of mineral hydration, likely hydrated calcium sulfate, in several rock samples including the broken fragments of "Tintina" rock and "Sutton Inlier" rock as well as in veins and nodules in other rocks like "Knorr" rock and "Wernicke" rock. Analysis using the rover's DAN instrument provided evidence of subsurface water, amounting to as much as 4% water content, down to a depth of 60 cm (24 in), during the rover's traverse from the *Bradbury Landing* site to the *Yellowknife Bay* area in the *Glenelg* terrain. In September 2015, NASA announced that they had found conclusive evidence of hydrated brine flows on recurring slope lineae, based on spectrometer readings of the darkened areas of slopes. These observations provided confirmation of earlier hypotheses based on timing of formation and their rate of growth, that these dark streaks resulted from water flowing in the very shallow subsurface. The streaks contain hydrated salts, perchlorates, which have water molecules in their crystal structure. The streaks flow downhill in Martian summer, when the temperature is above –23 degrees Celsius, and freeze at lower temperatures.

Researchers believe that much of the low northern plains of the planet were covered with an ocean hundreds of meters deep, though this remains controversial. In March 2015, scientists stated that such an ocean might have been the size of Earth's Arctic Ocean. This finding was derived from the ratio of water to deuterium in the modern Martian atmosphere compared to that ratio on Earth. The amount of Martian deuterium is eight times the amount that exists on Earth, suggesting that ancient Mars had significantly higher levels of water. Results from the *Curiosity* rover had previously found a high ratio of deuterium in Gale Crater, though not significantly high enough to suggest the former presence of an ocean. Other scientists caution that these results have not been confirmed, and point out that Martian climate models have not yet shown that the planet was warm enough in the past to support bodies of liquid water.

Polar caps

North polar early summer ice cap (1999)

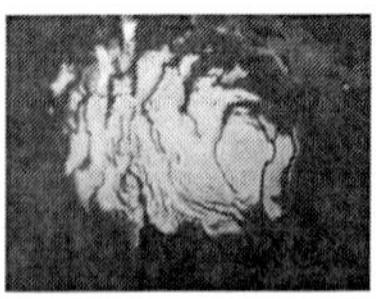

South polar midsummer ice cap (2000)

Mars has two permanent polar ice caps. During a pole's winter, it lies in continuous darkness, chilling the surface and causing the deposition of 25–30% of the atmosphere into slabs of CO_2 ice (dry ice). When the poles are again exposed to sunlight, the frozen CO_2 sublimes, creating enormous winds that sweep off the poles as fast as 400 km/h (250 mph). These seasonal actions transport large amounts of dust and water vapor, giving rise to Earth-like frost and large cirrus clouds. Clouds of water-ice were photographed by the *Opportunity* rover in 2004.

The caps at both poles consist primarily (70%) of water ice. Frozen carbon dioxide accumulates as a comparatively thin layer about one metre thick on the north cap in the northern winter only, whereas the south cap has a permanent dry ice cover about eight metres thick. This permanent dry ice cover at the south pole is peppered by flat floored, shallow, roughly circular pits, which

Figure 112: *A MOLA-based topographic map showing highlands (red and orange) dominating the southern hemisphere of Mars, lowlands (blue) the northern. Volcanic plateaus delimit regions of the northern plains, whereas the highlands are punctuated by several large impact basins.*

repeat imaging shows are expanding by meters per year; this suggests that the permanent CO_2 cover over the south pole water ice is degrading over time. The northern polar cap has a diameter of about 1,000 km (620 mi) during the northern Mars summer, and contains about 1.6 million cubic kilometres (380,000 cu mi) of ice, which, if spread evenly on the cap, would be 2 km (1.2 mi) thick. (This compares to a volume of 2.85 million cubic kilometres (680,000 cu mi) for the Greenland ice sheet.) The southern polar cap has a diameter of 350 km (220 mi) and a thickness of 3 km (1.9 mi). The total volume of ice in the south polar cap plus the adjacent layered deposits has been estimated at 1.6 million cubic km. Both polar caps show spiral troughs, which recent analysis of SHARAD ice penetrating radar has shown are a result of katabatic winds that spiral due to the Coriolis Effect.

The seasonal frosting of areas near the southern ice cap results in the formation of transparent 1-metre-thick slabs of dry ice above the ground. With the arrival of spring, sunlight warms the subsurface and pressure from subliming CO_2 builds up under a slab, elevating and ultimately rupturing it. This leads to geyser-like eruptions of CO_2 gas mixed with dark basaltic sand or dust. This process is rapid, observed happening in the space of a few days, weeks or months, a rate of change rather unusual in geology – especially for Mars. The gas rushing underneath a slab to the site of a geyser carves a spiderweb-like pattern of radial channels under the ice, the process being the inverted equivalent of an erosion network formed by water draining through a single plughole.

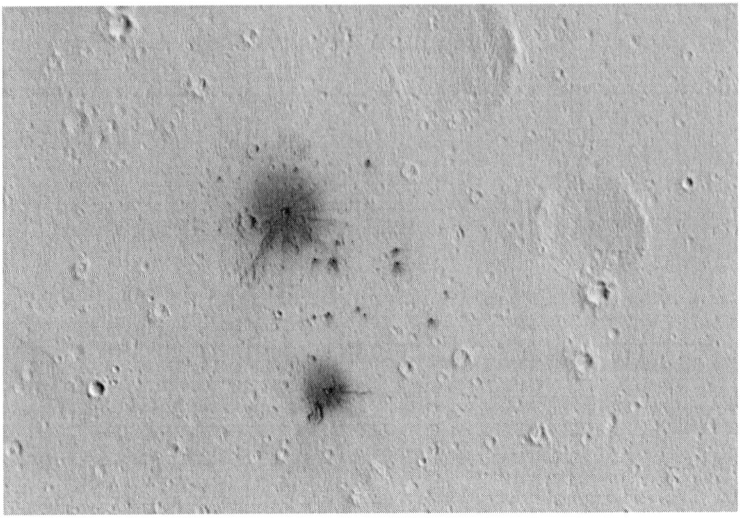

Figure 113: *These new impact craters on Mars occurred sometime between 2008 and 2014, as detected from orbit*

Geography and naming of surface features

Although better remembered for mapping the Moon, Johann Heinrich Mädler and Wilhelm Beer were the first "areographers". They began by establishing that most of Mars's surface features were permanent and by more precisely determining the planet's rotation period. In 1840, Mädler combined ten years of observations and drew the first map of Mars. Rather than giving names to the various markings, Beer and Mädler simply designated them with letters; Meridian Bay (Sinus Meridiani) was thus feature "*a*".

Today, features on Mars are named from a variety of sources. Albedo features are named for classical mythology. Craters larger than 60 km are named for deceased scientists and writers and others who have contributed to the study of Mars. Craters smaller than 60 km are named for towns and villages of the world with populations of less than 100,000. Large valleys are named for the word "Mars" or "star" in various languages; small valleys are named for rivers.

Large albedo features retain many of the older names, but are often updated to reflect new knowledge of the nature of the features. For example, *Nix Olympica* (the snows of Olympus) has become *Olympus Mons* (Mount Olympus). The surface of Mars as seen from Earth is divided into two kinds of areas, with differing albedo. The paler plains covered with dust and sand rich in reddish iron oxides were once thought of as Martian "continents" and given names like Arabia Terra (*land of Arabia*) or Amazonis Planitia (*Amazonian plain*). The dark features were thought to be seas, hence their names Mare Erythraeum, Mare Sirenum and Aurorae Sinus. The largest dark feature seen from Earth is Syrtis Major Planum. The permanent northern polar ice cap is named Planum Boreum, whereas the southern cap is called Planum Australe.

Mars's equator is defined by its rotation, but the location of its Prime Meridian was specified, as was Earth's (at Greenwich), by choice of an arbitrary point; Mädler and Beer selected a line for their first maps of Mars in 1830. After the spacecraft Mariner 9 provided extensive imagery of Mars in 1972, a small crater (later called Airy-0), located in the Sinus Meridiani ("Middle Bay" or "Meridian Bay"), was chosen for the definition of 0.0° longitude to coincide with the original selection.

Because Mars has no oceans and hence no "sea level", a zero-elevation surface had to be selected as a reference level; this is called the *areoid* of Mars, analogous to the terrestrial geoid. Zero altitude was defined by the height at which there is 610.5 Pa (6.105 mbar) of atmospheric pressure. This pressure corresponds to the triple point of water, and it is about 0.6% of the sea level surface pressure on Earth (0.006 atm). In practice, today this surface is defined directly from satellite gravity measurements.

Map of quadrangles

For mapping purposes, the United States Geological Survey divides the surface of Mars into thirty "quadrangles", each named for a prominent physiographic feature within that quadrangle. The quadrangles can be seen and explored via the interactive image map below.

File:MGS_MOC_Wide_Angle_Map_of_Mars_PIA03467.jpg

The thirty cartographic quadrangles of Mars, defined by the United States Geological Survey. The quadrangles are numbered with the prefix "MC" for "Mars Chart." Click on a quadrangle name link and you will be taken to the corresponding article. North is at the top; 0°N 180°W[160] is at the far left on the equator. The map images were taken by the Mars Global Surveyor.

[]

Figure 114: *Bonneville crater and Spirit rover's lander*

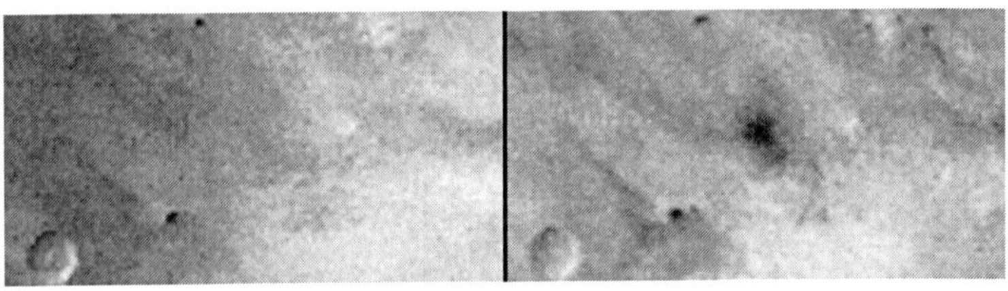

Figure 115: *Fresh asteroid impact on Mars at 3.34°N 219.38°E*[161]. *These before and after images of the same site were taken on the Martian afternoons of March 27 and 28, 2012 respectively (MRO)*

Impact topography

The dichotomy of Martian topography is striking: northern plains flattened by lava flows contrast with the southern highlands, pitted and cratered by ancient impacts. Research in 2008 has presented evidence regarding a theory proposed in 1980 postulating that, four billion years ago, the northern hemisphere of Mars was struck by an object one-tenth to two-thirds the size of Earth's Moon. If validated, this would make the northern hemisphere of Mars the site of an impact crater 10,600 by 8,500 km (6,600 by 5,300 mi) in size, or roughly the area of Europe, Asia, and Australia combined, surpassing the South Pole–Aitken basin as the largest impact crater in the Solar System.

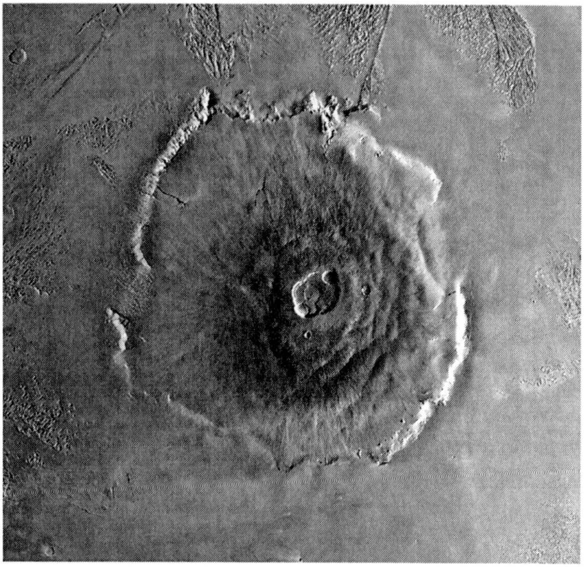

Figure 116: *Viking 1 image of Olympus Mons. The volcano and related terrain are approximately 550 km (340 mi) across.*

Mars is scarred by a number of impact craters: a total of 43,000 craters with a diameter of 5 km (3.1 mi) or greater have been found. The largest confirmed of these is the Hellas impact basin, a light albedo feature clearly visible from Earth. Due to the smaller mass of Mars, the probability of an object colliding with the planet is about half that of Earth. Mars is located closer to the asteroid belt, so it has an increased chance of being struck by materials from that source. Mars is more likely to be struck by short-period comets, *i.e.*, those that lie within the orbit of Jupiter. In spite of this, there are far fewer craters on Mars compared with the Moon, because the atmosphere of Mars provides protection against small meteors and surface modifying processes have erased some craters.

Martian craters can have a morphology that suggests the ground became wet after the meteor impacted.

Volcanoes

The shield volcano Olympus Mons (*Mount Olympus*) is an extinct volcano in the vast upland region Tharsis, which contains several other large volcanoes. Olympus Mons is roughly three times the height of Mount Everest, which in comparison stands at just over 8.8 km (5.5 mi). It is either the tallest or second-tallest mountain in the Solar System, depending on how it is measured, with

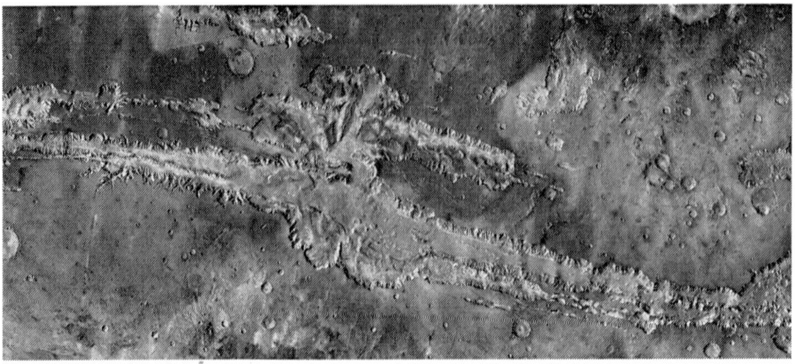

Figure 117: *Valles Marineris (2001 Mars Odyssey)*

various sources giving figures ranging from about 21 to 27 km (13 to 17 mi) high.

Tectonic sites

The large canyon, Valles Marineris (Latin for "Mariner Valleys", also known as Agathadaemon in the old canal maps), has a length of 4,000 km (2,500 mi) and a depth of up to 7 km (4.3 mi). The length of Valles Marineris is equivalent to the length of Europe and extends across one-fifth the circumference of Mars. By comparison, the Grand Canyon on Earth is only 446 km (277 mi) long and nearly 2 km (1.2 mi) deep. Valles Marineris was formed due to the swelling of the Tharsis area, which caused the crust in the area of Valles Marineris to collapse. In 2012, it was proposed that Valles Marineris is not just a graben, but a plate boundary where 150 km (93 mi) of transverse motion has occurred, making Mars a planet with possibly a two-tectonic plate arrangement.

Holes

Images from the Thermal Emission Imaging System (THEMIS) aboard NASA's Mars Odyssey orbiter have revealed seven possible cave entrances on the flanks of the volcano Arsia Mons. The caves, named after loved ones of their discoverers, are collectively known as the "seven sisters". Cave entrances measure from 100 to 252 m (328 to 827 ft) wide and they are estimated to be at least 73 to 96 m (240 to 315 ft) deep. Because light does not reach the floor of most of the caves, it is possible that they extend much deeper than these lower estimates and widen below the surface. "Dena" is the only exception; its floor is visible and was measured to be 130 m (430 ft) deep. The interiors of these caverns may be protected from micrometeoroids, UV radiation, solar flares and high energy particles that bombard the planet's surface.

Figure 118: *The tenuous atmosphere of Mars visible on the horizon*

Atmosphere

Mars lost its magnetosphere 4 billion years ago, possibly because of numerous asteroid strikes, so the solar wind interacts directly with the Martian ionosphere, lowering the atmospheric density by stripping away atoms from the outer layer. Both Mars Global Surveyor and Mars Express have detected ionised atmospheric particles trailing off into space behind Mars, and this atmospheric loss is being studied by the MAVEN orbiter. Compared to Earth, the atmosphere of Mars is quite rarefied. Atmospheric pressure on the surface today ranges from a low of 30 Pa (0.030 kPa) on Olympus Mons to over 1,155 Pa (1.155 kPa) in Hellas Planitia, with a mean pressure at the surface level of 600 Pa (0.60 kPa). The highest atmospheric density on Mars is equal to that found 35 km (22 mi) above Earth's surface. The resulting mean surface pressure is only 0.6% of that of Earth (101.3 kPa). The scale height of the atmosphere is about 10.8 km (6.7 mi), which is higher than Earth's, 6 km (3.7 mi), because the surface gravity of Mars is only about 38% of Earth's, an effect offset by both the lower temperature and 50% higher average molecular weight of the atmosphere of Mars.

The atmosphere of Mars consists of about 96% carbon dioxide, 1.93% argon and 1.89% nitrogen along with traces of oxygen and water. The atmosphere is quite dusty, containing particulates about 1.5 μm in diameter which give the

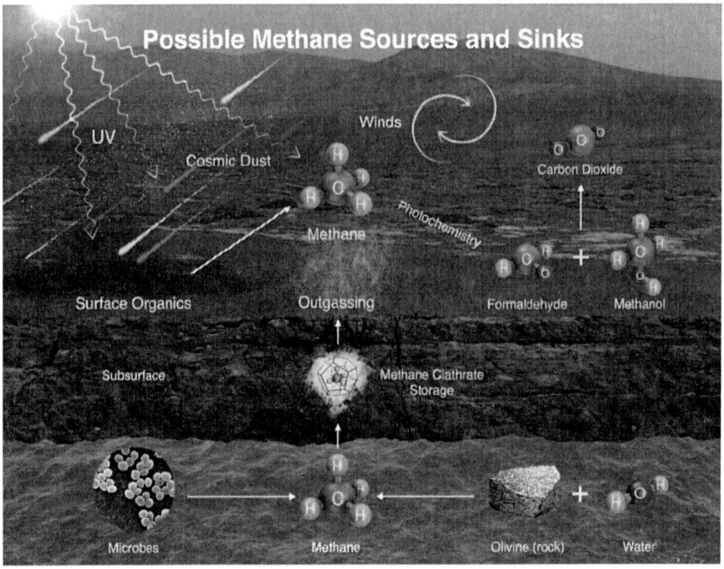

Figure 119: *Potential sources and sinks of methane (CH_4) on Mars*

Martian sky a tawny color when seen from the surface. It may take on a pink hue due to iron oxide particles suspended in it.

Methane has been detected in the Martian atmosphere with a concentration of about 30 ppb; it occurs in extended plumes, and the profiles imply that the methane was released from discrete regions. In northern midsummer, the principal plume contained 19,000 metric tons of methane, with an estimated source strength of 0.6 kilograms per second. The profiles suggest that there may be two local source regions, the first centered near 30°N 260°W[162] and the second near 0°N 310°W[163]. It is estimated that Mars must produce 270 tonnes per year of methane.

Methane can exist in the Martian atmosphere for only a limited period before it is destroyed—estimates of its lifetime range from 0.6–4 years. Its presence despite this short lifetime indicates that an active source of the gas must be present. Volcanic activity, cometary impacts, and the presence of methanogenic microbial life forms are among possible sources. Methane could be produced by a non-biological process called *serpentinization* involving water, carbon dioxide, and the mineral olivine, which is known to be common on Mars.

Figure 120: *Escaping atmosphere on Mars (carbon, oxygen, and hydrogen) by MAVEN in UV*

The *Curiosity* rover, which landed on Mars in August 2012, is able to make measurements that distinguish between different isotopologues of methane, but even if the mission is to determine that microscopic Martian life is the source of the methane, the life forms likely reside far below the surface, outside of the rover's reach. The first measurements with the Tunable Laser Spectrometer (TLS) indicated that there is less than 5 ppb of methane at the landing site at the point of the measurement. On September 19, 2013, NASA scientists, from further measurements by *Curiosity*, reported no detection of atmospheric methane with a measured value of 0.18±0.67 ppbv corresponding to an upper limit of only 1.3 ppbv (95% confidence limit) and, as a result, conclude that the probability of current methanogenic microbial activity on Mars is reduced.

The Mars Orbiter Mission by India is searching for methane in the atmosphere, while the ExoMars Trace Gas Orbiter, launched in 2016, would further study the methane as well as its decomposition products, such as formaldehyde and methanol.[164]

On December 16, 2014, NASA reported the *Curiosity* rover detected a "tenfold spike", likely localized, in the amount of methane in the Martian atmosphere. Sample measurements taken "a dozen times over 20 months" showed increases in late 2013 and early 2014, averaging "7 parts of methane per billion in the atmosphere." Before and after that, readings averaged around one-tenth that level.

Ammonia was tentatively detected on Mars by the Mars Express satellite, but with its relatively short lifetime, it is not clear what produced it. Ammonia is not stable in the Martian atmosphere and breaks down after a few hours. One possible source is volcanic activity.

In September 2017, NASA reported radiation levels on the surface of the planet Mars were temporarily doubled, and were associated with an aurora 25-times brighter than any observed earlier, due to a massive, and unexpected, solar storm in the middle of the month.

Aurora

In 1994, the European Space Agency's Mars Express found an ultraviolet glow coming from "magnetic umbrellas" in the southern hemisphere. Mars does not have a global magnetic field which guides charged particles entering the atmosphere. Mars has multiple umbrella-shaped magnetic fields mainly in the southern hemisphere, which are remnants of a global field that decayed billions of years ago.

In late December 2014, NASA's MAVEN spacecraft detected evidence of widespread auroras in Mars's northern hemisphere and descended to approximately 20–30 degrees North latitude of Mars's equator. The particles causing the aurora penetrated into the Martian atmosphere, creating auroras below 100 km above the surface, Earth's auroras range from 100 km to 500 km above the surface. Magnetic fields in the solar wind drape over Mars, into the atmosphere, and the charged particles follow the solar wind magnetic field lines into the atmosphere, causing auroras to occur outside the magnetic umbrellas.

On March 18, 2015, NASA reported the detection of an aurora that is not fully understood and an unexplained dust cloud in the atmosphere of Mars.

Climate

Dust storm on Mars

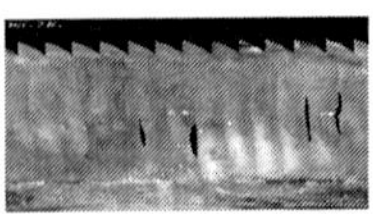

November 18, 2012

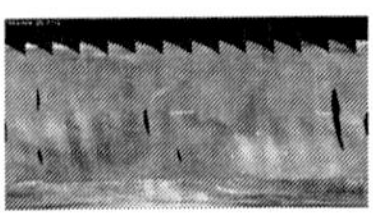

November 25, 2012

Opportunity and *Curiosity* rovers are noted.

Of all the planets in the Solar System, the seasons of Mars are the most Earth-like, due to the similar tilts of the two planets' rotational axes. The lengths of the Martian seasons are about twice those of Earth's because Mars's greater distance from the Sun leads to the Martian year being about two Earth years long. Martian surface temperatures vary from lows of about –143 °C

(–225 °F) at the winter polar caps to highs of up to 35 °C (95 °F) in equatorial summer. The wide range in temperatures is due to the thin atmosphere which cannot store much solar heat, the low atmospheric pressure, and the low thermal inertia of Martian soil. The planet is 1.52 times as far from the Sun as Earth, resulting in just 43% of the amount of sunlight.

If Mars had an Earth-like orbit, its seasons would be similar to Earth's because its axial tilt is similar to Earth's. The comparatively large eccentricity of the Martian orbit has a significant effect. Mars is near perihelion when it is summer in the southern hemisphere and winter in the north, and near aphelion when it is winter in the southern hemisphere and summer in the north. As a result, the seasons in the southern hemisphere are more extreme and the seasons in the northern are milder than would otherwise be the case. The summer temperatures in the south can be up to 30 K (30 °C; 54 °F) warmer than the equivalent summer temperatures in the north.

Mars has the largest dust storms in the Solar System. These can vary from a storm over a small area, to gigantic storms that cover the entire planet. They tend to occur when Mars is closest to the Sun, and have been shown to increase the global temperature.

Orbit and rotation

Mars's average distance from the Sun is roughly 230 million kilometres (143,000,000 mi), and its orbital period is 687 (Earth) days. The solar day (or sol) on Mars is only slightly longer than an Earth day: 24 hours, 39 minutes, and 35.244 seconds. A Martian year is equal to 1.8809 Earth years, or 1 year, 320 days, and 18.2 hours.

The axial tilt of Mars is 25.19 degrees relative to its orbital plane, which is similar to the axial tilt of Earth. As a result, Mars has seasons like Earth, though on Mars, they are nearly twice as long because its orbital period is that much longer. In the present day epoch, the orientation of the north pole of Mars is close to the star Deneb. Mars passed an aphelion in March 2010 and its perihelion in March 2011. The next aphelion came in February 2012 and the next perihelion came in January 2013.

Mars has a relatively pronounced orbital eccentricity of about 0.09; of the seven other planets in the Solar System, only Mercury has a larger orbital eccentricity. It is known that in the past, Mars has had a much more circular orbit. At one point, 1.35 million Earth years ago, Mars had an eccentricity of roughly 0.002, much less than that of Earth today. Mars's cycle of eccentricity is 96,000 Earth years compared to Earth's cycle of 100,000 years. Mars has a much longer cycle of eccentricity, with a period of 2.2 million Earth years, and this

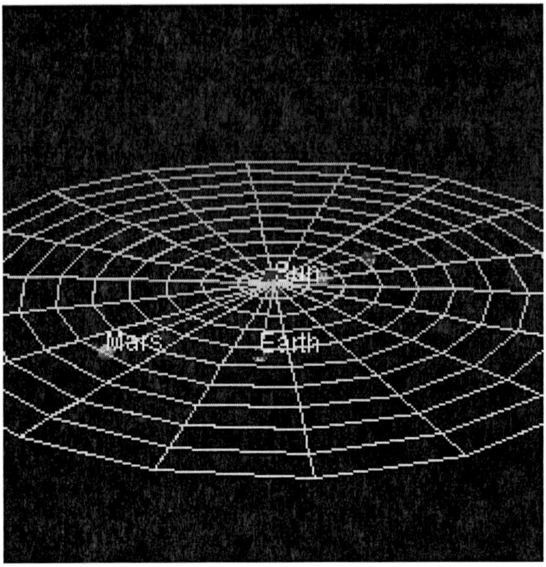

Figure 121: *Mars is about 230 million kilometres (143,000,000 mi) from the Sun; its orbital period is 687 (Earth) days, depicted in red. Earth's orbit is in blue.*

overshadows the 96,000-year cycle in the eccentricity graphs. For the last 35,000 years, the orbit of Mars has been getting slightly more eccentric because of the gravitational effects of the other planets. The closest distance between Earth and Mars will continue to mildly decrease for the next 25,000 years.

Habitability and search for life

Search for life

The current understanding of planetary habitability—the ability of a world to develop environmental conditions favorable to the emergence of life—favors planets that have liquid water on their surface. Most often this requires the orbit of a planet to lie within the habitable zone, which for the Sun extends from just beyond Venus to about the semi-major axis of Mars. During perihelion, Mars dips inside this region, but Mars's thin (low-pressure) atmosphere prevents liquid water from existing over large regions for extended periods. The past flow of liquid water demonstrates the planet's potential for habitability. Recent evidence has suggested that any water on the Martian surface may have been too salty and acidic to support regular terrestrial life.

Figure 122: *Viking 1 lander's sampling arm scooped up soil samples for tests (Chryse Planitia)*

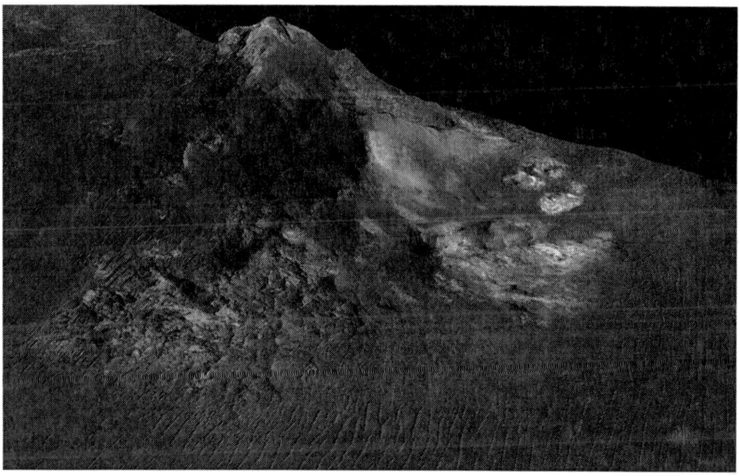

Figure 123: *Detection of impact glass deposits (green spots) at Alga crater, a possible site for preserved ancient life*

The lack of a magnetosphere and the extremely thin atmosphere of Mars are a challenge: the planet has little heat transfer across its surface, poor insulation against bombardment of the solar wind and insufficient atmospheric pressure to retain water in a liquid form (water instead sublimes to a gaseous state). Mars is nearly, or perhaps totally, geologically dead; the end of volcanic activity has apparently stopped the recycling of chemicals and minerals between the surface and interior of the planet.

In situ investigations have been performed on Mars by the *Viking* landers, *Spirit* and *Opportunity* rovers, *Phoenix* lander, and *Curiosity* rover. Evidence suggests that the planet was once significantly more habitable than it is today, but whether living organisms ever existed there remains unknown. The Viking probes of the mid-1970s carried experiments designed to detect microorganisms in Martian soil at their respective landing sites and had positive results, including a temporary increase of CO_2 production on exposure to water and nutrients. This sign of life was later disputed by scientists, resulting in a continuing debate, with NASA scientist Gilbert Levin asserting that Viking may have found life. A re-analysis of the Viking data, in light of modern knowledge of extremophile forms of life, has suggested that the Viking tests were not sophisticated enough to detect these forms of life. The tests could even have killed a (hypothetical) life form. Tests conducted by the Phoenix Mars lander have shown that the soil has a alkaline pH and it contains magnesium, sodium, potassium and chloride. The soil nutrients may be able to support life, but life would still have to be shielded from the intense ultraviolet light. A recent analysis of martian meteorite EETA79001 found 0.6 ppm ClO_4^-, 1.4 ppm ClO_3^-, and 16 ppm NO_3^-, most likely of Martian origin. The ClO_3^- suggests the presence of other highly oxidizing oxychlorines, such as ClO_2^- or ClO, produced both by UV oxidation of Cl and X-ray radiolysis of ClO_4^-. Thus, only highly refractory and/or well-protected (sub-surface) organics or life forms are likely to survive.

A 2014 analysis of the Phoenix WCL showed that the $Ca(ClO_4)_2$ in the Phoenix soil has not interacted with liquid water of any form, perhaps for as long as 600 Myr. If it had, the highly soluble $Ca(ClO_4)_2$ in contact with liquid water would have formed only $CaSO_4$. This suggests a severely arid environment, with minimal or no liquid water interaction.

Scientists have proposed that carbonate globules found in meteorite ALH84001, which is thought to have originated from Mars, could be fossilized microbes extant on Mars when the meteorite was blasted from the Martian surface by a meteor strike some 15 million years ago. This proposal has been met with skepticism, and an exclusively inorganic origin for the shapes has been proposed.

Small quantities of methane and formaldehyde detected by Mars orbiters are both claimed to be possible evidence for life, as these chemical compounds would quickly break down in the Martian atmosphere. Alternatively, these compounds may instead be replenished by volcanic or other geological means, such as serpentinization.

Impact glass, formed by the impact of meteors, which on Earth can preserve signs of life, has been found on the surface of the impact craters on Mars. Likewise, the glass in impact craters on Mars could have preserved signs of life if life existed at the site.

In May 2017, evidence of the earliest known life on land on Earth may have been found in 3.48-billion-year-old geyserite and other related mineral deposits (often found around hot springs and geysers) uncovered in the Pilbara Craton of Western Australia. These findings may be helpful in deciding where best to search for early signs of life on the planet Mars.

Moons

Enhanced-color HiRISE image of Phobos, showing a series of mostly parallel grooves and crater chains, with Stickney crater at right

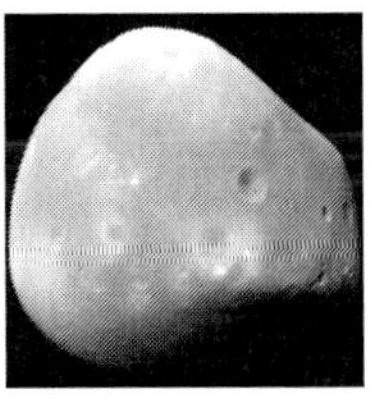

Enhanced-color HiRISE image of Deimos (not to scale), showing its smooth blanket of regolith

Mars has two relatively small natural moons, Phobos (about 22 km (14 mi) in diameter) and Deimos (about 12 km (7.5 mi) in diameter), which orbit close to the planet. Asteroid capture is a long-favored theory, but their origin remains uncertain. Both satellites were discovered in 1877 by Asaph Hall; they are named after the characters Phobos (panic/fear) and Deimos (terror/dread), who, in Greek mythology, accompanied their father Ares, god of war, into battle. Mars was the Roman counterpart of Ares. In modern Greek, though, the planet retains its ancient name *Ares* (Aris: Άρης).

From the surface of Mars, the motions of Phobos and Deimos appear different from that of the Moon. Phobos rises in the west, sets in the east, and rises

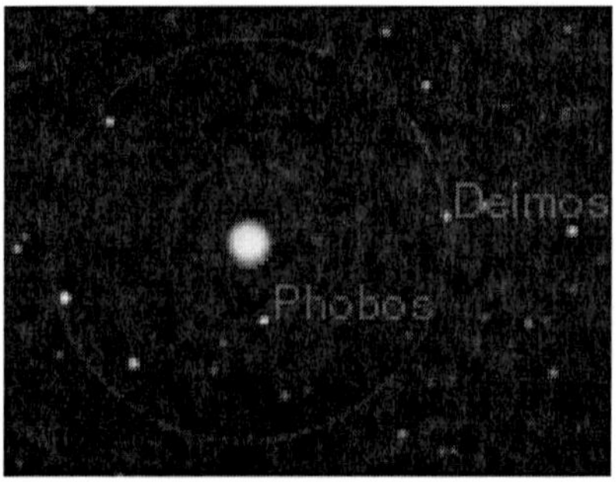

Figure 124: *Orbits of Phobos and Deimos (to scale)*

again in just 11 hours. Deimos, being only just outside synchronous orbit – where the orbital period would match the planet's period of rotation – rises as expected in the east but slowly. Despite the 30-hour orbit of Deimos, 2.7 days elapse between its rise and set for an equatorial observer, as it slowly falls behind the rotation of Mars.

Because the orbit of Phobos is below synchronous altitude, the tidal forces from the planet Mars are gradually lowering its orbit. In about 50 million years, it could either crash into Mars's surface or break up into a ring structure around the planet.

The origin of the two moons is not well understood. Their low albedo and carbonaceous chondrite composition have been regarded as similar to asteroids, supporting the capture theory. The unstable orbit of Phobos would seem to point towards a relatively recent capture. But both have circular orbits, near the equator, which is unusual for captured objects and the required capture dynamics are complex. Accretion early in the history of Mars is plausible, but would not account for a composition resembling asteroids rather than Mars itself, if that is confirmed.

A third possibility is the involvement of a third body or a type of impact disruption. More-recent lines of evidence for Phobos having a highly porous interior, and suggesting a composition containing mainly phyllosilicates and other minerals known from Mars, point toward an origin of Phobos from material ejected by an impact on Mars that reaccreted in Martian orbit, similar to the prevailing theory for the origin of Earth's moon. Although the VNIR spectra of the moons of Mars resemble those of outer-belt asteroids, the thermal infrared spectra of Phobos are reported to be inconsistent with chondrites of any class.

Mars may have moons smaller than 50 to 100 metres (160 to 330 ft) in diameter, and a dust ring is predicted to exist between Phobos and Deimos.

Figure 125: *Panorama of Gusev crater, where Spirit rover examined volcanic basalts*

Exploration

Dozens of crewless spacecraft, including orbiters, landers, and rovers, have been sent to Mars by the Soviet Union, the United States, Europe, and India to study the planet's surface, climate, and geology.

As of 2016[165], Mars is host to eight functioning spacecraft: six in orbit—*2001 Mars Odyssey*, *Mars Express*, *Mars Reconnaissance Orbiter*, MAVEN, Mars Orbiter Mission and ExoMars Trace Gas Orbiter—and two on the surface—Mars Exploration Rover *Opportunity* and the Mars Science Laboratory *Curiosity*. Observations by the *Mars Reconnaissance Orbiter* have revealed possible flowing water during the warmest months on Mars. In 2013, NASA's *Curiosity* rover discovered that Mars's soil contains between 1.5% and 3% water by mass (albeit attached to other compounds and thus not freely accessible). The public can request images of Mars via the *Mars Reconnaissance Orbiter*'s HiWish program.

The Mars Science Laboratory, named *Curiosity*, launched on November 26, 2011, and reached Mars on August 6, 2012 UTC. It is larger and more advanced than the Mars Exploration Rovers, with a movement rate up to 90 m (300 ft) per hour. Experiments include a laser chemical sampler that can deduce the make-up of rocks at a distance of 7 m (23 ft). On February 10, 2013, the *Curiosity* rover obtained the first deep rock samples ever taken from another planetary body, using its on-board drill.

On September 24, 2014, Mars Orbiter Mission (MOM), launched by the Indian Space Research Organisation, reached Mars orbit. ISRO launched MOM

Figure 126: *Mars Science Laboratory under parachute during its atmospheric entry at Mars*

on November 5, 2013, with the aim of analyzing the Martian atmosphere and topography. The Mars Orbiter Mission used a Hohmann transfer orbit to escape Earth's gravitational influence and catapult into a nine-month-long voyage to Mars. The mission is the first successful Asian interplanetary mission.

The European Space Agency, in collaboration with Roscosmos, launched the ExoMars Trace Gas Orbiter and *Schiaparelli* lander on March 14, 2016. While the Trace Gas Orbiter successfully entered Mars orbit on October 19, 2016, *Schiaparelli* crashed during its landing attempt.

Future

Planned for May 2018 is the launch of NASA's *InSight* lander, along with the twin MarCO CubeSats that will fly by Mars and provide a telemetry relay for the landing. The mission is expected to arrive at Mars in November 2018. NASA plans to launch its Mars 2020 astrobiology rover in July or August 2020.

The European Space Agency will launch the ExoMars rover and surface platform in July 2020.

The United Arab Emirates' *Mars Hope* orbiter is planned for launch in 2020, reaching Mars orbit in 2021. The probe will make a global study of the Martian atmosphere.

Several plans for a human mission to Mars have been proposed throughout the 20th century and into the 21st century, but no active plan has an arrival date sooner than the 2020s. SpaceX founder Elon Musk presented a plan in September 2016 to, optimistically, launch space tourists to Mars in 2024 at an estimated development cost of US$10 billion. In October 2016, President Barack Obama renewed U.S. policy to pursue the goal of sending humans to Mars in the 2030s, and to continue using the International Space Station as a technology incubator in that pursuit.

Astronomy on Mars

With the presence of various orbiters, landers, and rovers, it is possible to practice astronomy from Mars. Although Mars's moon Phobos appears about one-third the angular diameter of the full moon on Earth, Deimos appears more or less star-like, looking only slightly brighter than Venus does from Earth.

Various phenomena seen from Earth have also been observed from Mars, such as meteors and auroras. The apparent sizes of the moons Phobos and Deimos are sufficiently smaller than that of the Sun; thus, their partial "eclipses" of the Sun are best considered transits (see transit of Deimos and Phobos from Mars). Transits of Mercury and Venus have been observed from Mars. A transit of Earth will be seen from Mars on November 10, 2084.

On October 19, 2014, Comet Siding Spring passed extremely close to Mars, so close that the coma may have enveloped Mars.

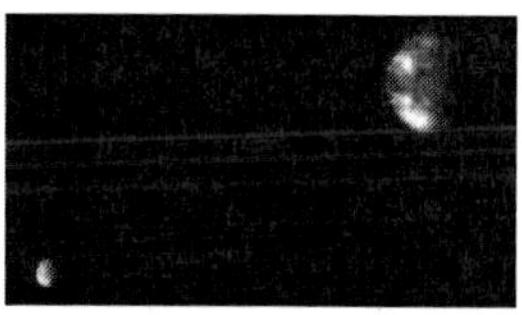

Earth and the Moon (*MRO* HiRISE, November 2016)

Phobos transits the Sun (*Opportunity*, March 10, 2004)

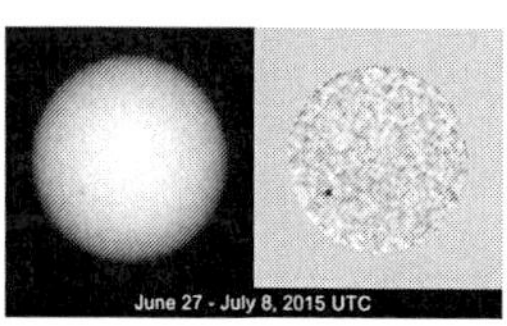

Tracking sunspots from Mars

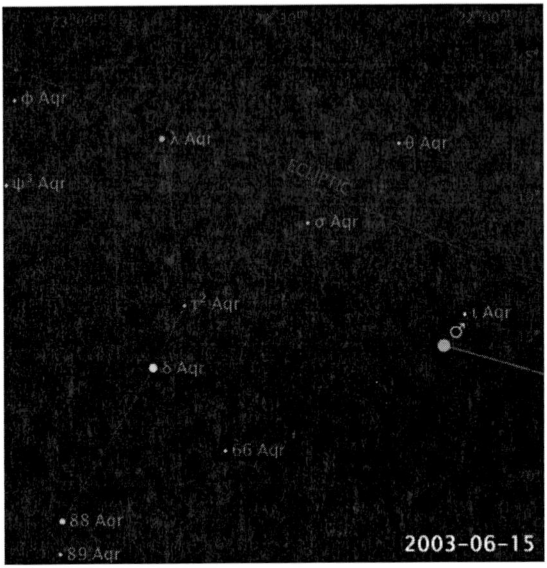

Figure 127: *Animation of the apparent retrograde motion of Mars in 2003 as seen from Earth*

Viewing

Because the orbit of Mars is eccentric, its apparent magnitude at opposition from the Sun can range from –3.0 to –1.4. The minimum brightness is magnitude +1.6 when the planet is in conjunction with the Sun. Mars usually appears distinctly yellow, orange, or red; the actual color of Mars is closer to butterscotch, and the redness seen is just dust in the planet's atmosphere. NASA's *Spirit* rover has taken pictures of a greenish-brown, mud-colored landscape with blue-grey rocks and patches of light red sand. When farthest away from Earth, it is more than seven times farther away than when it is closest. When least favorably positioned, it can be lost in the Sun's glare for months at a time. At its most favorable times—at 15- or 17-year intervals, and always between late July and late September—a lot of surface detail can be seen with a telescope. Especially noticeable, even at low magnification, are the polar ice caps.

As Mars approaches opposition, it begins a period of retrograde motion, which means it will appear to move backwards in a looping motion with respect to the background stars. The duration of this retrograde motion lasts for about 72 days, and Mars reaches its peak luminosity in the middle of this motion.

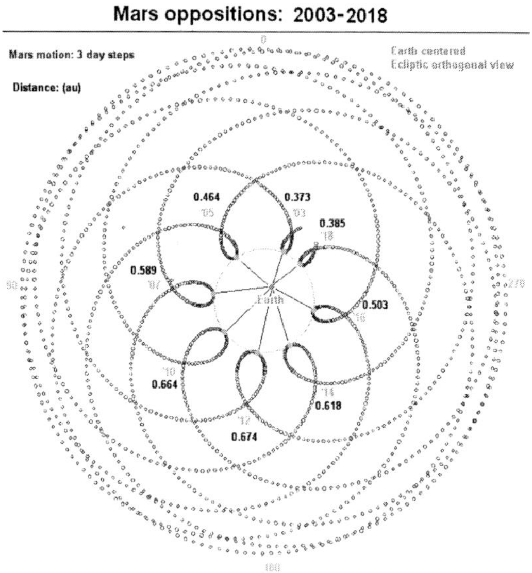

Figure 128: *Mars oppositions from 2003–2018, viewed from above the ecliptic with Earth centered*

Closest approaches

Relative

The point at which Mars's geocentric longitude is 180° different from the Sun's is known as opposition, which is near the time of closest approach to Earth. The time of opposition can occur as much as 8.5 days away from the closest approach. The distance at close approach varies between about 54 and about 103 million km due to the planets' elliptical orbits, which causes comparable variation in angular size. The last Mars opposition occurred on May 22, 2016 at a distance of about 76 million km. The next Mars opposition occurs on July 27, 2018 at a distance of about 58 million km. The average time between the successive oppositions of Mars, its synodic period, is 780 days; but the number of days between the dates of successive oppositions can range from 764 to 812.

As Mars approaches opposition it begins a period of retrograde motion, which makes it appear to move backwards in a looping motion relative to the background stars. The duration of this retrograde motion is about 72 days.

Absolute, around the present time

Mars made its closest approach to Earth and maximum apparent brightness in nearly 60,000 years, 55,758,006 km (0.37271925 AU; 34,646,419 mi), magnitude –2.88, on August 27, 2003 at 9:51:13 UT. This occurred when Mars was one day from opposition and about three days from its perihelion, making it particularly easy to see from Earth. The last time it came so close is estimated to have been on September 12, 57,617 BC, the next time being in 2287. This record approach was only slightly closer than other recent close approaches. For instance, the minimum distance on August 22, 1924 was 0.37285 AU, and the minimum distance on August 24, 2208 will be 0.37279 AU.

Historical observations

In 2005, radar data revealed the presence of large quantities of water ice at the poles and at mid-latitudes. The Mars rover *Spirit* sampled chemical compounds containing water molecules in March 2007. The *Phoenix* lander directly sampled water ice in shallow Martian soil on July 31, 2008. On September 28, 2015, NASA announced the presence of briny flowing salt water on the Martian surface.

The history of observations of Mars is marked by the oppositions of Mars, when the planet is closest to Earth and hence is most easily visible, which occur every couple of years. Even more notable are the perihelic oppositions of Mars, which occur every 15 or 17 years and are distinguished because Mars is close to perihelion, making it even closer to Earth.

Ancient and medieval observations

The ancient Sumerians believed that Mars was Nergal, the god of war and plague. During Sumerian times, Nergal was a minor deity of little significance, but, during later times, his main cult center was the city of Nineveh. In Mesopotamian texts, Mars is referred to as the "star of judgement of the fate of the dead". The existence of Mars as a wandering object in the night sky was recorded by the ancient Egyptian astronomers and, by 1534 BCE, they were familiar with the retrograde motion of the planet. By the period of the Neo-Babylonian Empire, the Babylonian astronomers were making regular records of the positions of the planets and systematic observations of their behavior. For Mars, they knew that the planet made 37 synodic periods, or 42 circuits of the zodiac, every 79 years. They invented arithmetic methods for making minor corrections to the predicted positions of the planets.

In the fourth century BCE, Aristotle noted that Mars disappeared behind the Moon during an occultation, indicating that the planet was farther away.

Ptolemy, a Greek living in Alexandria, attempted to address the problem of the orbital motion of Mars. Ptolemy's model and his collective work on astronomy was presented in the multi-volume collection *Almagest*, which became the authoritative treatise on Western astronomy for the next fourteen centuries. Literature from ancient China confirms that Mars was known by Chinese astronomers by no later than the fourth century BCE. In the fifth century CE, the Indian astronomical text *Surya Siddhanta* estimated the diameter of Mars. In the East Asian cultures, Mars is traditionally referred to as the "fire star" (火星), based on the Five elements.

During the seventeenth century, Tycho Brahe measured the diurnal parallax of Mars that Johannes Kepler used to make a preliminary calculation of the relative distance to the planet. When the telescope became available, the diurnal parallax of Mars was again measured in an effort to determine the Sun-Earth distance. This was first performed by Giovanni Domenico Cassini in 1672. The early parallax measurements were hampered by the quality of the instruments. The only occultation of Mars by Venus observed was that of October 13, 1590, seen by Michael Maestlin at Heidelberg. In 1610, Mars was viewed by Galileo Galilei, who was first to see it via telescope. The first person to draw a map of Mars that displayed any terrain features was the Dutch astronomer Christiaan Huygens.

Martian "canals"

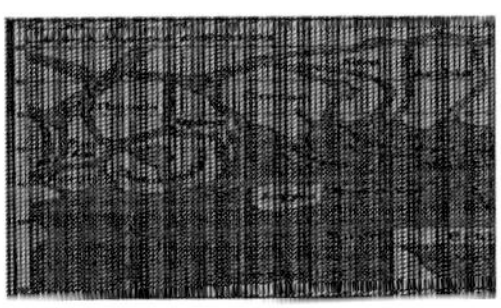

Map of Mars by Giovanni Schiaparelli

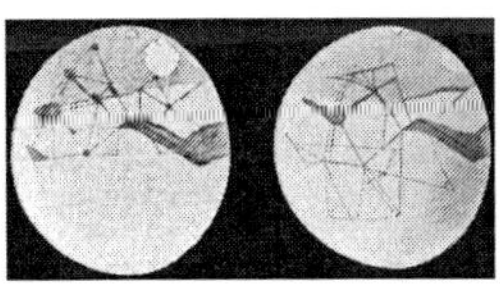

Mars sketched as observed by Lowell before 1914 (south on top)

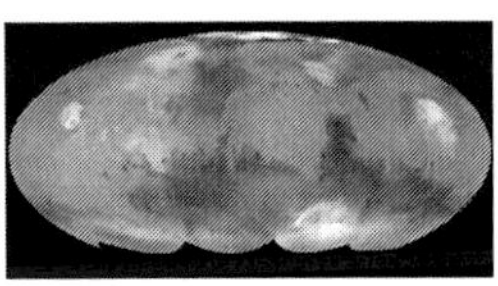

Map of Mars from the *Hubble Space Telescope* as seen near the 1999 opposition (north on top)

By the 19th century, the resolution of telescopes reached a level sufficient for surface features to be identified. A perihelic opposition of Mars occurred on September 5, 1877. In that year, the Italian astronomer Giovanni Schiaparelli used a 22 cm (8.7 in) telescope in Milan to help produce the first detailed map of Mars. These maps notably contained features he called *canali*, which were later shown to be an optical illusion. These *canali* were supposedly long, straight lines on the surface of Mars, to which he gave names of famous rivers on Earth. His term, which means "channels" or "grooves", was popularly mistranslated in English as "canals".

Influenced by the observations, the orientalist Percival Lowell founded an observatory which had 30 and 45 cm (12 and 18 in) telescopes. The observatory was used for the exploration of Mars during the last good opportunity in 1894 and the following less favorable oppositions. He published several books on Mars and life on the planet, which had a great influence on the public. The *canali* were independently found by other astronomers, like Henri Joseph Perrotin and Louis Thollon in Nice, using one of the largest telescopes of that time.

The seasonal changes (consisting of the diminishing of the polar caps and the dark areas formed during Martian summer) in combination with the canals led to speculation about life on Mars, and it was a long-held belief that Mars contained vast seas and vegetation. The telescope never reached the resolution required to give proof to any speculations. As bigger telescopes were used, fewer long, straight *canali* were observed. During an observation in 1909 by Flammarion with an 84 cm (33 in) telescope, irregular patterns were observed, but no *canali* were seen.

Even in the 1960s articles were published on Martian biology, putting aside explanations other than life for the seasonal changes on Mars. Detailed scenarios for the metabolism and chemical cycles for a functional ecosystem have been published.

Spacecraft visitation

Once spacecraft visited the planet during NASA's Mariner missions in the 1960s and 70s, these concepts were radically broken. The results of the Viking life-detection experiments aided an intermission in which the hypothesis of a hostile, dead planet was generally accepted.

Mariner 9 and Viking allowed better maps of Mars to be made using the data from these missions, and another major leap forward was the Mars Global Surveyor mission, launched in 1996 and operated until late 2006, that allowed complete, extremely detailed maps of the Martian topography, magnetic field and surface minerals to be obtained. These maps are available online; for

♂

example, at Google Mars. Mars Reconnaissance Orbiter and Mars Express continued exploring with new instruments, and supporting lander missions. NASA provides two online tools: Mars Trek, which provides visualizations of the planet using data from 50 years of exploration, and Experience Curiosity, which simulates traveling on Mars in 3-D with Curiosity.

In culture

Mars is named after the Roman god of war. In different cultures, Mars represents masculinity and youth. Its symbol, a circle with an arrow pointing out to the upper right, is used as a symbol for the male gender.

The many failures in Mars exploration probes resulted in a satirical counterculture blaming the failures on an Earth-Mars "Bermuda Triangle", a "Mars Curse", or a "Great Galactic Ghoul" that feeds on Martian spacecraft.

Intelligent "Martians"

The fashionable idea that Mars was populated by intelligent Martians exploded in the late 19th century. Schiaparelli's "canali" observations combined with Percival Lowell's books on the subject put forward the standard notion of a planet that was a drying, cooling, dying world with ancient civilizations constructing irrigation works.

Many other observations and proclamations by notable personalities added to what has been termed "Mars Fever". In 1899, while investigating atmospheric radio noise using his receivers in his Colorado Springs lab, inventor Nikola Tesla observed repetitive signals that he later surmised might have been radio communications coming from another planet, possibly Mars. In a 1901 interview Tesla said:

> *It was some time afterward when the thought flashed upon my mind that the disturbances I had observed might be due to an intelligent control. Although I could not decipher their meaning, it was impossible for me to think of them as having been entirely accidental. The feeling is constantly growing on me that I had been the first to hear the greeting of one planet to another.*

Tesla's theories gained support from Lord Kelvin who, while visiting the United States in 1902, was reported to have said that he thought Tesla had

Figure 129: *An 1893 soap ad playing on the popular idea that Mars was populated*

picked up Martian signals being sent to the United States. Kelvin "emphatically" denied this report shortly before departing America: "What I really said was that the inhabitants of Mars, if there are any, were doubtless able to see New York, particularly the glare of the electricity."

In a *New York Times* article in 1901, Edward Charles Pickering, director of the Harvard College Observatory, said that they had received a telegram from Lowell Observatory in Arizona that seemed to confirm that Mars was trying to communicate with Earth.

> *Early in December 1900, we received from Lowell Observatory in Arizona a telegram that a shaft of light had been seen to project from Mars (the Lowell observatory makes a specialty of Mars) lasting seventy minutes. I wired these facts to Europe and sent out neostyle copies through this country. The observer there is a careful, reliable man and there is no reason to doubt that the light existed. It was given as from a well-known geographical point on Mars. That was all. Now the story has gone the world over. In Europe it is stated that I have been in communication with Mars, and all sorts of exaggerations have spring up. Whatever the light was, we have no means of knowing. Whether it had intelligence or not, no one can say. It is absolutely inexplicable.*

Figure 130: *Martian tripod illustration from the 1906 French edition of The War of the Worlds by H. G. Wells*

Pickering later proposed creating a set of mirrors in Texas, intended to signal Martians.

In recent decades, the high-resolution mapping of the surface of Mars, culminating in Mars Global Surveyor, revealed no artifacts of habitation by "intelligent" life, but pseudoscientific speculation about intelligent life on Mars continues from commentators such as Richard C. Hoagland. Reminiscent of the *canali* controversy, these speculations are based on small scale features perceived in the spacecraft images, such as 'pyramids' and the 'Face on Mars'. Planetary astronomer Carl Sagan wrote:

> *Mars has become a kind of mythic arena onto which we have projected our Earthly hopes and fears.*

The depiction of Mars in fiction has been stimulated by its dramatic red color and by nineteenth century scientific speculations that its surface conditions might support not just life but intelligent life. Thus originated a large number of science fiction scenarios, among which is H. G. Wells' *The War of the Worlds*, published in 1898, in which Martians seek to escape their dying planet by invading Earth.

Influential works included Ray Bradbury's *The Martian Chronicles*, in which human explorers accidentally destroy a Martian civilization, Edgar Rice Burroughs' *Barsoom* series, C. S. Lewis' novel *Out of the Silent Planet* (1938), and a number of Robert A. Heinlein stories before the mid-sixties.

Jonathan Swift made reference to the moons of Mars, about 150 years before their actual discovery by Asaph Hall, detailing reasonably accurate descriptions of their orbits, in the 19th chapter of his novel *Gulliver's Travels*.

A comic figure of an intelligent Martian, Marvin the Martian, appeared on television in 1948 as a character in the Looney Tunes animated cartoons of Warner Brothers, and has continued as part of popular culture to the present.

After the Mariner and Viking spacecraft had returned pictures of Mars as it really is, an apparently lifeless and canal-less world, these ideas about Mars had to be abandoned, and a vogue for accurate, realist depictions of human colonies on Mars developed, the best known of which may be Kim Stanley Robinson's *Mars* trilogy. Pseudo-scientific speculations about the Face on Mars and other enigmatic landmarks spotted by space probes have meant that ancient civilizations continue to be a popular theme in science fiction, especially in film.

Notes

References

External links

- Mars[166] at DMOZ
- Mars Exploration Program[167] at NASA.gov
- Google Mars[168] and Google Mars 3D[169], interactive maps of the planet
- Geody Mars[170], mapping site that supports NASA World Wind, Celestia, and other applications

Images

- Mars images[171] by NASA's Planetary Photojournal
- Mars images[172] by NASA's Mars Exploration Program
- Mars images[173] by Malin Space Science Systems
- HiRISE image catalog[174] by the University of Arizona

Videos

- Rotating color globe of Mars[175] by the National Oceanic and Atmospheric Administration
- Rotating geological globe of Mars[176] by the United States Geological Survey

- NASA's *Curiosity* Finds Ancient Streambed – First Evidence of Water on Mars[177] on YouTube by The Science Channel (2012, 4:31)
- Flight Into Mariner Valley[178] by Arizona State University
- High resolution video[179] simulation of rotating Mars by Seán Doran, showing Arabia Terra, Valles Marineris and Tharsis (see album[180] for more)

Cartographic resources

- Mars nomenclature[181] and quadrangle maps with feature names[182] by the United States Geological Survey
- Geological map of Mars[183] by the United States Geological Survey
- Viking orbiter photomap[184] by Eötvös Loránd University
- Mars Global Surveyor topographical map[185] by Eötvös Loránd University

<indicator name="featured-star"> </indicator>

Asteroid belt

The **asteroid belt** is the circumstellar disc in the Solar System located roughly between the orbits of the planets Mars and Jupiter. It is occupied by numerous irregularly shaped bodies called asteroids or minor planets. The asteroid belt is also termed the **main asteroid belt** or **main belt** to distinguish it from other asteroid populations in the Solar System such as near-Earth asteroids and trojan asteroids. About half the mass of the belt is contained in the four largest asteroids: Ceres, Vesta, Pallas, and Hygiea. The total mass of the asteroid belt is approximately 4% that of the Moon, or 22% that of Pluto, and roughly twice that of Pluto's moon Charon (whose diameter is 1200 km).

Ceres, the asteroid belt's only dwarf planet, is about 950 km in diameter, whereas 4 Vesta, 2 Pallas, and 10 Hygiea have mean diameters of less than 600 km. The remaining bodies range down to the size of a dust particle. The asteroid material is so thinly distributed that numerous unmanned spacecraft have traversed it without incident. Nonetheless, collisions between large asteroids do occur, and these can produce an asteroid family whose members have similar orbital characteristics and compositions. Individual asteroids within the asteroid belt are categorized by their spectra, with most falling into three basic groups: carbonaceous (C-type), silicate (S-type), and metal-rich (M-type).

The asteroid belt formed from the primordial solar nebula as a group of planetesimals. Planetesimals are the smaller precursors of the protoplanets. Between Mars and Jupiter, however, gravitational perturbations from Jupiter imbued the protoplanets with too much orbital energy for them to accrete into a

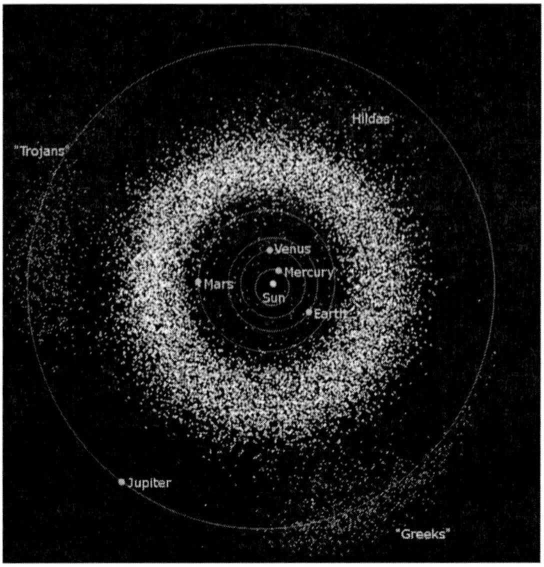

Figure 131:
The asteroids of the inner Solar System and Jupiter: The donut-shaped asteroid belt is located between the orbits of Jupiter and Mars.

planet. Collisions became too violent, and instead of fusing together, the planetesimals and most of the protoplanets shattered. As a result, 99.9% of the asteroid belt's original mass was lost in the first 100 million years of the Solar System's history. Some fragments eventually found their way into the inner Solar System, leading to meteorite impacts with the inner planets. Asteroid orbits continue to be appreciably perturbed whenever their period of revolution about the Sun forms an orbital resonance with Jupiter. At these orbital distances, a Kirkwood gap occurs as they are swept into other orbits.

Classes of small Solar System bodies in other regions are the near-Earth objects, the centaurs, the Kuiper belt objects, the scattered disc objects, the sednoids, and the Oort cloud objects.

On 22 January 2014, ESA scientists reported the detection, for the first definitive time, of water vapor on Ceres, the largest object in the asteroid belt. The detection was made by using the far-infrared abilities of the Herschel Space Observatory. The finding was unexpected because comets, not asteroids, are typically considered to "sprout jets and plumes". According to one of the scientists, "The lines are becoming more and more blurred between comets and asteroids."

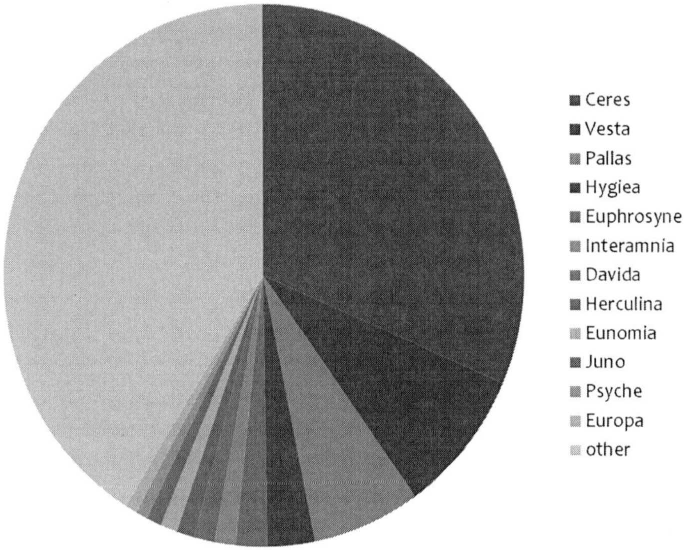

Figure 132: *The relative masses of the top twelve asteroids known compared to the remaining mass of all the other asteroids in the belt.*

History of observation

In 1596, Johannes Kepler predicted "Between Mars and Jupiter, I place a planet" in his *Mysterium Cosmographicum.* While analyzing Tycho Brahe's data, Kepler thought that there was too large a gap between the orbits of Mars and Jupiter.

In an anonymous footnote to his 1766 translation of Charles Bonnet's *Contemplation de la Nature*, the astronomer Johann Daniel Titius of Wittenberg noted an apparent pattern in the layout of the planets. If one began a numerical sequence at 0, then included 3, 6, 12, 24, 48, etc., doubling each time, and added four to each number and divided by 10, this produced a remarkably close approximation to the radii of the orbits of the known planets as measured in astronomical units *provided* one allowed for a "missing planet" (equivalent to 24 in the sequence) between the orbits of Mars (12) and Jupiter (48). In his footnote, Titius declared "But should the Lord Architect have left that space empty? Not at all."

When William Herschel discovered Uranus in 1781, the planet's orbit matched the law almost perfectly, leading astronomers to conclude that there had to be a planet between the orbits of Mars and Jupiter.

Figure 133: *By far the largest object within the belt is Ceres. The total mass of the asteroid belt is significantly less than Pluto's, and approximately twice that of Pluto's moon Charon.*

On January 1, 1801, Giuseppe Piazzi, chair of astronomy at the University of Palermo, Sicily, found a tiny moving object in an orbit with exactly the radius predicted by this pattern. He dubbed it "Ceres", after the Roman goddess of the harvest and patron of Sicily. Piazzi initially believed it to be a comet, but its lack of a coma suggested it was a planet.

Thus, the aforementioned pattern, now known as the Titius–Bode law, predicted the semi-major axes of all eight planets of the time (Mercury, Venus, Earth, Mars, Ceres, Jupiter, Saturn and Uranus).

Fifteen months later, Heinrich Olbers discovered a second object in the same region, Pallas. Unlike the other known planets, Ceres and Pallas remained points of light even under the highest telescope magnifications instead of resolving into discs. Apart from their rapid movement, they appeared indistinguishable from stars.

Accordingly, in 1802, William Herschel suggested they be placed into a separate category, named "asteroids", after the Greek *asteroeides*, meaning "star-like". Upon completing a series of observations of Ceres and Pallas, he concluded,

Figure 134: *Johannes Kepler, who first noticed in 1596 that there was something strange about the orbits of Mars and Jupiter.*

> *Neither the appellation of planets, nor that of comets, can with any propriety of language be given to these two stars ... They resemble small stars so much as hardly to be distinguished from them. From this, their asteroidal appearance, if I take my name, and call them Asteroids; reserving for myself however the liberty of changing that name, if another, more expressive of their nature, should occur.*

By 1807, further investigation revealed two new objects in the region: Juno and Vesta. The burning of Lilienthal in the Napoleonic wars, where the main body of work had been done,[186] brought this first period of discovery to a close.

Despite Herschel's coinage, for several decades it remained common practice to refer to these objects as planets and to prefix their names with numbers representing their date of discovery: 1 Ceres, 2 Pallas, 3 Juno, 4 Vesta. However, in 1845 astronomers detected a fifth object (5 Astraea) and, shortly thereafter, new objects were found at an accelerating rate. Counting them among the planets became increasingly cumbersome. Eventually, they were dropped from the planet list (as first suggested by Alexander von Humboldt in the early 1850s) and Herschel's choice of nomenclature, "asteroids", gradually came into common use.

Figure 135: *Giuseppe Piazzi, discoverer of Ceres, the largest object in the asteroid belt. For several decades after its discovery Ceres was known as a planet, after which it was reclassified as asteroid. In 2006, it was designated as a dwarf planet.*

The discovery of Neptune in 1846 led to the discrediting of the Titius–Bode law in the eyes of scientists, because its orbit was nowhere near the predicted position. To date, there is no scientific explanation for the law, and astronomers' consensus regards it as a coincidence.

The expression "asteroid belt" came into use in the very early 1850s, although it is hard to pinpoint who coined the term. The first English use seems to be in the 1850 translation (by E. C. Otté) of Alexander von Humboldt's *Cosmos*: "[...] and the regular appearance, about the 13th of November and the 11th of August, of shooting stars, which probably form part of a belt of asteroids intersecting the Earth's orbit and moving with planetary velocity". Another early appearance occurred in Robert James Mann's *A Guide to the Knowledge of the Heavens*:[187] "The orbits of the asteroids are placed in a wide belt of space, extending between the extremes of [...]". The American astronomer Benjamin Peirce seems to have adopted that terminology and to have been one of its promoters.[188]

One hundred asteroids had been located by mid-1868, and in 1891 the introduction of astrophotography by Max Wolf accelerated the rate of discovery still further. A total of 1,000 asteroids had been found by 1921, 10,000 by

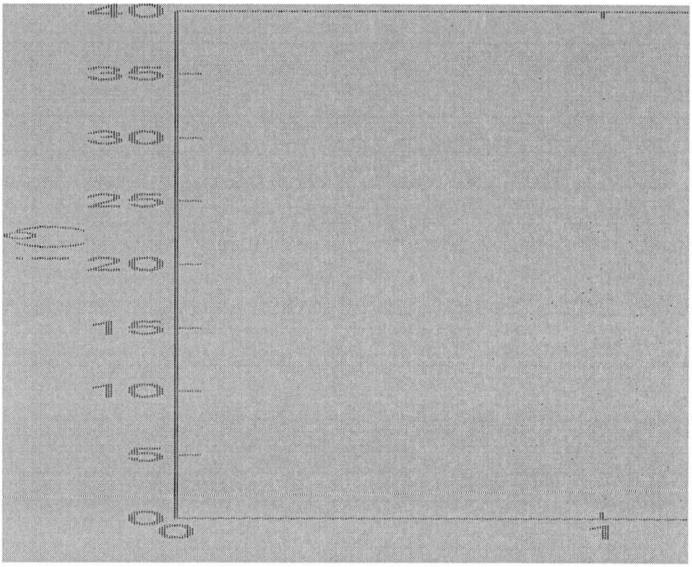

Figure 136: *The asteroid belt showing the orbital inclinations versus distances from the Sun, with asteroids in the core region of the asteroid belt in red and other asteroids in blue*

1981, and 100,000 by 2000. Modern asteroid survey systems now use automated means to locate new minor planets in ever-increasing quantities.

Origin

Formation

In 1802, shortly after discovering Pallas, Olbers suggested to Herschel that Ceres and Pallas were fragments of a much larger planet that once occupied the Mars–Jupiter region, this planet having suffered an internal explosion or a cometary impact many million years before. Over time, however, this hypothesis has fallen from favor. The large amount of energy required to destroy a planet, combined with the belt's low combined mass, which is only about 4% of the mass of the Moon, do not support the hypothesis. Further, the significant chemical differences between the asteroids become difficult to explain if they come from the same planet. Today, most scientists accept that, rather than fragmenting from a progenitor planet, the asteroids never formed a planet at all.

In general, in the Solar System, planetary formation is thought to have occurred via a process comparable to the long-standing nebular hypothesis: a cloud of interstellar dust and gas collapsed under the influence of gravity to form a rotating disc of material that then further condensed to form the Sun and planets. During the first few million years of the Solar System's history, an accretion process of sticky collisions caused the clumping of small particles, which gradually increased in size. Once the clumps reached sufficient mass, they could draw in other bodies through gravitational attraction and become planetesimals. This gravitational accretion led to the formation of the planets.

Planetesimals within the region which would become the asteroid belt were too strongly perturbed by Jupiter's gravity to form a planet. Instead they continued to orbit the Sun as before, occasionally colliding. In regions where the average velocity of the collisions was too high, the shattering of planetesimals tended to dominate over accretion, preventing the formation of planet-sized bodies. Orbital resonances occurred where the orbital period of an object in the belt formed an integer fraction of the orbital period of Jupiter, perturbing the object into a different orbit; the region lying between the orbits of Mars and Jupiter contains many such orbital resonances. As Jupiter migrated inward following its formation, these resonances would have swept across the asteroid belt, dynamically exciting the region's population and increasing their velocities relative to each other.

During the early history of the Solar System, the asteroids melted to some degree, allowing elements within them to be partially or completely differentiated by mass. Some of the progenitor bodies may even have undergone periods of explosive volcanism and formed magma oceans. However, because of the relatively small size of the bodies, the period of melting was necessarily brief (compared to the much larger planets), and had generally ended about 4.5 billion years ago, in the first tens of millions of years of formation. In August 2007, a study of zircon crystals in an Antarctic meteorite believed to have originated from 4 Vesta suggested that it, and by extension the rest of the asteroid belt, had formed rather quickly, within ten million years of the Solar System's origin.

Evolution

The asteroids are not samples of the primordial Solar System. They have undergone considerable evolution since their formation, including internal heating (in the first few tens of millions of years), surface melting from impacts, space weathering from radiation, and bombardment by micrometeorites. Although some scientists refer to the asteroids as residual planetesimals, other scientists consider them distinct.

The current asteroid belt is believed to contain only a small fraction of the mass of the primordial belt. Computer simulations suggest that the original asteroid belt may have contained mass equivalent to the Earth. Primarily because of gravitational perturbations, most of the material was ejected from the belt within about a million years of formation, leaving behind less than 0.1% of the original mass. Since their formation, the size distribution of the asteroid belt has remained relatively stable: there has been no significant increase or decrease in the typical dimensions of the main-belt asteroids.

The 4:1 orbital resonance with Jupiter, at a radius 2.06 AU, can be considered the inner boundary of the asteroid belt. Perturbations by Jupiter send bodies straying there into unstable orbits. Most bodies formed inside the radius of this gap were swept up by Mars (which has an aphelion at 1.67 AU) or ejected by its gravitational perturbations in the early history of the Solar System. The Hungaria asteroids lie closer to the Sun than the 4:1 resonance, but are protected from disruption by their high inclination.

When the asteroid belt was first formed, the temperatures at a distance of 2.7 AU from the Sun formed a "snow line" below the freezing point of water. Planetesimals formed beyond this radius were able to accumulate ice. In 2006 it was announced that a population of comets had been discovered within the asteroid belt beyond the snow line, which may have provided a source of water for Earth's oceans. According to some models, there was insufficient outgassing of water during the Earth's formative period to form the oceans, requiring an external source such as a cometary bombardment.

Characteristics

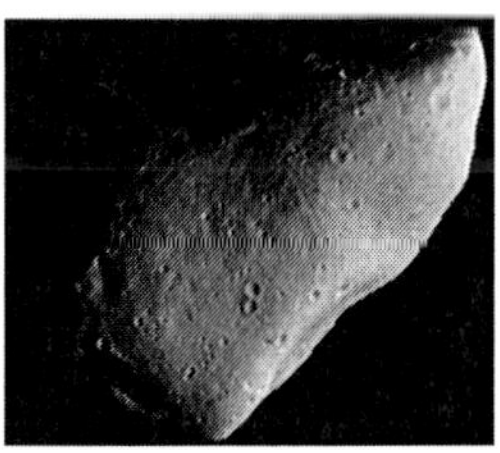

951 Gaspra, the first asteroid imaged by a spacecraft, as viewed during *Galileo*'s 1991 flyby; colors are exaggerated

Fragment of the Allende meteorite, a carbonaceous chondrite that fell to Earth in Mexico in 1969

Contrary to popular imagery, the asteroid belt is mostly empty. The asteroids are spread over such a large volume that it would be improbable to reach an asteroid without aiming carefully. Nonetheless, hundreds of thousands of asteroids are currently known, and the total number ranges in the millions or more, depending on the lower size cutoff. Over 200 asteroids are known to be larger than 100 km,[189] and a survey in the infrared wavelengths has shown that the asteroid belt has 0.7–1.7 million asteroids with a diameter of 1 km or more. The apparent magnitudes of most of the known asteroids are 11–19, with the median at about 16.

The total mass of the asteroid belt is estimated to be 2.8×10^{21} to 3.2×10^{21} kilograms, which is just 4% of the mass of the Moon. The four largest objects, Ceres, 4 Vesta, 2 Pallas, and 10 Hygiea, account for half of the belt's total mass, with almost one-third accounted for by Ceres alone.[190]

Composition

The current belt consists primarily of three categories of asteroids: C-type or carbonaceous asteroids, S-type or silicate asteroids, and M-type or metallic asteroids.

Carbonaceous asteroids, as their name suggests, are carbon-rich. They dominate the asteroid belt's outer regions. Together they comprise over 75% of the visible asteroids. They are redder in hue than the other asteroids and have a very low albedo. Their surface composition is similar to carbonaceous chondrite meteorites. Chemically, their spectra match the primordial composition of the early Solar System, with only the lighter elements and volatiles removed.

S-type (silicate-rich) asteroids are more common toward the inner region of the belt, within 2.5 AU of the Sun. The spectra of their surfaces reveal the presence of silicates and some metal, but no significant carbonaceous compounds. This indicates that their materials have been significantly modified from their primordial composition, probably through melting and reformation. They have a relatively high albedo, and form about 17% of the total asteroid population.

M-type (metal-rich) asteroids form about 10% of the total population; their spectra resemble that of iron-nickel. Some are believed to have formed from the metallic cores of differentiated progenitor bodies that were disrupted through collision. However, there are also some silicate compounds that can produce a similar appearance. For example, the large M-type asteroid 22 Kalliope does not appear to be primarily composed of metal. Within the asteroid belt, the number distribution of M-type asteroids peaks at a semi-major

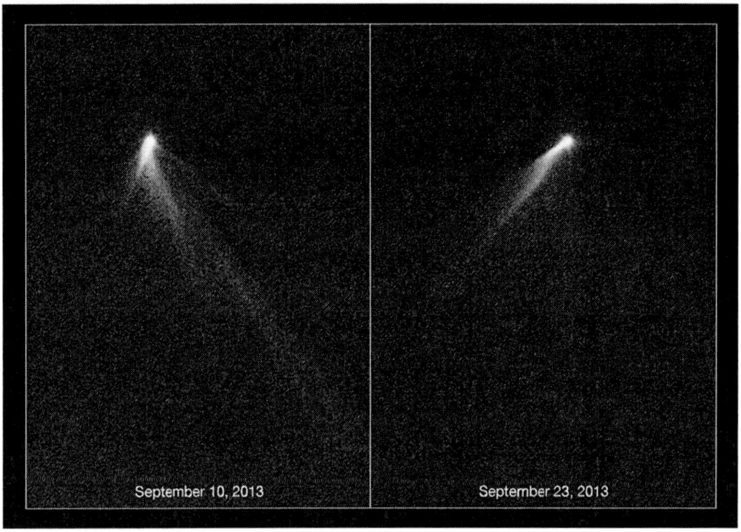

Figure 137: *Hubble views extraordinary multi-tailed asteroid P/2013 P5.*

axis of about 2.7 AU. It is not yet clear whether all M-types are compositionally similar, or whether it is a label for several varieties which do not fit neatly into the main C and S classes.

One mystery of the asteroid belt is the relative rarity of V-type, or basaltic asteroids. Theories of asteroid formation predict that objects the size of Vesta or larger should form crusts and mantles, which would be composed mainly of basaltic rock, resulting in more than half of all asteroids being composed either of basalt or olivine. Observations, however, suggest that 99 percent of the predicted basaltic material is missing. Until 2001, most basaltic bodies discovered in the asteroid belt were believed to originate from the asteroid Vesta (hence their name V-type). However, the discovery of the asteroid 1459 Magnya revealed a slightly different chemical composition from the other basaltic asteroids discovered until then, suggesting a different origin. This hypothesis was reinforced by the further discovery in 2007 of two asteroids in the outer belt, 7472 Kumakiri and (10537) 1991 RY$_{16}$, with differing basaltic composition that could not have originated from Vesta. These latter two are the only V-type asteroids discovered in the outer belt to date.

The temperature of the asteroid belt varies with the distance from the Sun. For dust particles within the belt, typical temperatures range from 200 K (–73 °C) at 2.2 AU down to 165 K (–108 °C) at 3.2 AU However, due to rotation, the surface temperature of an asteroid can vary considerably as the sides are alternately exposed to solar radiation and then to the stellar background.

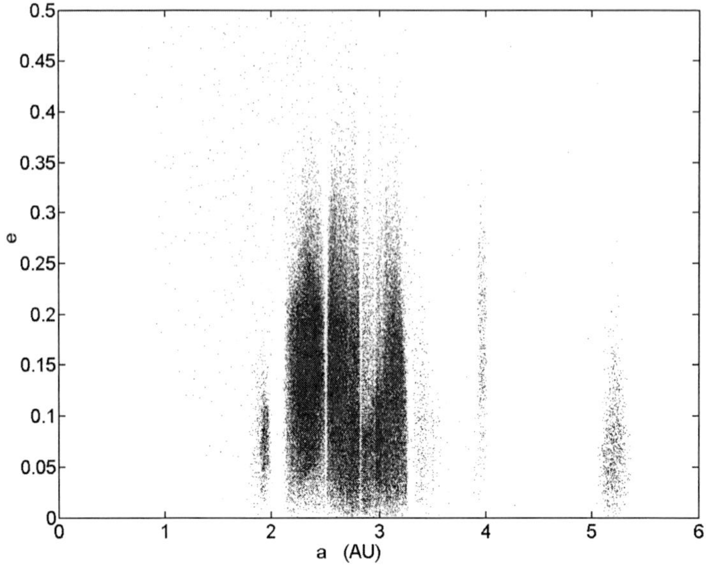

Figure 138: *The asteroid belt (showing eccentricities), with the asteroid belt in red and blue ("core" region in red)*

Main-belt comets

Several otherwise unremarkable bodies in the outer belt show cometary activity. Because their orbits cannot be explained through capture of classical comets, it is thought that many of the outer asteroids may be icy, with the ice occasionally exposed to sublimation through small impacts. Main-belt comets may have been a major source of the Earth's oceans, because the deuterium–hydrogen ratio is too low for classical comets to have been the principal source.

Orbits

Most asteroids within the asteroid belt have orbital eccentricities of less than 0.4, and an inclination of less than 30°. The orbital distribution of the asteroids reaches a maximum at an eccentricity of around 0.07 and an inclination below 4°. Thus although a typical asteroid has a relatively circular orbit and lies near the plane of the ecliptic, some asteroid orbits can be highly eccentric or travel well outside the ecliptic plane.

Sometimes, the term *main belt* is used to refer only to the more compact "core" region where the greatest concentration of bodies is found. This lies between

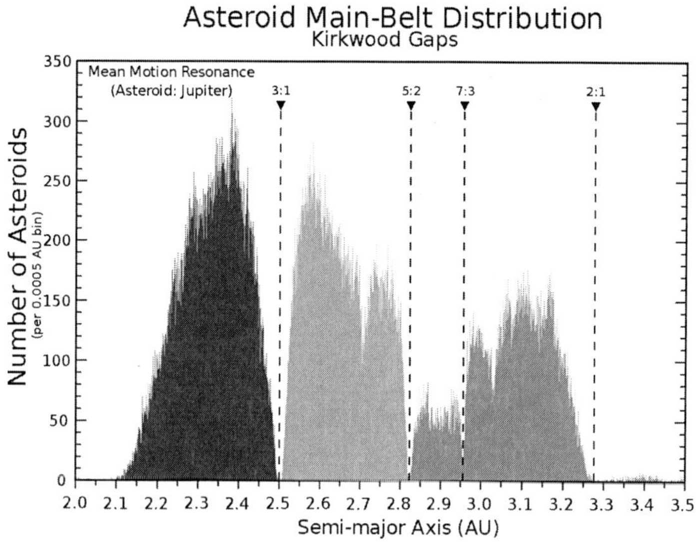

Figure 139: *Number of asteroids in the asteroid belt as a function of their semi-major axis. The dashed lines indicate the Kirkwood gaps, where orbital resonances with Jupiter destabilize orbits. The color gives a possible division into three zones:*
Zone I: inner main-belt ($a < 2.5$ AU)
Zone II: middle main-belt (2.5 AU $< a <$ 2.82 AU)
Zone III: outer main-belt ($a >$ 2.82 AU)

the strong 4:1 and 2:1 Kirkwood gaps at 2.06 and 3.27 AU, and at orbital eccentricities less than roughly 0.33, along with orbital inclinations below about 20°. As of 2006[191], this "core" region contained 93% of all discovered and numbered minor planets within the Solar System.[192]

Kirkwood gaps

The semi-major axis of an asteroid is used to describe the dimensions of its orbit around the Sun, and its value determines the minor planet's orbital period. In 1866, Daniel Kirkwood announced the discovery of gaps in the distances of these bodies' orbits from the Sun. They were located at positions where their period of revolution about the Sun was an integer fraction of Jupiter's orbital period. Kirkwood proposed that the gravitational perturbations of the planet led to the removal of asteroids from these orbits.

When the mean orbital period of an asteroid is an integer fraction of the orbital period of Jupiter, a mean-motion resonance with the gas giant is created that is sufficient to perturb an asteroid to new orbital elements. Asteroids that become located in the gap orbits (either primordially because of the migration of Jupiter's orbit, or due to prior perturbations or collisions) are gradually nudged into different, random orbits with a larger or smaller semi-major axis.

The gaps are not seen in a simple snapshot of the locations of the asteroids at any one time because asteroid orbits are elliptical, and many asteroids still cross through the radii corresponding to the gaps. The actual spatial density of asteroids in these gaps does not differ significantly from the neighboring regions.

The main gaps occur at the 3:1, 5:2, 7:3, and 2:1 mean-motion resonances with Jupiter. An asteroid in the 3:1 Kirkwood gap would orbit the Sun three times for each Jovian orbit, for instance. Weaker resonances occur at other semi-major axis values, with fewer asteroids found than nearby. (For example, an 8:3 resonance for asteroids with a semi-major axis of 2.71 AU.)

The main or core population of the asteroid belt is sometimes divided into three zones, based on the most prominent Kirkwood gaps:

- Zone I lies between the 4:1 resonance (2.06 AU) and 3:1 resonance (2.5 AU) Kirkwood gaps.
- Zone II continues from the end of Zone I out to the 5:2 resonance gap (2.82 AU).
- Zone III extends from the outer edge of Zone II to the 2:1 resonance gap (3.28 AU).

The asteroid belt may also be divided into the inner and outer belts, with the inner belt formed by asteroids orbiting nearer to Mars than the 3:1 Kirkwood gap (2.5 AU), and the outer belt formed by those asteroids closer to Jupiter's orbit. (Some authors subdivide the inner and outer belts at the 2:1 resonance gap (3.3 AU), whereas others suggest inner, middle, and outer belts; also see diagram).

Collisions

The high population of the asteroid belt makes for a very active environment, where collisions between asteroids occur frequently (on astronomical time scales). Collisions between main-belt bodies with a mean radius of 10 km are expected to occur about once every 10 million years. A collision may fragment an asteroid into numerous smaller pieces (leading to the formation of a new asteroid family). Conversely, collisions that occur at low relative speeds may also join two asteroids. After more than 4 billion years of such processes, the members of the asteroid belt now bear little resemblance to the original population.

Figure 140: *The zodiacal light, a minor part of which is created by dust from collisions in the asteroid belt*

Along with the asteroid bodies, the asteroid belt also contains bands of dust with particle radii of up to a few hundred micrometres. This fine material is produced, at least in part, from collisions between asteroids, and by the impact of micrometeorites upon the asteroids. Due to the Poynting–Robertson effect, the pressure of solar radiation causes this dust to slowly spiral inward toward the Sun.

The combination of this fine asteroid dust, as well as ejected cometary material, produces the zodiacal light. This faint auroral glow can be viewed at night extending from the direction of the Sun along the plane of the ecliptic. Asteroid particles that produce the visible zodiacal light average about 40 μm in radius. The typical lifetimes of main-belt zodiacal cloud particles are about 700,000 years. Thus, to maintain the bands of dust, new particles must be steadily produced within the asteroid belt. It was once thought that collisions of asteroids form a major component of the zodiacal light. However, computer simulations by Nesvorný and colleagues attributed 85 percent of the zodiacal-light dust to fragmentations of Jupiter-family comets, rather than to comets and collisions between asteroids in the asteroid belt. At most 10 percent of the dust is attributed to the asteroid belt.[193]

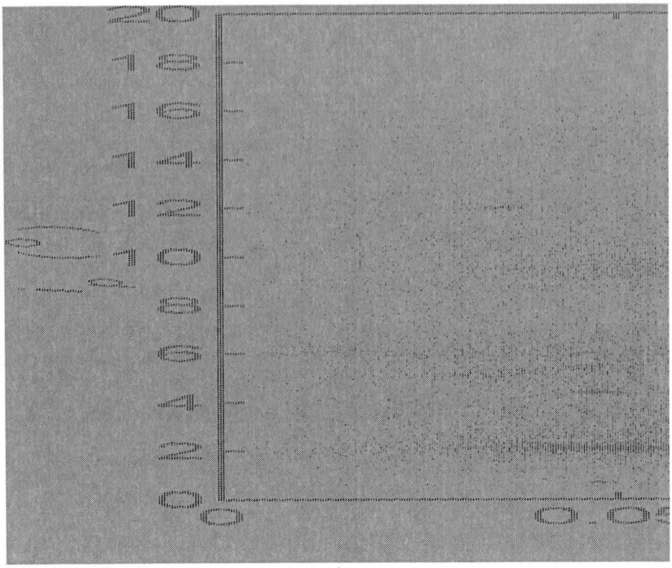

Figure 141: *This plot of orbital inclination* (i_p) *versus eccentricity* (e_p) *for the numbered main-belt asteroids clearly shows clumpings representing asteroid families.*

Meteorites

Some of the debris from collisions can form meteoroids that enter the Earth's atmosphere. Of the 50,000 meteorites found on Earth to date, 99.8 percent are believed to have originated in the asteroid belt.

Families and groups

In 1918, the Japanese astronomer Kiyotsugu Hirayama noticed that the orbits of some of the asteroids had similar parameters, forming families or groups.

Approximately one-third of the asteroids in the asteroid belt are members of an asteroid family. These share similar orbital elements, such as semi-major axis, eccentricity, and orbital inclination as well as similar spectral features, all of which indicate a common origin in the breakup of a larger body. Graphical displays of these elements, for members of the asteroid belt, show concentrations indicating the presence of an asteroid family. There are about 20–30 associations that are almost certainly asteroid families. Additional groupings have been found that are less certain. Asteroid families can be confirmed when the members display common spectral features. Smaller associations of asteroids are called groups or clusters.

Some of the most prominent families in the asteroid belt (in order of increasing semi-major axes) are the Flora, Eunoma, Koronis, Eos, and Themis families. The Flora family, one of the largest with more than 800 known members, may have formed from a collision less than a billion years ago. The largest asteroid to be a true member of a family (as opposed to an interloper in the case of Ceres with the Gefion family) is 4 Vesta. The Vesta family is believed to have formed as the result of a crater-forming impact on Vesta. Likewise, the HED meteorites may also have originated from Vesta as a result of this collision.

Three prominent bands of dust have been found within the asteroid belt. These have similar orbital inclinations as the Eos, Koronis, and Themis asteroid families, and so are possibly associated with those groupings.

Periphery

Skirting the inner edge of the belt (ranging between 1.78 and 2.0 AU, with a mean semi-major axis of 1.9 AU) is the Hungaria family of minor planets. They are named after the main member, 434 Hungaria; the group contains at least 52 named asteroids. The Hungaria group is separated from the main body by the 4:1 Kirkwood gap and their orbits have a high inclination. Some members belong to the Mars-crossing category of asteroids, and gravitational perturbations by Mars are likely a factor in reducing the total population of this group.

Another high-inclination group in the inner part of the asteroid belt is the Phocaea family. These are composed primarily of S-type asteroids, whereas the neighboring Hungaria family includes some E-types. The Phocaea family orbit between 2.25 and 2.5 AU from the Sun.

Skirting the outer edge of the asteroid belt is the Cybele group, orbiting between 3.3 and 3.5 AU. These have a 7:4 orbital resonance with Jupiter. The Hilda family orbit between 3.5 and 4.2 AU, and have relatively circular orbits and a stable 3:2 orbital resonance with Jupiter. There are few asteroids beyond 4.2 AU, until Jupiter's orbit. Here the two families of Trojan asteroids can be found, which, at least for objects larger than 1 km, are approximately as numerous as the asteroids of the asteroid belt.

New families

Some asteroid families have formed recently, in astronomical terms. The Karin Cluster apparently formed about 5.7 million years ago from a collision with a progenitor asteroid 33 km in radius. The Veritas family formed about 8.3 million years ago; evidence includes interplanetary dust recovered from ocean sediment.

Figure 142: *Artist's concept of the Dawn spacecraft with Vesta and Ceres*

More recently, the Datura cluster appears to have formed about 530 thousand years ago from a collision with a main-belt asteroid. The age estimate is based on the probability of the members having their current orbits, rather than from any physical evidence. However, this cluster may have been a source for some zodiacal dust material. Other recent cluster formations, such as the Iannini cluster (*circa* 1–5 million years ago), may have provided additional sources of this asteroid dust.

Exploration

The first spacecraft to traverse the asteroid belt was *Pioneer 10*, which entered the region on 16 July 1972. At the time there was some concern that the debris in the belt would pose a hazard to the spacecraft, but it has since been safely traversed by 12 spacecraft without incident. *Pioneer 11*, *Voyagers 1* and *2* and *Ulysses* passed through the belt without imaging any asteroids. *Galileo* imaged 951 Gaspra in 1991 and 243 Ida in 1993, NEAR imaged 253 Mathilde in 1997, *Cassini* imaged 2685 Masursky in 2000, *Stardust* imaged 5535 Annefrank in 2002, *New Horizons* imaged 132524 APL in 2006, *Rosetta* imaged 2867 Šteins in September 2008 and 21 Lutetia in July 2010, and *Dawn* orbited Vesta between July 2011 and September 2012 and has orbited Ceres since March 2015. On its way to Jupiter, *Juno* traversed the asteroid belt without collecting science data. Due to the low density of materials within the belt, the odds of a probe running into an asteroid are now estimated at less than one in a billion.

Most belt asteroids imaged to date have come from brief flyby opportunities by probes headed for other targets. Only the *Dawn*, NEAR and *Hayabusa* missions have studied asteroids for a protracted period in orbit and at the surface. *Dawn* explored Vesta from July 2011 to September 2012, and has been orbiting Ceres since March 2015.

Further reading

- Elkins-Tanton, Linda T. (2006). *Asteroids, Meteorites, and Comets* (First ed.). New York: Chelsea House. ISBN 0-8160-5195-X.

External links

<indicator name="spoken-icon"> </indicator>

Wikimedia Commons has media related to ***Main Belt asteroids***.

- Arnett, William A. (February 26, 2006). "Asteroids"[194]. The Nine Planets. Archived[195] from the original on 18 April 2007. Retrieved 2007-04-20.
- Asteroids Page[196] at NASA's Solar System Exploration[197]
- Cain, Fraser. "The Asteroid Belt"[198]. Universe Today. Archived[199] from the original on 7 March 2008. Retrieved 2008-04-01.
- "Main Asteroid Belt"[200]. Sol Company. Archived[201] from the original on 15 May 2007. Retrieved 2007-04-20.
- Munsell, Kirk (September 16, 2005). "Asteroids: Overview"[196]. NASA's Solar System Exploration. Archived[202] from the original on 24 May 2007. Retrieved 2007-05-26.
- Plots of eccentricity vs. semi-major axis[203] and inclination vs. semi-major axis[204] at Asteroid Dynamic Site
- Staff (October 31, 2006). "Asteroids"[205]. NASA. Archived[206] from the original on 11 April 2007. Retrieved 2007-04-20.
- Staff (2007). "Space Topics: Asteroids and Comets"[207]. The Planetary Society. Archived[208] from the original on 28 April 2007. Retrieved 2007-04-20.

<indicator name="featured-star"> </indicator>

Ceres (dwarf planet)

Ceres

A view of Ceres in natural color, pictured by the *Dawn* spacecraft in May 2015.[209]

Discovery	
Discovered by	Giuseppe Piazzi
Discovery date	1 January 1801
Designations	
MPC designation	**1 Ceres**
Pronunciation	/ˈsɪəriːz/
Named after	Cerēs
Alternative names	A899 OF; 1943 XB
Minor planet category	Dwarf planet Asteroid belt
Adjectives	Cererian /sɪˈrɪəriən/, rarely Cererean /sɛrɪˈriːən/
Orbital characteristics	
Epoch 2014-Dec-09 (JD 2,457,000.5)	
Aphelion	2.9773 AU (445,410,000 km)
Perihelion	2.5577 AU (382,620,000 km)
Semi-major axis	2.7675 AU (414,010,000 km)
Eccentricity	0.075823
Orbital period	4.60 yr 1,681.63 d
Synodic period	466.6 d 1.278 yr

Average orbital speed	17.905 km/s
Mean anomaly	95.9891°
Inclination	10.593° to ecliptic 9.20° to invariable plane[210]
Longitude of ascending node	80.3293°
Argument of perihelion	72.5220°
Satellites	None
Proper orbital elements	
Proper semi-major axis	2.7670962 AU
Proper eccentricity	0.1161977
Proper inclination	9.6474122°
Proper mean motion	78.193318 deg / yr
Proper orbital period	4.60397 yr (1681.601 d)
Precession of perihelion	54.070272 arcsec / yr
Precession of the ascending node	–59.170034 arcsec / yr
Physical characteristics	
Dimensions	(965.2 × 961.2 × 891.2) ± 2.0 km
Mean radius	473 km
Surface area	2,770,000 km^2[211]
Volume	421,000,000 km^3
Mass	$(9.393 \pm 0.005) \times 10^{20}$ kg 0.00015 Earths 0.0128 Moons
Mean density	2.161±0.009 g/cm^3
Surface gravity	0.28 m/s^2 0.029 *g*
Moment of inertia factor	0.37[212] (estimate)
Escape velocity	0.51 km/s
Sidereal rotation period	0.3781 d 9.074170±0.000002 h
Equatorial rotation velocity	92.61 m/s
Axial tilt	4°
North pole right ascension	291.42744°
North pole declination	66.764°
Albedo	0.090±0.0033 (V-band geometric)

Surface temp.	min	mean	max
Kelvin	?	≈ 168 K	235 K

Spectral type	C
Apparent magnitude	6.64 to 9.34[213]
Absolute magnitude *(H)*	3.36±0.02
Angular diameter	0.854” to 0.339”

Ceres (/ˈsɪəriːz/; minor-planet designation: **1 Ceres**) is the largest object in the asteroid belt that lies between the orbits of Mars and Jupiter. Its diameter is approximately 945 kilometers (587 miles), making it the largest of the minor planets within the orbit of Neptune. The 33rd-largest known body in the Solar System, it is the only dwarf planet within the orbit of Neptune.[214] Composed of rock and ice, Ceres is estimated to compose approximately one third of the mass of the entire asteroid belt. Ceres is the only object in the asteroid belt known to be rounded by its own gravity (though detailed analysis was required to exclude 4 Vesta). From Earth, the apparent magnitude of Ceres ranges from 6.7 to 9.3, and hence even at its brightest it is too dim to be seen with the naked eye except under extremely dark skies.

Ceres was the first asteroid to be discovered (by Giuseppe Piazzi at Palermo on 1 January 1801). It was originally considered a planet, but was reclassified as an asteroid in the 1850s after many other objects in similar orbits were discovered.

Ceres appears to be differentiated into a rocky core and an icy mantle, and may have a remnant internal ocean of liquid water under the layer of ice. The surface is probably a mixture of water ice and various hydrated minerals such as carbonates and clay. In January 2014, emissions of water vapor were detected from several regions of Ceres.[215] This was unexpected because large bodies in the asteroid belt typically do not emit vapor, a hallmark of comets.

The robotic NASA spacecraft *Dawn* entered orbit around Ceres on 6 March 2015. Pictures with a resolution previously unattained were taken during imaging sessions starting in January 2015 as *Dawn* approached Ceres, showing a cratered surface. Two distinct bright spots (or high-albedo features) inside a crater (different from the bright spots observed in earlier *Hubble* images) were seen in a 19 February 2015 image, leading to speculation about a possible cryovolcanic origin or outgassing. On 3 March 2015, a NASA spokesperson said the spots are consistent with highly reflective materials containing ice or salts, but that cryovolcanism is unlikely. However, on 2 September 2016, NASA scientists released a paper in *Science* that claimed that a massive ice volcano called Ahuna Mons is the strongest evidence yet for the existence of these mysterious ice volcanoes. On 11 May 2015, NASA released a higher-resolution image showing that, instead of one or two spots, there are actually several. On 9 December 2015, NASA scientists reported that the bright spots on Ceres may be related to a type of salt, particularly a form of brine containing magnesium sulfate hexahydrite ($MgSO_4 \cdot 6H_2O$); the spots were also found to be associated with ammonia-rich clays. In June 2016, near-infrared spectra of these bright areas were found to be consistent with a large amount of sodium carbonate (Na_2CO_3), implying that recent geologic activity was probably involved in the creation of the bright spots.

In October 2015, NASA released a true color portrait of Ceres made by *Dawn*. In February 2017, organics were reported to have been detected on Ceres in Ernutet crater (see image).

History

Discovery

Johann Elert Bode, in 1772, first suggested that an undiscovered planet could exist between the orbits of Mars and Jupiter. Kepler had already noticed the gap between Mars and Jupiter in 1596. Bode based his idea on the Titius–Bode law which is a now-discredited hypothesis that was first proposed in 1766. Bode observed that there was a regular pattern in the semi-major axes of the orbits of known planets, and that the pattern was marred only by the large gap between Mars and Jupiter. The pattern predicted that the missing planet ought to have an orbit with a semi-major axis near 2.8 astronomical units (AU). William Herschel's discovery of Uranus in 1781 near the predicted distance for the next body beyond Saturn increased faith in the law of Titius and Bode, and in 1800, a group headed by Franz Xaver von Zach, editor of the *Monatliche Correspondenz*, sent requests to twenty-four experienced astronomers (whom he dubbed the "celestial police"), asking that they combine their efforts and begin a methodical search for the expected planet. Although they did not discover Ceres, they later found several large asteroids.

One of the astronomers selected for the search was Giuseppe Piazzi, a Catholic priest at the Academy of Palermo, Sicily. Before receiving his invitation to join the group, Piazzi discovered Ceres on 1 January 1801. He was searching for "the 87th [star] of the Catalogue of the Zodiacal stars of Mr la Caille", but found that "it was preceded by another". Instead of a star, Piazzi had found a moving star-like object, which he first thought was a comet. Piazzi observed Ceres a total of 24 times, the final time on 11 February 1801, when illness

DELLA SCOPERTA

DEL NUOVO PIANETA

CERERE FERDINANDEA

OTTAVO TRA I PRIMARJ DEL NOSTRO SISTEMA SOLARE.

PALERMO

1802

NELLA STAMPERIA REALE.

Figure 143: *Piazzi's book Della scoperta del nuovo pianeta Cerere Ferdinandea outlining the discovery of Ceres, dedicated the new "planet" to Ferdinand I of the Two Sicilies.*

interrupted his observations. He announced his discovery on 24 January 1801 in letters to only two fellow astronomers, his compatriot Barnaba Oriani of Milan and Bode of Berlin. He reported it as a comet but "since its movement is so slow and rather uniform, it has occurred to me several times that it might be something better than a comet". In April, Piazzi sent his complete observations to Oriani, Bode, and Jérôme Lalande in Paris. The information was published in the September 1801 issue of the *Monatliche Correspondenz*.

By this time, the apparent position of Ceres had changed (mostly due to Earth's orbital motion), and was too close to the Sun's glare for other astronomers to confirm Piazzi's observations. Toward the end of the year, Ceres should have been visible again, but after such a long time it was difficult to predict its exact position. To recover Ceres, Carl Friedrich Gauss, then 24 years old, developed an efficient method of orbit determination. In only a few weeks, he predicted the path of Ceres and sent his results to von Zach. On 31 December 1801, von Zach and Heinrich W. M. Olbers found Ceres near the predicted position and thus recovered it.

The early observers were only able to calculate the size of Ceres to within an order of magnitude. Herschel underestimated its diameter as 260 km in 1802, whereas in 1811 Johann Hieronymus Schröter overestimated it as 2,613 km.[216]

Name

Piazzi originally suggested the name *Cerere Ferdinandea* for his discovery, after the goddess Ceres (Roman goddess of agriculture, *Cerere* in Italian, who was believed to have originated in Sicily and whose oldest temple was there) and King Ferdinand of Sicily. "Ferdinandea", however, was not acceptable to other nations and was dropped. Ceres was called *Hera* for a short time in Germany. In Greece, it is called Demeter (Δήμητρα), after the Greek equivalent of the Roman *Cerēs*;[217] in English, that name is used for the asteroid 1108 Demeter.

The regular adjectival forms of the name are *Cererian* and *Cererean*, derived from the Latin genitive *Cereris*, but *Ceresian* is occasionally seen for the goddess (as in the sickle-shaped Ceresian Lake), as is the shorter form *Cerean*.

The old astronomical symbol of Ceres is a sickle, □⚳□ (⚳),[218] similar to Venus' symbol □♀□ but with a break in the circle. It has a variant □ ⚳ □, reversed under the influence of the initial letter 'C' of 'Ceres'. These were later replaced with the generic asteroid symbol of a numbered disk, □①□.

Cerium, a rare-earth element discovered in 1803, was named after Ceres.[219] </ref> In the same year another element was also initially named after Ceres, but when cerium was named, its discoverer changed the name to palladium, after the second asteroid, 2 Pallas.

Classification

The categorization of Ceres has changed more than once and has been the subject of some disagreement. Johann Elert Bode believed Ceres to be the "missing planet" he had proposed to exist between Mars and Jupiter, at a distance of 419 million km (2.8 AU) from the Sun. Ceres was assigned a planetary symbol, and remained listed as a planet in astronomy books and tables (along with 2 Pallas, 3 Juno, and 4 Vesta) for half a century.

As other objects were discovered in the neighborhood of Ceres, it was realized that Ceres represented the first of a new class of objects. In 1802, with the discovery of 2 Pallas, William Herschel coined the term *asteroid* ("star-like") for these bodies, writing that "they resemble small stars so much as hardly to be distinguished from them, even by very good telescopes". As the first such body to be discovered, Ceres was given the designation 1 Ceres under the modern system of minor-planet designations. By the 1860s, the existence of a fundamental difference between asteroids such as Ceres and the major planets was widely accepted, though a precise definition of "planet" was never formulated.

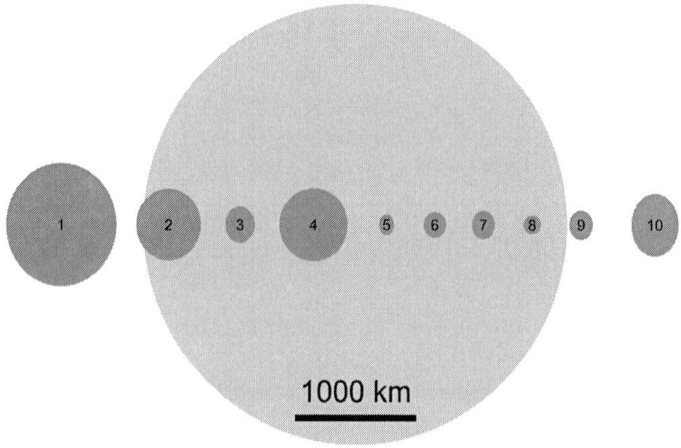

Figure 144: *Sizes of the first ten main-belt objects discovered profiled against the Moon. Ceres is far left (1).*

Ceres (bottom left), the Moon and Earth, shown to scale

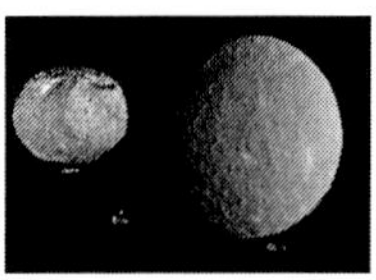

Size comparison of Vesta, Ceres and Eros

The 2006 debate surrounding Pluto and what constitutes a planet led to Ceres being considered for reclassification as a planet. A proposal before the International Astronomical Union for the definition of a planet would have defined a planet as "a celestial body that (a) has sufficient mass for its self-gravity to overcome rigid-body forces so that it assumes a hydrostatic equilibrium (nearly round) shape, and (b) is in orbit around a star, and is neither a star nor a satellite of a planet". Had this resolution been adopted, it would have made Ceres the fifth planet in order from the Sun. This never happened, however, and on 24 August 2006 a modified definition was adopted, carrying the additional requirement that a planet must have "cleared the neighborhood around its orbit". By this definition, Ceres is not a planet because it does not dominate its orbit, sharing it as it does with the thousands of other asteroids in the asteroid belt and constituting only about a third of the mass of the belt. Bodies that met the first proposed definition but not the second, such as Ceres, were instead classified as dwarf planets.

Ceres is the largest object in the asteroid belt. It is sometimes assumed that Ceres has been *re*classified as a dwarf planet, and that it is therefore no longer considered an asteroid. For example, a news update at Space.com spoke of "Pallas, the largest asteroid, and Ceres, the dwarf planet formerly classified as an asteroid",[220] whereas an IAU question-and-answer posting states, "Ceres is (or now we can say it was) the largest asteroid", though it then speaks of "other asteroids" crossing Ceres' path and otherwise implies that Ceres is still considered an asteroid. The Minor Planet Center notes that such bodies may have dual designations. The 2006 IAU decision that classified Ceres as a dwarf planet never addressed whether it is or is not an asteroid. Indeed, the IAU has never defined the word 'asteroid' at all, having preferred the term 'minor planet' until 2006, and preferring the terms 'small Solar System body' and 'dwarf planet' after 2006. Lang (2011) comments "the [IAU has] added a new designation to Ceres, classifying it as a dwarf planet. ... By [its] definition, Eris, Haumea, Makemake and Pluto, as well as the largest asteroid, 1 Ceres, are all dwarf planets", and describes it elsewhere as "the dwarf planet–asteroid 1 Ceres". NASA continues to refer to Ceres as an asteroid,[221] as do various academic textbooks.

Orbit

Proper (long-term mean) orbital elements compared to osculating (instant) orbital elements for Ceres:

Element type	a (in AU)	e	i	Period (in days)
Proper	2.7671	0.116198	9.647435	1,681.60
Osculating (Epoch 23 July 2010)	2.7653	0.079138	10.586821	1,679.66
Difference	0.0018	0.03706	0.939386	1.94

Ceres follows an orbit between Mars and Jupiter, within the asteroid belt, with a period of 4.6 Earth years. The orbit is moderately inclined ($i = 10.6°$ compared to 7° for Mercury and 17° for Pluto) and moderately eccentric ($e = 0.08$ compared to 0.09 for Mars).

The diagram illustrates the orbits of Ceres (blue) and several planets (white and gray). The segments of orbits below the ecliptic are plotted in darker colors, and the orange plus sign is the Sun's location. The top left diagram is a polar

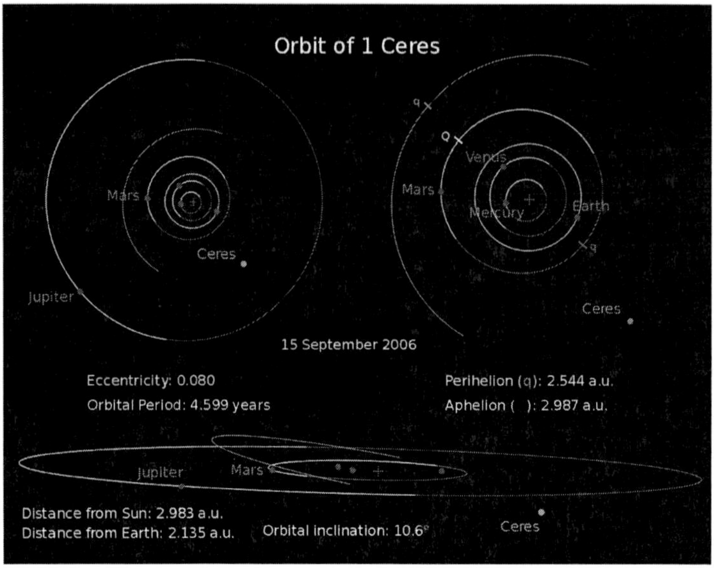

Figure 145: *Orbit of Ceres*

view that shows the location of Ceres in the gap between Mars and Jupiter. The top right is a close-up demonstrating the locations of the perihelia (q) and aphelia (Q) of Ceres and Mars. In this diagram (but not in general), the perihelion of Mars is on the opposite side of the Sun from those of Ceres and several of the large main-belt asteroids, including 2 Pallas and 10 Hygiea. The bottom diagram is a side view showing the inclination of the orbit of Ceres compared to the orbits of Mars and Jupiter.

Ceres was once thought to be a member of an asteroid family. The asteroids of this family share similar proper orbital elements, which may indicate a common origin through an asteroid collision some time in the past. Ceres was later found to have spectral properties different from other members of the family, which is now called the Gefion family after the next-lowest-numbered family member, 1272 Gefion. Ceres appears to be merely an interloper in the Gefion family, coincidentally having similar orbital elements but not a common origin.

Resonances

Ceres is in a near-1:1 mean-motion orbital resonance with Pallas (their proper orbital periods differ by 0.2%). However, a true resonance between the two would be unlikely; due to their small masses relative to their large separations, such relationships among asteroids are very rare. Nevertheless, Ceres is able to capture other asteroids into temporary 1:1 resonant orbital relationships (making them temporary trojans) for periods up to 2 million years or more; fifty such objects have been identified.

Transits of planets from Ceres

Mercury, Venus, Earth, and Mars can all appear to cross the Sun, or transit it, from a vantage point on Ceres. The most common transits are those of Mercury, which usually happen every few years, most recently in 2006 and 2010. The most recent transit of Venus was in 1953, and the next will be in 2051; the corresponding dates are 1814 and 2081 for transits of Earth, and 767 and 2684 for transits of Mars.

Rotation and axial tilt

The rotation period of Ceres (the Cererian day) is 9 hours and 4 minutes. It has an axial tilt of 4°. This is small enough for Ceres's polar regions to contain permanently shadowed craters that are expected to act as cold traps and accumulate water ice over time, similar to the situation on the Moon and Mercury. About 0.14% of water molecules released from the surface are expected to end up in the traps, hopping an average of 3 times before escaping or being trapped.

Geology

Ceres has a mass of 9.39×10^{20} kg as determined from the *Dawn* spacecraft. With this mass Ceres composes approximately a third of the estimated total $3.0 \pm 0.2 \times 10^{21}$ kg mass of the asteroid belt, which is in turn approximately 4% of the mass of the Moon. Ceres is massive enough to give it a nearly spherical, equilibrium shape. Among Solar System bodies, Ceres is intermediate in size between the smaller Vesta and the larger Tethys. Its surface area is approximately the same as the land area of India or Argentina.[222]

Surface

The surface composition of Ceres is broadly similar to that of C-type asteroids. Some differences do exist. The ubiquitous features in Ceres' IR spectrum are those of hydrated materials, which indicate the presence of significant amounts of water in its interior. Other possible surface constituents include iron-rich clay minerals (cronstedtite) and carbonate minerals (dolomite and siderite), which are common minerals in carbonaceous chondrite meteorites. The spectral features of carbonates and clay minerals are usually absent in the spectra of other C-type asteroids. Sometimes Ceres is classified as a G-type asteroid.

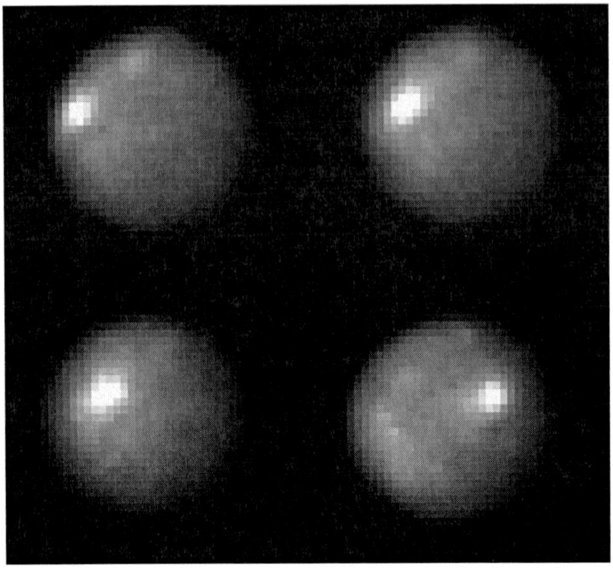

Figure 146: *HST images taken over a span of 2 hours and 20 minutes in 2004*

Ceres' surface is relatively warm. The maximum temperature with the Sun overhead was estimated from measurements to be 235 K (approximately –38 °C, –36 °F) on 5 May 1991. Ice is unstable at this temperature. Material left behind by the sublimation of surface ice could explain the dark surface of Ceres compared to the icy moons of the outer Solar System.

Studies by the Hubble Space Telescope reveal that graphite, sulfur, and sulfur dioxide are present on Ceres's surface. The former is evidently the result of space weathering on Ceres's older surfaces; the latter two are volatile under Cerean conditions and would be expected to either escape quickly or settle in cold traps, and are evidently associated with areas with recent geological activity.

Observations prior to *Dawn*

Prior to the *Dawn* mission, only a few surface features had been unambiguously detected on Ceres. High-resolution ultraviolet *Hubble Space Telescope* images taken in 1995 showed a dark spot on its surface, which was nicknamed "Piazzi" in honor of the discoverer of Ceres. This was thought to be a crater. Later near-infrared images with a higher resolution taken over a whole rotation with the Keck telescope using adaptive optics showed several bright and dark features moving with Ceres' rotation. Two dark features had circular shapes and were presumed to be craters; one of them was observed to have a bright central region, whereas another was identified as the "Piazzi" feature. Visible-light *Hubble Space Telescope* images of a full rotation taken in 2003 and 2004 showed eleven recognizable surface features, the natures of which were then undetermined. One of these features corresponds to the "Piazzi" feature observed earlier.

These last observations indicated that the north pole of Ceres pointed in the direction of right ascension 19 h 24 min (291°), declination +59°, in the constellation Draco, resulting in an axial tilt of approximately 3°. *Dawn* later determined that the north polar axis actually points at right ascension 19 h 25 m 40.3 s (291.418°), declination +66° 45' 50" (about 1.5 degrees from Delta Draconis), which means an axial tilt of 4°.

Observations by *Dawn*

Dawn revealed that Ceres has a heavily cratered surface; nevertheless, Ceres does not have as many large craters as expected, likely due to past geological processes. An unexpectedly large number of Cererian craters have central pits, perhaps due to cryovolcanic processes, and many have central peaks. Ceres has one prominent mountain, Ahuna Mons; this peak appears to be a cryovolcano and has few craters, suggesting a maximum age of no more than a few hundred million years. A later computer simulation has suggested that there were originally other cryovolcanoes on Ceres that are now unrecognisable due to viscous relaxation. Several bright spots have been observed by *Dawn*, the brightest spot ("Spot 5") located in the middle of an 80-kilometer (50 mi) crater called Occator. From images taken of Ceres on 4 May 2015, the secondary bright spot was revealed to actually be a group of scattered bright areas, possibly as many as ten. These bright features have an albedo of approximately 40% that are caused by a substance on the surface, possibly ice or salts, reflecting sunlight. A haze periodically appears above Spot 5, the best known bright spot, supporting the hypothesis that some sort of outgassing or sublimating ice formed the bright spots. In March 2016, *Dawn* found definitive evidence of water molecules on the surface of Ceres at Oxo crater.

On 9 December 2015, NASA scientists reported that the bright spots on Ceres may be related to a type of salt, particularly a form of brine containing magnesium sulfate hexahydrite ($MgSO_4{\cdot}6H_2O$); the spots were also found to be associated with ammonia-rich clays. Near-infrared spectra of these bright areas were reported in 2017 to be consistent with a large amount of sodium carbonate, (Na
2CO
3) and smaller amounts of ammonium chloride (NH
4Cl) or ammonium bicarbonate (NH

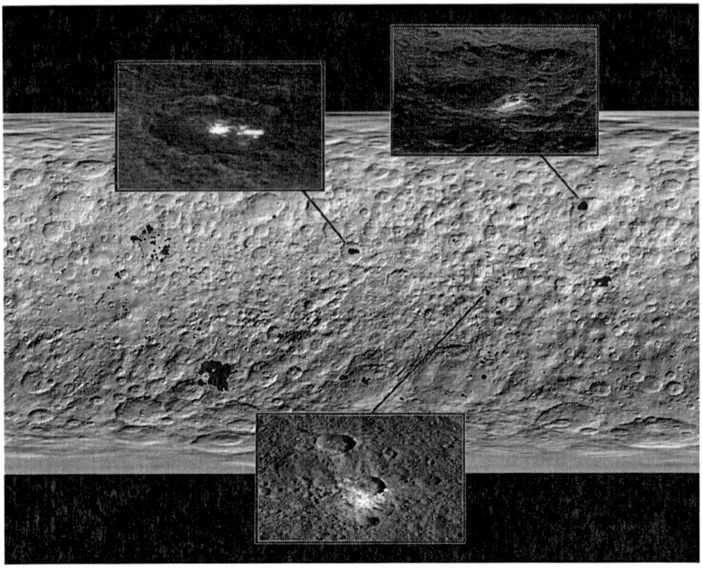

Figure 147:
Map of bright spots on Ceres (released 10 December 2015).

4HCO
3).[223] These materials have been suggested to originate from the recent crystallization of brines that reached the surface from below.

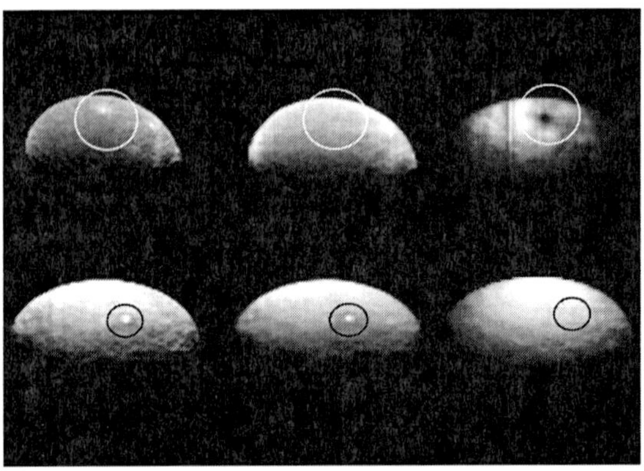

Bright spots on Ceres in visible and infrared:

"Spot 1" (top row) ("cooler" than surroundings);

"Spot 5" (bottom) ("similar in temperature" as surroundings) (April 2015)

"Bright Spot 5" in the crater Occator. Imaged by *Dawn* from 385 km (239 mi) (LAMO)

Ahuna Mons is an estimated 5 km (3 mi) high on its steepest side. Imaged by *Dawn* from 385 km (239 mi) in December 2015.

Internal structure

Ceres' oblateness is consistent with a differentiated body, a rocky core overlain with an icy mantle. This 100-kilometer thick mantle (23%–28% of Ceres by mass; 50% by volume)[224] contains up to 200 million cubic kilometers of water, which would be more than the amount of fresh water on Earth. This result is supported by the observations made by the Keck telescope in 2002 and by evolutionary modeling. Also, some characteristics of its surface and history (such as its distance from the Sun, which weakened solar radiation enough to allow some fairly low-freezing-point components to be incorporated during its formation), point to the presence of volatile materials in the interior of Ceres. It has been suggested that a remnant layer of liquid water may have survived to the present under a layer of ice.

Shape and gravity field measurements by *Dawn* confirm Ceres is a body in hydrostatic equilibrium with partial differentiation and isostatic compensation,

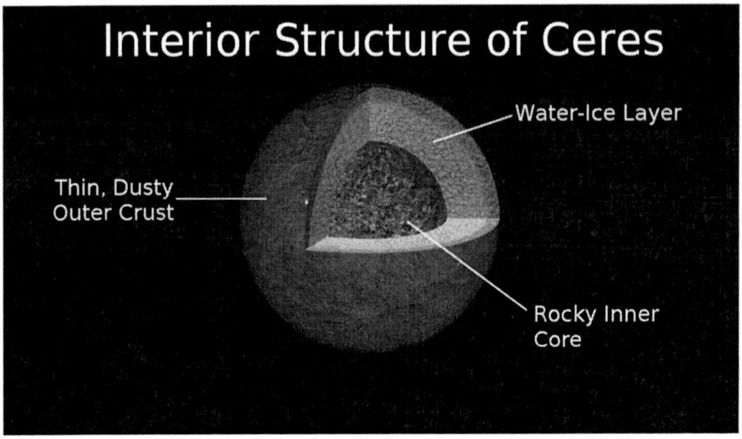

Figure 148: *Diagram showing a possible internal structure of Ceres*

with a mean moment of inertia of 0.37 (which is similar to that of Callisto at ~0.36). The densities of the core and outer layer are estimated to be 2.46–2.90 and 1.68–1.95 g/cm^3, with the latter being about 70–190 km thick. Only partial dehydration of the core is expected. The high density of the outer layer (relative to water ice) reflects its enrichment in silicates and salts. Ceres is the smallest object confirmed to be in hydrostatic equilibrium, being 600 km smaller and less than half the mass of Saturn's moon Rhea, the next smallest such object. Modeling has suggested Ceres could have a small metallic core from partial differentiation of its rocky fraction.

Atmosphere

There are indications that Ceres has a tenuous water vapor atmosphere outgassing from water ice on the surface.[225,226]

Surface water ice is unstable at distances less than 5 AU from the Sun, so it is expected to sublime if it is exposed directly to solar radiation. Water ice can migrate from the deep layers of Ceres to the surface, but escapes in a very short time. As a result, it is difficult to detect water vaporization. Water escaping from polar regions of Ceres was possibly observed in the early 1990s but this has not been unambiguously demonstrated. It may be possible to detect escaping water from the surroundings of a fresh impact crater or from cracks in the subsurface layers of Ceres. Ultraviolet observations by the IUE spacecraft detected statistically significant amounts of hydroxide ions near Ceres' north pole, which is a product of water vapor dissociation by ultraviolet solar radiation.

In early 2014, using data from the *Herschel Space Observatory*, it was discovered that there are several localized (not more than 60 km in diameter) mid-latitude sources of water vapor on Ceres, which each give off approximately 10^{26} molecules (or 3 kg) of water per second.[227] and 7000 kg/s, respectively.</ref> Two potential source regions, designated Piazzi (123°E, 21°N) and Region A (231°E, 23°N), have been visualized in the near infrared as dark areas (Region A also has a bright center) by the W. M. Keck Observatory. Possible mechanisms for the vapor release are sublimation from approximately 0.6 km^2 of exposed surface ice, or cryovolcanic eruptions resulting from radiogenic internal heat or from pressurization of a subsurface ocean due to growth of an overlying layer of ice. Surface sublimation would be expected to be lower when Ceres is farther from the Sun in its orbit, whereas internally powered emissions should not be affected by its orbital position. The limited data available was more consistent with cometary-style sublimation; however, subsequent evidence from Dawn strongly suggests ongoing geologic activity could be at least partially responsible.

Studies using *Dawn's* gamma ray and neutron detector (GRaND) reveal that Ceres is accelerating electrons from the solar wind regularly; although there are several possibilities as to what is causing this, the most accepted is that these electrons are being accelerated by collisions between the solar wind and a tenuous water vapor exosphere.

In 2017, *Dawn* confirmed that Ceres has a transient atmosphere that appears to be linked to solar activity. Ice on Ceres can sublimate when energetic particles from the Sun hit exposed ice within craters.

Origin and evolution

Ceres is possibly a surviving protoplanet (planetary embryo), which formed 4.57 billion years ago in the asteroid belt. Although the majority of inner Solar System protoplanets (including all lunar- to Mars-sized bodies) either merged with other protoplanets to form terrestrial planets or were ejected from the Solar System by Jupiter, Ceres is thought to have survived relatively intact. An alternative theory proposes that Ceres formed in the Kuiper belt and later migrated to the asteroid belt.[228] The discovery of ammonia salts in Occator crater supports an origin in the outer Solar System. Another possible protoplanet, Vesta, is less than half the size of Ceres; it suffered a major impact after solidifying, losing $\sim$1% of its mass.

The geological evolution of Ceres was dependent on the heat sources available during and after its formation: friction from planetesimal accretion, and decay of various radionuclides (possibly including short-lived extinct radionuclides such as aluminium-26). These are thought to have been sufficient to allow

Ceres to differentiate into a rocky core and icy mantle soon after its formation. This process may have caused resurfacing by water volcanism and tectonics, erasing older geological features. Ceres's relatively warm surface temperature implies that any of the resulting ice on its surface would have gradually sublimated, leaving behind various hydrated minerals like clay minerals and carbonates.

Today, Ceres has become considerably less geologically active, with a surface sculpted chiefly by impacts; nevertheless, evidence from *Dawn* reveals that internal processes have continued to sculpt Ceres's surface to a significant extent, in stark contrast to Vesta and of previous expectations that Ceres would have become geologically dead early in its history due to its small size. The presence of significant amounts of water ice in its composition and evidence of recent geological resurfacing, raises the possibility that Ceres has a layer of liquid water in its interior. This hypothetical layer is often called an ocean. If such a layer of liquid water exists, it is hypothesized to be located between the rocky core and ice mantle like that of the theorized ocean on Europa. The existence of an ocean is more likely if solutes (i.e. salts), ammonia, sulfuric acid or other antifreeze compounds are dissolved in the water.

Potential habitability

Although not as actively discussed as a potential home for microbial extraterrestrial life as Mars, Titan, Europa or Enceladus, there is evidence that Ceres' icy mantle was once a watery subterranean ocean, and that has led to speculations that life could have existed there, and that hypothesized ejecta bearing microorganisms could have come from Ceres to Earth.[229]

Observation and exploration

Observation

When Ceres has an opposition near the perihelion, it can reach a visual magnitude of +6.7. This is generally regarded as too dim to be seen with the naked eye, but under exceptional viewing conditions a very sharp-sighted person may be able to see it. The only other asteroids that can reach a similarly bright magnitude are 4 Vesta, and, during rare oppositions near perihelion, 2 Pallas and 7 Iris.[230] At a conjunction Ceres has a magnitude of around +9.3, which corresponds to the faintest objects visible with 10×50 binoculars. It can thus be seen with binoculars whenever it is above the horizon of a fully dark sky.

Some notable observations and milestones for Ceres include:

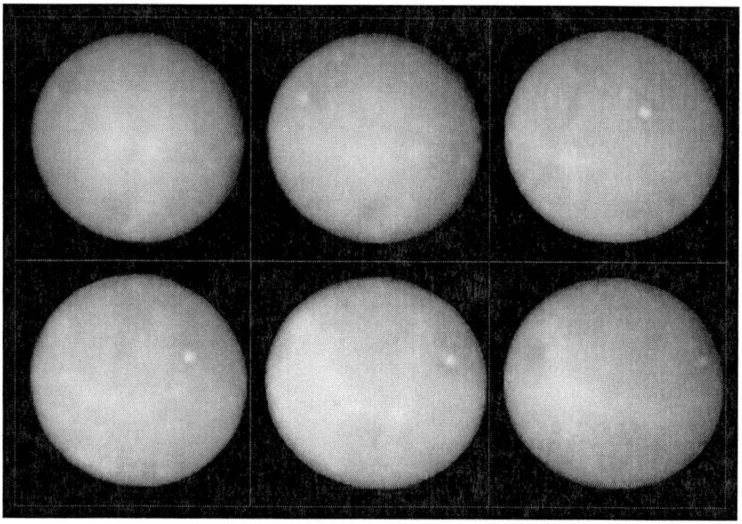

Figure 149: *Polarimetric map of Ceres*

- 1984 November 13: An occultation of a star by Ceres observed in Mexico, Florida and across the Caribbean.
- 1995 June 25: Ultraviolet *Hubble Space Telescope* images with 50-kilometer resolution.
- 2002: Infrared images with 30-km resolution taken with the Keck telescope using adaptive optics.
- 2003 and 2004: Visible light images with 30-km resolution (the best prior to the *Dawn* mission) taken using *Hubble*.
- 2012 December 22: Ceres occulted the star TYC 1865-00446-1 over parts of Japan, Russia, and China. Ceres' brightness was magnitude 6.9 and the star, 12.2.
- 2014. Ceres was found to have an atmosphere with water vapor, confirmed by the *Herschel* space telescope.
- 2015: The NASA *Dawn* spacecraft approached and orbited Ceres, sending detailed images and scientific data back to Earth.

Exploration

In 1981, a proposal for an asteroid mission was submitted to the European Space Agency (ESA). Named the Asteroidal Gravity Optical and Radar Analysis (AGORA), this spacecraft was to launch some time in 1990–1994 and perform two flybys of large asteroids. The preferred target for this mission was Vesta. AGORA would reach the asteroid belt either by a gravitational slingshot

Figure 150: *Artist's conception of Dawn, travelling from Vesta to Ceres*

trajectory past Mars or by means of a small ion engine. However, the proposal was refused by ESA. A joint NASA–ESA asteroid mission was then drawn up for a Multiple Asteroid Orbiter with Solar Electric Propulsion (MAOSEP), with one of the mission profiles including an orbit of Vesta. NASA indicated they were not interested in an asteroid mission. Instead, ESA set up a technological study of a spacecraft with an ion drive. Other missions to the asteroid belt were proposed in the 1980s by France, Germany, Italy, and the United States, but none were approved. Exploration of Ceres by fly-by and impacting penetrator was the second main target of the second plan of the multiaimed Soviet Vesta mission, developed in cooperation with European countries for realisation in 1991–1994 but canceled due to the Soviet Union disbanding.

In the early 1990s, NASA initiated the Discovery Program, which was intended to be a series of low-cost scientific missions. In 1996, the program's study team recommended as a high priority a mission to explore the asteroid belt using a spacecraft with an ion engine. Funding for this program remained problematic for several years, but by 2004 the *Dawn* vehicle had passed its critical design review.

It was launched on 27 September 2007, as the space mission to make the first visits to both Vesta and Ceres. On 3 May 2011, *Dawn* acquired its first targeting image 1.2 million kilometers from Vesta. After orbiting Vesta for 13

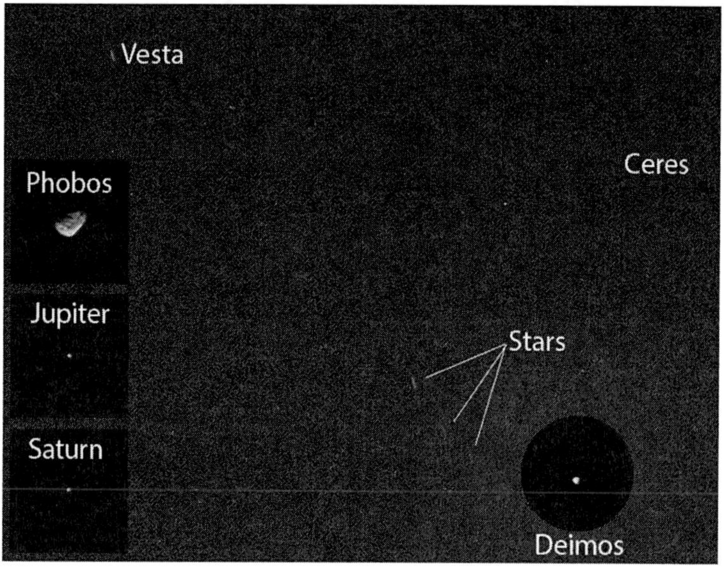

Figure 151: *First asteroid image (Ceres and Vesta) from Mars – viewed by Curiosity (20 April 2014)*

months, *Dawn* used its ion engine to depart for Ceres, with gravitational capture occurring on 6 March 2015 at a separation of 61,000 km, four months prior to the *New Horizons* flyby of Pluto.

Dawn's mission profile calls for it to study Ceres from a series of circular polar orbits at successively lower altitudes. It entered its first observational orbit ("RC3") around Ceres at an altitude of 13,500 km on 23 April 2015, staying for only approximately one orbit (fifteen days). The spacecraft will subsequently reduce its orbital distance to 4,400 km for its second observational orbit ("survey") for three weeks, then down to 1,470 km ("HAMO;" high altitude mapping orbit) for two months and then down to its final orbit at 375 km ("LAMO;" low altitude mapping orbit) for at least three months. The spacecraft instrumentation includes a framing camera, a visual and infrared spectrometer, and a gamma-ray and neutron detector. These instruments will examine Ceres' shape and elemental composition. On 13 January 2015, *Dawn* took the first images of Ceres at near-*Hubble* resolution, revealing impact craters and a small high-albedo spot on the surface, near the same location as that observed previously. Additional imaging sessions, at increasingly better resolution took place on 25 January, 4, 12, 19, and 25 February, 1 March, and 10 and 15 April.

Dawn's arrival in a stable orbit around Ceres was delayed after, close to reaching Ceres, it was hit by a cosmic ray, making it take another, longer route around Ceres in back, instead of a direct spiral towards it.

The Chinese Space Agency is designing a sample retrieval mission from Ceres that would take place during the 2020s.[231]

Maps

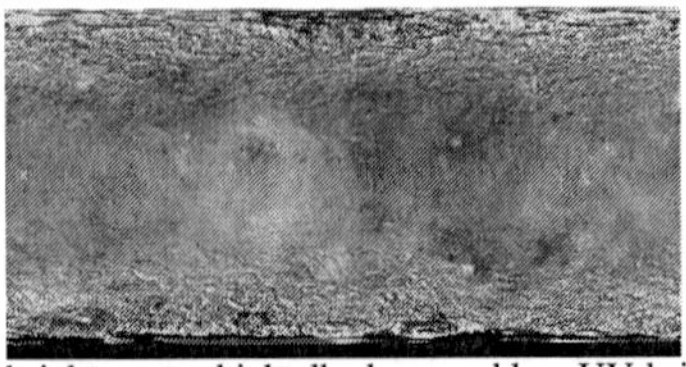

Map of Ceres (red=IR-bright;green=high albedo areas;blue=UV-bright) (September 2015)

Map of Ceres (centered on 180° longitude; color; March 2015)

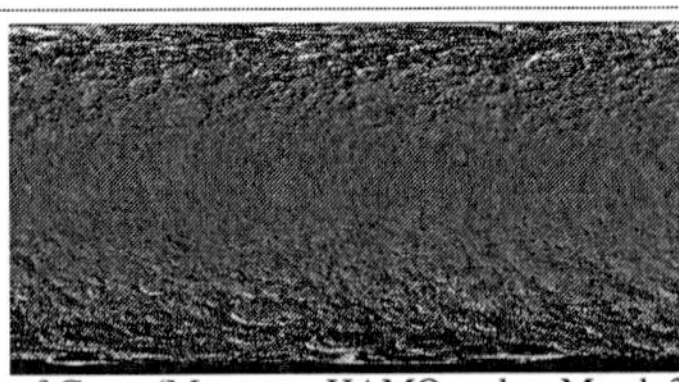

Map of Ceres (Mercator; HAMO; color; March 2016)

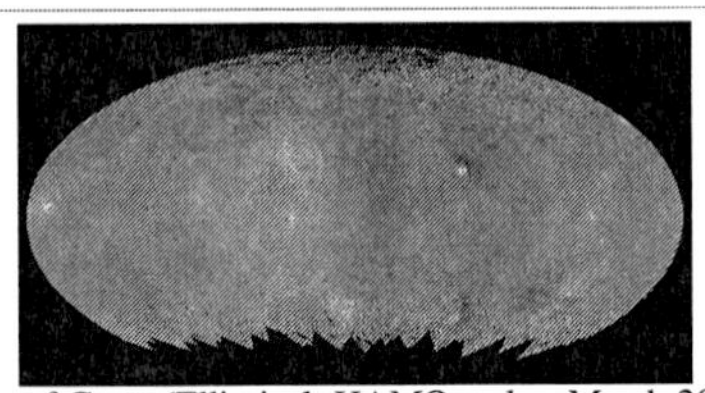

Map of Ceres (Elliptical; HAMO; color; March 2016)

Black-and-white photographic map of Ceres, centered on 180° longitude, with official nomenclature (September 2017)

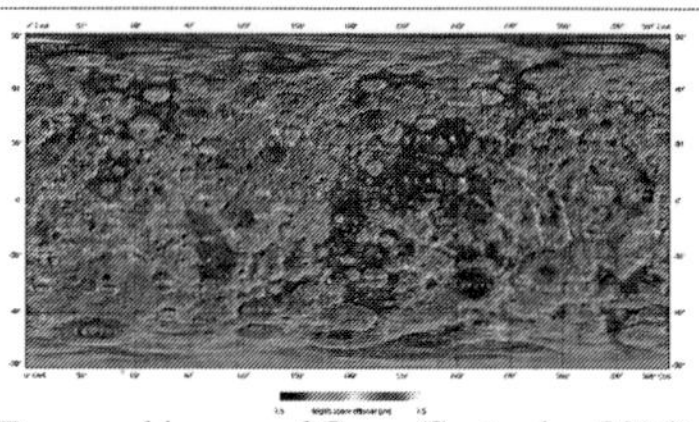

Topographic map of Ceres (September 2016).
15 km (10 mi) of elevation separate the lowest crater floors (indigo) from the highest peaks (white).

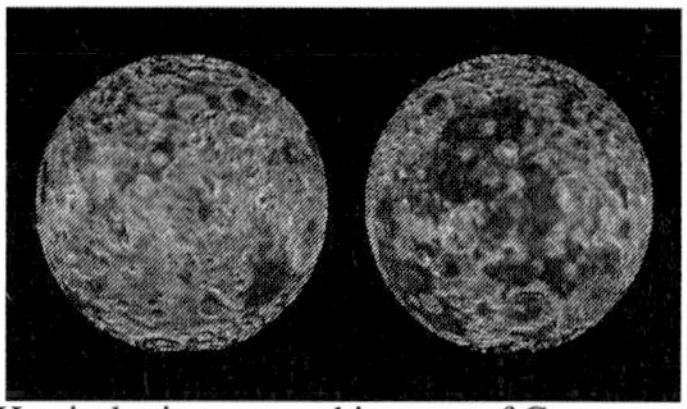

Hemispheric topographic maps of Ceres, centered on 60° and 240° east longitude (July 2015).

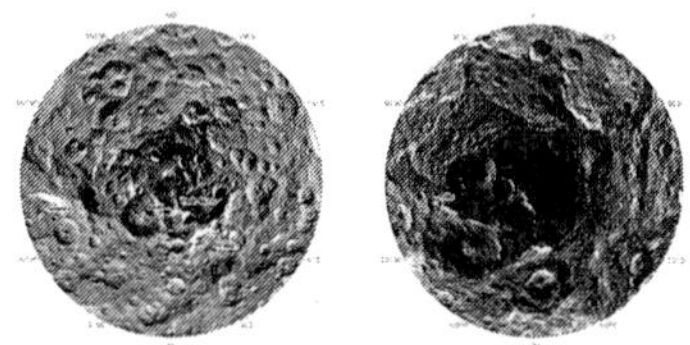

Ceres, polar regions (November 2015): North (left); south (right).

Ceres – Survey Maps (June 2015)

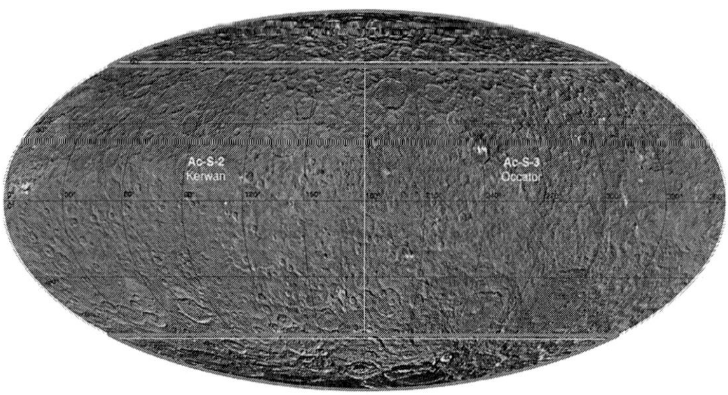

Overall

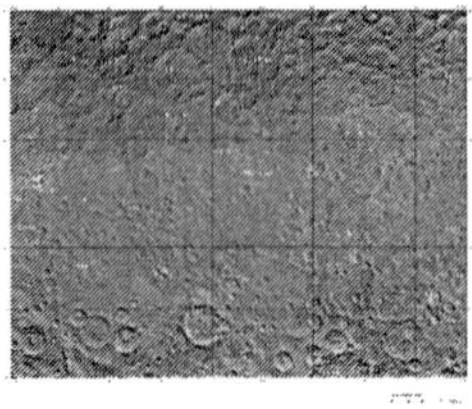

Kerwan section

(PDF version[232])

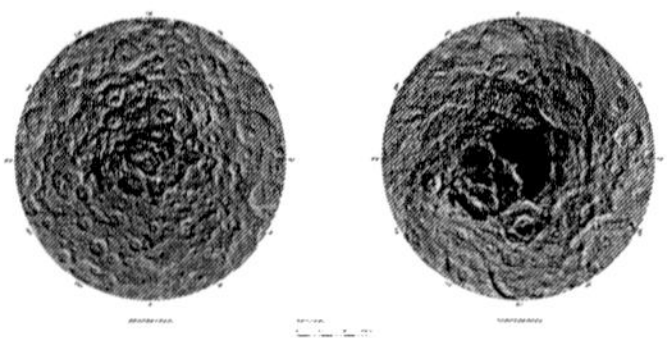

Asari-Zadeni section

(PDF version[233])

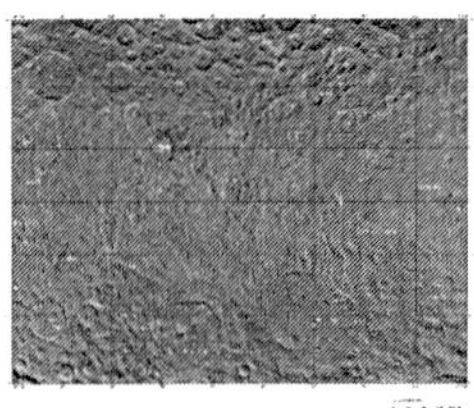

Occator section

(PDF version[234])

Map of quadrangles

The following imagemap of the dwarf planet Ceres is divided into 15 quadrangles. They are named after the first craters whose names the IAU approved in July 2015. The map image(s) were taken by the *Dawn* space probe.

File:20150319-Ceres-DwarfPlanet-Dawn-OpNav5HIPASS-QuadNames-blank.png

Topographic map of Ceres as of February 2015. Darker areas represent lower elevations, and brighter areas represent higher elevations.

[]

Gallery

Ceres in half shadow from 40,000 km (25 February 2015)

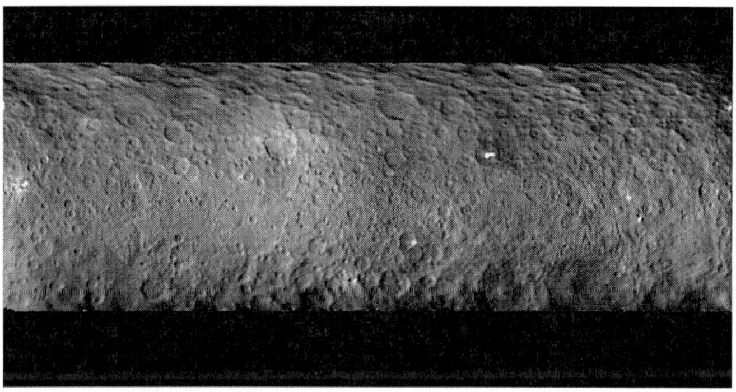

Dawn Ceres mosaic – 19 February 2015

Ceres from *Dawn*, 47,000 kilometers (29,000 mi) away. At this distance, Ceres is approximately the apparent size of the full moon (19 February 2015).

The large impact basin in the lower portion of the left image appears relatively young.

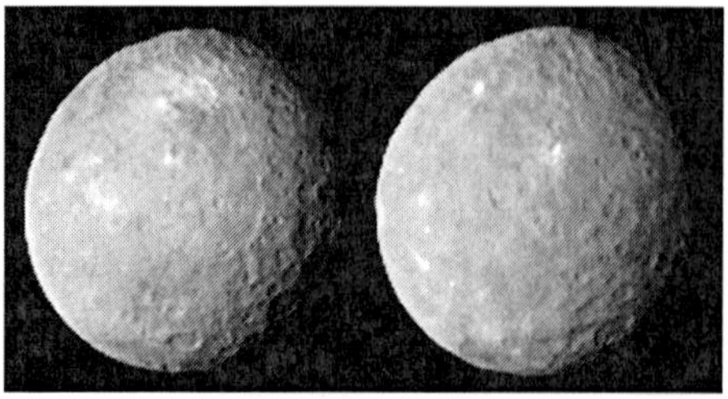

Ceres at 84,000 kilometers (52,000 mi) away (12 February 2015), at half the apparent size of the full moon. Relative to these images, those at left were taken at similar longitudes but a more northerly latitude, and are rotated approximately 45° clockwise.

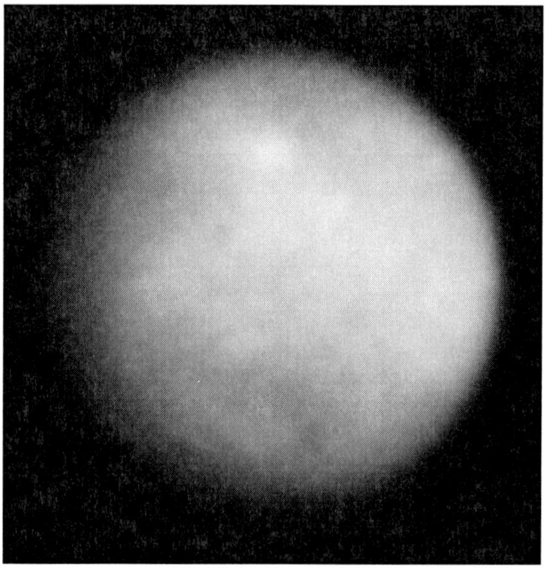

Figure 152: *2004 Hubble Space Telescope*

Figure 153: *14 April 2015; Dawn*

Figure 154: *6 May 2015; Dawn*

Figure 155: *16 May 2015; Dawn*

Figure 156: *22 May 2015; Dawn*

Mapping orbits and resolution – Photos of Ceres by *Dawn* on Commons

Orbit phase	No.	Dates	Altitude (km; mi)	Orbital period	Reso-lution (km/px)	Im-prove-ment over *Hubble*	Notes
RC3	1st	23 April 2015 – 9 May 2015	13,500 km (8,400 mi)	15 days	1.3	24×	
Survey	2nd	6 June 2015 – 30 June 2015	4,400 km (2,700 mi)	3.1 days	0.41	73×	
HAMO	3rd	17 August 2015 – 23 October 2015	1,450 km (900 mi)	19 hours	0.14 (140 m)	217×	
LAMO/-XMO1	4th	16 December 2015 – 2 September 2016	375 km (233 mi)	5.5 hours	0.035 (35 m)	850×	
XMO2	5th	5 October 2016 – 4 November 2016	1,480 km (920 mi)	19 hours	0.14 (140 m)	217×	
XMO3	6th	5 December 2016 –	7,520–9,350 km (4,670–5,810 mi)	≈8 days	0.9 (est)	34× (est)	

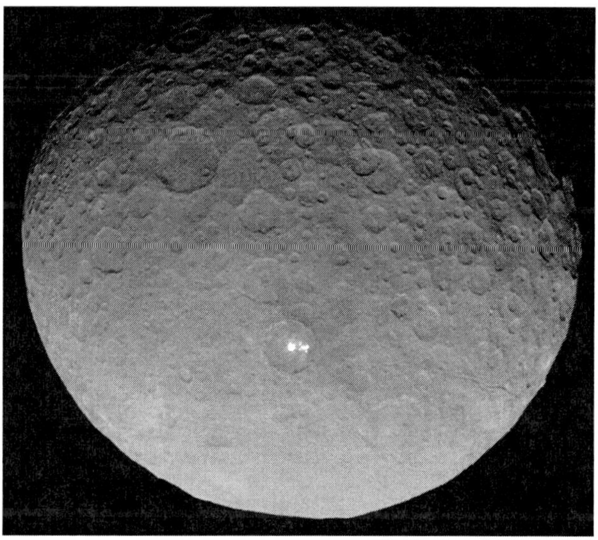

Figure 157: *23 April 2015; Dawn*
1st Map Orbit - RC3

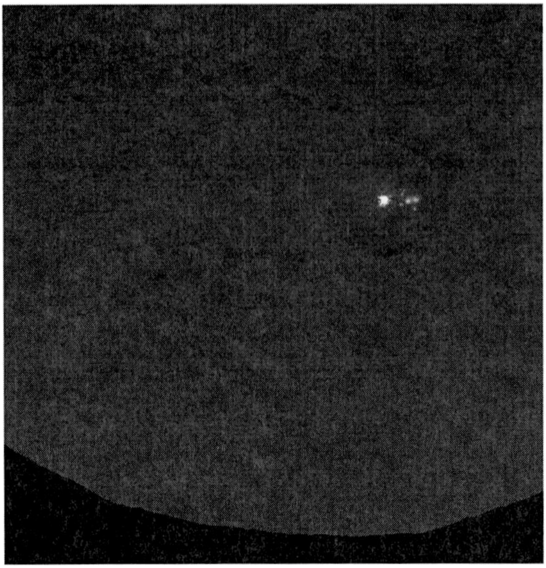

Figure 158: *6 June 2015: Dawn*
2nd Map Orbit - SRVY

Figure 159: *17 August 2015: Dawn*
3rd Map Orbit - HAMO

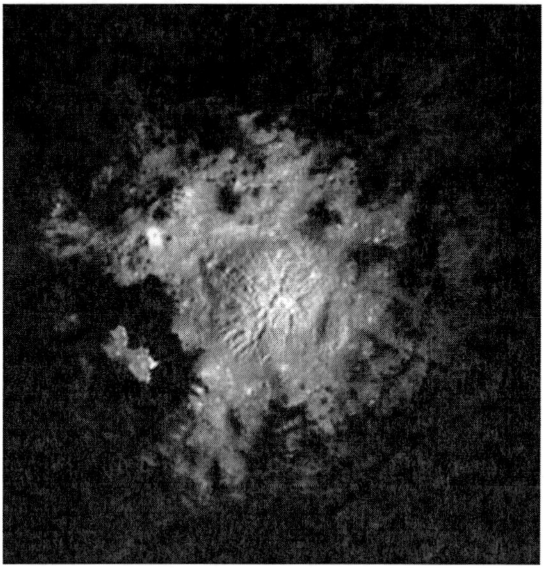

Figure 160: *10 Dec 2015: Dawn*
4th Map Orbit - LAMOa

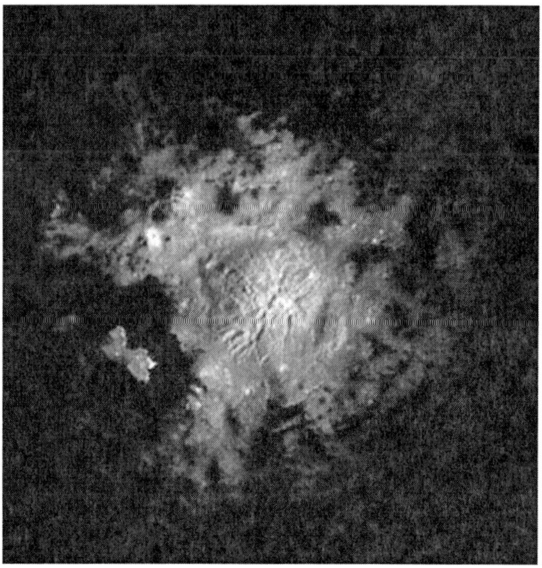

Figure 161: *10 Dec 2015: Dawn*
4th Map Orbit - LAMOb

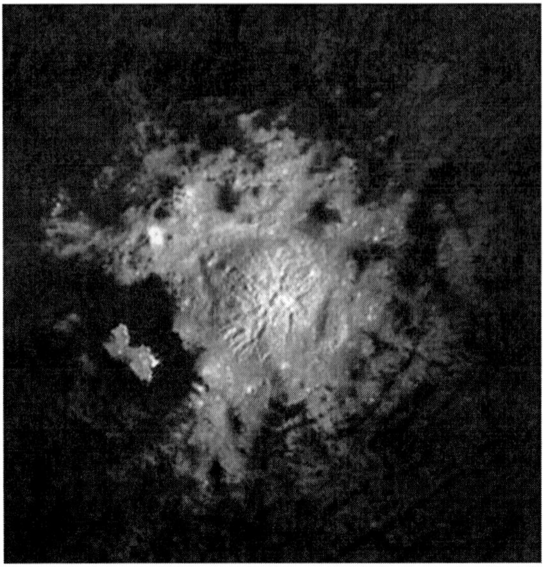

Figure 162: *10 Dec 2015: Dawn*
4th Map Orbit - LAMOc

Animations

Figure 163: *4 February 2015; Dawn*

Figure 164: *29 April 2017: Dawn*

Figure 165: *4 May 2015; Dawn*

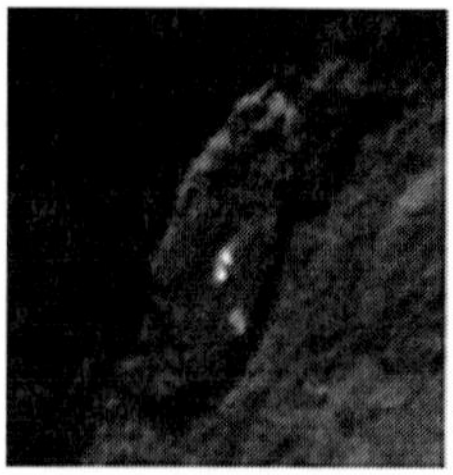

Figure 166: *4 May 2015; Dawn bright spots*

Ceres flyover animations

Surface features exaggerated

(simulated; 01:15; 8 June 2015)

Focus on Occator Crater

(false colors; 01:12; 9 December 2015)

Flight over dwarf planet Ceres

(color; 03:43; 29 January 2016)

External links

Wikimedia Commons has media related to ***Ceres (dwarf planet)***.

- Destination Ceres:Breakfast at Dawn – NASA[235]
- *Dawn* mission home page[236] at JPL
- A simulation of the orbit of Ceres[237]
- JPL Ephemeris[238]
- How Gauss determined the orbit of Ceres[239] from keplersdiscovery.com

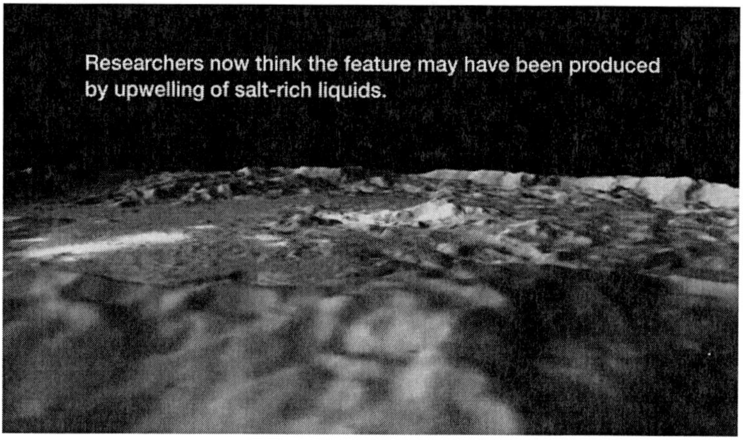

Figure 167:
Ceres – Occator Crater – Flyover (animation; 02:20; 15 December 2016)

- Hilton, James L. (1999). ”U.S. Naval Observatory Ephemerides of the Largest Asteroids”[240]. *The Astronomical Journal.* **117** (2): 1077–1086. Bibcode: 1999AJ....117.1077H[241]. doi: 10.1086/300728[242].
- Animated reprojected colorized map of Ceres[243] (22 February 2015)
- Rotating relief model[244] of Ceres by Seán Doran (about 60% of a full rotation; starts with Occator midway above center)

<indicator name=”featured-star”> </indicator>

Outer planets

The **outer planets** are those planets in the Solar System beyond the asteroid belt, and hence refers to the gas giants, which are in order of their distance from the Sun:

- Jupiter is the largest planet in the Solar System. It has four very large satellites (moons).
- Saturn is the second-largest planet, with a large and bright ring system.
- Uranus is the third-largest planet and the least massive of the four outer planets. It is tilted almost onto the plane of its orbit.
- Neptune is the fourth-largest planet, as smallest of the four outer planets, but third-most massive. It has one big retrograde moon and many small ones.

Figure 168: *From top to bottom: Neptune, Uranus, Saturn, and Jupiter to approximate scale and color.*

The outer planets all have ring systems, although all but Saturn's are faint when viewed from Earth.[245]

Another aspect common to the gas giants is their many natural satellites (moons), two of which are larger than the planet Mercury (Jupiter's Ganymede and Saturn's Titan). That pair and Io, Callisto, Europa, and Triton, are larger than Pluto and Eris.

This region of space is also occupied by centaurs, various fields of trojans, and many comets.

Pluto was considered to be an outer planet from its discovery in 1930 until its reclassification as a dwarf planet in 2006 (see also: Kuiper belt).

Exploration

In situ exploration by spacecraft includes *Pioneer 10*, *Pioneer 11*, *Voyager 1*, *Voyager 2*, *Ulysses*, *Galileo*, *Cassini–Huygens*, and *New Horizons*. Planned missions include *Juno* Jupiter Orbiter and possibly the Outer Planet Flagship Missions; there are various proposals too, such as Uranus orbiter and probe.

Figure 169: *The Galileo Probe plunged deep into Jupiter in 1995. It was carried to the Jovian system by the Galileo spacecraft, where it was released and survived what was then the highest-velocity atmospheric entry yet attempted.*

Ongoing missions for the outer planets as of 2011 include Cassini probe, orbiting Saturn, New Horizons, headed for Pluto, and Juno, headed for Jupiter. Cassini and New Horizons also visited Jupiter with a flyby.

One of the breakthroughs that made exploration of the outer planets much easier, was the concept of the planetary gravity assist.[246] Discovered in the 1960s, a spacecraft approaches a planet like Jupiter in such a way as to be accelerated to a higher speed. This allows a much smaller rocket to be used for a given launch.

Another promising technology, tested on *Deep Space 1*, is the ion engine. Ion engines can make much more efficient use of propellant than existing chemical rockets.

Spacecraft Exploration Summary

Spacecraft	Launch Year	Jupiter	Saturn	Uranus	Neptune	End Year
Pioneer 10	1972	flyby				2003
Pioneer 11	1973	flyby	flyby			1995
Voyager 1	1977	flyby	flyby			—
Voyager 2	1977	flyby	flyby	flyby	flyby	—
Galileo	1989	orbiter				2003
Galileo Probe	1989	entry				1995
Ulysses	1990	flyby				2009
Cassini	1997	flyby	orbiter			—
New Horizons	2006	flyby				—
Juno	2011	orbiter				—

Mission concepts

List of mission concepts:

- Argo[247] - Jupiter, Saturn, Neptune, Kuiper belt flyby
- AVIATR (Titan Airplane)[248]
- Chiron Orbiter Mission Concept Study
- Enceladus Flyby & Sample Return Concept Studies
- Enceladus Orbiter Concept Study
- Europa Flyby
- Europa Flyby Study[249]
- Europa Lander
- Europa Lander Study
- Europa Orbiter[250]
- Exploration of Jovian Atmosphere Using Nuclear Ramjet Flyer[251]
- Ganymede Orbiter Concept Study
- Io Observer Concept Study
- Journey to Enceladus and Titan (JET)[252]
- Jupiter Europa Orbiter Study[253]
- Life Investigation For Enceladus (LIFE) - flyby sample return for Enceladus and Saturn's E-ring.[254]
- Neptune-Triton-Kuiper Belt Objects Mission Concept Study
- Saturn Atmospheric Entry Probe Study
- Saturn Ring Observer Concept Study
- Titan Lake Probe Concept Study

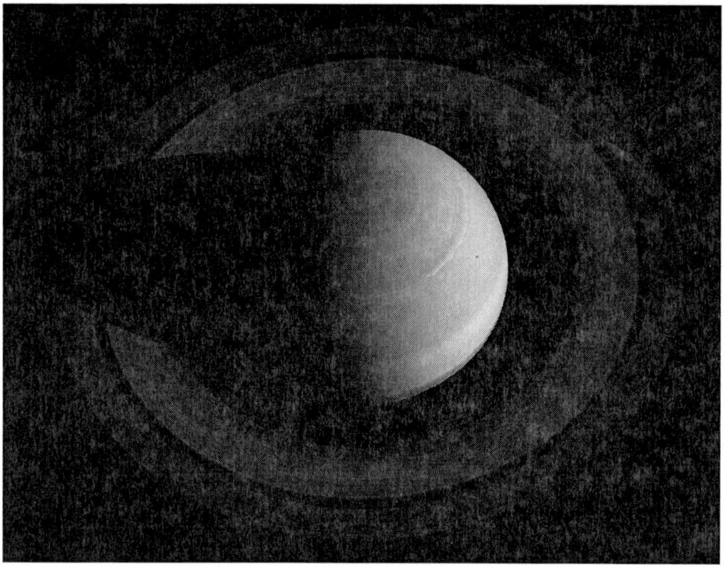

Figure 170: *Real image of the northern hemisphere of Saturn, as taken by the Cassini–Huygens mission*

Figure 171: *Jupiter by Voyager 1 on February 25, 1979*

- Titan Saturn System Mission
- Titan Rover[255]
- Trojan Tour Concept Study
- Uranus and Neptune Orbiter and Probe Concept Studies
- Uranus orbiter and probe
- Uranus Pathfinder[256]

Further reading

- Solar Power for Outer Planets Study (2007) – NASA Glenn Research Center[257]
- Human Outer Planet Exploration (2003) – NASA Langley Research Center and Princeton University[258]
- Future Exploration of the Outer Planet Satellites: A Decadal Perspective (2009)[259]

External links

- Outer Planets Assessment Group (NASA)[260]
  - OPAG Reports[261]
- U.S. Centennial of Flight Commission - Exploration of the Outer Planets[262]
- Space Probes to the Outer Planets[263]

Giant planet

A **giant planet** is any massive planet. They are usually primarily composed of low-boiling-point materials (gases or ices), rather than rock or other solid matter, but massive solid planets can also exist. There are four known giant planets in the Solar System: Jupiter, Saturn, Uranus and Neptune. Many extrasolar giant planets have been identified orbiting other stars.

Giant planets are also sometimes called **jovian planets**, after Jupiter. They are also sometimes known as gas giants. However, many astronomers apply the latter term only to Jupiter and Saturn, classifying Uranus and Neptune, which have different compositions, as ice giants.[264] Both names are potentially misleading: all of the giant planets consist primarily of fluids above their critical points, where distinct gas and liquid phases do not exist. The principal components are hydrogen and helium in the case of Jupiter and Saturn, and water, ammonia and methane in the case of Uranus and Neptune.

Figure 172: *The Solar System's four giant planets against the Sun, to scale*

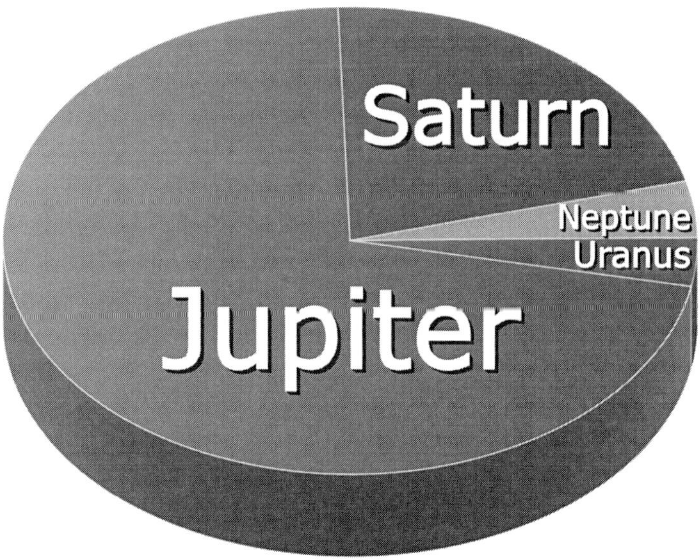

Figure 173: *Relative masses of the giant planets of the Solar System*

The defining differences between a very low-mass brown dwarf and a gas giant ($\sim$13 M_J) are debated. One school of thought is based on formation; the other, on the physics of the interior. Part of the debate concerns whether "brown dwarfs" must, by definition, have experienced nuclear fusion at some point in their history.

Terminology

The term *gas giant* was coined in 1952 by the science fiction writer James Blish and was originally used to refer to all giant planets. Arguably it is something of a misnomer, because throughout most of the volume of these planets the pressure is so high that matter is not in gaseous form. Other than solids in the core and the upper layers of the atmosphere, all matter is above the critical point, where there is no distinction between liquids and gases. *Fluid planet* would be a more accurate term. Jupiter also has metallic hydrogen near its center, but much of its volume is hydrogen, helium, and traces of other gases above their critical points. The observable atmospheres of all these planets (at less than unit optical depth) are quite thin compared to their radii, only extending perhaps one percent of the way to the center. Thus the observable portions are gaseous (in contrast to Mars and Earth, which have gaseous atmospheres through which the crust can be seen).

The rather misleading term has caught on because planetary scientists typically use "rock", "gas", and "ice" as shorthands for classes of elements and compounds commonly found as planetary constituents, irrespective of what phase the matter may appear in. In the outer Solar System, hydrogen and helium are referred to as "gases"; water, methane, and ammonia as "ices"; and silicates and metals as "rock". When deep planetary interiors are considered, it may not be far off to say that, by "ice" astronomers mean oxygen and carbon, by "rock" they mean silicon, and by "gas" they mean hydrogen and helium. The many ways in which Uranus and Neptune differ from Jupiter and Saturn have led some to use the term only for the planets similar to the latter two. With this terminology in mind, some astronomers have started referring to Uranus and Neptune as ice giants to indicate the predominance of the "ices" (in fluid form) in their interior composition.

The alternative term *jovian planet* refers to the Roman god Jupiter—the genitive form of which is *Jovis*, hence *Jovian*—and was intended to indicate that all of these planets were similar to Jupiter.

Objects large enough to start deuterium fusion (above 13 Jupiter masses for solar composition) are called brown dwarfs, and these occupy the mass range between that of large giant planets and the lowest-mass stars. The 13-Jupiter-mass (M_J) cutoff is a rule of thumb rather than something of precise physical

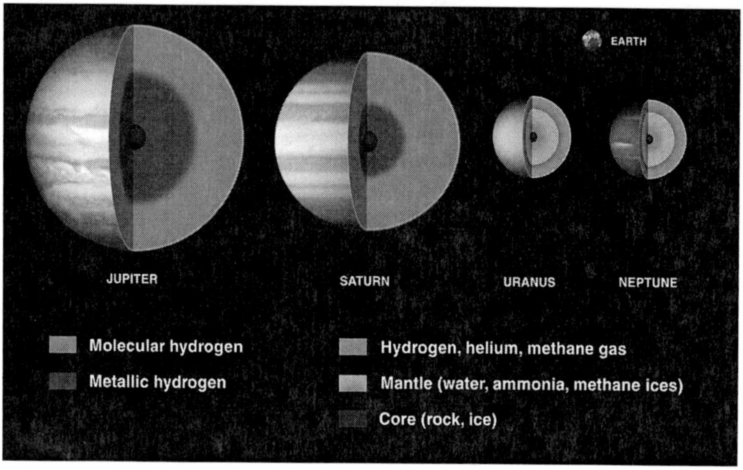

Figure 174: *These cut-aways illustrate interior models of the giant planets. Jupiter is shown with a rocky core overlaid by a deep layer of metallic hydrogen.*

significance. Larger objects will burn most of their deuterium and smaller ones will burn only a little, and the 13 M_J value is somewhere in between. The amount of deuterium burnt depends not only on the mass but also on the composition of the planet, especially on the amount of helium and deuterium present.[265] The Extrasolar Planets Encyclopaedia includes objects up to 25 Jupiter masses, and the Exoplanet Data Explorer up to 24 Jupiter masses.

Description

A giant planet is a massive planet and has a thick atmosphere of hydrogen and helium. They may have a dense molten core of rocky elements, or the core may have completely dissolved and dispersed throughout the planet if the planet is hot enough.[266] In "traditional" giant planets such as Jupiter and Saturn (the gas giants) hydrogen and helium constitute most of the mass of the planet, whereas they only make up an outer envelope on Uranus and Neptune, which are instead mostly composed of water, ammonia, and methane and therefore increasingly referred to as "ice giants".

Extrasolar giant planets that orbit very close to their stars are the exoplanets that are easiest to detect. These are called hot Jupiters and hot Neptunes because they have very high surface temperatures. Hot Jupiters were, until the advent of space-borne telescopes, the most common form of exoplanet known, due to the relative ease of detecting them with ground-based instruments.

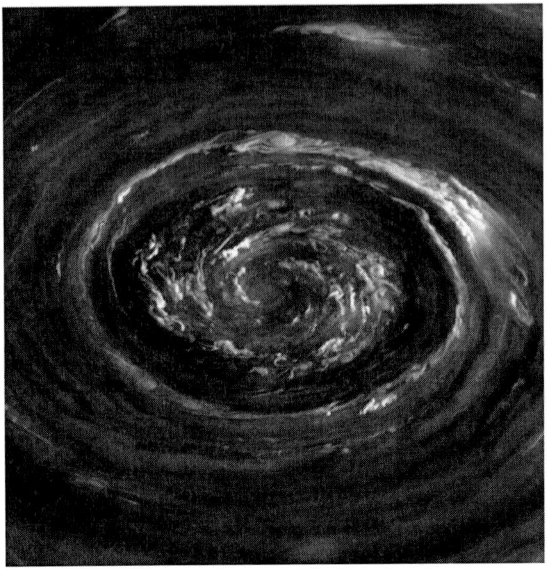

Figure 175: *Saturn's north polar vortex*

Giant planets are commonly said to lack solid surfaces, but it is more accurate to say that they lack surfaces altogether since the gases that constitute them simply become thinner and thinner with increasing distance from the planets' centers, eventually becoming indistinguishable from the interplanetary medium. Therefore, landing on a giant planet may or may not be possible, depending on the size and composition of its core.

Subtypes

Gas giants

Gas giants consist mostly of hydrogen and helium. The Solar System's gas giants, Jupiter and Saturn, have heavier elements making up between 3 and 13 percent of their mass.[267] Gas giants are thought to consist of an outer layer of molecular hydrogen, surrounding a layer of liquid metallic hydrogen, with a probable molten core with a rocky composition.

Jupiter and Saturn's outermost portion of the hydrogen atmosphere has many layers of visible clouds that are mostly composed of water and ammonia. The layer of metallic hydrogen makes up the bulk of each planet, and is referred to as "metallic" because the very high pressure turns hydrogen into an electrical conductor. The core is thought to consist of heavier elements at such high temperatures (20,000 K) and pressures that their properties are poorly understood.

Ice giants

Ice giants have distinctly different interior compositions from gas giants. The Solar System's ice giants, Uranus and Neptune, have a hydrogen-rich atmosphere that extends from the cloud tops down to about 80% (Uranus) or 85% (Neptune) of their radius. Below this, they are predominantly "icy", i.e. consist mostly of water, methane, and ammonia. There is also some rock and gas, but various proportions of ice–rock–gas could mimic pure ice, so that the exact proportions are unknown.

Uranus and Neptune have very hazy atmospheric layers with small amounts of methane, giving them aquamarine colors; light blue and ultramarine respectively.Wikipedia:Please clarify Both have magnetic fields that are sharply inclined to their axes of rotation.

Unlike the other giant planets, Uranus has an extreme tilt that causes its seasons to be severely pronounced. The two planets also have other subtle but important differences. Uranus has more hydrogen and helium than Neptune despite being less massive overall. Neptune is therefore denser and has much more internal heat and a more active atmosphere. The Nice model, in fact, suggests that Neptune formed closer to the Sun than Uranus did, and should therefore have more heavy elements.

Massive solid planets

Massive solid planets can also exist. For example, Kepler-10c is composed of mostly of rock with possibly up to 20% high-pressure water ice, but does have a hydrogen-dominated envelope.

Solid planets up to thousands of Earth masses may be able to form around massive stars (B-type and O-type stars; 5–120 solar masses), where the protoplanetary disk would contain enough heavy elements. Also, these stars have high UV radiation and winds that could photoevaporate the gas in the disk, leaving just the heavy elements. For comparison, Neptune's mass equals 17 Earth masses, Jupiter has 318 Earth masses, and the 13 Jupiter-mass limit used in the IAU's working definition of an exoplanet equals approximately 4000 Earth masses.

Figure 176: *An artist's conception of 79 Ceti b, the first extra-solar giant planet found with a minimum mass less than Saturn.*

Extrasolar giant planets

Because of the limited techniques currently available to detect exoplanets, many of those found to date have been of a size associated, in the Solar System, with giant planets. Because these large planets are inferred to share more in common with Jupiter than with the other giant planets, some have claimed that "jovian planet" is a more accurate term for them. Many of the exoplanets are much closer to their parent stars and hence much hotter than the giant planets in the Solar System, making it possible that some of those planets are a type not observed in the Solar System. Considering the relative abundances of the elements in the universe (approximately 98% hydrogen and helium) it would be surprising to find a predominantly rocky planet more massive than Jupiter. On the other hand, models of planetary-system formation have suggested that giant planets would be inhibited from forming as close to their stars as many of the extrasolar giant planets have been observed to orbit.

Atmospheres

The bands seen in the atmosphere of Jupiter are due to counter-circulating streams of material called zones and belts, encircling the planet parallel to its equator. The zones are the lighter bands, and are at higher altitudes in the atmosphere. They have an internal updraft and are high-pressure regions. The

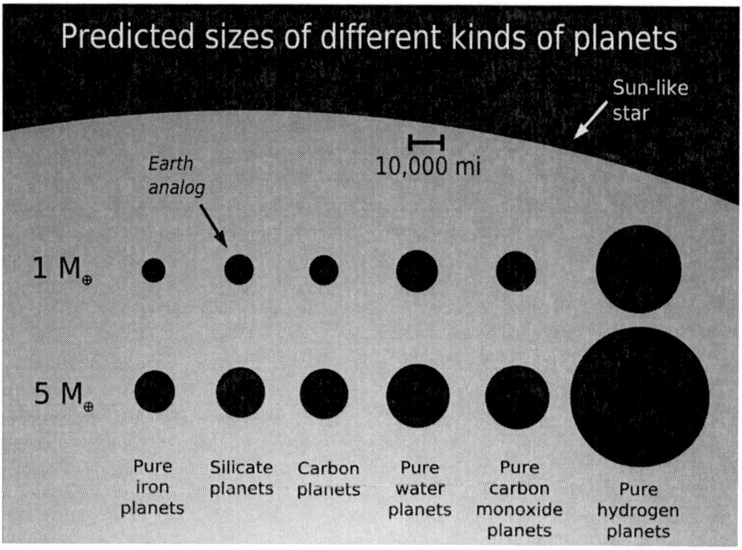

Figure 177: *Comparison of sizes of planets of a given mass with different compositions*

belts are the darker bands, are lower in the atmosphere, and have an internal downdraft. They are low-pressure regions. These structures are somewhat analogous to the high and low-pressure cells in Earth's atmosphere, but they have a very different structure—latitudinal bands that circle the entire planet, as opposed to small confined cells of pressure. This appears to be a result of the rapid rotation and underlying symmetry of the planet. There are no oceans or landmasses to cause local heating and the rotation speed is much higher than that of Earth.

There are smaller structures as well: spots of different sizes and colors. On Jupiter, the most noticeable of these features is the Great Red Spot, which has been present for at least 300 years. These structures are huge storms. Some such spots are thunderheads as well.

Bibliography

- SPACE.com: Q&A: The IAU's Proposed Planet Definition, 16 August 2006, 2:00 AM ET
- BBC News: Q&A New planets proposal Wednesday, 16 August 2006, 13:36 GMT 14:36 UK

External links

Wikimedia Commons has media related to ***Gas giant***.

Look up ***gas giant*** or ***Jovian planet***in Wiktionary, the free dictionary.

- SPACE.com: Q&A: The IAU's Proposed Planet Definition[268] 16 August 2006 2:00 am ET
- BBC News: Q&A New planets proposal[269] Wednesday, 16 August 2006, 13:36 GMT 14:36 UK
- Gas Giants in Science Fiction:[270]
- Episode "Giants" on The Science Channel TV show *Planets*

Saturn

<indicator name="pp-default"> </indicator>

Saturn

Saturn in natural color approaching equinox, photographed by *Cassini* in July 2008. The dot in the bottom left corner is Titan.

Designations	
Pronunciation	/ˈsætərn/ Media:en-us-Saturn.ogg
Named after	Saturn
Adjectives	Saturnian, Cronian
Orbital characteristics	
Epoch J2000.0	
Aphelion	1,514.50 million km (10.1238 AU)
Perihelion	1,352.55 million km (9.0412 AU)
Semi-major axis	1,433.53 million km (9.5826 AU)
Eccentricity	0.0565
Orbital period	• 29.4571 yr • 10,759.22 d • 24,491.07 Saturnian solar days
Synodic period	378.09 days
Average orbital speed	9.68 km/s (6.01 mi/s)
Mean anomaly	317.020°
Inclination	• 2.485° to ecliptic • 5.51° to Sun's equator • 0.93° to invariable plane
Longitude of ascending node	113.665°
Argument of perihelion	339.392°
Known satellites	62 with formal designations; innumerable additional moonlets.
Physical characteristics	
Mean radius	58,232 km (36,184 mi)

Equatorial radius	• 60,268 km (37,449 mi)[271] • 9.449 Earths
Polar radius	• 54,364 km (33,780 mi) • 8.552 Earths
Flattening	0.09796
Surface area	• 4.27×10^{10} km^2 (1.65×10^{10} sq mi) • 83.703 Earths
Volume	• 8.2713×10^{14} km^3 (1.9844×10^{14} cu mi) • 763.59 Earths
Mass	• 5.6834×10^{26} kg (1.2530×10^{27} lb) • 95.159 Earths
Mean density	0.687 g/cm^3 (0.0248 lb/cu in)[272] (less than water)
Surface gravity	• 10.44 m/s^2 (34.3 ft/s^2) • 1.065 *g*
Moment of inertia factor	0.210 I/MR^2 estimate
Escape velocity	35.5 km/s (22.1 mi/s)
Sidereal rotation period	10.55 hours (10 hr 33 min)
Equatorial rotation velocity	9.87 km/s (6.13 mi/s; 35,500 km/h)
Axial tilt	26.73° (to orbit)
North pole right ascension	40.589°; 2^h 42^m 21^s
North pole declination	83.537°
Albedo	• 0.342 (Bond) • 0.499 (geometric)

Surface temp.	min	mean	max
1 bar		134 K (–139 °C)	
0.1 bar		84 K (–189 °C)	

Apparent magnitude	+1.47 to –0.24
Angular diameter	14.5" to 20.1" (excludes rings)
Atmosphere	
Surface pressure	140 kPa
Scale height	59.5 km (37.0 mi)

Composition by volume	by volume:	
	96.3±2.4%	hydrogen (H_2)
	3.25±2.4%	helium (He)
	0.45±0.2%	methane (CH_4)
	0.0125±0.0075%	ammonia (NH_3)
	0.0110±0.0058%	hydrogen deuteride (HD)
	0.0007±0.00015%	ethane (C_2H_6)
	Ices: • ammonia (NH_3) • water (H_2O) • ammonium hydrosulfide (NH_4SH)	

Saturn is the sixth planet from the Sun and the second-largest in the Solar System, after Jupiter. It is a gas giant with an average radius about nine times that of Earth. It has only one-eighth the average density of Earth, but with its larger volume Saturn is over 95 times more massive. Saturn is named after the Roman god of agriculture; its astronomical symbol (♄) represents the god's sickle.

Saturn's interior is probably composed of a core of iron–nickel and rock (silicon and oxygen compounds). This core is surrounded by a deep layer of metallic hydrogen, an intermediate layer of liquid hydrogen and liquid helium, and finally a gaseous outer layer. Saturn has a pale yellow hue due to ammonia crystals in its upper atmosphere. Electrical current within the metallic hydrogen layer is thought to give rise to Saturn's planetary magnetic field, which is weaker than Earth's, but has a magnetic moment 580 times that of Earth due to Saturn's larger size. Saturn's magnetic field strength is around one-twentieth of Jupiter's. The outer atmosphere is generally bland and lacking in contrast, although long-lived features can appear. Wind speeds on Saturn can reach 1,800 km/h (500 m/s), higher than on Jupiter, but not as high as those on Neptune.

The planet's most famous feature is its prominent ring system that is composed mostly of ice particles, with a smaller amount of rocky debris and dust. At least 62 moons are known to orbit Saturn, of which 53 are officially named. This does not include the hundreds of moonlets in the rings. Titan, Saturn's largest moon, and the second-largest in the Solar System, is larger than the planet Mercury, although less massive, and is the only moon in the Solar System to have a substantial atmosphere.

Figure 178: *Composite image comparing the sizes of Saturn and Earth*

Physical characteristics

Saturn is a gas giant because it is predominantly composed of hydrogen and helium. It lacks a definite surface, though it may have a solid core. Saturn's rotation causes it to have the shape of an oblate spheroid; that is, it is flattened at the poles and bulges at its equator. Its equatorial and polar radii differ by almost 10%: 60,268 km versus 54,364 km. Jupiter, Uranus, and Neptune, the other giant planets in the Solar System, are also oblate but to a lesser extent. Saturn is the only planet of the Solar System that is less dense than water—about 30% less. Although Saturn's core is considerably denser than water, the average specific density of the planet is 0.69 g/cm^3 due to the atmosphere. Jupiter has 318 times the Earth's mass, and Saturn is 95 times the mass of the Earth. Together, Jupiter and Saturn hold 92% of the total planetary mass in the Solar System.

On 8 January 2015, NASA reported that a team of scientists from the Jet Propulsion Laboratory determined the barycenter of the planet Saturn and its family of moons to within 4 km (2.5 mi) using data from an experiment conducted with the Cassini spacecraft and the Very Large Baseline Array.

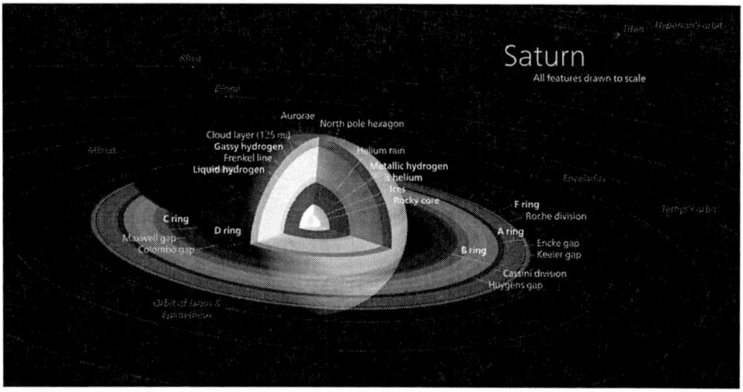

Figure 179: *Diagram of Saturn, to scale*

Internal structure

Despite consisting mostly of hydrogen and helium, most of Saturn's mass is not in the gas phase, because hydrogen becomes a non-ideal liquid when the density is above 0.01 g/cm^3, which is reached at a radius containing 99.9% of Saturn's mass. The temperature, pressure, and density inside Saturn all rise steadily toward the core, which causes hydrogen to be a metal in the deeper layers.

Standard planetary models suggest that the interior of Saturn is similar to that of Jupiter, having a small rocky core surrounded by hydrogen and helium with trace amounts of various volatiles. This core is similar in composition to the Earth, but more dense. Examination of Saturn's gravitational moment, in combination with physical models of the interior, has allowed constraints to be placed on the mass of Saturn's core. In 2004, scientists estimated that the core must be 9–22 times the mass of the Earth, which corresponds to a diameter of about 25,000 km. This is surrounded by a thicker liquid metallic hydrogen layer, followed by a liquid layer of helium-saturated molecular hydrogen that gradually transitions to a gas with increasing altitude. The outermost layer spans 1,000 km and consists of gas.

Saturn has a hot interior, reaching 11,700 °C at its core, and it radiates 2.5 times more energy into space than it receives from the Sun. Jupiter's thermal energy is generated by the Kelvin–Helmholtz mechanism of slow gravitational compression, but such a process alone may not be sufficient to explain heat production for Saturn, because it is less massive. An alternative or additional mechanism may be generation of heat through the "raining out" of droplets of helium deep in Saturn's interior. As the droplets descend through the lower-density hydrogen, the process releases heat by friction and leaves Saturn's outer

Figure 180: *Methane bands circle Saturn. The moon Dione hangs below the rings to the right.*

layers depleted of helium. These descending droplets may have accumulated into a helium shell surrounding the core. Rainfalls of diamonds have been suggested to occur on Saturn, as well as on Jupiter and ice giants Uranus and Neptune.

Atmosphere

The outer atmosphere of Saturn contains 96.3% molecular hydrogen and 3.25% helium by volume.[273] The proportion of helium is significantly deficient compared to the abundance of this element in the Sun. The quantity of elements heavier than helium (metallicity) is not known precisely, but the proportions are assumed to match the primordial abundances from the formation of the Solar System. The total mass of these heavier elements is estimated to be 19–31 times the mass of the Earth, with a significant fraction located in Saturn's core region.

Trace amounts of ammonia, acetylene, ethane, propane, phosphine and methane have been detected in Saturn's atmosphere. The upper clouds are composed of ammonia crystals, while the lower level clouds appear to consist of either ammonium hydrosulfide (NH_4SH) or water. Ultraviolet radiation from the Sun causes methane photolysis in the upper atmosphere, leading to

Figure 181: *A global storm girdles the planet in 2011. The head of the storm (bright area) passes the tail circling around the left limb.*

a series of hydrocarbon chemical reactions with the resulting products being carried downward by eddies and diffusion. This photochemical cycle is modulated by Saturn's annual seasonal cycle.

Cloud layers

Saturn's atmosphere exhibits a banded pattern similar to Jupiter's, but Saturn's bands are much fainter and are much wider near the equator. The nomenclature used to describe these bands is the same as on Jupiter. Saturn's finer cloud patterns were not observed until the flybys of the *Voyager* spacecraft during the 1980s. Since then, Earth-based telescopy has improved to the point where regular observations can be made.

The composition of the clouds varies with depth and increasing pressure. In the upper cloud layers, with the temperature in the range 100–160 K and pressures extending between 0.5–2 bar, the clouds consist of ammonia ice. Water ice clouds begin at a level where the pressure is about 2.5 bar and extend down to 9.5 bar, where temperatures range from 185–270 K. Intermixed in this layer is a band of ammonium hydrosulfide ice, lying in the pressure range 3–6 bar with temperatures of 190–235 K. Finally, the lower layers, where pressures are between 10–20 bar and temperatures are 270–330 K, contains a region of water droplets with ammonia in aqueous solution.

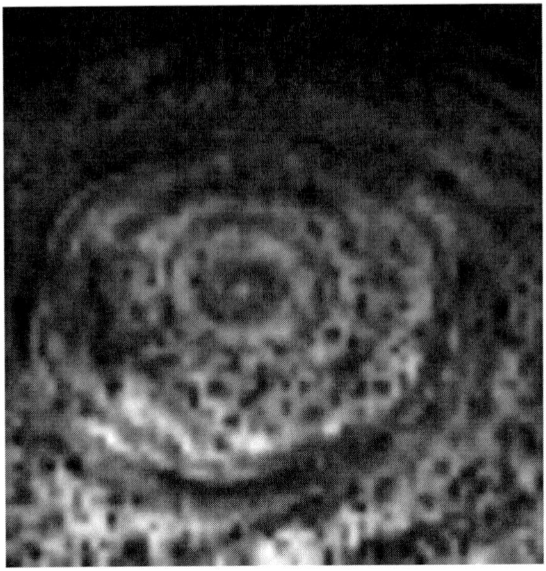

Figure 182: *Saturn's north pole (IR;animation)*

Saturn's usually bland atmosphere occasionally exhibits long-lived ovals and other features common on Jupiter. In 1990, the Hubble Space Telescope imaged an enormous white cloud near Saturn's equator that was not present during the *Voyager* encounters, and in 1994 another smaller storm was observed. The 1990 storm was an example of a Great White Spot, a unique but short-lived phenomenon that occurs once every Saturnian year, roughly every 30 Earth years, around the time of the northern hemisphere's summer solstice. Previous Great White Spots were observed in 1876, 1903, 1933 and 1960, with the 1933 storm being the most famous. If the periodicity is maintained, another storm will occur in about 2020.

The winds on Saturn are the second fastest among the Solar System's planets, after Neptune's. Voyager data indicate peak easterly winds of 500 m/s (1800 km/h). In images from the *Cassini* spacecraft during 2007, Saturn's northern hemisphere displayed a bright blue hue, similar to Uranus. The color was most likely caused by Rayleigh scattering. Thermography has shown that Saturn's south pole has a warm polar vortex, the only known example of such a phenomenon in the Solar System. Whereas temperatures on Saturn are normally –185 °C, temperatures on the vortex often reach as high as –122 °C, suspected to be the warmest spot on Saturn.

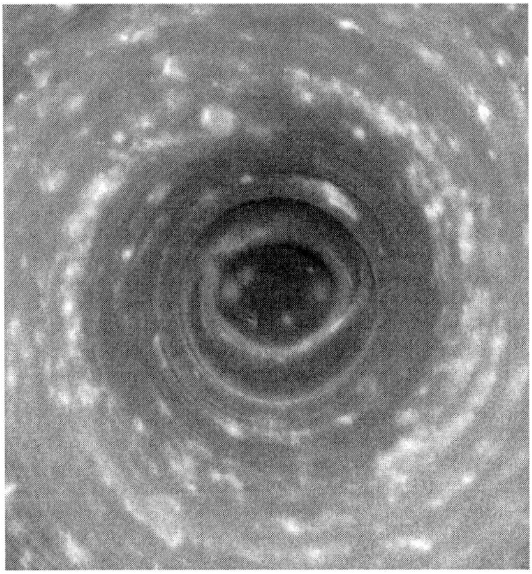

Figure 183: *Saturn's south pole*

North pole hexagonal cloud pattern

A persisting hexagonal wave pattern around the north polar vortex in the atmosphere at about 78°N was first noted in the Voyager images. The sides of the hexagon are each about 13,800 km (8,600 mi) long, which is longer than the diameter of the Earth. The entire structure rotates with a period of 10h 39m 24s (the same period as that of the planet's radio emissions) which is assumed to be equal to the period of rotation of Saturn's interior. The hexagonal feature does not shift in longitude like the other clouds in the visible atmosphere. The pattern's origin is a matter of much speculation. Most scientists think it is a standing wave pattern in the atmosphere. Polygonal shapes have been replicated in the laboratory through differential rotation of fluids.[274,275]

South pole vortex

HST imaging of the south polar region indicates the presence of a jet stream, but no strong polar vortex nor any hexagonal standing wave. NASA reported in November 2006 that *Cassini* had observed a "hurricane-like" storm locked to the south pole that had a clearly defined eyewall. Eyewall clouds had not previously been seen on any planet other than Earth. For example, images from the *Galileo* spacecraft did not show an eyewall in the Great Red Spot of Jupiter.

Figure 184: *Polar aurorae on Saturn*

The south pole storm may have been present for billions of years. This vortex is comparable to the size of Earth, and it has winds of 550 km/h.

Other features

Cassini has observed a series of cloud features nicknamed "String of Pearls" found in northern latitudes. These features are cloud clearings that reside in deeper cloud layers.

Magnetosphere

Saturn has an intrinsic magnetic field that has a simple, symmetric shape – a magnetic dipole. Its strength at the equator – 0.2 gauss (20 μT) – is approximately one twentieth of that of the field around Jupiter and slightly weaker than Earth's magnetic field. As a result, Saturn's magnetosphere is much smaller than Jupiter's. When *Voyager 2* entered the magnetosphere, the solar wind pressure was high and the magnetosphere extended only 19 Saturn radii, or 1.1 million km (712,000 mi), although it enlarged within several hours, and remained so for about three days. Most probably, the magnetic field is generated similarly to that of Jupiter – by currents in the liquid metallic-hydrogen layer called a metallic-hydrogen dynamo. This magnetosphere is efficient at deflecting the solar wind particles from the Sun. The moon Titan orbits within the

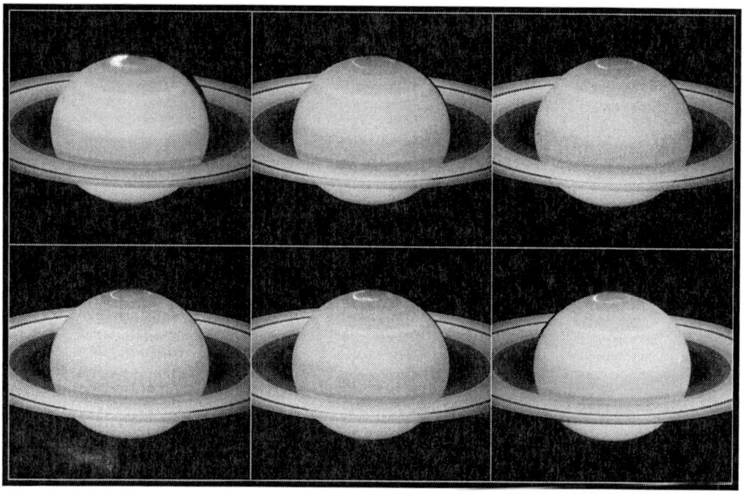

Figure 185: *Auroral lights at Saturn's north pole*

outer part of Saturn's magnetosphere and contributes plasma from the ionized particles in Titan's outer atmosphere. Saturn's magnetosphere, like Earth's, produces aurorae.

Orbit and rotation

The average distance between Saturn and the Sun is over 1.4 billion kilometers (9 AU). With an average orbital speed of 9.68 km/s, it takes Saturn 10,759 Earth days (or about 29 $^1/_2$ years), to finish one revolution around the Sun. The elliptical orbit of Saturn is inclined 2.48° relative to the orbital plane of the Earth. The perihelion and aphelion distances are, respectively, 9.195 and 9.957 AU, on average.[276] The visible features on Saturn rotate at different rates depending on latitude and multiple rotation periods have been assigned to various regions (as in Jupiter's case).

Astronomers use three different systems for specifying the rotation rate of Saturn. *System I* has a period of 10 hr 14 min 00 sec (844.3°/d) and encompasses the Equatorial Zone, the South Equatorial Belt and the North Equatorial Belt. The polar regions are considered to have rotation rates similar to *System I*. All other Saturnian latitudes, excluding the north and south polar regions, are indicated as *System II* and have been assigned a rotation period of 10 hr 38 min 25.4 sec (810.76°/d). *System III* refers to Saturn's internal rotation rate. Based on radio emissions from the planet in the period of the Voyager flybys, it has

Figure 186: *Saturn and rings as viewed by the Cassini spacecraft (28 October 2016)*

been assigned a rotation period of 10 hr 39 min 22.4 sec (810.8°/d). Because it is close to System II, it has largely superseded it.

A precise value for the rotation period of the interior remains elusive. While approaching Saturn in 2004, *Cassini* found that the radio rotation period of Saturn had increased appreciably, to approximately 10 hr 45 min 45 sec (± 36 sec). The latest estimate of Saturn's rotation (as an indicated rotation rate for Saturn as a whole) based on a compilation of various measurements from the *Cassini*, *Voyager* and Pioneer probes was reported in September 2007 is 10 hr 32 min 35 sec.

In March 2007, it was found that the variation of radio emissions from the planet did not match Saturn's rotation rate. This variance may be caused by geyser activity on Saturn's moon Enceladus. The water vapor emitted into Saturn's orbit by this activity becomes charged and creates a drag upon Saturn's magnetic field, slowing its rotation slightly relative to the rotation of the planet.

Figure 187: *A montage of Saturn and its principal moons (Dione, Tethys, Mimas, Enceladus, Rhea and Titan; Iapetus not shown). This famous image was created from photographs taken in November 1980 by the Voyager 1 spacecraft.*

Natural satellites

Saturn has 62 known moons, 53 of which have formal names. In addition, there is evidence of dozens to hundreds of moonlets with diameters of 40–500 meters in Saturn's rings, which are not considered to be true moons. Titan, the largest moon, comprises more than 90% of the mass in orbit around Saturn, including the rings. Saturn's second-largest moon, Rhea, may have a tenuous ring system of its own, along with a tenuous atmosphere.

Many of the other moons are small: 34 are less than 10 km in diameter and another 14 between 10 and 50 km in diameter. Traditionally, most of Saturn's moons have been named after Titans of Greek mythology. Titan is the only satellite in the Solar System with a major atmosphere, in which a complex organic chemistry occurs. It is the only satellite with hydrocarbon lakes.

On 6 June 2013, scientists at the IAA-CSIC reported the detection of polycyclic aromatic hydrocarbons in the upper atmosphere of Titan, a possible precursor for life. On 23 June 2014, NASA claimed to have strong evidence that nitrogen in the atmosphere of Titan came from materials in the Oort cloud, associated with comets, and not from the materials that formed Saturn in earlier times.

Figure 188: *Possible beginning of a new moon (white dot) of Saturn (image taken by Cassini on 15 April 2013)*

Saturn's moon Enceladus, which seems similar in chemical makeup to comets, has often been regarded as a potential habitat for microbial life. Evidence of this possibility includes the satellite's salt-rich particles having an "ocean-like" composition that indicates most of Enceladus's expelled ice comes from the evaporation of liquid salt water. A 2015 flyby by *Cassini* through a plume on Enceladus found most of the ingredients to sustain life forms that live by methanogenesis.

In April 2014, NASA scientists reported the possible beginning of a new moon within the A Ring, which was imaged by *Cassini* on 15 April 2013.

Planetary rings

The rings of Saturn (imaged here by *Cassini* in 2007) are the most massive and conspicuous in the Solar System.

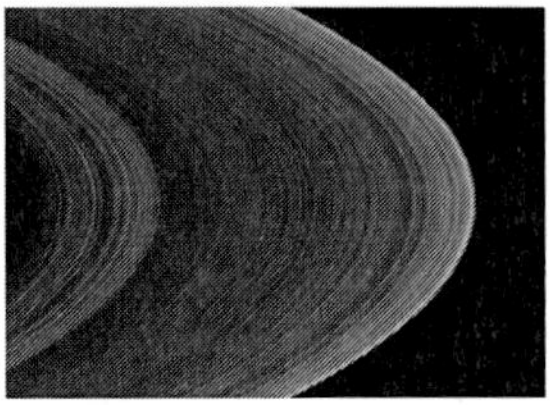

False-color UV image of Saturn's outer B and A rings; dirtier ringlets in the Cassini Division and Encke Gap show up red.

Saturn is probably best known for the system of planetary rings that makes it visually unique. The rings extend from 6,630 km to 120,700 km outward from Saturn's equator, average approximately 20 meters in thickness and are composed of 93% water ice with traces of tholin impurities and 7% amorphous carbon. The particles that make up the rings range in size from specks of dust up to 10 m. While the other gas giants also have ring systems, Saturn's is the largest and most visible.

There are two main hypotheses regarding the origin of the rings. One hypothesis is that the rings are remnants of a destroyed moon of Saturn. The second hypothesis is that the rings are left over from the original nebular material from which Saturn formed. Some ice in the E ring comes from the moon Enceladus's geysers.

In the past, astronomers once thought the rings formed alongside the planet when it formed billions of years ago. Instead, the age of these planetary rings is probably some hundreds of millions of years.

Beyond the main rings at a distance of 12 million km from the planet is the sparse Phoebe ring, which is tilted at an angle of 27° to the other rings and, like Phoebe, orbits in retrograde fashion.

Some of the moons of Saturn, including Pandora and Prometheus, act as shepherd moons to confine the rings and prevent them from spreading out. Pan and Atlas cause weak, linear density waves in Saturn's rings that have yielded more reliable calculations of their masses.

History of observation and exploration

There have been three main phases in the observation and exploration of Saturn. The first era was ancient observations (such as with the naked eye), before the invention of the modern telescopes. Starting in the 17th century progressively more advanced telescopic observations from Earth have been made. The other type is visitation by spacecraft, either by orbiting or flyby. In the 21st century observations continue from the Earth (or Earth-orbiting observatories) and from the Cassini orbiter at Saturn.

Ancient observations

Saturn has been known since prehistoric times. In ancient times, it was the most distant of the known planets in the Solar System and thus a major character in various mythologies. Babylonian astronomers systematically observed and recorded the movements of Saturn. In ancient Roman mythology, the god Saturnus, from which the planet takes its name, was the god of agriculture. The Romans considered Saturnus the equivalent of the Greek god Cronus. The Greeks had made the outermost planet sacred to Cronus, and the Romans followed suit. (In modern Greek, the planet retains its ancient name *Cronus*—Κρόνος: *Kronos*.)[277]

The Greek scientist Ptolemy based his calculations of Saturn's orbit on observations he made while it was in opposition. In Hindu astrology, there are nine astrological objects, known as Navagrahas. Saturn is known as "Shani" and judges everyone based on the good and bad deeds performed in life. Ancient Chinese and Japanese culture designated the planet Saturn as the "earth star" (土星). This was based on Five Elements which were traditionally used to classify natural elements.

In ancient Hebrew, Saturn is called 'Shabbathai'. Its angel is Cassiel. Its intelligence or beneficial spirit is Agiel (layga) and its spirit (darker aspect) is Zazel (lzaz). In Ottoman Turkish, Urdu and Malay, its name is 'Zuhal', derived from Arabic زحل.

European observations (17th–19th centuries)

Saturn's rings require at least a 15-mm-diameter telescope to resolve and thus were not known to exist until Galileo first saw them in 1610. He thought of them as two moons on Saturn's sides. It was not until Christiaan Huygens used greater telescopic magnification that this notion was refuted. Huygens discovered Saturn's moon Titan; Giovanni Domenico Cassini later discovered four other moons: Iapetus, Rhea, Tethys and Dione. In 1675, Cassini discovered the gap now known as the Cassini Division.

No further discoveries of significance were made until 1789 when William Herschel discovered two further moons, Mimas and Enceladus. The irregularly shaped satellite Hyperion, which has a resonance with Titan, was discovered in 1848 by a British team.

In 1899 William Henry Pickering discovered Phoebe, a highly irregular satellite that does not rotate synchronously with Saturn as the larger moons do. Phoebe was the first such satellite found and it takes more than a year to orbit Saturn in a retrograde orbit. During the early 20th century, research on Titan led to the confirmation in 1944 that it had a thick atmosphere – a feature unique among the Solar System's moons.

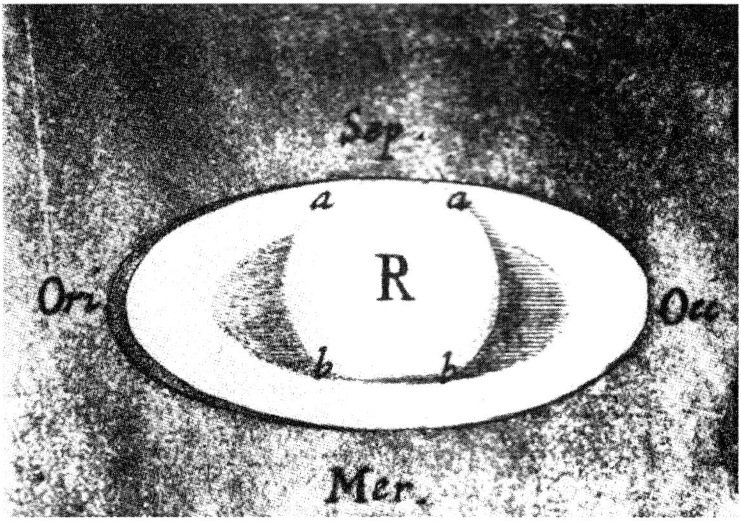

Figure 189: *Robert Hooke noted the shadows (a and b) cast by both the globe and the rings on each other in this drawing of Saturn in 1666.*

Modern NASA and ESA probes

Pioneer 11 flyby

Pioneer 11 made the first flyby of Saturn in September 1979, when it passed within 20,000 km of the planet's cloud tops. Images were taken of the planet and a few of its moons, although their resolution was too low to discern surface detail. The spacecraft also studied Saturn's rings, revealing the thin F-ring and the fact that dark gaps in the rings are bright when viewed at high phase angle (towards the Sun), meaning that they contain fine light-scattering material. In addition, Pioneer 11 measured the temperature of Titan.

Voyager flybys

In November 1980, the *Voyager 1* probe visited the Saturn system. It sent back the first high-resolution images of the planet, its rings and satellites. Surface features of various moons were seen for the first time. *Voyager 1* performed a close flyby of Titan, increasing knowledge of the atmosphere of the moon. It proved that Titan's atmosphere is impenetrable in visible wavelengths; therefore no surface details were seen. The flyby changed the spacecraft's trajectory out from the plane of the Solar System.

Almost a year later, in August 1981, *Voyager 2* continued the study of the Saturn system. More close-up images of Saturn's moons were acquired, as

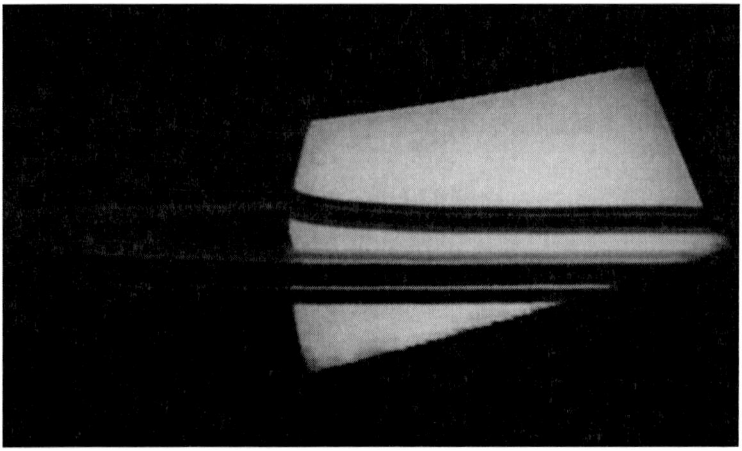

Figure 190: *Pioneer 11 image of Saturn*

well as evidence of changes in the atmosphere and the rings. Unfortunately, during the flyby, the probe's turnable camera platform stuck for a couple of days and some planned imaging was lost. Saturn's gravity was used to direct the spacecraft's trajectory towards Uranus.

The probes discovered and confirmed several new satellites orbiting near or within the planet's rings, as well as the small Maxwell Gap (a gap within the C Ring) and Keeler gap (a 42 km wide gap in the A Ring).

Cassini–Huygens spacecraft

On 1 July 2004, the *Cassini–Huygens* space probe performed the SOI (Saturn Orbit Insertion) maneuver and entered orbit around Saturn. Before the SOI, *Cassini* had already studied the system extensively. In June 2004, it had conducted a close flyby of Phoebe, sending back high-resolution images and data. *Cassini*'s flyby of Saturn's largest moon, Titan, has captured radar images of large lakes and their coastlines with numerous islands and mountains. The orbiter completed two Titan flybys before releasing the *Huygens* probe on 25 December 2004. *Huygens* descended onto the surface of Titan on 14 January 2005, sending a flood of data during the atmospheric descent and after the landing. *Cassini* has since conducted multiple flybys of Titan and other icy satellites.

Starting in early 2005, scientists used *Cassini* to track lightning on Saturn. The power of the lightning is approximately 1,000 times that of lightning on Earth.

In 2006, NASA reported that *Cassini* had found evidence of liquid water reservoirs that erupt in geysers on Saturn's moon Enceladus. Images had shown jets

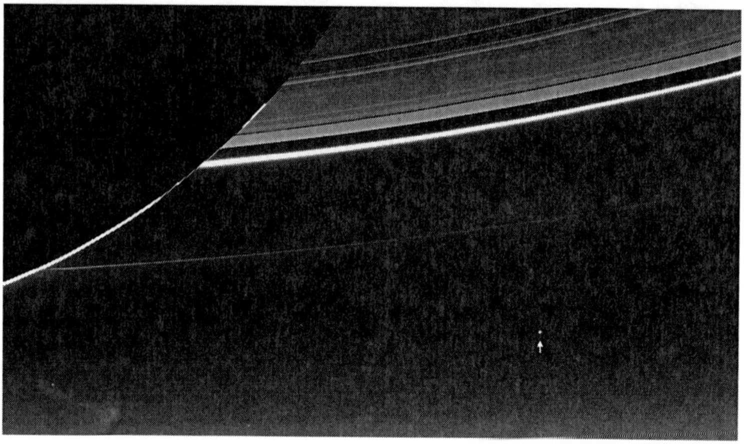

Figure 191: *Earth and Moon by Cassini (19 July 2013).*

Figure 192: *Earth and Moon viewed from the rings of Saturn (12 April 2017)*

Figure 193: *At Enceladus's south pole geysers spray water from many locations along the tiger stripes.*

of icy particles being emitted into orbit around Saturn from vents in the moon's south polar region. According to Andrew Ingersoll, California Institute of Technology, "Other moons in the Solar System have liquid-water oceans covered by kilometers of icy crust. What's different here is that pockets of liquid water may be no more than tens of meters below the surface." Over 100 geysers have been identified on Enceladus. In May 2011, NASA scientists at an Enceladus Focus Group Conference reported that Enceladus "is emerging as the most habitable spot beyond Earth in the Solar System for life as we know it".

Cassini photographs have led to other significant discoveries. They have revealed a previously undiscovered planetary ring, outside the brighter main rings of Saturn and inside the G and E rings. The source of this ring is hypothesized to be the crashing of a meteoroid off Janus and Epimetheus. In July 2006, *Cassini* images provided evidence of hydrocarbon lakes near Titan's north pole, the presence of which were confirmed in January 2007. In March 2007, additional images near Titan's north pole revealed hydrocarbon seas, the largest of which is almost the size of the Caspian Sea. In October 2006, the probe detected an 8,000 km diameter cyclone-like storm with an eyewall at Saturn's south pole.

From 2004 to 2 November 2009, the probe discovered and confirmed eight new satellites. Its primary mission ended in 2008 when the spacecraft had completed 74 orbits around the planet. The probe's mission was extended to

Figure 194: *Amateur telescopic view of Saturn*

September 2010 and then extended again to 2017, to study a full period of Saturn's seasons.

In April 2013 *Cassini* sent back images of a hurricane at the planet's north pole 20 times larger than those found on Earth, with winds faster than 530 km/h.

On 19 July 2013, "The Day the Earth Smiled", *Cassini* was pointed towards Earth to capture an image of the Earth and the Moon (and, as well, Venus and Mars) as part of a natural light, multi-image portrait of the entire Saturn system. It was the first time NASA informed the people of Earth that a long-distance photo was being taken in advance.

On 15 September 2017, the *Cassini-Huygens* spacecraft performed the "Grand Finale" of its mission: a number of passes through gaps between Saturn and Saturn's inner rings. The atmospheric entry of *Cassini* ended the mission.

Possible future missions

The continued exploration of Saturn is still considered to be a viable option for NASA as part of their ongoing New Frontiers program of missions. NASA previously requested for plans to be put forward for a mission to Saturn that included an atmospheric entry probe and possible investigations into the habitability and possible discovery of life on Saturn's moons Titan and Enceladus.

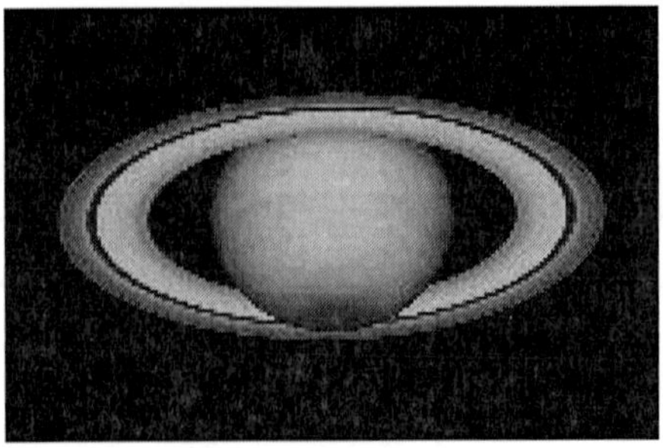

Figure 195: *Simulated appearance of Saturn as seen from Earth (at opposition) during an orbit of Saturn, 2001-2029*

Observation

Saturn is the most distant of the five planets easily visible to the naked eye from Earth, the other four being Mercury, Venus, Mars and Jupiter. (Uranus and occasionally 4 Vesta are visible to the naked eye in dark skies.) Saturn appears to the naked eye in the night sky as a bright, yellowish point of light with an apparent magnitude of usually between +1 and 0. It takes approximately 29.5 years for the planet to complete an entire circuit of the ecliptic against the background constellations of the zodiac. Most people will require an optical aid (very large binoculars or a small telescope) that magnifies at least 30 times to achieve an image of Saturn's rings, in which clear resolution is present. Twice every Saturnian year (roughly every 15 Earth years), the rings briefly disappear from view, due to the way in which they are angled and because they are so thin. Such a "disappearance" will next occur in 2025, but Saturn will be too close to the Sun for any ring-crossing observation to be possible.

Saturn and its rings are best seen when the planet is at, or near, opposition, the configuration of a planet when it is at an elongation of 180°, and thus appears opposite the Sun in the sky. A Saturnian opposition occurs every year—approximately every 378 days—and results in the planet appearing at its brightest. Both the Earth and Saturn orbit the Sun on eccentric orbits, which means their distances from the Sun vary over time, and therefore so do their distances from each other, hence varying the brightness of Saturn from one opposition to the next. Saturn also appears brighter when the rings are angled such that they are more visible. For example, during the opposition of 17 December 2002, Saturn appeared at its brightest due to a favorable orientation of its rings relative to the Earth, even though Saturn was closer to the Earth and Sun in late 2003.

From time to time Saturn is occulted by the Moon (that is, the Moon covers up Saturn in the sky). As with all the planets in the Solar System, occultations of

Figure 196: *Saturn eclipses the Sun, as seen from Cassini. The rings are visible, including the F Ring.*

Saturn occur in "seasons". Saturnian occultations will take place 12 or more times over a 12-month period, followed by about a five-year period in which no such activity is registered. Australian astronomy experts Hill and Horner explain the seasonal nature of Saturnian occultations:

> *This is the result of the fact that the moon's orbit around the Earth is tilted to the orbit of the Earth around the Sun – and so most of the time, the moon will pass above or below Saturn in the sky, and no occultation will occur. It is only when Saturn lies near the point that the moon's orbit crosses the "plane of the ecliptic" that occultations can happen – and then they occur every time the moon swings by, until Saturn moves away from the crossing point.*

Further reading

- Alexander, Arthur Francis O'Donel (1962). *The Planet Saturn - A History of Observation, Theory and Discovery*.
- Gore, Rick (July 1981). "Voyager 1 at Saturn: Riddles of the Rings". *National Geographic*. Vol. 160 no. 1. pp. 3–31. ISSN 0027-9358[278]. OCLC 643483454[279].
- Lovett, L.; et al. (2006). *Saturn: A New View*. New York: Harry N. Abrams, Inc. ISBN 978-0-8109-3090-2.
- Karttunen, H.; et al. (2007). *Fundamental Astronomy* (5th ed.). New York: Springer. ISBN 978-3-540-34143-7.

- Seidelmann, P. Kenneth; et al. (2007). "Report of the IAU/IAG Working Group on cartographic coordinates and rotational elements: 2006". *Celestial Mechanics and Dynamical Astronomy*. **98** (3): 155–180. Bibcode: 2007CeMDA..98..155S[280]. doi: 10.1007/s10569-007-9072-y[281].
- de Pater, Imke; Lissauer, Jack J. (2015). *Planetary Sciences*[282] (2nd updated ed.). New York: Cambridge University Press. p. 250. ISBN 978-0521853712.

External links

<indicator name="spoken-icon"> </indicator>

- Saturn profile[283] at NASA's Solar System Exploration site
- Saturn Fact Sheet[284], by NASA
- Gazetteer of Planetary Nomenclature – Saturn (USGS)[285]
- *Cassini–Huygens* mission[286] to Saturn, by NASA
- General information[287] about Saturn
- Astronomy Cast: Saturn[288]
- Saturn in Daytime (12 inch telescope)[289]
- Saturn 'Rev 175' Raw Preview[290]
- Haese, Paul. "Capturing Saturn"[291]. *Sixty Symbols*. Brady Haran for the University of Nottingham.
- BBC *In Our Time* radio program[292]

<indicator name="featured-star"> </indicator>

Uranus

<indicator name="pp-default"> </indicator>

Uranus

Uranus as a featureless disc, photographed by *Voyager 2* in 1986

Discovery	
Discovered by	William Herschel
Discovery date	March 13, 1781
Designations	
Pronunciation	/ˈjʊərənəs/ Media:uranus.ogg or /jʊˈreɪnəs/ Media:en-us-Uranus.ogg
Adjectives	Uranian
Orbital characteristics[293]	
Epoch J2000	
Aphelion	20.11 AU (3,008 Gm)
Perihelion	18.33 AU (2,742 Gm)
Semi-major axis	19.2184 AU (2,875.04 Gm)
Eccentricity	0.046381
Orbital period	• 84.0205 yr • 30,688.5 d • 42,718 Uranian solar days
Synodic period	369.66 days
Average orbital speed	6.80 km/s

Mean anomaly	142.238600°
Inclination	0.773° to ecliptic 6.48° to Sun's equator 1.02° to invariable plane
Longitude of ascend-ing node	74.006°
Argument of perihelion	96.998857°
Known satellites	27
Physical characteristics	
Mean radius	25,362±7 km
Equatorial radius	25,559±4 km 4.007 Earths
Polar radius	24,973±20 km 3.929 Earths
Flattening	0.0229±0.0008[294]
Circum-ference	159,354.1 km
Surface area	8.1156×10^9 km^2 15.91 Earths
Volume	6.833×10^{13} km^3 63.086 Earths
Mass	$(8.6810 \pm 0.0013) \times 10^{25}$ kg 14.536 Earths GM=5,793,939±13 km^3/s^2
Mean density	1.27 g/cm^3
Surface gravity	8.69 m/s^2 0.886 g
Moment of inertia factor	0.23 (estimate)
Escape velocity	21.3 km/s
Sidereal rotation period	–0.71833 d (retrograde) 17 h 14 min 24 s
Equatorial ro-tation velocity	2.59 km/s 9,320 km/h
Axial tilt	97.77° (to orbit)
North pole right ascension	$17^h\ 9^m\ 15^s$ 257.311°
North pole declination	–15.175°

<table>
<tr><td>Albedo</td><td>0.300 (Bond)
0.51 (geom.)</td></tr>
<tr><td></td><td><table>
<tr><th>Surface temp.</th><th>min</th><th>mean</th><th>max</th></tr>
<tr><td>1 bar level</td><td></td><td>76 K (−197.2 °C)</td><td></td></tr>
<tr><td>0.1 bar (tropopause)</td><td>47 K</td><td>53 K</td><td>57 K</td></tr>
</table></td></tr>
<tr><td>Apparent magnitude</td><td>5.9 to 5.32</td></tr>
<tr><td>Angular diameter</td><td>3.3” to 4.1”</td></tr>
<tr><td colspan="2">Atmosphere[295]</td></tr>
<tr><td>Scale height</td><td>27.7 km</td></tr>
<tr><td>Composition by volume</td><td>(Below 1.3 bar)
Gases:
• 83 ± 3% hydrogen (H_2)
• 15 ± 3% helium (He)
• 2.3% methane (CH_4)
• 0.009% (0.007–0.015%) hydrogen deuteride (HD)
Ices:
• ammonia (NH_3)
• water (H_2O)
• ammonium hydrosulfide (NH_4SH)
• methane hydrate</td></tr>
<tr><td colspan="2"></td></tr>
</table>

Uranus is the seventh planet from the Sun. It has the third-largest planetary radius and fourth-largest planetary mass in the Solar System. Uranus is similar in composition to Neptune, and both have different bulk chemical composition from that of the larger gas giants Jupiter and Saturn. For this reason, scientists often classify Uranus and Neptune as "ice giants" to distinguish them from the gas giants. Uranus's atmosphere is similar to Jupiter's and Saturn's in its primary composition of hydrogen and helium, but it contains more "ices" such as water, ammonia, and methane, along with traces of other hydrocarbons. It is the coldest planetary atmosphere in the Solar System, with a minimum temperature of 49 K (−224 °C, −371 °F), and has a complex, layered cloud structure with water thought to make up the lowest clouds and methane the uppermost layer of clouds. The interior of Uranus is mainly composed of ices and rock.

Uranus is the only planet whose name is derived directly from a figure from Greek mythology, from the Latinised version of the Greek god of the sky Ouranos. Like the other giant planets, Uranus has a ring system, a magnetosphere, and numerous moons. The Uranian system has a unique configuration among those of the planets because its axis of rotation is tilted sideways, nearly into the plane of its solar orbit. Its north and south poles, therefore,

Figure 197: *William Herschel, discoverer of Uranus*

lie where most other planets have their equators. In 1986, images from *Voyager 2* showed Uranus as an almost featureless planet in visible light, without the cloud bands or storms associated with the other giant planets. Observations from Earth have shown seasonal change and increased weather activity as Uranus approached its equinox in 2007. Wind speeds can reach 250 metres per second (900 km/h; 560 mph).

History

Like the classical planets, Uranus is visible to the naked eye, but it was never recognised as a planet by ancient observers because of its dimness and slow orbit. Sir William Herschel announced its discovery on 13 March 1781, expanding the known boundaries of the Solar System for the first time in history and making Uranus the first planet discovered with a telescope.

Discovery

Uranus had been observed on many occasions before its recognition as a planet, but it was generally mistaken for a star. Possibly the earliest known observation was by Hipparchos, who in 128 BCE might have recorded it as a star for his star catalogue that was later incorporated into Ptolemy's Almagest.

Figure 198: *Replica of the telescope used by Herschel to discover Uranus*

The earliest definite sighting was in 1690, when John Flamsteed observed it at least six times, cataloguing it as 34 Tauri. The French astronomer Pierre Charles Le Monnier observed Uranus at least twelve times between 1750 and 1769, including on four consecutive nights.

Sir William Herschel observed Uranus on 13 March 1781 from the garden of his house at 19 New King Street in Bath, Somerset, England (now the Herschel Museum of Astronomy), and initially reported it (on 26 April 1781) as a comet. Herschel "engaged in a series of observations on the parallax of the fixed stars",[296] using a telescope of his own design.

Herschel recorded in his journal. "In the quartile near ζ Tauri ... either [a] Nebulous star or perhaps a comet."[297] On 17 March he noted: "I looked for the Comet or Nebulous Star and found that it is a Comet, for it has changed its place."[298] When he presented his discovery to the Royal Society, he continued to assert that he had found a comet, but also implicitly compared it to a planet:

> *The power I had on when I first saw the comet was 227. From experience I know that the diameters of the fixed stars are not proportionally magnified with higher powers, as planets are; therefore I now put the powers at 460 and 932, and found that the diameter of the comet increased in proportion to the power, as it ought to be, on the supposition of its not being a fixed star, while the diameters of the stars to which I compared it were not*

increased in the same ratio. Moreover, the comet being magnified much beyond what its light would admit of, appeared hazy and ill-defined with these great powers, while the stars preserved that lustre and distinctness which from many thousand observations I knew they would retain. The sequel has shown that my surmises were well-founded, this proving to be the Comet we have lately observed.

Herschel notified the Astronomer Royal, Nevil Maskelyne, of his discovery and received this flummoxed reply from him on 23 April 1781: "I don't know what to call it. It is as likely to be a regular planet moving in an orbit nearly circular to the sun as a Comet moving in a very eccentric ellipsis. I have not yet seen any coma or tail to it."[299]

Although Herschel continued to describe his new object as a comet, other astronomers had already begun to suspect otherwise. Finnish-Swedish astronomer Anders Johan Lexell, working in Russia, was the first to compute the orbit of the new object. Its nearly circular orbit led him to a conclusion that it was a planet rather than a comet. Berlin astronomer Johann Elert Bode described Herschel's discovery as "a moving star that can be deemed a hitherto unknown planet-like object circulating beyond the orbit of Saturn".[300] Bode concluded that its near-circular orbit was more like a planet than a comet.[301]

The object was soon universally accepted as a new planet. By 1783, Herschel acknowledged this to Royal Society president Joseph Banks: "By the observation of the most eminent Astronomers in Europe it appears that the new star, which I had the honour of pointing out to them in March 1781, is a Primary Planet of our Solar System." In recognition of his achievement, King George III gave Herschel an annual stipend of £200 on condition that he move to Windsor so that the Royal Family could look through his telescopes.

Name

The name of Uranus references the ancient Greek deity of the sky Uranus (Ancient Greek: Οὐρανός), the father of Cronus (Saturn) and grandfather of Zeus (Jupiter), which in Latin became "Ūranus" (Latin pronunciation: [ˈuranʊs]). It is the only planet whose name is derived directly from a figure of Greek mythology. The adjectival form of Uranus is "Uranian". The pronunciation of the name *Uranus* preferred among astronomers is /ˈjʊərənəs/, with stress on the first syllable as in Latin *Ūranus,* in contrast to /jʊˈreɪnəs/, with stress on the second syllable and a long *a*, though both are considered acceptable.[302] </ ref>

Consensus on the name was not reached until almost 70 years after the planet's discovery. During the original discussions following discovery, Maskelyne asked Herschel to "do the astronomical world the faver [*sic*] to give a name to your planet, which is entirely your own, [and] which we are so much obliged to you for the discovery of".[303] In response to Maskelyne's request, Herschel decided to name the object *Georgium Sidus* (George's Star), or the "Georgian Planet" in honour of his new patron, King George III. He explained this decision in a letter to Joseph Banks:

> *In the fabulous ages of ancient times the appellations of Mercury, Venus, Mars, Jupiter and Saturn were given to the Planets, as being the names of their principal heroes and divinities. In the present more philosophical era it would hardly be allowable to have recourse to the same method and call it Juno, Pallas, Apollo or Minerva, for a name to our new heavenly body. The first consideration of any particular event, or remarkable incident, seems to be its chronology: if in any future age it should be asked, when this last-found Planet was discovered? It would be a very satisfactory answer to say, 'In the reign of King George the Third'.*

Herschel's proposed name was not popular outside Britain, and alternatives were soon proposed. Astronomer Jérôme Lalande proposed that it be named *Herschel* in honour of its discoverer. Swedish astronomer Erik Prosperin proposed the name *Neptune*, which was supported by other astronomers who liked the idea to commemorate the victories of the British Royal Naval fleet in the course of the American Revolutionary War by calling the new planet even *Neptune George III* or *Neptune Great Britain*.

In a March 1782 treatise, Bode proposed *Uranus*, the Latinised version of the Greek god of the sky, Ouranos.[304] Bode argued that the name should follow the mythology so as not to stand out as different from the other planets, and that Uranus was an appropriate name as the father of the first generation of the Titans. He also noted that elegance of the name in that just as Saturn was the father of Jupiter, the new planet should be named after the father of Saturn. In 1789, Bode's Royal Academy colleague Martin Klaproth named his newly discovered element uranium in support of Bode's choice. Ultimately, Bode's suggestion became the most widely used, and became universal in 1850 when HM Nautical Almanac Office, the final holdout, switched from using *Georgium Sidus* to *Uranus*.

Uranus has two astronomical symbols. The first to be proposed, ♅, was suggested by Lalande in 1784. In a letter to Herschel, Lalande described it as "un globe surmonté par la première lettre de votre nom" ("a globe surmounted by the first letter of your surname"). A later proposal, □, is a hybrid of the symbols for Mars and the Sun because Uranus was the Sky in Greek mythology, which was thought to be dominated by the combined powers of the Sun and Mars.

Uranus is called by a variety of translations in other languages. In Chinese, Japanese, Korean, and Vietnamese, its name is literally translated as the "sky

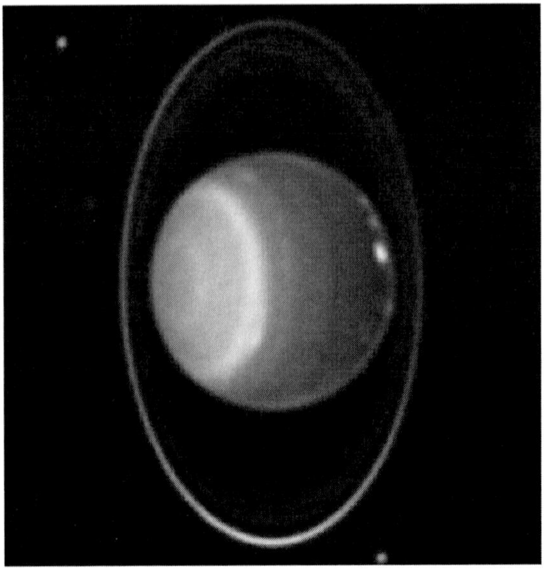

Figure 199: *A 1998 false-colour near-infrared image of Uranus showing cloud bands, rings, and moons obtained by the Hubble Space Telescope's NICMOS camera.*

king star" (天王星). In Thai, its official name is *Dao Yurenat* (ดาวยูเรนัส), as in English. Its other name in Thai is *Dao Maritayu* (ดาวมฤตยู, Star of Mṛtyu), after the Sanskrit word for "death", Mrtyu (मृत्यु). In Mongolian, its name is *Tengeriin Van* (Тэнгэрийн ван), translated as "King of the Sky", reflecting its namesake god's role as the ruler of the heavens. In Hawaiian, its name is *Hele'ekala*. In Māori, its name is *Whērangi*. It is also named in Māori as *Rangipō*. In Nahuatl, Uranus is known as "Ilhuicateōcītlalli", named after the word for "sky", "ilhuicatl". It is also named in Nahuatl as "Xiuhtēuccītlalli", after the god Xiuhtecuhtli.

Orbit and rotation

Uranus orbits the Sun once every 84 years. Its average distance from the Sun is roughly 20 AU (3 billion km; 2 billion mi). The difference between its minimum and maximum distance from the Sun is 1.8 AU, larger than that of any other planet, though not as large as that of dwarf planet Pluto.[305] The intensity of sunlight varies inversely with the square of distance, and so on Uranus (at about 20 times the distance from the Sun compared to Earth) it is about 1/400 the intensity of light on Earth. Its orbital elements were first calculated in 1783 by Pierre-Simon Laplace. With time, discrepancies began to appear between the predicted and observed orbits, and in 1841, John Couch Adams first proposed that the differences might be due to the gravitational tug of an unseen planet. In 1845, Urbain Le Verrier began his own independent research into Uranus's orbit. On 23 September 1846, Johann Gottfried Galle located a new planet, later named Neptune, at nearly the position predicted by Le Verrier.

The rotational period of the interior of Uranus is 17 hours, 14 minutes. As on all the giant planets, its upper atmosphere experiences strong winds in the direction of rotation. At some latitudes, such as about 60 degrees south, visible features of the atmosphere move much faster, making a full rotation in as little as 14 hours.

Axial tilt

The Uranian axis of rotation is approximately parallel with the plane of the Solar System, with an axial tilt of 97.77° (as defined by prograde rotation). This gives it seasonal changes completely unlike those of the other planets. Near the solstice, one pole faces the Sun continuously and the other faces away. Only a narrow strip around the equator experiences a rapid day–night cycle, but with the Sun low over the horizon. At the other side of Uranus's orbit the orientation of the poles towards the Sun is reversed. Each pole gets around 42 years of continuous sunlight, followed by 42 years of darkness. Near the time of the equinoxes, the Sun faces the equator of Uranus giving a period of day–night cycles similar to those seen on most of the other planets. In contrast to the other planets, whose motions around the Sun resemble that of spinning tops, Uranus's motion can be visualised as that of a ball rolling on the ecliptic plane near solstices and of a spinning rifle bullet near equinoxes.

Uranus reached its most recent equinox on 7 December 2007.

Northern hemisphere	Year	Southern hemisphere
Winter solstice	1902, 1986	Summer solstice
Vernal equinox	1923, 2007	Autumnal equinox
Summer solstice	1944, 2028	Winter solstice
Autumnal equinox	1965, 2049	Vernal equinox

One result of this axis orientation is that, averaged over the Uranian year, the polar regions of Uranus receive a greater energy input from the Sun than its equatorial regions. Nevertheless, Uranus is hotter at its equator than at its poles. The underlying mechanism that causes this is unknown. The reason for Uranus's unusual axial tilt is also not known with certainty, but the usual speculation is that during the formation of the Solar System, an Earth-sized protoplanet collided with Uranus, causing the skewed orientation. Uranus's south pole was pointed almost directly at the Sun at the time of Voyager 2's flyby in 1986. The labelling of this pole as "south" uses the definition currently endorsed by the International Astronomical Union, namely that the north pole of a planet or satellite is the pole that points above the invariable plane of the Solar System, regardless of the direction the planet is spinning. A different convention is sometimes used, in which a body's north and south poles are defined according to the right-hand rule in relation to the direction of rotation.

Visibility

From 1995 to 2006, Uranus's apparent magnitude fluctuated between +5.6 and +5.9, placing it just within the limit of naked eye visibility at +6.5. Its angular diameter is between 3.4 and 3.7 arcseconds, compared with 16 to 20 arcseconds for Saturn and 32 to 45 arcseconds for Jupiter. At opposition, Uranus is visible to the naked eye in dark skies, and becomes an easy target even in urban conditions with binoculars. In larger amateur telescopes with an objective diameter of between 15 and 23 cm, Uranus appears as a pale cyan disk with distinct limb darkening. With a large telescope of 25 cm or wider, cloud patterns, as well as some of the larger satellites, such as Titania and Oberon, may be visible.

Physical characteristics

Internal structure

Uranus's mass is roughly 14.5 times that of Earth, making it the least massive of the giant planets. Its diameter is slightly larger than Neptune's at roughly four times that of Earth. A resulting density of 1.27 g/cm^3 makes Uranus the second least dense planet, after Saturn. This value indicates that it is made primarily of various ices, such as water, ammonia, and methane. The total mass of ice in Uranus's interior is not precisely known, because different figures emerge depending on the model chosen; it must be between 9.3 and 13.5 Earth masses. Hydrogen and helium constitute only a small part of the total, with between 0.5 and 1.5 Earth masses. The remainder of the non-ice mass (0.5 to 3.7 Earth masses) is accounted for by rocky material.

The standard model of Uranus's structure is that it consists of three layers: a rocky (silicate/iron–nickel) core in the centre, an icy mantle in the middle and an outer gaseous hydrogen/helium envelope. The core is relatively small, with a mass of only 0.55 Earth masses and a radius less than 20% of Uranus's;

Figure 200: *Size comparison of Earth and Uranus*

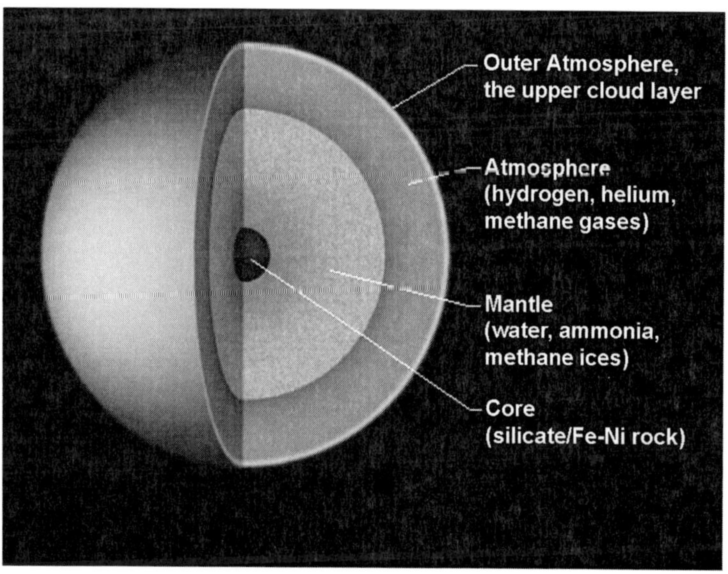

Figure 201: *Diagram of the interior of Uranus*

the mantle comprises its bulk, with around 13.4 Earth masses, and the upper atmosphere is relatively insubstantial, weighing about 0.5 Earth masses and extending for the last 20% of Uranus's radius. Uranus's core density is around 9 g/cm^3, with a pressure in the centre of 8 million bars (800 GPa) and a temperature of about 5000 K. The ice mantle is not in fact composed of ice in the conventional sense, but of a hot and dense fluid consisting of water, ammonia and other volatiles. This fluid, which has a high electrical conductivity, is sometimes called a water–ammonia ocean.

The extreme pressure and temperature deep within Uranus may break up the methane molecules, with the carbon atoms condensing into crystals of diamond that rain down through the mantle like hailstones. Very-high-pressure experiments at the Lawrence Livermore National Laboratory suggest that the base of the mantle may comprise an ocean of liquid diamond, with floating solid 'diamond-bergs'.

The bulk compositions of Uranus and Neptune are different from those of Jupiter and Saturn, with ice dominating over gases, hence justifying their separate classification as ice giants. There may be a layer of ionic water where the water molecules break down into a soup of hydrogen and oxygen ions, and deeper down superionic water in which the oxygen crystallises but the hydrogen ions move freely within the oxygen lattice.[306]

Although the model considered above is reasonably standard, it is not unique; other models also satisfy observations. For instance, if substantial amounts of hydrogen and rocky material are mixed in the ice mantle, the total mass of ices in the interior will be lower, and, correspondingly, the total mass of rocks and hydrogen will be higher. Presently available data does not allow a scientific determination which model is correct. The fluid interior structure of Uranus means that it has no solid surface. The gaseous atmosphere gradually transitions into the internal liquid layers. For the sake of convenience, a revolving oblate spheroid set at the point at which atmospheric pressure equals 1 bar (100 kPa) is conditionally designated as a "surface". It has equatorial and polar radii of 25,559 ± 4 km (15,881.6 ± 2.5 mi) and 24,973 ± 20 km (15,518 ± 12 mi), respectively. This surface is used throughout this article as a zero point for altitudes.

Internal heat

Uranus's internal heat appears markedly lower than that of the other giant planets; in astronomical terms, it has a low thermal flux. Why Uranus's internal temperature is so low is still not understood. Neptune, which is Uranus's near twin in size and composition, radiates 2.61 times as much energy into space as it receives from the Sun, but Uranus radiates hardly any excess heat at all. The total power radiated by Uranus in the far infrared (i.e. heat) part of the spectrum is 1.06±0.08 times the solar energy absorbed in its atmosphere. Uranus's heat flux is only 0.042±0.047 W/m^2, which is lower than the internal heat flux of Earth of about 0.075 W/m^2. The lowest temperature recorded in Uranus's tropopause is 49 K (–224.2 °C; –371.5 °F), making Uranus the coldest planet in the Solar System.

One of the hypotheses for this discrepancy suggests that when Uranus was hit by a supermassive impactor, which caused it to expel most of its primordial heat, it was left with a depleted core temperature. Another hypothesis is that some form of barrier exists in Uranus's upper layers that prevents the core's heat from reaching the surface. For example, convection may take place in a set of compositionally different layers, which may inhibit the upward heat transport; perhaps double diffusive convection is a limiting factor.

Atmosphere

Although there is no well-defined solid surface within Uranus's interior, the outermost part of Uranus's gaseous envelope that is accessible to remote sensing is called its atmosphere. Remote-sensing capability extends down to roughly 300 km below the 1 bar (100 kPa) level, with a corresponding pressure around 100 bar (10 MPa) and temperature of 320 K (47 °C; 116 °F). The tenuous thermosphere extends over two planetary radii from the nominal surface, which is defined to lie at a pressure of 1 bar. The Uranian atmosphere can be divided into three layers: the troposphere, between altitudes of –300 and 50 km (–186 and 31 mi) and pressures from 100 to 0.1 bar (10 MPa to 10 kPa); the stratosphere, spanning altitudes between 50 and 4,000 km (31 and 2,485 mi) and pressures of between 0.1 and 10^{-10} bar (10 kPa to 10 μPa); and the thermosphere extending from 4,000 km to as high as 50,000 km from the surface. There is no mesosphere.

Composition

The composition of Uranus's atmosphere is different from its bulk, consisting mainly of molecular hydrogen and helium. The helium molar fraction, i.e. the number of helium atoms per molecule of gas, is 0.15±0.03 in the upper troposphere, which corresponds to a mass fraction 0.26±0.05. This value is close to the protosolar helium mass fraction of 0.275±0.01, indicating that helium has not settled in its centre as it has in the gas giants. The third-most-abundant component of Uranus's atmosphere is methane (CH_4). Methane has prominent absorption bands in the visible and near-infrared (IR), making Uranus aquamarine or cyan in colour. Methane molecules account for 2.3% of the atmosphere by molar fraction below the methane cloud deck at the pressure level of 1.3 bar (130 kPa); this represents about 20 to

30 times the carbon abundance found in the Sun. The mixing ratio[307] is much lower in the upper atmosphere due to its extremely low temperature, which lowers the saturation level and causes excess methane to freeze out. The abundances of less volatile compounds such as ammonia, water, and hydrogen sulfide in the deep atmosphere are poorly known. They are probably also higher than solar values. Along with methane, trace amounts of various hydrocarbons are found in the stratosphere of Uranus, which are thought to be produced from methane by photolysis induced by the solar ultraviolet (UV) radiation. They include ethane (C_2H_6), acetylene (C_2H_2), methylacetylene (CH_3C_2H), and diacetylene (C_2HC_2H). Spectroscopy has also uncovered traces of water vapour, carbon monoxide and carbon dioxide in the upper atmosphere, which can only originate from an external source such as infalling dust and comets.

Troposphere

The troposphere is the lowest and densest part of the atmosphere and is characterised by a decrease in temperature with altitude. The temperature falls from about 320 K (47 °C; 116 °F) at the base of the nominal troposphere at –300 km to 53 K (–220 °C; –364 °F) at 50 km. The temperatures in the coldest upper region of the troposphere (the tropopause) actually vary in the range between 49 and 57 K (–224 and –216 °C; –371 and –357 °F) depending on planetary latitude. The tropopause region is responsible for the vast majority of Uranus's thermal far infrared emissions, thus determining its effective temperature of 59.1 ± 0.3 K (–214.1 ± 0.3 °C; –353.3 ± 0.5 °F).

The troposphere is thought to have a highly complex cloud structure; water clouds are hypothesised to lie in the pressure range of 50 to 100 bar (5 to 10 MPa), ammonium hydrosulfide clouds in the range of 20 to 40 bar (2 to 4 MPa), ammonia or hydrogen sulfide clouds at between 3 and 10 bar (0.3 and 1 MPa) and finally directly detected thin methane clouds at 1 to 2 bar (0.1 to 0.2 MPa). The troposphere is a dynamic part of the atmosphere, exhibiting strong winds, bright clouds and seasonal changes.

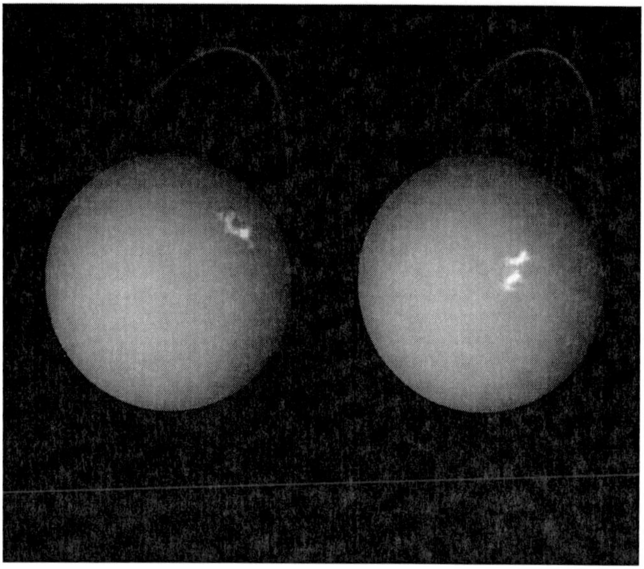

Figure 202: *Aurorae on Uranus taken by the Space Telescope Imaging Spectrograph (STIS) installed on Hubble.*

Upper atmosphere

The middle layer of the Uranian atmosphere is the stratosphere, where temperature generally increases with altitude from 53 K (–220 °C; –364 °F) in the tropopause to between 800 and 850 K (527 and 577 °C; 980 and 1,070 °F) at the base of the thermosphere. The heating of the stratosphere is caused by absorption of solar UV and IR radiation by methane and other hydrocarbons, which form in this part of the atmosphere as a result of methane photolysis. Heat is also conducted from the hot thermosphere. The hydrocarbons occupy a relatively narrow layer at altitudes of between 100 and 300 km corresponding to a pressure range of 10 to 0.1 mbar (10.00 to 0.10 hPa) and temperatures of between 75 and 170 K (–198 and –103 °C; –325 and –154 °F). The most abundant hydrocarbons are methane, acetylene and ethane with mixing ratios of around 10^{-7} relative to hydrogen. The mixing ratio of carbon monoxide is similar at these altitudes. Heavier hydrocarbons and carbon dioxide have mixing ratios three orders of magnitude lower. The abundance ratio of water is around 7×10^{-9}. Ethane and acetylene tend to condense in the colder lower part of stratosphere and tropopause (below 10 mBar level) forming haze layers, which may be partly responsible for the bland appearance of Uranus. The concentration of hydrocarbons in the Uranian stratosphere above the haze is significantly lower than in the stratospheres of the other giant planets.

The outermost layer of the Uranian atmosphere is the thermosphere and corona, which has a uniform temperature around 800 to 850 K. The heat sources necessary to sustain such a high level are not understood, as neither the solar UV nor the auroral activity can provide the necessary energy to maintain these temperatures. The weak cooling efficiency due to the lack of hydrocarbons in the stratosphere above 0.1 mBar pressure level may contribute too. In addition to molecular hydrogen, the thermosphere-corona contains many free hydrogen atoms. Their small mass and high temperatures explain why the corona extends as far as 50,000 km (31,000 mi), or two Uranian radii, from its surface. This extended corona is a unique feature of Uranus. Its effects include a drag on small particles orbiting Uranus, causing a general depletion of dust in the Uranian rings. The Uranian thermosphere, together with the upper part of the stratosphere, corresponds to the ionosphere of Uranus. Observations show that the ionosphere occupies altitudes from 2,000 to 10,000 km (1,200 to 6,200 mi). The Uranian ionosphere is denser than that of either Saturn or Neptune, which may arise from the low concentration of hydrocarbons in the stratosphere. The ionosphere is mainly sustained by solar UV radiation and its density depends on the solar activity. Auroral activity is insignificant as compared to Jupiter and Saturn.

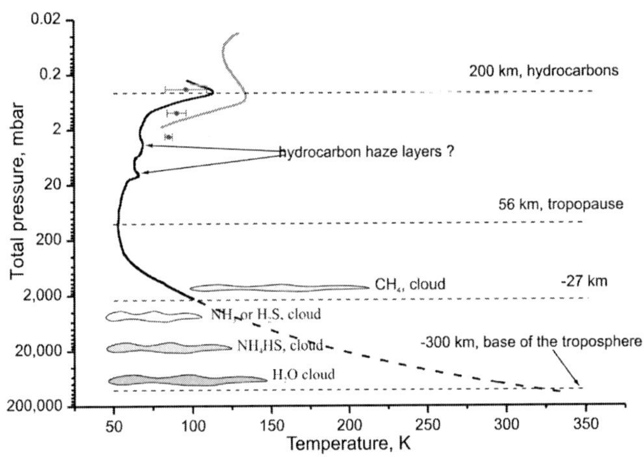

Figure 203: *Temperature profile of the Uranian troposphere and lower stratosphere. Cloud and haze layers are also indicated.*

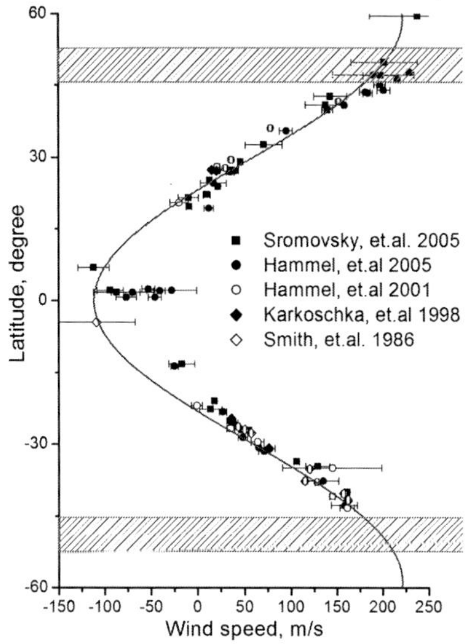

Figure 204: *Zonal wind speeds on Uranus. Shaded areas show the southern collar and its future northern counterpart. The red curve is a symmetrical fit to the data.*

Magnetosphere

Before the arrival of *Voyager 2*, no measurements of the Uranian magnetosphere had been taken, so its nature remained a mystery. Before 1986, scientists had expected the magnetic field of Uranus to be in line with the solar wind, because it would then align with Uranus's poles that lie in the ecliptic.

Voyager's observations revealed that Uranus's magnetic field is peculiar, both because it does not originate from its geometric centre, and because it is tilted at 59° from the axis of rotation. In fact the magnetic dipole is shifted from the Uranus's centre towards the south rotational pole by as much as one third of the planetary radius. This unusual geometry results in a highly asymmetric magnetosphere, where the magnetic field strength on the surface in the southern hemisphere can be as low as 0.1 gauss (10 μT), whereas in the northern hemisphere it can be as high as 1.1 gauss (110 μT). The average field at the surface is 0.23 gauss (23 μT). Studies of *Voyager 2* data in 2017 suggest that this asymmetry causes Uranus's magnetosphere to connect with the solar wind once a Uranian day, opening the planet to the Sun's particles. In comparison, the magnetic field of Earth is roughly as strong at either pole, and its "magnetic equator" is roughly parallel with its geographical equator. The dipole moment of Uranus is 50 times that of Earth. Neptune has a similarly displaced and

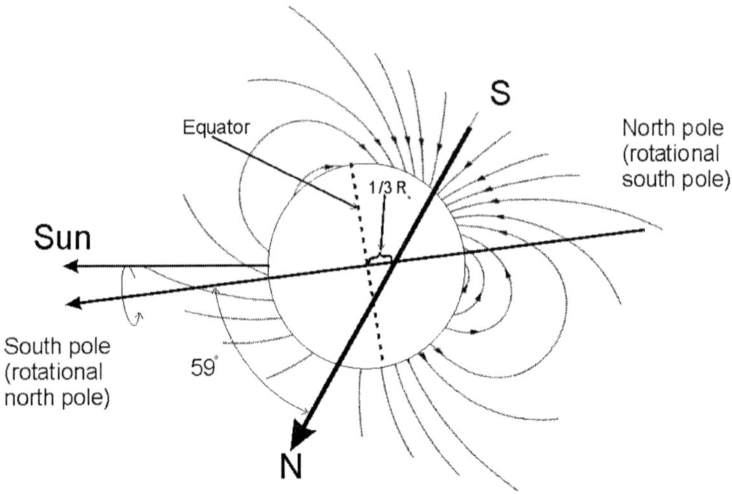

Figure 205: *The magnetic field of Uranus as observed by Voyager 2 in 1986. S and N are magnetic south and north poles.*

tilted magnetic field, suggesting that this may be a common feature of ice giants. One hypothesis is that, unlike the magnetic fields of the terrestrial and gas giants, which are generated within their cores, the ice giants' magnetic fields are generated by motion at relatively shallow depths, for instance, in the water–ammonia ocean. Another possible explanation for the magnetosphere's alignment is that there are oceans of liquid diamond in Uranus's interior that would deter the magnetic field.

Despite its curious alignment, in other respects the Uranian magnetosphere is like those of other planets: it has a bow shock at about 23 Uranian radii ahead of it, a magnetopause at 18 Uranian radii, a fully developed magnetotail, and radiation belts. Overall, the structure of Uranus's magnetosphere is different from Jupiter's and more similar to Saturn's. Uranus's magnetotail trails behind it into space for millions of kilometres and is twisted by its sideways rotation into a long corkscrew.

Uranus's magnetosphere contains charged particles: mainly protons and electrons, with a small amount of H_2^+ ions. No heavier ions have been detected. Many of these particles probably derive from the thermosphere. The ion and electron energies can be as high as 4 and 1.2 megaelectronvolts, respectively. The density of low-energy (below 1 kiloelectronvolt) ions in the inner magnetosphere is about 2 cm^{-3}. The particle population is strongly affected by the

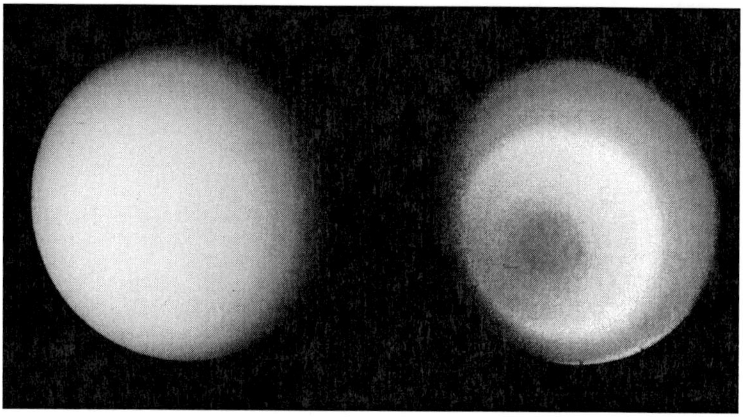

Figure 206: *Uranus's southern hemisphere in approximate natural colour (left) and in shorter wavelengths (right), showing its faint cloud bands and atmospheric "hood" as seen by Voyager 2*

Uranian moons, which sweep through the magnetosphere, leaving noticeable gaps. The particle flux is high enough to cause darkening or space weathering of their surfaces on an astronomically rapid timescale of 100,000 years. This may be the cause of the uniformly dark colouration of the Uranian satellites and rings. Uranus has relatively well developed aurorae, which are seen as bright arcs around both magnetic poles. Unlike Jupiter's, Uranus's aurorae seem to be insignificant for the energy balance of the planetary thermosphere.

Climate

At ultraviolet and visible wavelengths, Uranus's atmosphere is bland in comparison to the other giant planets, even to Neptune, which it otherwise closely resembles. When *Voyager 2* flew by Uranus in 1986, it observed a total of ten cloud features across the entire planet. One proposed explanation for this dearth of features is that Uranus's internal heat appears markedly lower than that of the other giant planets. The lowest temperature recorded in Uranus's tropopause is 49 K (–224 °C; –371 °F), making Uranus the coldest planet in the Solar System.

Figure 207: *The first dark spot observed on Uranus. Image obtained by the HST ACS in 2006.*

Banded structure, winds and clouds

In 1986, *Voyager 2* found that the visible southern hemisphere of Uranus can be subdivided into two regions: a bright polar cap and dark equatorial bands. Their boundary is located at about –45° of latitude. A narrow band straddling the latitudinal range from –45 to –50° is the brightest large feature on its visible surface. It is called a southern "collar". The cap and collar are thought to be a dense region of methane clouds located within the pressure range of 1.3 to 2 bar (see above). Besides the large-scale banded structure, Voyager 2 observed ten small bright clouds, most lying several degrees to the north from the collar. In all other respects Uranus looked like a dynamically dead planet in 1986. Voyager 2 arrived during the height of Uranus's southern summer and could not observe the northern hemisphere. At the beginning of the 21st century, when the northern polar region came into view, the Hubble Space Telescope (HST) and Keck telescope initially observed neither a collar nor a polar cap in the northern hemisphere. So Uranus appeared to be asymmetric: bright near the south pole and uniformly dark in the region north of the southern collar. In 2007, when Uranus passed its equinox, the southern collar almost disappeared, and a faint northern collar emerged near 45° of latitude.

In the 1990s, the number of the observed bright cloud features grew considerably partly because new high-resolution imaging techniques became available. Most were found in the northern hemisphere as it started to become visible. An early explanation—that bright clouds are easier to identify in its dark part, whereas in the southern hemisphere the bright collar masks them – was shown to be incorrect. Nevertheless there are differences between the clouds of each hemisphere. The northern clouds are smaller, sharper and brighter. They appear to lie at a higher altitude. The lifetime of clouds spans several orders of magnitude. Some small clouds live for hours; at least one southern cloud may have persisted since the *Voyager 2* flyby. Recent observation also discovered that cloud features on Uranus have a lot in common with those on Neptune. For example, the dark spots common on Neptune had never been observed on Uranus before 2006, when the first such feature dubbed Uranus Dark Spot was imaged. The speculation is that Uranus is becoming more Neptune-like during its equinoctial season.

The tracking of numerous cloud features allowed determination of zonal winds blowing in the upper troposphere of Uranus. At the equator winds are retrograde, which means that they blow in the reverse direction to the planetary rotation. Their speeds are from –360 to –180 km/h (–220 to –110 mph). Wind speeds increase with the distance from the equator, reaching zero values near ±20° latitude, where the troposphere's temperature minimum is located. Closer to the poles, the winds shift to a prograde direction, flowing with Uranus's rotation. Wind speeds continue to increase reaching maxima at ±60° latitude before falling to zero at the poles. Wind speeds at –40° latitude range from 540 to 720 km/h (340 to 450 mph). Because the collar obscures all clouds below that parallel, speeds between it and the southern pole are impossible to measure. In contrast, in the northern hemisphere maximum speeds as high as 860 km/h (540 mph) are observed near +50° latitude.

Seasonal variation

For a short period from March to May 2004, large clouds appeared in the Uranian atmosphere, giving it a Neptune-like appearance. Observations included record-breaking wind speeds of 820 km/h (510 mph) and a persistent thunderstorm referred to as "Fourth of July fireworks". On 23 August 2006, researchers at the Space Science Institute (Boulder, Colorado) and the University of Wisconsin observed a dark spot on Uranus's surface, giving scientists more insight into Uranus's atmospheric activity. Why this sudden upsurge in activity occurred is not fully known, but it appears that Uranus's extreme axial tilt results in extreme seasonal variations in its weather. Determining the nature of this seasonal variation is difficult because good data on Uranus's atmosphere have existed for less than 84 years, or one full Uranian year. Photometry over

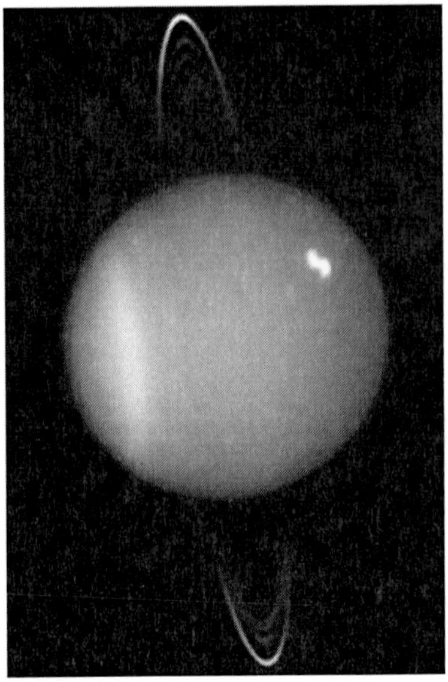

Figure 208: *Uranus in 2005. Rings, southern collar and a bright cloud in the northern hemisphere are visible (HST ACS image).*

the course of half a Uranian year (beginning in the 1950s) has shown regular variation in the brightness in two spectral bands, with maxima occurring at the solstices and minima occurring at the equinoxes. A similar periodic variation, with maxima at the solstices, has been noted in microwave measurements of the deep troposphere begun in the 1960s. Stratospheric temperature measurements beginning in the 1970s also showed maximum values near the 1986 solstice. The majority of this variability is thought to occur owing to changes in the viewing geometry.

There are some indications that physical seasonal changes are happening in Uranus. Although Uranus is known to have a bright south polar region, the north pole is fairly dim, which is incompatible with the model of the seasonal change outlined above. During its previous northern solstice in 1944, Uranus displayed elevated levels of brightness, which suggests that the north pole was not always so dim. This information implies that the visible pole brightens some time before the solstice and darkens after the equinox. Detailed analysis of the visible and microwave data revealed that the periodical changes of brightness are not completely symmetrical around the solstices, which also indicates a change in the meridional albedo patterns. In the 1990s, as Uranus moved away from its solstice, Hubble and ground-based telescopes revealed that the south polar cap darkened noticeably (except the southern collar, which remained bright), whereas the northern hemisphere demonstrated increasing activity, such as cloud formations and stronger winds, bolstering expectations that it should brighten soon. This indeed happened in 2007 when it passed an equinox: a faint northern polar collar arose, and the southern collar became nearly invisible, although the zonal wind profile remained slightly asymmetric, with northern winds being somewhat slower than southern.

The mechanism of these physical changes is still not clear. Near the summer and winter solstices, Uranus's hemispheres lie alternately either in full glare of the Sun's rays or facing deep space. The brightening of the sunlit hemisphere is thought to result from the local thickening of the methane clouds and haze layers located in the troposphere. The bright collar at –45° latitude is also connected with methane clouds. Other changes in the southern polar region can be explained by changes in the lower cloud layers. The variation of the microwave emission from Uranus is probably caused by changes in the deep tropospheric circulation, because thick polar clouds and haze may inhibit convection. Now that the spring and autumn equinoxes are arriving on Uranus, the dynamics are changing and convection can occur again.

Formation

Many argue that the differences between the ice giants and the gas giants extend to their formation. The Solar System is hypothesised to have formed from a giant rotating ball of gas and dust known as the presolar nebula. Much of the nebula's gas, primarily hydrogen and helium, formed the Sun, and the dust grains collected together to form the first protoplanets. As the planets grew, some of them eventually accreted enough matter for their gravity to hold on to the nebula's leftover gas. The more gas they held onto, the larger they became; the larger they became, the more gas they held onto until a critical point was reached, and their size began to increase exponentially. The ice giants, with only a few Earth masses of nebular gas, never reached that critical point. Recent simulations of planetary migration have suggested that both ice giants formed closer to the Sun than their present positions, and moved outwards after formation (the Nice model).

Moons

Uranus has 27 known natural satellites. The names of these satellites are chosen from characters in the works of Shakespeare and Alexander Pope. The five main satellites are Miranda, Ariel, Umbriel, Titania, and Oberon. The Uranian satellite system is the least massive among those of the giant planets; the combined mass of the five major satellites would be less than half that

Figure 209: *Major moons of Uranus in order of increasing distance (left to right), at their proper relative sizes and albedos (collage of Voyager 2 photographs)*

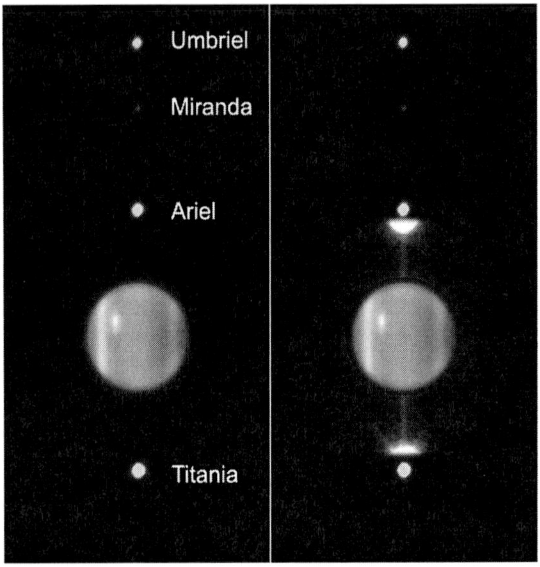

Figure 210: *The Uranus System (NACO/VLT image)*

of Triton (largest moon of Neptune) alone. The largest of Uranus's satellites, Titania, has a radius of only 788.9 km (490.2 mi), or less than half that of the Moon, but slightly more than Rhea, the second-largest satellite of Saturn, making Titania the eighth-largest moon in the Solar System. Uranus's satellites have relatively low albedos; ranging from 0.20 for Umbriel to 0.35 for Ariel (in green light). They are ice–rock conglomerates composed of roughly 50% ice and 50% rock. The ice may include ammonia and carbon dioxide.

Among the Uranian satellites, Ariel appears to have the youngest surface with the fewest impact craters and Umbriel's the oldest. Miranda has fault canyons

20 km (12 mi) deep, terraced layers, and a chaotic variation in surface ages and features. Miranda's past geologic activity is thought to have been driven by tidal heating at a time when its orbit was more eccentric than currently, probably as a result of a former 3:1 orbital resonance with Umbriel. Extensional processes associated with upwelling diapirs are the likely origin of Miranda's 'racetrack'-like coronae. Ariel is thought to have once been held in a 4:1 resonance with Titania.

Uranus has at least one horseshoe orbiter occupying the Sun–Uranus L_3 Lagrangian point—a gravitationally unstable region at 180° in its orbit, 83982 Crantor. Crantor moves inside Uranus's co-orbital region on a complex, temporary horseshoe orbit. 2010 EU65 is also a promising Uranus horseshoe librator candidate.

Planetary rings

The Uranian rings are composed of extremely dark particles, which vary in size from micrometres to a fraction of a metre. Thirteen distinct rings are presently known, the brightest being the ε ring. All except two rings of Uranus are extremely narrow – they are usually a few kilometres wide. The rings are probably quite young; the dynamics considerations indicate that they did not form with Uranus. The matter in the rings may once have been part of a moon (or moons) that was shattered by high-speed impacts. From numerous pieces of debris that formed as a result of those impacts, only a few particles survived, in stable zones corresponding to the locations of the present rings.

William Herschel described a possible ring around Uranus in 1789. This sighting is generally considered doubtful, because the rings are quite faint, and in the two following centuries none were noted by other observers. Still, Herschel made an accurate description of the epsilon ring's size, its angle relative to Earth, its red colour, and its apparent changes as Uranus travelled around the Sun. The ring system was definitively discovered on 10 March 1977 by James L. Elliot, Edward W. Dunham, and Jessica Mink using the Kuiper Airborne Observatory. The discovery was serendipitous; they planned to use the occultation of the star SAO 158687 (also known as HD 128598) by Uranus to study its atmosphere. When their observations were analysed, they found that the star had disappeared briefly from view five times both before and after it disappeared behind Uranus. They concluded that there must be a ring system around Uranus. Later they detected four additional rings. The rings were directly imaged when Voyager 2 passed Uranus in 1986. Voyager 2 also discovered two additional faint rings, bringing the total number to eleven.

In December 2005, the Hubble Space Telescope detected a pair of previously unknown rings. The largest is located twice as far from Uranus as the previously known rings. These new rings are so far from Uranus that they are called the "outer" ring system. Hubble also spotted two small satellites, one of which, Mab, shares its orbit with the outermost newly discovered ring. The new rings bring the total number of Uranian rings to 13. In April 2006, images of the new rings from the Keck Observatory yielded the colours of the outer rings: the outermost is blue and the other one red. One hypothesis concerning the outer ring's blue colour is that it is composed of minute particles of water ice from the surface of Mab that are small enough to scatter blue light. In contrast, Uranus's inner rings appear grey.

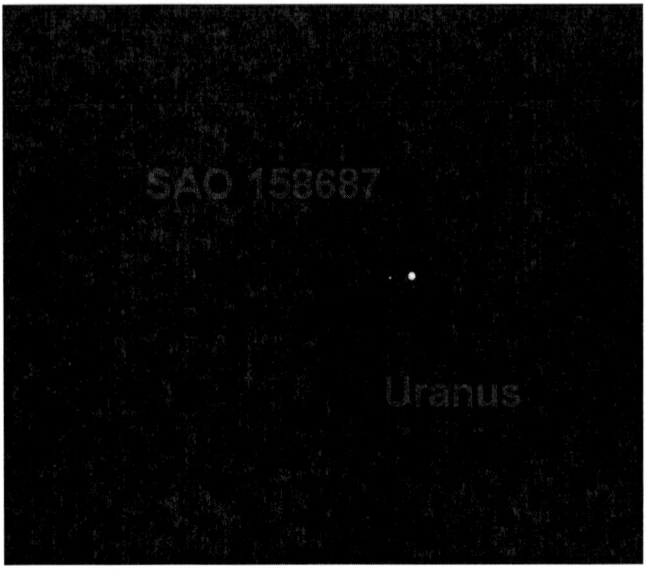

Figure 211: *Animation about the discovering occultation in 1977. (Click on it to start)*

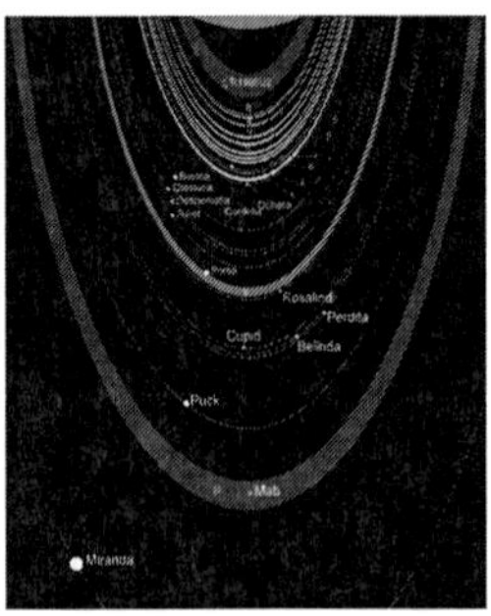

Figure 212: *Uranus has a complicated planetary ring system, which was the second such system to be discovered in the Solar System after Saturn's. UNIQ-ref-1-3951097dbf03c794-QINU*

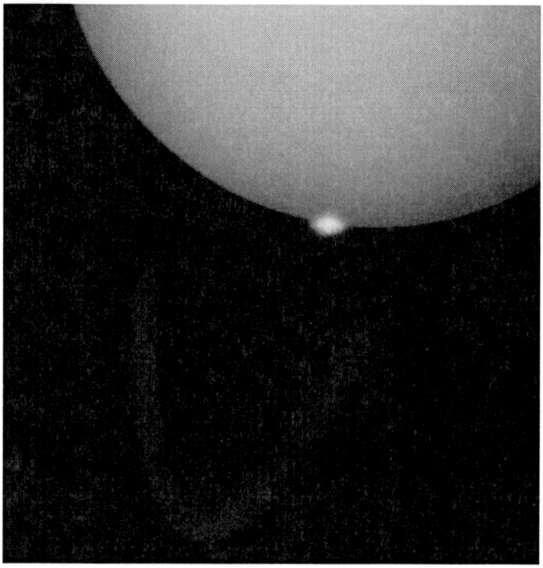

Figure 213: *Uranus's aurorae against its equatorial rings, imaged by the Hubble telescope. Unlike the aurorae of Earth and Jupiter, those of Uranus are not in line with its poles, due to its lopsided magnetic field.*

Exploration

In 1986, NASA's *Voyager 2* interplanetary probe encountered Uranus. This flyby remains the only investigation of Uranus carried out from a short distance and no other visits are planned. Launched in 1977, *Voyager 2* made its closest approach to Uranus on 24 January 1986, coming within 81,500 km (50,600 mi) of the cloudtops, before continuing its journey to Neptune. The spacecraft studied the structure and chemical composition of Uranus's atmosphere, including its unique weather, caused by its axial tilt of 97.77°. It made the first detailed investigations of its five largest moons and discovered 10 new ones. It examined all nine of the system's known rings and discovered two more. It also studied the magnetic field, its irregular structure, its tilt and its unique corkscrew magnetotail caused by Uranus's sideways orientation.

Voyager 1 was unable to visit Uranus because investigation of Saturn's moon Titan was considered a priority. This trajectory took *Voyager 1* out the plane of the ecliptic, ending its planetary science mission.[:118]

The possibility of sending the *Cassini* spacecraft from Saturn to Uranus was evaluated during a mission extension planning phase in 2009, but was ultimately rejected in favour of destroying it in the Saturnian atmosphere. It would

Figure 214: *Crescent Uranus as imaged by Voyager 2 while en route to Neptune*

have taken about twenty years to get to the Uranian system after departing Saturn. A Uranus orbiter and probe was recommended by the 2013–2022 Planetary Science Decadal Survey published in 2011; the proposal envisages launch during 2020–2023 and a 13-year cruise to Uranus. A Uranus entry probe could use Pioneer Venus Multiprobe heritage and descend to 1–5 atmospheres. The ESA evaluated a "medium-class" mission called Uranus Pathfinder.[308] A New Frontiers Uranus Orbiter has been evaluated and recommended in the study, *The Case for a Uranus Orbiter*. Such a mission is aided by the ease with which a relatively big mass can be sent to the system—over 1500 kg with an Atlas 521 and 12-year journey. For more concepts see Proposed Uranus missions.

In culture

In astrology, the planet Uranus (♅) is the ruling planet of Aquarius. Because Uranus is cyan and Uranus is associated with electricity, the colour electric blue, which is close to cyan, is associated with the sign Aquarius (see Uranus in astrology).

The chemical element uranium, discovered in 1789 by the German chemist Martin Heinrich Klaproth, was named after the newly discovered planet Uranus.

"Uranus, the Magician" is a movement in Gustav Holst's orchestral suite *The Planets*, written between 1914 and 1916.

Operation Uranus was the successful military operation in World War II by the Soviet army to take back Stalingrad and marked the turning point in the land war against the Wehrmacht.

The lines "Then felt I like some watcher of the skies/When a new planet swims into his ken", from John Keats's "On First Looking Into Chapman's Homer", are a reference to Herschel's discovery of Uranus.

Many references to Uranus in popular culture and news involve humor about one pronunciation of its name resembling that of the phrase "your anus".

Further reading

- Alexander, Arthur Francis O'Donel (1965). *The Planet Uranus - A History of Observation, Theory and Discovery.*
- Miner, Ellis D. (1998). *Uranus: The Planet, Rings and Satellites.* New York: John Wiley and Sons. ISBN 978-0-471-97398-0.
- Bode, Johann Elert (1784). *Von dem neu entdeckten Planeten*[309]. Verfasser.
- Gore, Rick (August 1986). "Uranus — Voyager Visits a Dark Planet". *National Geographic*. Vol. 170 no. 2. pp. 178–194. ISSN 0027-9358[310]. OCLC 643483454[311].

External links

Wikiquote has quotations related to: ***Uranus***

- Uranus[312] at European Space Agency
- NASA's Uranus fact sheet[313]
- Uranus Profile[314] at NASA's Solar System Exploration site[315]
- Planets – Uranus[316] A kid's guide to Uranus.
- Uranus[317] at Jet Propulsion Laboratory's planetary photojournal. (photos)
- Voyager at Uranus[318] (photos)
- Uranus (Astronomy Cast homepage)[319] (blog)
- Uranian system montage[320] (photo)
- Gray, Meghan; Merrifield, Michael (2010). "Uranus"[321]. *Sixty Symbols*. Brady Haran for the University of Nottingham.
- "How to Pronounce Uranus"[322] by CGP Grey

<indicator name="featured-star"> </indicator>

Neptune

<indicator name="pp-default"> </indicator>

Neptune

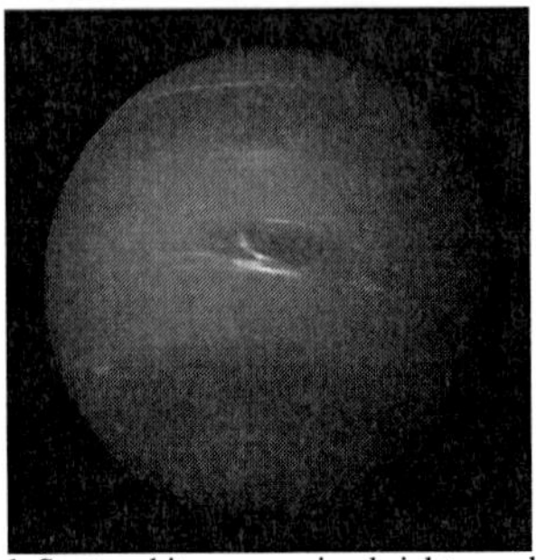

Neptune's Great Dark Spot and its companion bright smudge; on the west limb the fast moving bright feature called Scooter and the little dark spot are visible.

Discovery	
Discovered by	• Johann Galle • Urbain Le Verrier
Discovery date	23 September 1846
Designations	
Pronunciation	/ˈnɛptuːn/ Media:en-us-Neptune.ogg /ˈnɛptjuːn/
Adjectives	Neptunian
Orbital characteristics[323,324]	
Epoch J2000	
Aphelion	30.33 AU (4.54 billion km)
Perihelion	29.81 AU (4.46 billion km)
Semi-major axis	30.110387 AU (4.50 billion km)
Eccentricity	0.009456
Orbital period	• 164.8 yr • 60,182 days • 89,666 Neptunian solar days
Synodic period	367.49 days
Average orbital speed	5.43 km/s
Mean anomaly	256.228°

<table>
<tr><td>Inclination</td><td>1.767975° to ecliptic
6.43° to Sun’s equator
0.72° to invariable plane[325]</td></tr>
<tr><td>Longitude of ascending node</td><td>131.784°</td></tr>
<tr><td>Argument of perihelion</td><td>276.336°</td></tr>
<tr><td>Known satellites</td><td>14</td></tr>
<tr><td colspan="2">Physical characteristics</td></tr>
<tr><td>Mean radius</td><td>24,622±19 km</td></tr>
<tr><td>Equatorial radius</td><td>24,764±15 km[326]
3.883 Earths</td></tr>
<tr><td>Polar radius</td><td>24,341±30 km
3.829 Earths</td></tr>
<tr><td>Flattening</td><td>0.0171±0.0013</td></tr>
<tr><td>Surface area</td><td>7.6183×10^9 km^2
14.98 Earths</td></tr>
<tr><td>Volume</td><td>6.254×10^{13} km^3
57.74 Earths</td></tr>
<tr><td>Mass</td><td>1.0243×10^{26} kg
17.147 Earths
5.15×10^{-5} Suns</td></tr>
<tr><td>Mean density</td><td>1.638 g/cm^3[327]</td></tr>
<tr><td>Surface gravity</td><td>11.15 m/s^2
1.14 g</td></tr>
<tr><td>Moment of inertia factor</td><td>0.23 (estimate)</td></tr>
<tr><td>Escape velocity</td><td>23.5 km/s</td></tr>
<tr><td>Sidereal rotation period</td><td>0.6713 day
16 h 6 min 36 s</td></tr>
<tr><td>Equatorial rotation velocity</td><td>2.68 km/s (9,650 km/h)</td></tr>
<tr><td>Axial tilt</td><td>28.32° (to orbit)</td></tr>
<tr><td>North pole right ascension</td><td>$19^h\,57^m\,20^s$
299.3°</td></tr>
<tr><td>North pole declination</td><td>42.950°</td></tr>
<tr><td>Albedo</td><td>0.290 (bond)
0.41 (geom.)</td></tr>
<tr><td colspan="2">
<table>
<tr><th>Surface temp.</th><th>min</th><th>mean</th><th>max</th></tr>
<tr><td>1 bar level</td><td></td><td>72 K (–201 °C)</td><td></td></tr>
<tr><td>0.1 bar (10 kPa)</td><td></td><td>55 K (–218 °C)</td><td></td></tr>
</table>
</td></tr>
<tr><td>Apparent magnitude</td><td>8.02 to 7.78</td></tr>
</table>

Angular diameter	2.2–2.4”
Atmosphere	
Scale height	19.7±0.6 km
Composition by volume	• **Gases**: • 80%±3.2% hydrogen (H_2) • 19%±3.2% helium (He) • 1.5%±0.5% methane (CH_4) • ∼0.019% hydrogen deuteride (HD) • ∼0.00015% ethane (C_2H_6) • **Ices**: • ammonia (NH_3) • water (H_2O) • ammonium hydrosulfide (NH_4SH) • methane ice (?) ($CH_4 \cdot 5.75H_2O$)

Neptune is the eighth and farthest known planet from the Sun in the Solar System. In the Solar System, it is the fourth-largest planet by diameter, the third-most-massive planet, and the densest giant planet. Neptune is 17 times the mass of Earth and is slightly more massive than its near-twin Uranus, which is 15 times the mass of Earth and slightly larger than Neptune.[328] Neptune orbits the Sun once every 164.8 years at an average distance of 30.1 astronomical units (4.50×10^9 km). It is named after the Roman god of the sea and has the astronomical symbol ♆, a stylised version of the god Neptune's trident.

Neptune is not visible to the unaided eye and is the only planet in the Solar System found by mathematical prediction rather than by empirical observation. Unexpected changes in the orbit of Uranus led Alexis Bouvard to deduce that its orbit was subject to gravitational perturbation by an unknown planet. Neptune was subsequently observed with a telescope on 23 September 1846 by Johann Galle within a degree of the position predicted by Urbain Le Verrier. Its largest moon, Triton, was discovered shortly thereafter, though none of the planet's remaining known 13 moons were located telescopically until the 20th century. The planet's distance from Earth gives it a very small apparent size, making it challenging to study with Earth-based telescopes. Neptune was visited by *Voyager 2*, when it flew by the planet on 25 August 1989. The advent of the *Hubble Space Telescope* and large ground-based telescopes with adaptive optics has recently allowed for additional detailed observations from afar.

Like Jupiter and Saturn, Neptune's atmosphere is composed primarily of hydrogen and helium, along with traces of hydrocarbons and possibly nitrogen, but it contains a higher proportion of "ices" such as water, ammonia, and methane. However, its interior, like that of Uranus, is primarily composed of ices and rock, which is why Uranus and Neptune are normally considered "ice giants" to emphasise this distinction. Traces of methane in the outermost regions in part account for the planet's blue appearance.

In contrast to the hazy, relatively featureless atmosphere of Uranus, Neptune's atmosphere has active and visible weather patterns. For example, at the time of the *Voyager 2* flyby in 1989, the planet's southern hemisphere had a Great Dark Spot comparable to the Great Red Spot on Jupiter. These weather patterns are driven by the strongest sustained winds of any planet in the Solar System, with recorded wind speeds as high as 2,100 kilometres per hour (580 m/s; 1,300 mph). Because of its great distance from the Sun, Neptune's outer atmosphere is one of the coldest places in the Solar System, with temperatures at its cloud tops approaching 55 K (–218 °C). Temperatures at the planet's centre are approximately 5,400 K (5,100 °C). Neptune has a faint and fragmented ring system (labelled "arcs"), which was discovered in 1982, then later confirmed by *Voyager 2*.

History

Discovery

Some of the earliest recorded observations ever made through a telescope, Galileo's drawings on 28 December 1612 and 27 January 1613, contain plotted points that match up with what is now known to be the position of Neptune. On both occasions, Galileo seems to have mistaken Neptune for a fixed star when it appeared close—in conjunction—to Jupiter in the night sky; hence, he is not credited with Neptune's discovery. At his first observation in December 1612, Neptune was almost stationary in the sky because it had just turned retrograde that day. This apparent backward motion is created when Earth's orbit takes it past an outer planet. Because Neptune was only beginning its yearly retrograde cycle, the motion of the planet was far too slight to be detected with Galileo's small telescope. In July 2009, University of Melbourne physicist David Jamieson announced new evidence suggesting that Galileo was at least aware that the "star" he had observed had moved relative to the fixed stars.

In 1821, Alexis Bouvard published astronomical tables of the orbit of Neptune's neighbour Uranus. Subsequent observations revealed substantial deviations from the tables, leading Bouvard to hypothesise that an unknown body was perturbing the orbit through gravitational interaction. In 1843, John

Figure 215: *Urbain Le Verrier*

Couch Adams began work on the orbit of Uranus using the data he had. Via Cambridge Observatory director James Challis, he requested extra data from Sir George Airy, the Astronomer Royal, who supplied it in February 1844. Adams continued to work in 1845–46 and produced several different estimates of a new planet.

In 1845–46, Urbain Le Verrier, independently of Adams, developed his own calculations but aroused no enthusiasm in his compatriots. In June 1846, upon seeing Le Verrier's first published estimate of the planet's longitude and its similarity to Adams's estimate, Airy persuaded Challis to search for the planet. Challis vainly scoured the sky throughout August and September.

Meanwhile, Le Verrier by letter urged Berlin Observatory astronomer Johann Gottfried Galle to search with the observatory's refractor. Heinrich d'Arrest, a student at the observatory, suggested to Galle that they could compare a recently drawn chart of the sky in the region of Le Verrier's predicted location with the current sky to seek the displacement characteristic of a planet, as opposed to a fixed star. On the evening of 23 September 1846, the day Galle received the letter, he discovered Neptune within 1° of where Le Verrier had predicted it to be, about 12° from Adams' prediction. Challis later realised that he had observed the planet twice, on 4 and 12 August, but did not recognise it as a planet because he lacked an up-to-date star map and was distracted by his concurrent work on comet observations.

In the wake of the discovery, there was much nationalistic rivalry between the French and the British over who deserved credit for the discovery. Eventually, an international consensus emerged that both Le Verrier and Adams jointly deserved credit. Since 1966, Dennis Rawlins has questioned the credibility of Adams's claim to co-discovery, and the issue was re-evaluated by historians with the return in 1998 of the "Neptune papers" (historical documents) to the Royal Observatory, Greenwich. After reviewing the documents, they suggest that "Adams does not deserve equal credit with Le Verrier for the discovery of Neptune. That credit belongs only to the person who succeeded both in predicting the planet's place and in convincing astronomers to search for it."

Naming

Shortly after its discovery, Neptune was referred to simply as "the planet exterior to Uranus" or as "Le Verrier's planet". The first suggestion for a name came from Galle, who proposed the name *Janus*. In England, Challis put forward the name *Oceanus*.[329]

Claiming the right to name his discovery, Le Verrier quickly proposed the name *Neptune* for this new planet, though falsely stating that this had been officially approved by the French Bureau des Longitudes. In October, he sought to name the planet *Le Verrier*, after himself, and he had loyal support in this from the observatory director, François Arago. This suggestion met with stiff resistance outside France. French almanacs quickly reintroduced the name *Herschel* for Uranus, after that planet's discoverer Sir William Herschel, and *Leverrier* for the new planet.

Struve came out in favour of the name *Neptune* on 29 December 1846, to the Saint Petersburg Academy of Sciences. Soon, *Neptune* became the internationally accepted name. In Roman mythology, Neptune was the god of the sea, identified with the Greek Poseidon. The demand for a mythological name seemed to be in keeping with the nomenclature of the other planets, all of which, except for Earth, were named for deities in Greek and Roman mythology.

Most languages today, even in countries that have no direct link to Greco-Roman culture, use some variant of the name "Neptune" for the planet. However, in Chinese, Japanese, and Korean, the planet's name was translated as "sea king star" (海王星), because Neptune was the god of the sea. In Mongolian, Neptune is called *Dalain Van* (Далайн ван), reflecting its namesake god's role as the ruler of the sea. In modern Greek the planet is called *Poseidon* (Ποσειδώνας, *Poseidonas*), the Greek counterpart of Neptune.[330] In Hebrew, "Rahab" ,(רהב) from a Biblical sea monster mentioned in the Book of Psalms, was selected in a vote managed by the Academy of the Hebrew

Language in 2009 as the official name for the planet, even though the existing Latin term ”Neptun” (נפטון) is commonly used.[331,332] In Māori, the planet is called *Tangaroa*, named after the Māori god of the sea.[333] In Nahuatl, the planet is called *Tlāloccītlalli*, named after the rain god Tlāloc. In Thai, Neptune is referred both by its Westernised name *Dao Nepjun* (ดาวเนปจูน), and is also named *Dao Ketu* (ดาวเกตุ, ”Star of Ketu”), after the descending lunar node Ketu (केतु) who plays a role in Hindu astrology.

Status

From its discovery in 1846 until the subsequent discovery of Pluto in 1930, Neptune was the farthest known planet. When Pluto was discovered, it was considered a planet, and Neptune thus became the second-farthest known planet, except for a 20-year period between 1979 and 1999 when Pluto's elliptical orbit brought it closer than Neptune to the Sun. The discovery of the Kuiper belt in 1992 led many astronomers to debate whether Pluto should be considered a planet or as part of the Kuiper belt. In 2006, the International Astronomical Union defined the word ”planet” for the first time, reclassifying Pluto as a ”dwarf planet” and making Neptune once again the outermost known planet in the Solar System.

Physical characteristics

Neptune's mass of 1.0243×10^{26} kg is intermediate between Earth and the larger gas giants: it is 17 times that of Earth but just 1/19th that of Jupiter.[334] Its gravity at 1 bar is 11.15 m/s^2, 1.14 times the surface gravity of Earth, and surpassed only by Jupiter. Neptune's equatorial radius of 24,764 km is nearly four times that of Earth. Neptune, like Uranus, is an ice giant, a subclass of giant planet, due to their smaller size and higher concentrations of volatiles relative to Jupiter and Saturn. In the search for extrasolar planets, Neptune has been used as a metonym: discovered bodies of similar mass are often referred to as ”Neptunes”, just as scientists refer to various extrasolar bodies as ”Jupiters”.

Internal structure

Neptune's internal structure resembles that of Uranus. Its atmosphere forms about 5% to 10% of its mass and extends perhaps 10% to 20% of the way towards the core, where it reaches pressures of about 10 GPa, or about 100,000 times that of Earth's atmosphere. Increasing concentrations of methane, ammonia and water are found in the lower regions of the atmosphere.

The mantle is equivalent to 10 to 15 Earth masses and is rich in water, ammonia and methane. As is customary in planetary science, this mixture is

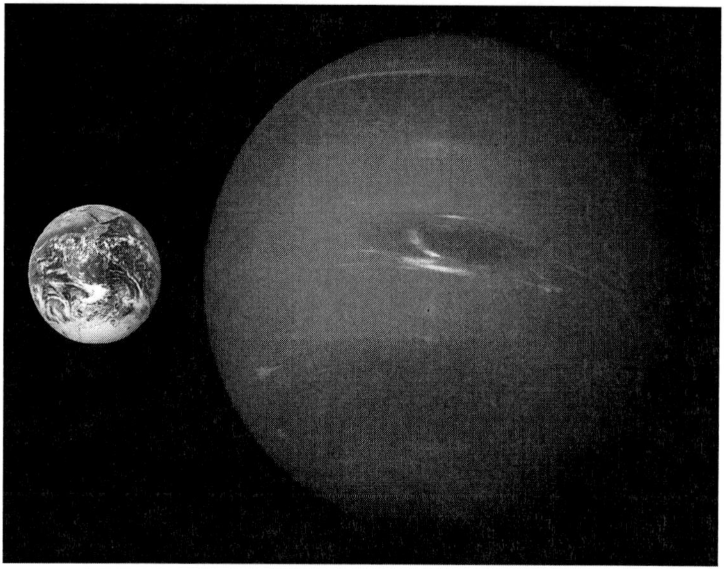

Figure 216: *A size comparison of Neptune and Earth*

referred to as icy even though it is a hot, dense fluid. This fluid, which has a high electrical conductivity, is sometimes called a water–ammonia ocean. The mantle may consist of a layer of ionic water in which the water molecules break down into a soup of hydrogen and oxygen ions, and deeper down superionic water in which the oxygen crystallises but the hydrogen ions float around freely within the oxygen lattice.[335] At a depth of 7,000 km, the conditions may be such that methane decomposes into diamond crystals that rain downwards like hailstones. Very-high-pressure experiments at the Lawrence Livermore National Laboratory suggest that the base of the mantle may be an ocean of liquid carbon with floating solid 'diamonds'.

The core of Neptune is likely composed of iron, nickel and silicates, with an interior model giving a mass about 1.2 times that of Earth. The pressure at the centre is 7 Mbar (700 GPa), about twice as high as that at the centre of Earth, and the temperature may be 5,400 K.

Atmosphere

At high altitudes, Neptune's atmosphere is 80% hydrogen and 19% helium. A trace amount of methane is also present. Prominent absorption bands of methane exist at wavelengths above 600 nm, in the red and infrared portion

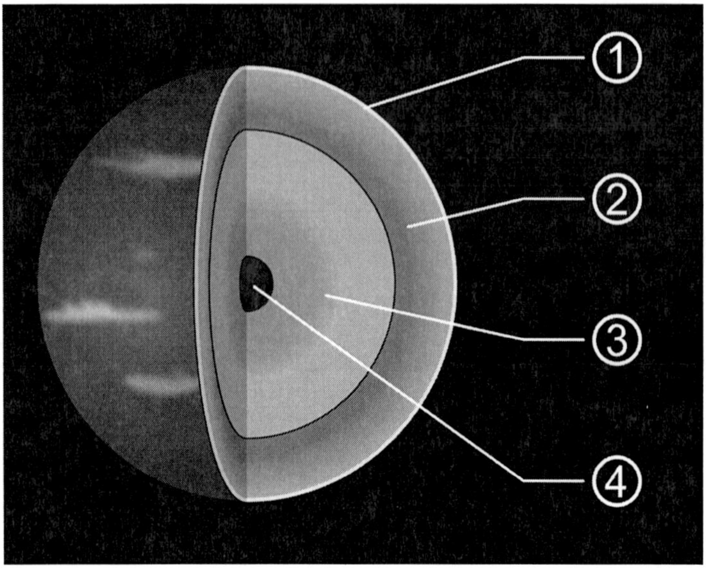

Figure 217:
The internal structure of Neptune:
1. *Upper atmosphere, top clouds*
2. *Atmosphere consisting of hydrogen, helium and methane gas*
3. *Mantle consisting of water, ammonia and methane ices*
4. *Core consisting of rock (silicates and nickel–iron)*

of the spectrum. As with Uranus, this absorption of red light by the atmospheric methane is part of what gives Neptune its blue hue, although Neptune's vivid azure differs from Uranus's milder cyan. Because Neptune's atmospheric methane content is similar to that of Uranus, some unknown atmospheric constituent is thought to contribute to Neptune's colour.

Neptune's atmosphere is subdivided into two main regions: the lower troposphere, where temperature decreases with altitude, and the stratosphere, where temperature increases with altitude. The boundary between the two, the tropopause, lies at a pressure of 0.1 bars (10 kPa). The stratosphere then gives way to the thermosphere at a pressure lower than 10^{-5} to 10^{-4} bars (1 to 10 Pa). The thermosphere gradually transitions to the exosphere.

Models suggest that Neptune's troposphere is banded by clouds of varying compositions depending on altitude. The upper-level clouds lie at pressures below one bar, where the temperature is suitable for methane to condense. For pressures between one and five bars (100 and 500 kPa), clouds of ammonia and hydrogen sulfide are thought to form. Above a pressure of five bars,

Figure 218: *Combined colour and near-infrared image of Neptune, showing bands of methane in its atmosphere, and four of its moons, Proteus, Larissa, Galatea, and Despina.*

Figure 219: *A time-lapse video of Neptune and its moons*

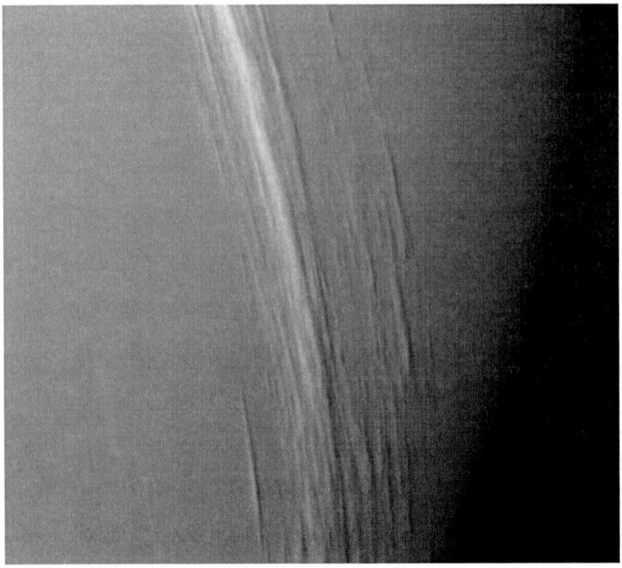

Figure 220: *Bands of high-altitude clouds cast shadows on Neptune's lower cloud deck*

the clouds may consist of ammonia, ammonium sulfide, hydrogen sulfide and water. Deeper clouds of water ice should bc found at pressures of about 50 bars (5.0 MPa), where the temperature reaches 273 K (0 °C). Underneath, clouds of ammonia and hydrogen sulfide may be found.

High-altitude clouds on Neptune have been observed casting shadows on the opaque cloud deck below. There are also high-altitude cloud bands that wrap around the planet at constant latitude. These circumferential bands have widths of 50–150 km and lie about 50–110 km above the cloud deck. These altitudes are in the layer where weather occurs, the troposphere. Weather does not occur in the higher stratosphere or thermosphere. Unlike Uranus, Neptune's composition has a higher volume of ocean, whereas Uranus has a smaller mantle.

Neptune's spectra suggest that its lower stratosphere is hazy due to condensation of products of ultraviolet photolysis of methane, such as ethane and ethyne. The stratosphere is also home to trace amounts of carbon monoxide and hydrogen cyanide. The stratosphere of Neptune is warmer than that of Uranus due to the elevated concentration of hydrocarbons.

For reasons that remain obscure, the planet's thermosphere is at an anomalously high temperature of about 750 K. The planet is too far from the Sun for this heat to be generated by ultraviolet radiation. One candidate for a heating mechanism is atmospheric interaction with ions in the planet's magnetic field. Other candidates are gravity waves from the interior that dissipate in the atmosphere. The thermosphere contains traces of carbon dioxide and water, which may have been deposited from external sources such as meteorites and dust.

Magnetosphere

Neptune resembles Uranus in its magnetosphere, with a magnetic field strongly tilted relative to its rotational axis at 47° and offset at least 0.55 radii, or about 13,500 km from the planet's physical centre. Before *Voyager 2*'s arrival at Neptune, it was hypothesised that Uranus's tilted magnetosphere was the result of its sideways rotation. In comparing the magnetic fields of the two planets, scientists now think the extreme orientation may be characteristic of flows in the planets' interiors. This field may be generated by convective fluid motions in a thin spherical shell of electrically conducting liquids (probably a combination of ammonia, methane and water) resulting in a dynamo action.

The dipole component of the magnetic field at the magnetic equator of Neptune is about 14 microteslas (0.14 G). The dipole magnetic moment of Neptune is about 2.2×10^{17} T·m^3 (14 μT·$R_N{}^3$, where R_N is the radius of Neptune). Neptune's magnetic field has a complex geometry that includes relatively large contributions from non-dipolar components, including a strong quadrupole moment that may exceed the dipole moment in strength. By contrast, Earth, Jupiter and Saturn have only relatively small quadrupole moments, and their fields are less tilted from the polar axis. The large quadrupole moment of Neptune may be the result of offset from the planet's centre and geometrical constraints of the field's dynamo generator.

Neptune's bow shock, where the magnetosphere begins to slow the solar wind, occurs at a distance of 34.9 times the radius of the planet. The magnetopause, where the pressure of the magnetosphere counterbalances the solar wind, lies at a distance of 23–26.5 times the radius of Neptune. The tail of the magnetosphere extends out to at least 72 times the radius of Neptune, and likely much farther.

Climate

Neptune's weather is characterised by extremely dynamic storm systems, with winds reaching speeds of almost 600 m/s (2,200 km/h; 1,300 mph)—nearly reaching supersonic flow. More typically, by tracking the motion of persistent clouds, wind speeds have been shown to vary from 20 m/s in the easterly direction to 325 m/s westward. At the cloud tops, the prevailing winds range in

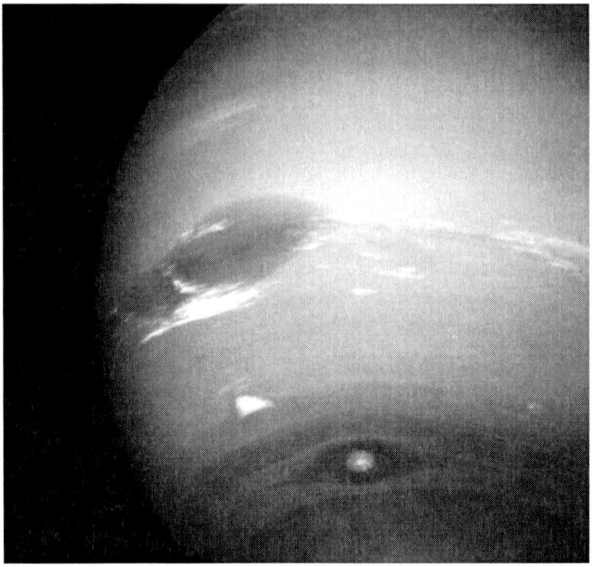

Figure 221: *The Great Dark Spot (top), Scooter (middle white cloud), and the Small Dark Spot (bottom), with contrast exaggerated.*

speed from 400 m/s along the equator to 250 m/s at the poles. Most of the winds on Neptune move in a direction opposite the planet's rotation.[336] The general pattern of winds showed prograde rotation at high latitudes vs. retrograde rotation at lower latitudes. The difference in flow direction is thought to be a "skin effect" and not due to any deeper atmospheric processes. At 70° S latitude, a high-speed jet travels at a speed of 300 m/s.

Neptune differs from Uranus in its typical level of meteorological activity. *Voyager 2* observed weather phenomena on Neptune during its 1989 flyby, but no comparable phenomena on Uranus during its 1986 fly-by.

The abundance of methane, ethane and acetylene at Neptune's equator is 10–100 times greater than at the poles. This is interpreted as evidence for upwelling at the equator and subsidence near the poles.Wikipedia:Please clarify

In 2007, it was discovered that the upper troposphere of Neptune's south pole was about 10 K warmer than the rest of its atmosphere, which averages approximately 73 K (–200 °C). The temperature differential is enough to let methane, which elsewhere is frozen in the troposphere, escape into the stratosphere near the pole. The relative "hot spot" is due to Neptune's axial tilt, which has exposed the south pole to the Sun for the last quarter of Neptune's year, or roughly

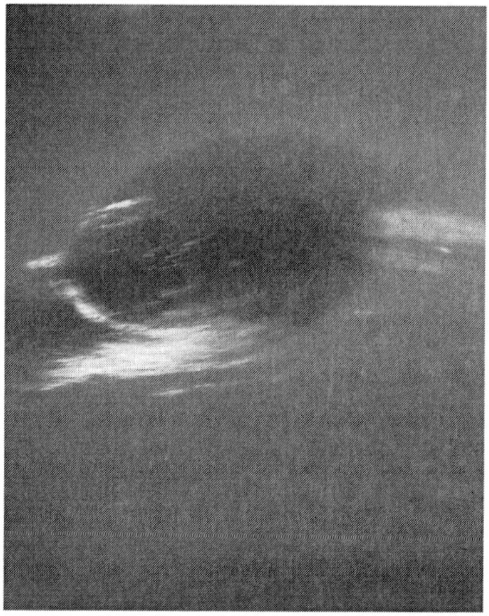

Figure 222: *The Great Dark Spot, as imaged by Voyager 2*

40 Earth years. As Neptune slowly moves towards the opposite side of the Sun, the south pole will be darkened and the north pole illuminated, causing the methane release to shift to the north pole.

Because of seasonal changes, the cloud bands in the southern hemisphere of Neptune have been observed to increase in size and albedo. This trend was first seen in 1980 and is expected to last until about 2020. The long orbital period of Neptune results in seasons lasting forty years.

Storms

In 1989, the Great Dark Spot, an anti-cyclonic storm system spanning 13,000 × 6,600 km, was discovered by NASA's *Voyager 2* spacecraft. The storm resembled the Great Red Spot of Jupiter. Some five years later, on 2 November 1994, the *Hubble Space Telescope* did not see the Great Dark Spot on the planet. Instead, a new storm similar to the Great Dark Spot was found in Neptune's northern hemisphere.

The Scooter is another storm, a white cloud group farther south than the Great Dark Spot. This nickname first arose during the months leading up to the *Voyager 2* encounter in 1989, when they were observed moving at speeds faster than the Great Dark Spot (and images acquired later would subsequently reveal the presence of clouds moving even faster than those that had initially been

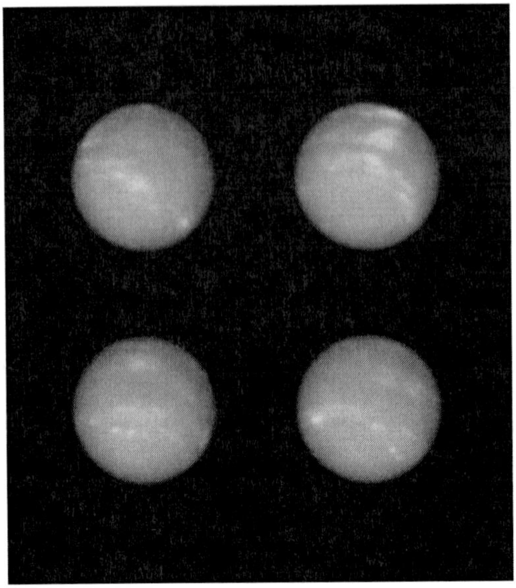

Figure 223: *Four images taken a few hours apart with the NASA/ESA Hubble Space Telescope's Wide Field Camera 3*

detected by *Voyager 2*). The Small Dark Spot is a southern cyclonic storm, the second-most-intense storm observed during the 1989 encounter. It was initially completely dark, but as *Voyager 2* approached the planet, a bright core developed and can be seen in most of the highest-resolution images.

Neptune's dark spots are thought to occur in the troposphere at lower altitudes than the brighter cloud features, so they appear as holes in the upper cloud decks. As they are stable features that can persist for several months, they are thought to be vortex structures. Often associated with dark spots are brighter, persistent methane clouds that form around the tropopause layer. The persistence of companion clouds shows that some former dark spots may continue to exist as cyclones even though they are no longer visible as a dark feature. Dark spots may dissipate when they migrate too close to the equator or possibly through some other unknown mechanism.

Internal heating

Neptune's more varied weather when compared to Uranus is due in part to its higher internal heating. Although Neptune lies over 50% farther from the Sun than Uranus, and receives only 40% its amount of sunlight, the two planets'

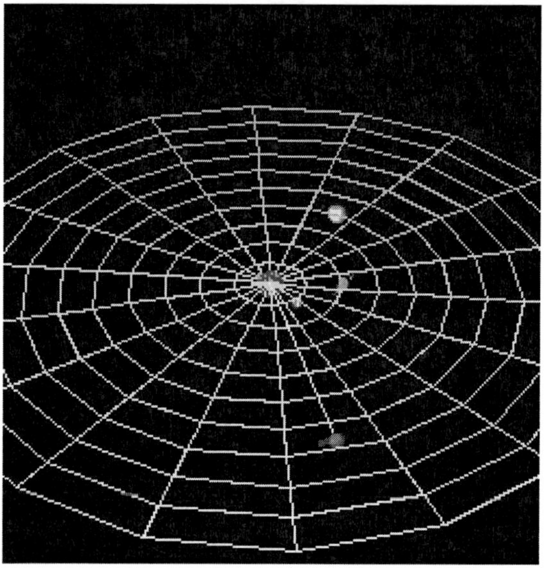

Figure 224: *Neptune (red arc) completes one orbit around the Sun (centre) for every 164.79 orbits of Earth. The light blue object represents Uranus.*

surface temperatures are roughly equal. The upper regions of Neptune's troposphere reach a low temperature of 51.8 K (–221.3 °C). At a depth where the atmospheric pressure equals 1 bar (100 kPa), the temperature is 72.00 K (–201.15 °C). Deeper inside the layers of gas, the temperature rises steadily. As with Uranus, the source of this heating is unknown, but the discrepancy is larger: Uranus only radiates 1.1 times as much energy as it receives from the Sun; whereas Neptune radiates about 2.61 times as much energy as it receives from the Sun. Neptune is the farthest planet from the Sun, yet its internal energy is sufficient to drive the fastest planetary winds seen in the Solar System. Depending on the thermal properties of its interior, the heat left over from Neptune's formation may be sufficient to explain its current heat flow, though it is more difficult to simultaneously explain Uranus's lack of internal heat while preserving the apparent similarity between the two planets.[337]

Orbit and rotation

The average distance between Neptune and the Sun is 4.50 billion km (about 30.1 astronomical units (AU)), and it completes an orbit on average every 164.79 years, subject to a variability of around ±0.1 years. The perihelion distance is 29.81 AU; the aphelion distance is 30.33 AU.[338]

On 11 July 2011, Neptune completed its first full barycentric orbit since its discovery in 1846, although it did not appear at its exact discovery position in the sky, because Earth was in a different location in its 365.26-day orbit. Because of the motion of the Sun in relation to the barycentre of the Solar System, on 11 July Neptune was also not at its exact discovery position in relation to the Sun; if the more common heliocentric coordinate system is used, the discovery longitude was reached on 12 July 2011.[339,340]

The elliptical orbit of Neptune is inclined 1.77° compared to that of Earth.

The axial tilt of Neptune is 28.32°, which is similar to the tilts of Earth (23°) and Mars (25°). As a result, Neptune experiences similar seasonal changes to Earth. The long orbital period of Neptune means that the seasons last for forty Earth years. Its sidereal rotation period (day) is roughly 16.11 hours. Because its axial tilt is comparable to Earth's, the variation in the length of its day over the course of its long year is not any more extreme.

Because Neptune is not a solid body, its atmosphere undergoes differential rotation. The wide equatorial zone rotates with a period of about 18 hours, which is slower than the 16.1-hour rotation of the planet's magnetic field. By contrast, the reverse is true for the polar regions where the rotation period is 12 hours. This differential rotation is the most pronounced of any planet in the Solar System, and it results in strong latitudinal wind shear.

Orbital resonances

Neptune's orbit has a profound impact on the region directly beyond it, known as the Kuiper belt. The Kuiper belt is a ring of small icy worlds, similar to the asteroid belt but far larger, extending from Neptune's orbit at 30 AU out to about 55 AU from the Sun. Much in the same way that Jupiter's gravity dominates the asteroid belt, shaping its structure, so Neptune's gravity dominates the Kuiper belt. Over the age of the Solar System, certain regions of the Kuiper belt became destabilised by Neptune's gravity, creating gaps in the Kuiper belt's structure. The region between 40 and 42 AU is an example.

There do exist orbits within these empty regions where objects can survive for the age of the Solar System. These resonances occur when Neptune's orbital period is a precise fraction of that of the object, such as 1:2, or 3:4. If, say, an object orbits the Sun once for every two Neptune orbits, it will only complete half an orbit by the time Neptune returns to its original position. The most heavily populated resonance in the Kuiper belt, with over 200 known objects, is the 2:3 resonance. Objects in this resonance complete 2 orbits for every 3 of Neptune, and are known as plutinos because the largest of the known Kuiper belt objects, Pluto, is among them. Although Pluto crosses Neptune's orbit

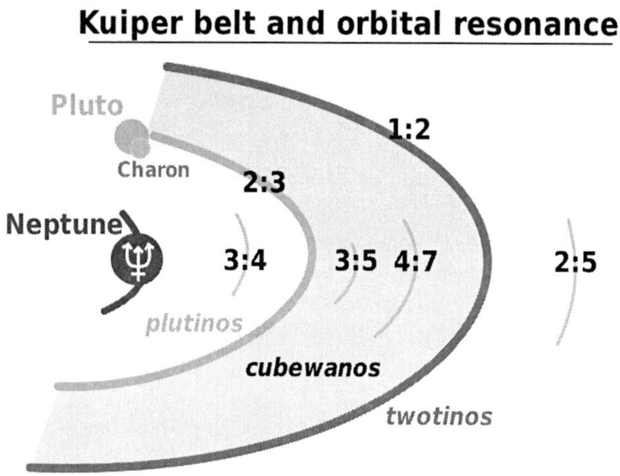

Figure 225: *A diagram showing the major orbital resonances in the Kuiper belt caused by Neptune: the highlighted regions are the 2:3 resonance (plutinos), the nonresonant "classical belt" (cubewanos), and the 1:2 resonance (twotinos).*

regularly, the 2:3 resonance ensures they can never collide. The 3:4, 3:5, 4:7 and 2:5 resonances are less populated.

Neptune has a number of known trojan objects occupying both the Sun–Neptune L_4 and L_5 Lagrangian points—gravitationally stable regions leading and trailing Neptune in its orbit, respectively. Neptune trojans can be viewed as being in a 1:1 resonance with Neptune. Some Neptune trojans are remarkably stable in their orbits, and are likely to have formed alongside Neptune rather than being captured. The first object identified as associated with Neptune's trailing L_5 Lagrangian point was 2008 LC18. Neptune also has a temporary quasi-satellite, (309239) 2007 RW10. The object has been a quasi-satellite of Neptune for about 12,500 years and it will remain in that dynamical state for another 12,500 years.

Formation and migration

The formation of the ice giants, Neptune and Uranus, has proven difficult to model precisely. Current models suggest that the matter density in the outer regions of the Solar System was too low to account for the formation of such

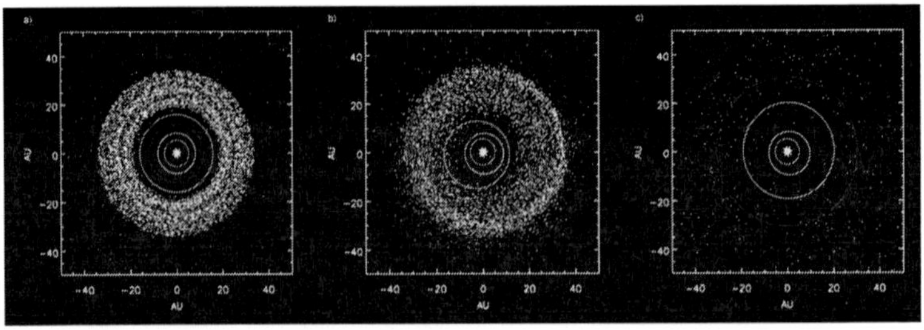

Figure 226: *A simulation showing the outer planets and Kuiper belt: a) before Jupiter and Saturn reached a 2:1 resonance; b) after inward scattering of Kuiper belt objects following the orbital shift of Neptune; c) after ejection of scattered Kuiper belt bodies by Jupiter*

large bodies from the traditionally accepted method of core accretion, and various hypotheses have been advanced to explain their formation. One is that the ice giants were not formed by core accretion but from instabilities within the original protoplanetary disc and later had their atmospheres blasted away by radiation from a nearby massive OB star.

An alternative concept is that they formed closer to the Sun, where the matter density was higher, and then subsequently migrated to their current orbits after the removal of the gaseous protoplanetary disc. This hypothesis of migration after formation is favoured, due to its ability to better explain the occupancy of the populations of small objects observed in the trans-Neptunian region. The current most widely accepted explanation of the details of this hypothesis is known as the Nice model, which explores the effect of a migrating Neptune and the other giant planets on the structure of the Kuiper belt.

Moons

Neptune has 14 known moons.[341] Triton is the largest Neptunian moon, comprising more than 99.5% of the mass in orbit around Neptune,[342] and it is the only one massive enough to be spheroidal. Triton was discovered by William Lassell just 17 days after the discovery of Neptune itself. Unlike all other large planetary moons in the Solar System, Triton has a retrograde orbit, indicating that it was captured rather than forming in place; it was probably once a dwarf planet in the Kuiper belt. It is close enough to Neptune to be locked into a synchronous rotation, and it is slowly spiralling inward because of tidal acceleration. It will eventually be torn apart, in about 3.6 billion years, when it reaches the Roche limit. In 1989, Triton was the coldest object that had

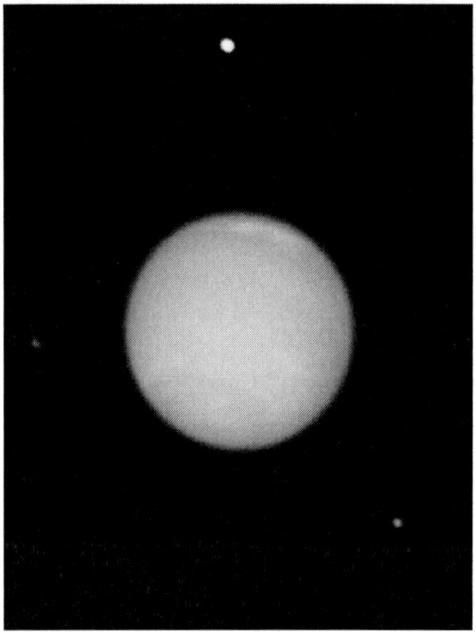

Figure 227: *Natural-colour view of Neptune with Proteus (top), Larissa (lower right), and Despina (left), from the Hubble Space Telescope*

yet been measured in the Solar System, with estimated temperatures of 38 K (–235 °C).

Neptune's second known satellite (by order of discovery), the irregular moon Nereid, has one of the most eccentric orbits of any satellite in the Solar System. The eccentricity of 0.7512 gives it an apoapsis that is seven times its periapsis distance from Neptune.[343]

From July to September 1989, *Voyager 2* discovered six moons of Neptune. Of these, the irregularly shaped Proteus is notable for being as large as a body of its density can be without being pulled into a spherical shape by its own gravity. Although the second-most-massive Neptunian moon, it is only 0.25% the mass of Triton. Neptune's innermost four moons—Naiad, Thalassa, Despina and Galatea—orbit close enough to be within Neptune's rings. The next-farthest out, Larissa, was originally discovered in 1981 when it had occulted a star. This occultation had been attributed to ring arcs, but when *Voyager 2* observed Neptune in 1989, Larissa was found to have caused it. Five new irregular moons discovered between 2002 and 2003 were announced in 2004. A new moon and the smallest yet, S/2004 N 1, was found in 2013. Because Neptune was the Roman god of the sea, Neptune's moons have been named after lesser sea gods.

Figure 228: *Neptune's moon Proteus*

Planetary rings

Neptune has a planetary ring system, though one much less substantial than that of Saturn. The rings may consist of ice particles coated with silicates or carbon-based material, which most likely gives them a reddish hue. The three main rings are the narrow Adams Ring, 63,000 km from the centre of Neptune, the Le Verrier Ring, at 53,000 km, and the broader, fainter Galle Ring, at 42,000 km. A faint outward extension to the Le Verrier Ring has been named Lassell; it is bounded at its outer edge by the Arago Ring at 57,000 km.

The first of these planetary rings was detected in 1968 by a team led by Edward Guinan. In the early 1980s, analysis of this data along with newer observations led to the hypothesis that this ring might be incomplete. Evidence that the rings might have gaps first arose during a stellar occultation in 1984 when the rings obscured a star on immersion but not on emersion. Images from *Voyager 2* in 1989 settled the issue by showing several faint rings.

The outermost ring, Adams, contains five prominent arcs now named *Courage*, *Liberté*, *Egalité 1*, *Egalité 2* and *Fraternité* (Courage, Liberty, Equality and Fraternity). The existence of arcs was difficult to explain because the laws of motion would predict that arcs would spread out into a uniform ring over short timescales. Astronomers now estimate that the arcs are corralled into their

Figure 229: *Neptune's rings*

current form by the gravitational effects of Galatea, a moon just inward from the ring.

Earth-based observations announced in 2005 appeared to show that Neptune's rings are much more unstable than previously thought. Images taken from the W. M. Keck Observatory in 2002 and 2003 show considerable decay in the rings when compared to images by *Voyager 2*. In particular, it seems that the *Liberté* arc might disappear in as little as one century.

Observation

With an apparent magnitude between +7.7 and +8.0, Neptune is never visible to the naked eye and can be outshone by Jupiter's Galilean moons, the dwarf planet Ceres and the asteroids 4 Vesta, 2 Pallas, 7 Iris, 3 Juno, and 6 Hebe.[344] A telescope or strong binoculars will resolve Neptune as a small blue disk, similar in appearance to Uranus.[345]

Because of the distance of Neptune from Earth, its angular diameter only ranges from 2.2 to 2.4 arcseconds, the smallest of the Solar System planets. Its small apparent size makes it challenging to study it visually. Most telescopic data was fairly limited until the advent of the *Hubble Space Telescope* and large ground-based telescopes with adaptive optics (AO).[346] The first scientifically useful observation of Neptune from ground-based telescopes using adaptive optics, was commenced in 1997 from Hawaii.[347] Neptune is currently entering its spring and summer season and has been shown to be heating up, with increased atmospheric activity and brightness as a consequence. Combined with technological advancements, ground-based telescopes with adaptive optics are recording increasingly more detailed images of it. Both *Hubble* and the adaptive-optics telescopes on Earth have made many new discoveries within the Solar System since the mid-1990s, with a large increase in the number of known satellites and moons around the outer planet, among others. In 2004 and 2005, five new small satellites of Neptune with diameters between 38 and 61 kilometres were discovered.[348]

From Earth, Neptune goes through apparent retrograde motion every 367 days, resulting in a looping motion against the background stars during each opposition. These loops carried it close to the 1846 discovery coordinates in April and July 2010 and again in October and November 2011.

Observation of Neptune in the radio-frequency band shows that it is a source of both continuous emission and irregular bursts. Both sources are thought to originate from its rotating magnetic field. In the infrared part of the spectrum, Neptune's storms appear bright against the cooler background, allowing the size and shape of these features to be readily tracked.

Exploration

Voyager 2 is the only spacecraft that has visited Neptune. The spacecraft's closest approach to the planet occurred on 25 August 1989. Because this was the last major planet the spacecraft could visit, it was decided to make a close flyby of the moon Triton, regardless of the consequences to the trajectory, similarly to what was done for *Voyager 1*'s encounter with Saturn and its moon Titan. The images relayed back to Earth from *Voyager 2* became the basis of a 1989 PBS all-night program, *Neptune All Night*.

During the encounter, signals from the spacecraft required 246 minutes to reach Earth. Hence, for the most part, *Voyager 2*'s mission relied on preloaded commands for the Neptune encounter. The spacecraft performed a near-encounter with the moon Nereid before it came within 4,400 km of Neptune's atmosphere on 25 August, then passed close to the planet's largest moon Triton later the same day.[349]

The spacecraft verified the existence of a magnetic field surrounding the planet and discovered that the field was offset from the centre and tilted in a manner similar to the field around Uranus. Neptune's rotation period was determined

Figure 230: *A Voyager 2 mosaic of Triton*

using measurements of radio emissions and *Voyager 2* also showed that Neptune had a surprisingly active weather system. Six new moons were discovered, and the planet was shown to have more than one ring.

The flyby also provided the first accurate measurement of Neptune's mass which was found to be 0.5 percent less than previously calculated. The new figure disproved the hypothesis that an undiscovered Planet X acted upon the orbits of Neptune and Uranus.[350]

After the *Voyager 2* flyby mission, the next step in scientific exploration of the Neptunian system, is considered to be a Flagship orbital mission. Such a hypothetical mission is envisioned to be possible in the late 2020s or early 2030s. However, there have been discussions to launch Neptune missions sooner. In 2003, there was a proposal in NASA's "Vision Missions Studies" for a "Neptune Orbiter with Probes" mission that does *Cassini*-level science. Another, more recent proposal was for *Argo*, a flyby spacecraft to be launched in 2019, that would visit Jupiter, Saturn, Neptune, and a Kuiper belt object. The focus would be on Neptune and its largest moon Triton to be investigated around 2029. The proposed *New Horizons 2* mission (which was later scrapped) might also have done a close flyby of the Neptunian system.

Bibliography

- Burgess, Eric (1991). *Far Encounter: The Neptune System*. Columbia University Press. ISBN 978-0-231-07412-4.
- Moore, Patrick (2000). *The Data Book of Astronomy*. CRC Press. ISBN 978-0-7503-0620-1.

Further reading

- Miner, Ellis D.; Wessen, Randii R. (2002). *Neptune: The Planet, Rings, and Satellites*. Springer-Verlag. ISBN 978-1-85233-216-7.
- Standage, Tom (2001). *The Neptune File*. Penguin. ISBN 978-0-8027-1363-6.

External links

- NASA's Neptune fact sheet[351]
- Neptune[352] from Bill Arnett's nineplanets.org
- Neptune[353] Astronomy Cast episode No. 63, includes full transcript.
- Neptune Profile[354] at NASA's Solar System Exploration site[355]
- Planets – Neptune[356] A children's guide to Neptune.
- Merrifield, Michael; Bauer, Amanda (2010). "Neptune"[357]. *Sixty Symbols*. Brady Haran for the University of Nottingham.
- Neptune by amateur[358] (The Planetary Society)

<indicator name="featured-star"> </indicator>

Centaur (minor planet)

Types of distant minor planets

- Cis-Neptunian objects
  - Centaurs
  - Neptune trojans
- Trans-Neptunian objects (TNOs)‡
  - Kuiper belt objects (KBOs)
    - Classical KBOs (cubewanos)
    - Resonant KBOs
      - Plutinos (2:3 resonance)
  - Scattered disc objects (SDOs)
    - Resonant SDOs
  - Detached objects
  - Sednoids
  - Oort cloud objects (ICO/OCOs)

‡ Trans-Neptunian dwarf planets are called "plutoids"
• v • t • e[359]

Centaurs are small solar system bodies with a semi-major axis between those of the outer planets. They have unstable orbits because they cross or have crossed the orbits of one or more of the giant planets; almost all their orbits have dynamic lifetimes of only a few million years. Centaurs typically behave with characteristics of both asteroids and comets. They are named after the mythological centaurs that were a mixture of horse and human. It has been estimated that there are around 44,000 centaurs in the Solar System with diameters larger than 1 km.

The first centaur to be discovered, under the definition of the Jet Propulsion Laboratory (JPL) and the one used here, was 944 Hidalgo in 1920. However, they were not recognized as a distinct population until the discovery of 2060 Chiron in 1977. The largest confirmed centaur is 10199 Chariklo, which at 260 km in diameter is as big as a mid-sized main-belt asteroid, and is known to have a system of rings. It was discovered in 1997. However, the lost centaur 1995 SN55 may be somewhat larger.

No centaur has been photographed up close, although there is evidence that Saturn's moon Phoebe, imaged by the *Cassini* probe in 2004, may be a captured centaur that originated in the Kuiper belt. In addition, the Hubble Space Telescope has gleaned some information about the surface features of 8405 Asbolus.

As of 2008[360], three centaurs have been found to display comet-like comas: Chiron, 60558 Echeclus, and 166P/NEAT. Chiron and Echeclus are therefore classified as both asteroids and comets. Other centaurs, such as 52872 Okyrhoe and 2012 CG, are suspected of having shown comas. Any centaur that is perturbed close enough to the Sun is expected to become a comet.

Classification

The generic definition of a centaur is a small body that orbits the Sun between Jupiter and Neptune and crosses the orbits of one or more of the giant planets. Due to the inherent long-term instability of orbits in this region, even centaurs such as 2000 GM137 and 2001 XZ255, which do not currently cross the orbit of any planet, are in gradually changing orbits that will be perturbed until they start to cross the orbit of one or more of the giant planets.

However, different institutions have different criteria for classifying borderline objects, based on particular values of their orbital elements:

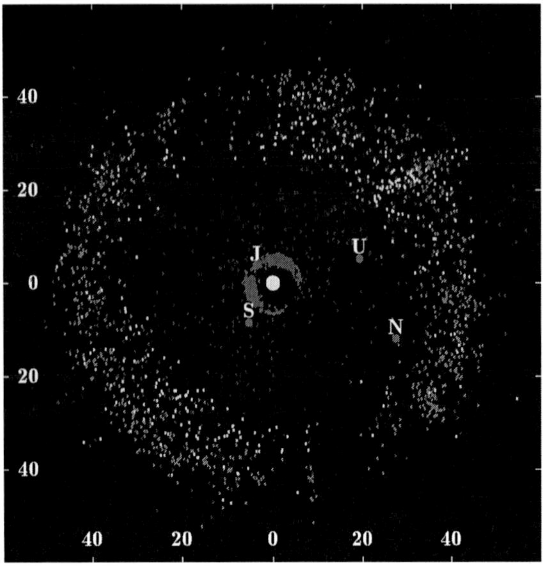

Figure 231:

Positions of known outer Solar System objects. The centaurs lie generally inwards of the Kuiper belt and outside the Jupiter trojans.

(scale in AU; epoch as of January 2015; # of objects in parenthesis)

- The Minor Planet Center (MPC) defines centaurs as having a perihelion beyond the orbit of Jupiter ($q > 5.2$ AU) and a semi-major axis less than that of Neptune ($a < 30.1$ AU). Though nowadays the MPC often lists centaurs and scattered disc objects together as a single group.
- The Jet Propulsion Laboratory (JPL) similarly defines centaurs as having a semi-major axis, a, between those of Jupiter (5.5 AU $< a$) and Neptune ($a < 30.1$ AU).
- In contrast, the Deep Ecliptic Survey (DES) defines centaurs using a dynamical classification scheme. These classifications are based on the simulated change in behavior of the present orbit when extended over 10 million years. The DES defines centaurs as non-resonant objects whose instantaneous (osculating) perihelia are less than the osculating semi-major axis of Neptune at any time during the simulation. This definition is intended to be synonymous with planet-crossing orbits and to suggest comparatively short lifetimes in the current orbit.

The collection *The Solar System Beyond Neptune* (2008) defines objects with a semi-major axis between those of Jupiter and Neptune and a Jupiter – Tisserand's parameter above 3.05 – as centaurs, classifying the objects with a Jupiter Tisserand's parameter below this and, to exclude Kuiper belt objects, an arbitrary perihelion cut-off half-way to Saturn ($q < 7.35$) as Jupiter-family comets[361] and classifying those objects on unstable orbits with a semi-major axis larger than Neptune's as members of the scattered disc. Other astronomers prefer to define centaurs as objects that are non-resonant with a perihelion inside the orbit of Neptune that can be shown to likely cross the Hill sphere of a gas giant within the next 10 million years, so that centaurs can be thought of as objects scattered inwards and that interact more strongly and scatter more quickly than typical scattered-disc objects.

The JPL Small-Body Database lists 406 centaurs. There are an additional 97 trans-Neptunian objects (objects with a semi-major axis further than Neptune's, i.e. $a > 30.1$ AU) with a perihelion closer than the orbit of Uranus ($q < 19.2$ AU). The Committee on Small Body Nomenclature of the International Astronomical Union has not formally weighed in on either side of the debate. Instead, it has adopted the following naming convention for such objects: Befitting their centaur-like transitional orbits between TNOs and comets, "objects on unstable, non-resonant, giant-planet-crossing orbits with semimajor axes greater than Neptune's" are to be named for other hybrid and shape-shifting mythical creatures. Thus far, only the binary objects Ceto and Phorcys and Typhon and Echidna have been named according to the new policy.

Other objects caught between these differences in classification methods include 944 Hidalgo which was discovered in 1920 and is listed as a centaur in the JPL Small-Body Database. (44594) 1999 OX3, which has a semi-major axis of 32 AU but crosses the orbits of both Uranus and Neptune is listed as an outer centaur by the *Deep Ecliptic Survey* (DES). Among the inner centaurs, (434620) 2005 VD, with a perihelion distance very near Jupiter, is listed as a centaur by both JPL and DES.

Centaurs with measured diameters listed as possible dwarf planets according to Mike Brown's website include 10199 Chariklo, 2060 Chiron, and 54598 Bienor.

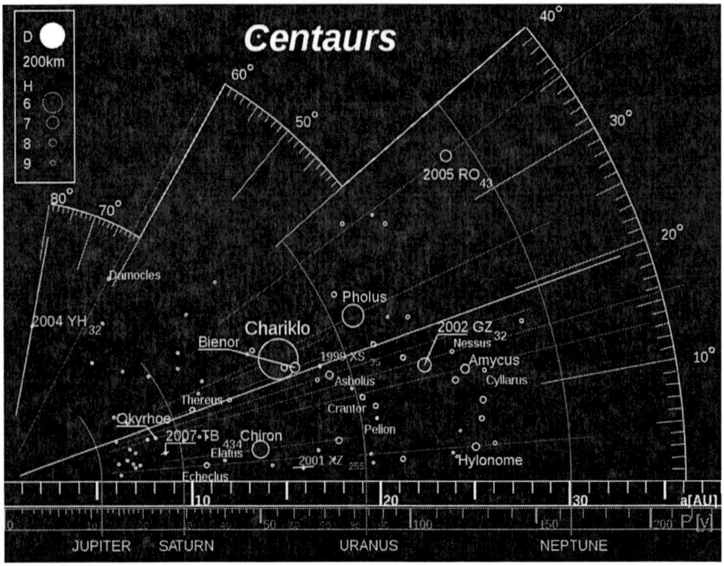

Figure 232: *Orbits of known centaurs*[362]

Orbits

Distribution

The diagram illustrates the orbits of known centaurs in relation to the orbits of the planets. For selected objects, the eccentricity of the orbits is represented by red segments (extending from perihelion to aphelion).

The orbits of centaurs show a wide range of eccentricity, from highly eccentric (Pholus, Asbolus, Amycus, Nessus) to more circular (Chariklo and the Saturn-crossers Thereus and Okyrhoe).

To illustrate the range of the orbits' parameters, the diagram shows a few objects with very unusual orbits, plotted in yellow :

- 1999 XS35 (Apollo asteroid) follows an extremely eccentric orbit (e = 0.947), leading it from inside Earth's orbit (0.94 AU) to well beyond Neptune (> 34 AU)
- 2007 TB_{434} follows a quasi-circular orbit ($e < 0.026$)
- 2001 XZ_{255} has the lowest inclination ($i < 3°$).
- Damocles is among a few centaurs on orbits with extreme inclination (prograde $i > 70°$, e.g. 2007 DA_{61}, 2004 YH_{32}; retrograde $i < 120°$, e.g. 2005 JT_{50}; not shown)

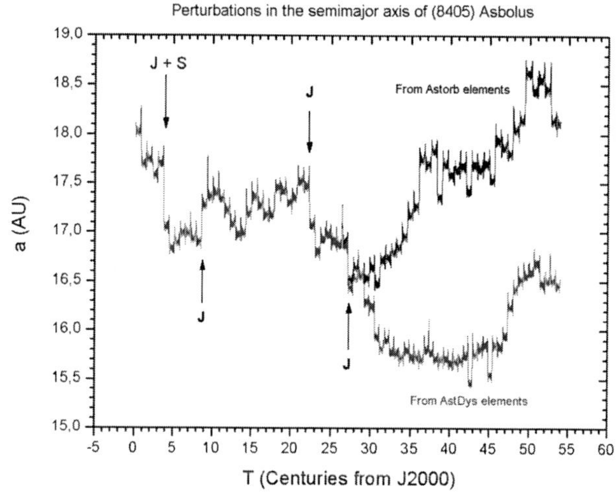

Figure 233: *The semi-major axis of Asbolus during the next 5500 years, using two slightly different estimates of present-day orbital elements. After the Jupiter encounter of year 4713 the two calculations diverge.*[363]

- 2004 YH_{32} follows such a highly inclined orbit (nearly 80°) that, while it crosses from the distance of the asteroid belt from the Sun to past the distance of Saturn, if its orbit is projected onto the plane of Jupiter's orbit, it does not even go out as far as Jupiter.

A dozen known centaurs, including Dioretsa ("asteroid" spelled backwards), follow retrograde orbits.

Changing orbits

Because the centaurs are not protected by orbital resonances, their orbits are unstable within a timescale of 10^6–10^7 years. For example, 55576 Amycus is in an unstable orbit near the 3:4 resonance of Uranus. Dynamical studies of their orbits indicate that being a centaur is probably an intermediate orbital state of objects transitioning from the Kuiper belt to the Jupiter family of short-period comets. Objects may be perturbed from the Kuiper belt, whereupon they become Neptune-crossing and interact gravitationally with that planet (see theories of origin). They then become classed as centaurs, but their orbits are chaotic, evolving relatively rapidly as the centaur makes repeated close approaches to one or more of the outer planets. Some centaurs will evolve into Jupiter-crossing orbits whereupon their perihelia may become

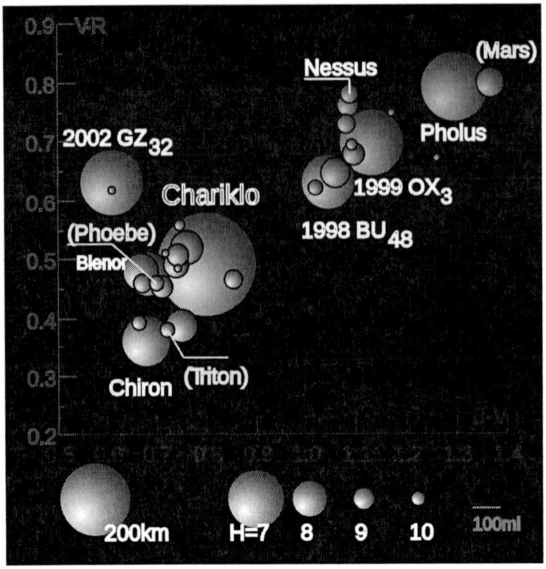

Figure 234: *Colour distribution of centaurs*

reduced into the inner Solar System and they may be reclassified as active comets in the Jupiter family if they display cometary activity. Centaurs will thus ultimately collide with the Sun or a planet or else they may be ejected into interstellar space after a close approach to one of the planets, particularly Jupiter.

Physical characteristics

The relatively small size of centaurs precludes surface observations, but colour indices and spectra can indicate possible surface composition and can provide insight into the origin of the bodies.[364]

Colours

The colours of centaurs are very diverse, which challenges any simple model of surface composition. In the side-diagram, the colour indices are measures of apparent magnitude of an object through blue (B), visible (V) (i.e. green-yellow) and red (R) filters. The diagram illustrates these differences (in exaggerated colours) for all centaurs with known colour indices. For reference, two moons: Triton and Phoebe, and planet Mars are plotted (yellow labels, size not to scale).

Centaurs appear to be grouped into two classes:

- very red – for example 5145 Pholus
- blue (or blue-grey, according to some authors) – for example 2060 Chiron

There are numerous theories to explain this colour difference, but they can be divided broadly into two categories:

- The colour difference results from a difference in the origin and/or composition of the centaur (see origin below)
- The colour difference reflects a different level of space-weathering from radiation and/or cometary activity.

As examples of the second category, the reddish colour of Pholus has been explained as a possible mantle of irradiated red organics, whereas Chiron has instead had its ice exposed due to its periodic cometary activity, giving it a blue/grey index. The correlation with activity and color is not certain, however, as the active centaurs span the range of colors from blue (Chiron) to red (166P/NEAT).[365] Alternatively, Pholus may have been only recently expelled from the Kuiper belt, so that surface transformation processes have not yet taken place.

Delsanti et al. suggest multiple competing processes: reddening by the radiation, and blushing by collisions.[366]

Spectra

The interpretation of spectra is often ambiguous, related to particle sizes and other factors, but the spectra offer an insight into surface composition. As with the colours, the observed spectra can fit a number of models of the surface.

Water ice signatures have been confirmed on a number of centaurs (including 2060 Chiron, 10199 Chariklo and 5145 Pholus). In addition to the water ice signature, a number of other models have been put forward:

- Chariklo's surface has been suggested to be a mixture of tholins (like those detected on Titan and Triton) with amorphous carbon.
- Pholus has been suggested to be covered by a mixture of Titan-like tholins, carbon black, olivine[367] and methanol ice.
- The surface of 52872 Okyrhoe has been suggested to be a mixture of kerogens, olivines and small percentage of water ice.
- 8405 Asbolus has been suggested to be a mixture of 15% Triton-like tholins, 8% Titan-like tholin, 37% amorphous carbon and 40% ice tholin.

Chiron appears to be the most complex. The spectra observed vary depending on the period of the observation. Water ice signature was detected during a period of low activity and disappeared during high activity.

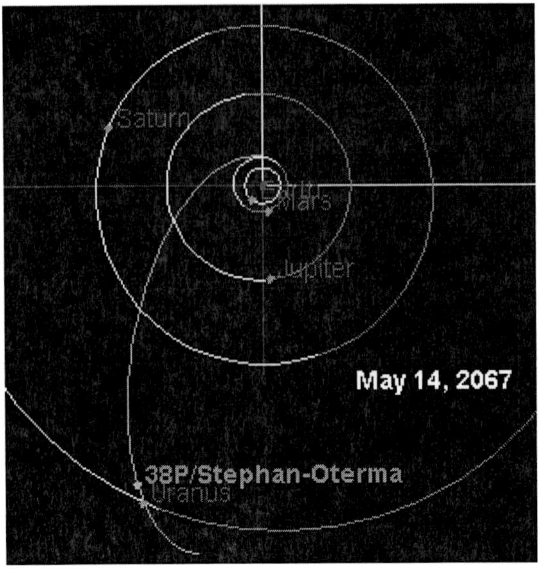

Figure 235: *Comet 38P exhibits centaur-like behavior by making close approaches to Jupiter, Saturn, and Uranus between 1982 and 2067.*

Similarities to comets

Observations of Chiron in 1988 and 1989 near its perihelion found it to display a coma (a cloud of gas and dust evaporating from its surface). It is thus now officially classified as both a comet and an asteroid, although it is far larger than a typical comet and there is some lingering controversy. Other centaurs are being monitored for comet-like activity: so far two, 60558 Echeclus, and 166P/NEAT have shown such behavior. 166P/NEAT was discovered while it exhibited a coma, and so is classified as a comet, though its orbit is that of a centaur. 60558 Echeclus was discovered without a coma but recently became active, and so it too is now classified as both a comet and an asteroid.

Carbon monoxide has been detected in 60558 Echeclus and Chiron in very small amounts, and the derived CO production rate was calculated to be sufficient to account for the observed coma. The calculated CO production rate from both 60558 Echeclus and Chiron is substantially lower than what is typically observed for 29P/Schwassmann–Wachmann, another distantly active comet often classified as a centaur.

There is no clear orbital distinction between centaurs and comets. Both 29P/ Schwassmann-Wachmann and 39P/Oterma have been referred to as centaurs since they have typical centaur orbits. The comet 39P/Oterma is currently inactive and was seen to be active only before it was perturbed into a centaur orbit by Jupiter in 1963. The faint comet 38P/Stephan–Oterma would probably not show a coma if it had a perihelion distance beyond Jupiter's orbit at 5 AU. By the year 2200, comet 78P/Gehrels will probably migrate outwards into a centaur-like orbit.

Theories of origin

The study of centaur development is rich in recent developments, but any conclusions are still hampered by limited physical data. Different models have been put forward for possible origin of centaurs.

Simulations indicate that the orbit of some Kuiper belt objects can be perturbed, resulting in the object's expulsion so that it becomes a centaur. Scattered disc objects would be dynamically the best candidates (For instance, the centaurs could be part of an "inner" scattered disc of objects perturbed inwards from the Kuiper belt.) for such expulsions, but their colours do not fit the bicoloured nature of the centaurs. Plutinos are a class of Kuiper belt object that display a similar bicoloured nature, and there are suggestions that not all plutinos' orbits are as stable as initially thought, due to perturbation by Pluto. Further developments are expected with more physical data on Kuiper belt objects.

Notable centaurs

Name	Year	Discoverer	Half-life (forward)	Class[368]
55576 Amycus	2002	NEAT at Palomar	11.1 Ma	UK
54598 Bienor	2000	Marc W. Buie et al.	?	U
10370 Hylonome	1995	Mauna Kea Observatory	6.3 Ma	UN
10199 Chariklo	1997	Spacewatch	10.3 Ma	U
8405 Asbolus	1995	Spacewatch (James V. Scotti)	0.86 Ma	SN
7066 Nessus	1993	Spacewatch (David L. Rabinowitz)	4.9 Ma	SK
5145 Pholus	1992	Spacewatch (David L. Rabinowitz)	1.28 Ma	SN
2060 Chiron	1977	Charles T. Kowal	1.03 Ma	SU

^ the class is defined by the perihelion and aphelion distance of the object: S indicates a perihelion/aphelion near Saturn, U near Uranus, N near Neptune, and K in the Kuiper belt.

External links

Wikimedia Commons has media related to ***Centaurs (planetoids)***.

- List of Centaurs and Scattered-Disk Objects[369]
- Centaurs from The Encyclopedia of Astrobiology Astronomy and Spaceflight[370]
- Horner, Jonathan; Lykawka, Patryk Sofia (2010). "Planetary Trojans – the main source of short period comets?". *International Journal of Astrobiology*. **9** (4): 227–234. Bibcode: 2010IJAsB...9..227H[371]. arXiv: 1007.2541[372]. doi: 10.1017/S1473550410000212[373].
- NASA's WISE Finds Mysterious Centaurs May Be Comets (2013)[374]

Comet

<indicator name="pp-autoreview"> </indicator>

<indicator name="good-star"> </indicator>

Comets – nucleus, coma and tail:
- Top: 9P/Tempel (impactor collision: *Deep Impact*), 67P/Churyumov–Gerasimenko (*Rosetta*)
- Middle: 17P/Holmes and its blue ionized tail, and 81P/Wild (Wild 2) visited by *Stardust*
- Bottom: Hale–Bopp seen from Earth in 1997, and C/2011 W3 (Lovejoy) imaged from Earth orbit

A **comet** is an icy small Solar System body that, when passing close to the Sun, warms and begins to release gases, a process called outgassing. This produces a visible atmosphere or coma, and sometimes also a tail. These phenomena are due to the effects of solar radiation and the solar wind acting upon the nucleus of the comet. Comet nuclei range from a few hundred metres to tens of kilometres across and are composed of loose collections of ice, dust, and small rocky particles. The coma may be up to 15 times the Earth's diameter, while the tail may stretch one astronomical unit. If sufficiently bright, a comet may be seen from the Earth without the aid of a telescope and may subtend an arc of 30° (60 Moons) across the sky. Comets have been observed and recorded since ancient times by many cultures.

Comets usually have highly eccentric elliptical orbits, and they have a wide range of orbital periods, ranging from several years to potentially several millions of years. Short-period comets originate in the Kuiper belt or its associated scattered disc, which lie beyond the orbit of Neptune. Long-period comets are thought to originate in the Oort cloud, a spherical cloud of icy bodies extending from outside the Kuiper belt to halfway to the nearest star. Long-period comets are set in motion towards the Sun from the Oort cloud by gravitational perturbations caused by passing stars and the galactic tide. Hyperbolic comets may pass once through the inner Solar System before being flung to interstellar space. The appearance of a comet is called an apparition.

Comets are distinguished from asteroids by the presence of an extended, gravitationally unbound atmosphere surrounding their central nucleus. This atmosphere has parts termed the coma (the central part immediately surrounding the nucleus) and the tail (a typically linear section consisting of dust or gas blown out from the coma by the Sun's light pressure or outstreaming solar wind plasma). However, extinct comets that have passed close to the Sun many times have lost nearly all of their volatile ices and dust and may come to resemble small asteroids. Asteroids are thought to have a different origin from comets, having formed inside the orbit of Jupiter rather than in the outer Solar System. The discovery of main-belt comets and active centaur minor planets has blurred the distinction between asteroids and comets.

As of November 2014[375] there are 5,253 known comets, a number that is steadily increasing as they are discovered. However, this represents only a tiny fraction of the total potential comet population, as the reservoir of comet-like bodies in the outer Solar System (in the Oort cloud) is estimated to be one trillion. Roughly one comet per year is visible to the naked eye, though many of those are faint and unspectacular. Particularly bright examples are called "great comets". Comets have been visited by unmanned probes such as the European Space Agency's Rosetta, which became the first ever to land a robotic spacecraft on a comet,[376] and NASA's Deep Impact, which blasted a crater on Comet Tempel 1 to study its interior.

Etymology

The word *comet* derives from the Old English *cometa* from the Latin *comēta* or *comētēs*. That, in turn, is a latinisation of the Greek κομήτης ("wearing long hair"), and the *Oxford English Dictionary* notes that the term (ἀστὴρ) κομήτης already meant "long-haired star, comet" in Greek. Κομήτης was derived from κομᾶν ("to wear the hair long"), which was itself derived from κόμη ("the hair of the head") and was used to mean "the tail of a comet".

The astronomical symbol for comets is ☄ (in Unicode U+2604), consisting of a small disc with three hairlike extensions.

Physical characteristics

Nucleus

The solid, core structure of a comet is known as the nucleus. Cometary nuclei are composed of an amalgamation of rock, dust, water ice, and frozen gases such as carbon dioxide, carbon monoxide, methane, and ammonia. As such, they are popularly described as "dirty snowballs" after Fred Whipple's model. However, some comets may have a higher dust content, leading them to be called "icy dirtballs". Research conducted in 2014 suggests that comets are like "deep fried ice cream", in that their surfaces are formed of dense crystalline ice mixed with organic compounds, while the interior ice is colder and less dense.

The surface of the nucleus is generally dry, dusty or rocky, suggesting that the ices are hidden beneath a surface crust several metres thick. In addition to the gases already mentioned, the nuclei contain a variety of organic compounds, which may include methanol, hydrogen cyanide, formaldehyde, ethanol, and ethane and perhaps more complex molecules such as long-chain hydrocarbons and amino acids. In 2009, it was confirmed that the amino acid glycine had been found in the comet dust recovered by NASA's Stardust mission. In August 2011, a report, based on NASA studies of meteorites found on Earth, was published suggesting DNA and RNA components (adenine, guanine, and related organic molecules) may have been formed on asteroids and comets.

Figure 236: *Nucleus of 103P/Hartley as imaged during a spacecraft flyby. The nucleus is about 2 km in length.*

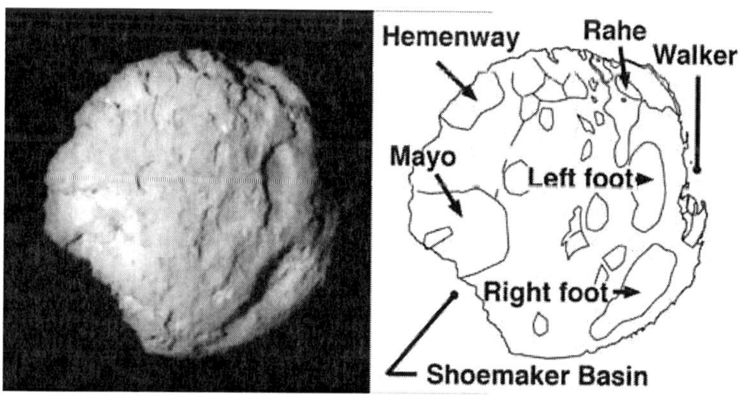

Figure 237: *Comet 81P/Wild exhibits jets on light side and dark side, stark relief, and is dry.*

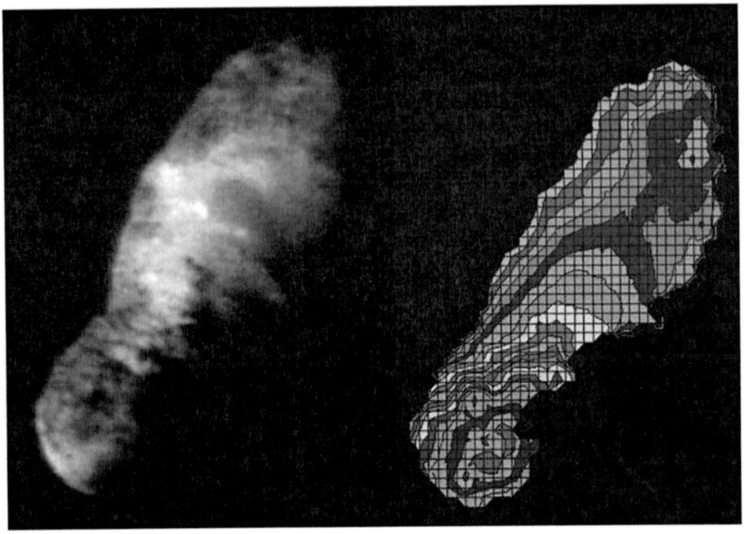

Figure 238: *Comet Borrelly exhibits jets, but has no surface ice.*

The outer surfaces of cometary nuclei have a very low albedo, making them among the least reflective objects found in the Solar System. The Giotto space probe found that the nucleus of Halley's Comet reflects about four percent of the light that falls on it, and Deep Space 1 discovered that Comet Borrelly's surface reflects less than 3.0% of the light that falls on it; by comparison, asphalt reflects seven percent. The dark surface material of the nucleus may consist of complex organic compounds. Solar heating drives off lighter volatile compounds, leaving behind larger organic compounds that tend to be very dark, like tar or crude oil. The low reflectivity of cometary surfaces causes them to absorb the heat that drives their outgassing processes.

Comet nuclei with radii of up to 30 kilometres (19 mi) have been observed, but ascertaining their exact size is difficult. The nucleus of 322P/SOHO is probably only 100–200 metres (330–660 ft) in diameter. A lack of smaller comets being detected despite the increased sensitivity of instruments has led some to suggest that there is a real lack of comets smaller than 100 metres (330 ft) across. Known comets have been estimated to have an average density of 0.6 g/cm^3 (0.35 oz/cu in). Because of their low mass, comet nuclei do not become spherical under their own gravity and therefore have irregular shapes.

Roughly six percent of the near-Earth asteroids are thought to be extinct nuclei of comets that no longer experience outgassing, including 14827 Hypnos and 3552 Don Quixote.

Results from the *Rosetta* and *Philae* spacecraft show that the nucleus of 67P/Churyumov–Gerasimenko has no magnetic field, which suggests that magnetism may not have played a role in the early formation of planetesimals. Further, the ALICE spectrograph on *Rosetta* determined that electrons (within 1 km (0.62 mi) above the comet nucleus) produced from photoionization of water molecules by solar radiation, and not photons from the Sun as thought earlier, are responsible for the degradation of water and carbon dioxide molecules released from the comet nucleus into its coma. Instruments on the *Philae* lander found at least sixteen organic compounds at the comet's surface, four of which (acetamide, acetone, methyl isocyanate and propionaldehyde) have been detected for the first time on a comet.

Properties of some comets

Name	Dimensions (km)	Density (g/cm^3)	Mass (kg)[377]	Refs
Halley's Comet	$15 \times 8 \times 8$	0.6	3×10^{14}	
Tempel 1	7.6×4.9	0.62	7.9×10^{13}	
19P/Borrelly	$8 \times 4 \times 4$	0.3	2.0×10^{13}	
81P/Wild	$5.5 \times 4.0 \times 3.3$	0.6	2.3×10^{13}	
67P/Churyumov–Gerasimenko	$4.1 \times 3.3 \times 1.8$	0.47	1.0×10^{13}	

Coma

The streams of dust and gas thus released form a huge and extremely thin atmosphere around the comet called the "coma". The force exerted on the coma by the Sun's radiation pressure and solar wind cause an enormous "tail" to form pointing away from the Sun.

The coma is generally made of H_2O and dust, with water making up to 90% of the volatiles that outflow from the nucleus when the comet is within 3 to 4 astronomical units (450,000,000 to 600,000,000 km; 280,000,000 to 370,000,000 mi) of the Sun. The H_2O parent molecule is destroyed primarily through photodissociation and to a much smaller extent photoionization, with the solar wind playing a minor role in the destruction of water compared to photochemistry. Larger dust particles are left along the comet's orbital path whereas smaller particles are pushed away from the Sun into the comet's tail by light pressure.

Although the solid nucleus of comets is generally less than 60 kilometres (37 mi) across, the coma may be thousands or millions of kilometres across, sometimes becoming larger than the Sun. For example, about a month after an outburst in October 2007, comet 17P/Holmes briefly had a tenuous dust

Figure 239: *Hubble image of Comet ISON shortly before perihelion.*

Figure 240: *C/2006 W3 (Chistensen) emitting carbon gas (IR image)*

atmosphere larger than the Sun. The Great Comet of 1811 also had a coma roughly the diameter of the Sun. Even though the coma can become quite large, its size can decrease about the time it crosses the orbit of Mars around 1.5 astronomical units (220,000,000 km; 140,000,000 mi) from the Sun. At this distance the solar wind becomes strong enough to blow the gas and dust away from the coma, and in doing so enlarging the tail. Ion tails have been observed to extend one astronomical unit (150 million km) or more.

Both the coma and tail are illuminated by the Sun and may become visible when a comet passes through the inner Solar System, the dust reflects Sunlight directly while the gases glow from ionisation. Most comets are too faint to be visible without the aid of a telescope, but a few each decade become bright enough to be visible to the naked eye. Occasionally a comet may experience a huge and sudden outburst of gas and dust, during which the size of the coma greatly increases for a period of time. This happened in 2007 to Comet Holmes.

In 1996, comets were found to emit X-rays. This greatly surprised astronomers because X-ray emission is usually associated with very high-temperature bodies. The X-rays are generated by the interaction between comets and the solar wind: when highly charged solar wind ions fly through a cometary atmosphere, they collide with cometary atoms and molecules, "stealing" one or more electrons from the atom in a process called "charge exchange". This exchange or transfer of an electron to the solar wind ion is followed by its de-excitation into the ground state of the ion by the emission of X-rays and far ultraviolet photons.

Tails

In the outer Solar System, comets remain frozen and inactive and are extremely difficult or impossible to detect from Earth due to their small size. Statistical detections of inactive comet nuclei in the Kuiper belt have been reported from observations by the Hubble Space Telescope but these detections have been questioned. As a comet approaches the inner Solar System, solar radiation causes the volatile materials within the comet to vaporize and stream out of the nucleus, carrying dust away with them.

The streams of dust and gas each form their own distinct tail, pointing in slightly different directions. The tail of dust is left behind in the comet's orbit in such a manner that it often forms a curved tail called the type II or dust tail. At the same time, the ion or type I tail, made of gases, always points directly away from the Sun because this gas is more strongly affected by the solar wind than is dust, following magnetic field lines rather than an orbital trajectory. On occasions - such as when the Earth passes through a comet's orbital plane, a

Figure 241: *Diagram of a comet showing the dust trail (or antitail), the dust tail, and the ion gas tail, which is formed by the solar wind flow.*

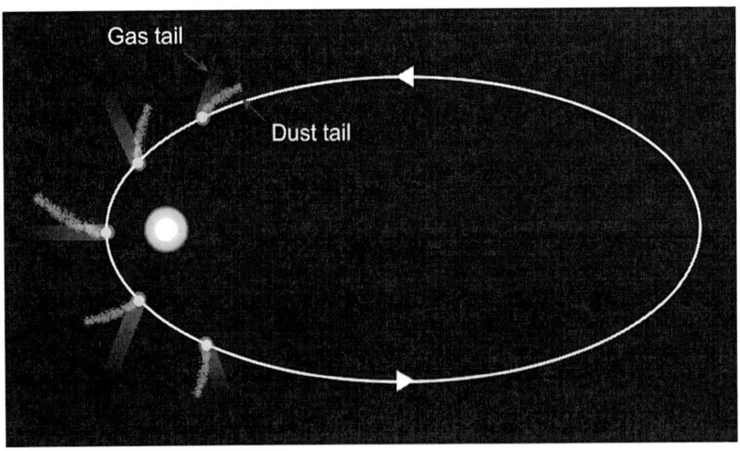

Figure 242: *Typical direction of tails over a comet's orbit near the Sun*

tail pointing in the opposite direction to the ion and dust tails called the antitail may be seen.

The observation of antitails contributed significantly to the discovery of solar wind. The ion tail is formed as a result of the ionisation by solar ultra-violet radiation of particles in the coma. Once the particles have been ionized, they attain a net positive electrical charge, which in turn gives rise to an "induced magnetosphere" around the comet. The comet and its induced magnetic field form an obstacle to outward flowing solar wind particles. Because the relative orbital speed of the comet and the solar wind is supersonic, a bow shock is formed upstream of the comet in the flow direction of the solar wind. In this bow shock, large concentrations of cometary ions (called "pick-up ions") congregate and act to "load" the solar magnetic field with plasma, such that the field lines "drape" around the comet forming the ion tail.

If the ion tail loading is sufficient, the magnetic field lines are squeezed together to the point where, at some distance along the ion tail, magnetic reconnection occurs. This leads to a "tail disconnection event". This has been observed on a number of occasions, one notable event being recorded on 20 April 2007, when the ion tail of Encke's Comet was completely severed while the comet passed through a coronal mass ejection. This event was observed by the STEREO space probe.

In 2013, ESA scientists reported that the ionosphere of the planet Venus streams outwards in a manner similar to the ion tail seen streaming from a comet under similar conditions."

Jets

Uneven heating can cause newly generated gases to break out of a weak spot on the surface of comet's nucleus, like a geyser.[378] These streams of gas and dust can cause the nucleus to spin, and even split apart. In 2010 it was revealed dry ice (frozen carbon dioxide) can power jets of material flowing out of a comet nucleus.[379] Infrared imaging of Hartley 2 shows such jets exiting and carrying with it dust grains into the coma.

Orbital characteristics

Most comets are small Solar System bodies with elongated elliptical orbits that take them close to the Sun for a part of their orbit and then out into the further reaches of the Solar System for the remainder. Comets are often classified according to the length of their orbital periods: The longer the period the more elongated the ellipse.

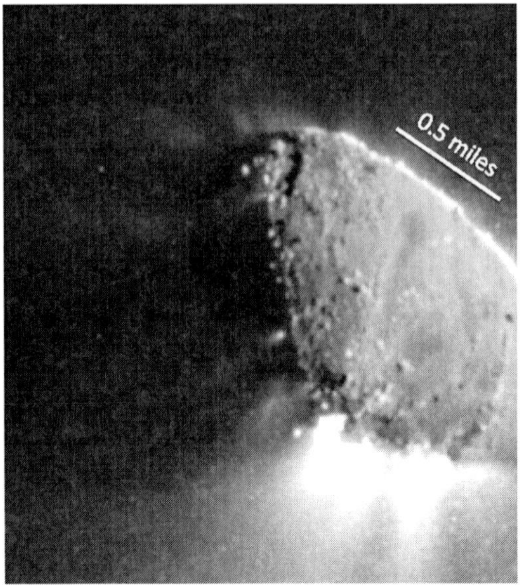

Figure 243: *Gas and snow jets of 103P/Hartley*

Short period

Periodic comets or short-period comets are generally defined as those having orbital periods of less than 200 years. They usually orbit more-or-less in the ecliptic plane in the same direction as the planets. Their orbits typically take them out to the region of the outer planets (Jupiter and beyond) at aphelion; for example, the aphelion of Halley's Comet is a little beyond the orbit of Neptune. Comets whose aphelia are near a major planet's orbit are called its "family". Such families are thought to arise from the planet capturing formerly long-period comets into shorter orbits.

At the shorter orbital period extreme, Encke's Comet has an orbit that does not reach the orbit of Jupiter, and is known as an **Encke-type comet**. Short-period comets with orbital periods less than 20 years and low inclinations (up to 30 degrees) to the ecliptic are called **Jupiter-family comets** (JFCs). Those like Halley, with orbital periods of between 20 and 200 years and inclinations extending from zero to more than 90 degrees, are called **Halley-type comets** (HTCs). As of 2017[375], only 87 HTCs have been observed, compared with 567 identified JFCs.

Recently discovered main-belt comets form a distinct class, orbiting in more circular orbits within the asteroid belt.

Because their elliptical orbits frequently take them close to the giant planets, comets are subject to further gravitational perturbations. Short-period comets have a tendency for their aphelia to coincide with a giant planet's semi-major axis, with the JFCs being the largest group. It is clear that comets coming in from the Oort cloud often have their orbits strongly influenced by the gravity of giant planets as a result of a close encounter. Jupiter is the source of the greatest perturbations, being more than twice as massive as all the other planets combined. These perturbations can deflect long-period comets into shorter orbital periods.

Based on their orbital characteristics, short-period comets are thought to originate from the centaurs and the Kuiper belt/scattered disc —a disk of objects in the trans-Neptunian region—whereas the source of long-period comets is thought to be the far more distant spherical Oort cloud (after the Dutch astronomer Jan Hendrik Oort who hypothesised its existence). Vast swarms of comet-like bodies are thought to orbit the Sun in these distant regions in roughly circular orbits. Occasionally the gravitational influence of the outer planets (in the case of Kuiper belt objects) or nearby stars (in the case of Oort cloud objects) may throw one of these bodies into an elliptical orbit that takes it inwards toward the Sun to form a visible comet. Unlike the return of periodic comets, whose orbits have been established by previous observations, the appearance of new comets by this mechanism is unpredictable.

Long period

Long-period comets have highly eccentric orbits and periods ranging from 200 years to thousands of years. An eccentricity greater than 1 when near perihelion does not necessarily mean that a comet will leave the Solar System. For example, Comet McNaught had a heliocentric osculating eccentricity of 1.000019 near its perihelion passage epoch in January 2007 but is bound to the Sun with roughly a 92,600-year orbit because the eccentricity drops below 1 as it moves further from the Sun. The future orbit of a long-period comet is properly obtained when the osculating orbit is computed at an epoch after leaving the planetary region and is calculated with respect to the center of mass of the Solar System. By definition long-period comets remain gravitationally bound to the Sun; those comets that are ejected from the Solar System due to close passes by major planets are no longer properly considered as having "periods". The orbits of long-period comets take them far beyond the outer planets at aphelia, and the plane of their orbits need not lie near the ecliptic. Long-period comets such as Comet West and C/1999 F1 can have aphelion distances of nearly 70,000 AU with orbital periods estimated around 6 million years.

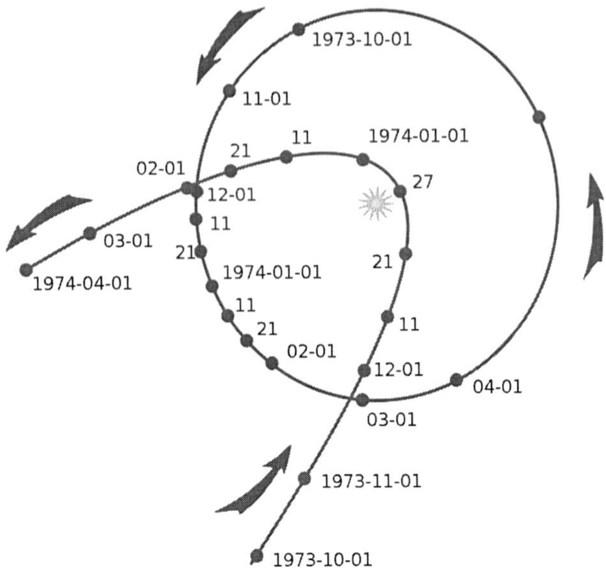

Figure 244: *Orbits of Comet Kohoutek (red) and the Earth (blue), illustrating the high eccentricity of its orbit and its rapid motion when close to the Sun.*

Figure 245: *Comets C/2012 F6 (Lemmon) (top) and C/2011 L4 (PANSTARRS) (bottom)*

Single-apparition or non-periodic comets are similar to long-period comets because they also have parabolic or slightly hyperbolic trajectories when near perihelion in the inner Solar System. However, gravitational perturbations from giant planets cause their orbits to change. Single-apparition comets have a hyperbolic or parabolic osculating orbit which allows them to permanently exit the Solar System after a single pass of the Sun. The Sun's Hill sphere has an unstable maximum boundary of 230,000 AU (1.1 parsecs (3.6 light-years)). Only a few hundred comets have been seen to reach a hyperbolic orbit ($e > 1$) when near perihelion that using a heliocentric unperturbed two-body best-fit suggests they may escape the Solar System.

No comets with an eccentricity significantly greater than one have been observed, so there are no confirmed observations of comets that are likely to have originated outside the Solar System. Comet C/1980 E1 had an orbital period of roughly 7.1 million years before the 1982 perihelion passage, but a 1980 encounter with Jupiter accelerated the comet giving it the largest eccentricity (1.057) of any known hyperbolic comet. Comets not expected to return to the inner Solar System include C/1980 E1, C/2000 U5, C/2001 Q4 (NEAT), C/2009 R1, C/1956 R1, and C/2007 F1 (LONEOS).

Some authorities use the term "periodic comet" to refer to any comet with a periodic orbit (that is, all short-period comets plus all long-period comets), whereas others use it to mean exclusively short-period comets. Similarly, although the literal meaning of "non-periodic comet" is the same as "single-apparition comet", some use it to mean all comets that are not "periodic" in the second sense (that is, to also include all comets with a period greater than 200 years).

Early observations have revealed a few genuinely hyperbolic (i.e. non-periodic) trajectories, but no more than could be accounted for by perturbations from Jupiter. If comets pervaded interstellar space, they would be moving with velocities of the same order as the relative velocities of stars near the Sun (a few tens of km per second). If such objects entered the Solar System, they would have positive specific orbital energy and would be observed to have genuinely hyperbolic trajectories. A rough calculation shows that there might be four hyperbolic comets per century within Jupiter's orbit, give or take one and perhaps two orders of magnitude.

Year	2007	2008	2009	2010	2011	2012	2013	2014	2015	2016
Number	12	7	8	4	13	10	16	10	15	4

|+ Hyperbolic comet discoveries

Oort cloud and Hills cloud

The Oort cloud is thought to occupy a vast space starting from between 2,000 and 5,000 AU (0.03 and 0.08 ly) to as far as 50,000 AU (0.79 ly) from the Sun. Some estimates place the outer edge at between 100,000 and 200,000 AU (1.58 and 3.16 ly). The region can be subdivided into a spherical outer Oort cloud of 20,000–50,000 AU (0.32–0.79 ly), and a doughnut-shaped inner cloud, the Hills cloud, of 2,000–20,000 AU (0.03–0.32 ly). The outer cloud is only weakly bound to the Sun and supplies the long-period (and possibly Halley-type) comets that fall to inside the orbit of Neptune. The inner Oort cloud is also known as the Hills cloud, named after J. G. Hills, who proposed its existence in 1981. Models predict that the inner cloud should have tens or hundreds of times as many cometary nuclei as the outer halo; it is seen as a possible source of new comets that resupply the relatively tenuous outer cloud as the latter's numbers are gradually depleted. The Hills cloud explains the continued existence of the Oort cloud after billions of years.

Exocomets

Exocomets beyond the Solar System have also been detected and may be common in the Milky Way. The first exocomet system detected was around Beta Pictoris, a very young A-type main-sequence star, in 1987. A total of 10 such exocomet systems have been identified as of 2013[375], using the absorption spectrum caused by the large clouds of gas emitted by comets when passing close to their star.

Effects of comets

Connection to meteor showers

As a result of outgassing, comets leave in their wake a trail of solid debris too large to be swept away by radiation pressure and the solar wind. If the Earth's orbit sends it though that debris, there are likely to be meteor showers as Earth passes through. The Perseid meteor shower, for example, occurs every year between 9 and 13 August, when Earth passes through the orbit of Comet Swift–Tuttle. Halley's Comet is the source of the Orionid shower in October.

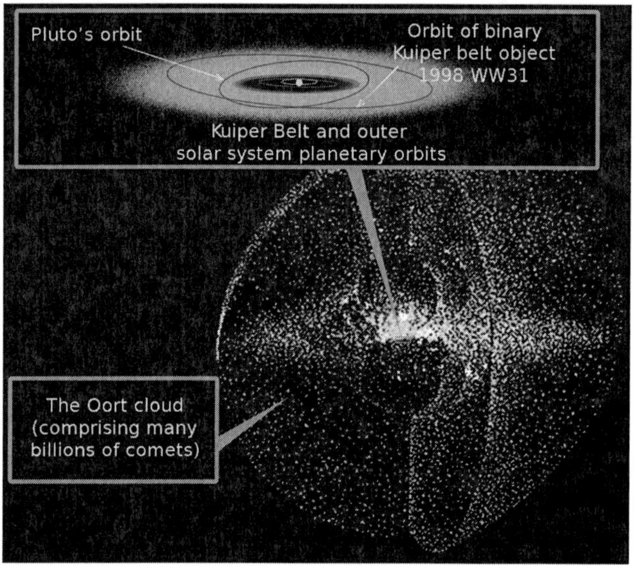

Figure 246: *The Oort cloud thought to surround the Solar System*

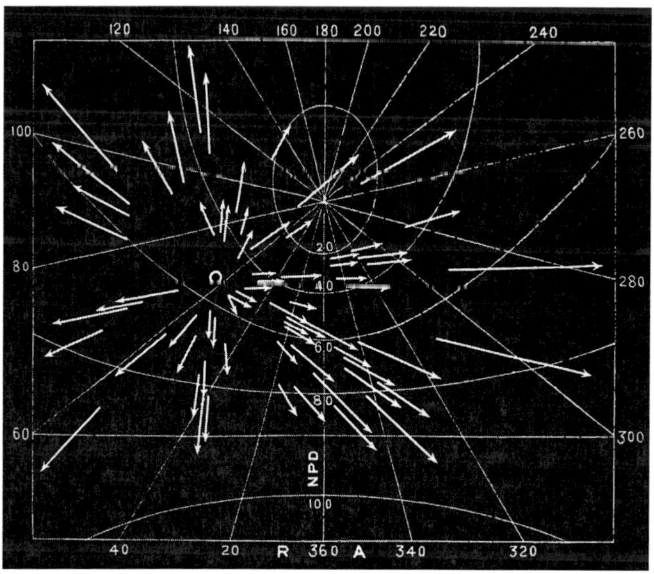

Figure 247: *Diagram of Perseids meteors*

Comets and impact on life

Many comets and asteroids collided with Earth in its early stages. Many scientists think that comets bombarding the young Earth about 4 billion years ago brought the vast quantities of water that now fill the Earth's oceans, or at least a significant portion of it. Others have cast doubt on this idea. The detection of organic molecules, including polycyclic aromatic hydrocarbons, in significant quantities in comets has led to speculation that comets or meteorites may have brought the precursors of life—or even life itself—to Earth. In 2013 it was suggested that impacts between rocky and icy surfaces, such as comets, had the potential to create the amino acids that make up proteins through shock synthesis. In 2015, scientists found significant amounts of molecular oxygen in the outgassings of comet 67P, suggesting that the molecule may occur more often than had been thought, and thus less an indicator of life as has been supposed.[380]

It is suspected that comet impacts have, over long timescales, also delivered significant quantities of water to the Earth's Moon, some of which may have survived as lunar ice. Comet and meteoroid impacts are also thought to be responsible for the existence of tektites and australites.

Fate of comets

Departure (ejection) from Solar System

If a comet is traveling fast enough, it may leave the Solar System. Such comets follow the open path of a hyperbola, and as such they are called hyperbolic comets. To date, comets are only known to be ejected by interacting with another object in the Solar System, such as Jupiter. An example of this is thought to be Comet C/1980 E1, which was shifted from a predicted orbit of 7.1 million years around the Sun, to a hyperbolic trajectory, after a 1980 close pass by the planet Jupiter.[381]

Volatiles exhausted

Jupiter-family comets and long-period comets appear to follow very different fading laws. The JFCs are active over a lifetime of about 10,000 years or $\sim$1,000 orbits whereas long-period comets fade much faster. Only 10% of the long-period comets survive more than 50 passages to small perihelion and only 1% of them survive more than 2,000 passages. Eventually most of the volatile material contained in a comet nucleus evaporates, and the comet becomes a small, dark, inert lump of rock or rubble that can resemble an asteroid. Some asteroids in elliptical orbits are now identified as extinct comets. Roughly six percent of the near-Earth asteroids are thought to be extinct comet nuclei.

Breakup and collisions

The nucleus of some comets may be fragile, a conclusion supported by the observation of comets splitting apart. A significant cometary disruption was that of Comet Shoemaker–Levy 9, which was discovered in 1993. A close encounter in July 1992 had broken it into pieces, and over a period of six days in July 1994, these pieces fell into Jupiter's atmosphere—the first time astronomers had observed a collision between two objects in the Solar System. Other splitting comets include 3D/Biela in 1846 and 73P/Schwassmann–Wachmann from 1995 to 2006. Greek historian Ephorus reported that a comet split apart as far back as the winter of 372–373 BC. Comets are suspected of splitting due to thermal stress, internal gas pressure, or impact.

Comets 42P/Neujmin and 53P/Van Biesbroeck appear to be fragments of a parent comet. Numerical integrations have shown that both comets had a rather close approach to Jupiter in January 1850, and that, before 1850, the two orbits were nearly identical.

Some comets have been observed to break up during their perihelion passage, including great comets West and Ikeya–Seki. Biela's Comet was one significant example, when it broke into two pieces during its passage through the perihelion in 1846. These two comets were seen separately in 1852, but never again afterward. Instead, spectacular meteor showers were seen in 1872 and 1885 when the comet should have been visible. A minor meteor shower, the Andromedids, occurs annually in November, and it is caused when the Earth crosses the orbit of Biela's Comet.

Some comets meet a more spectacular end – either falling into the Sun or smashing into a planet or other body. Collisions between comets and planets or moons were common in the early Solar System: some of the many craters on the Moon, for example, may have been caused by comets. A recent collision of a comet with a planet occurred in July 1994 when Comet Shoemaker–Levy 9 broke up into pieces and collided with Jupiter.

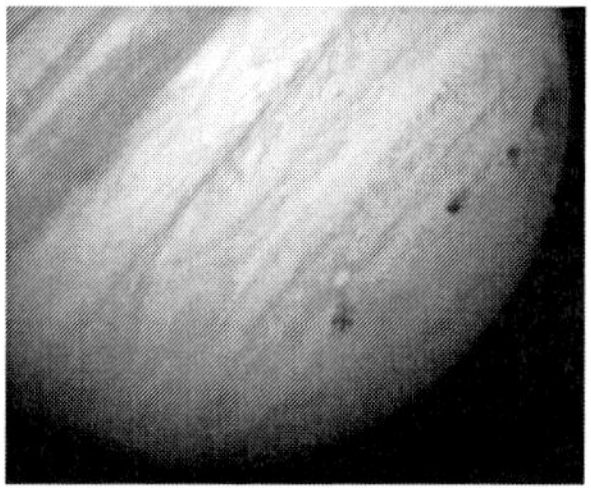

Brown spots mark impact sites of Comet Shoemaker–Levy 9 on Jupiter

The break up of 73P/Schwassmann–Wachmann within three days (1995)

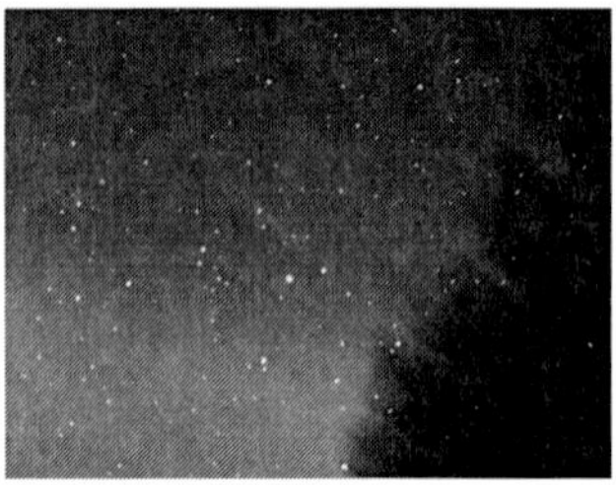

Ghost tail of C/2015 D1 (SOHO) after passage at the sun

Disintegration of P/2013 R3 (2014)

Nomenclature

The names given to comets have followed several different conventions over the past two centuries. Prior to the early 20th century, most comets were simply referred to by the year when they appeared, sometimes with additional adjectives for particularly bright comets; thus, the "Great Comet of 1680", the "Great Comet of 1882", and the "Great January Comet of 1910".

After Edmund Halley demonstrated that the comets of 1531, 1607, and 1682 were the same body and successfully predicted its return in 1759 by calculating its orbit, that comet became known as Halley's Comet. Similarly, the second and third known periodic comets, Encke's Comet and Biela's Comet, were named after the astronomers who calculated their orbits rather than their original discoverers. Later, periodic comets were usually named after their

Figure 248: *Halley's Comet in 1910*

discoverers, but comets that had appeared only once continued to be referred to by the year of their appearance.

In the early 20th century, the convention of naming comets after their discoverers became common, and this remains so today. A comet can be named after its discoverers, or an instrument or program that helped to find it.

History of study

Early observations and thought

From ancient sources, such as Chinese oracle bones, it is known that their appearances have been noticed by humans for millennia. Until the sixteenth century, comets were usually considered bad omens of deaths of kings or noble men, or coming catastrophes, or even interpreted as attacks by heavenly beings against terrestrial inhabitants.

Aristotle believed that comets were atmospheric phenomena, due to the fact that they could appear outside of the Zodiac and vary in brightness over the course of a few days. Pliny the Elder believed that comets were connected with political unrest and death.

Figure 249: *Halley's Comet appeared in 1066, prior to the Battle of Hastings (Bayeux Tapestry).*

In India, by the 6th century astronomers believed that comets were celestial bodies that re-appeared periodically. This was the view expressed in the 6th century by the astronomers Varāhamihira and Bhadrabahu, and the 10th-century astronomer Bhaṭṭotpala listed the names and estimated periods of certain comets, but it is not known how these figures were calculated or how accurate they were.

In the 16th century Tycho Brahe demonstrated that comets must exist outside the Earth's atmosphere by measuring the parallax of the Great Comet of 1577 from observations collected by geographically separated observers. Within the precision of the measurements, this implied the comet must be at least four times more distant than from the Earth to the Moon.

Orbital studies

Isaac Newton, in his *Principia Mathematica* of 1687, proved that an object moving under the influence of gravity must trace out an orbit shaped like one of the conic sections, and he demonstrated how to fit a comet's path through the sky to a parabolic orbit, using the comet of 1680 as an example.

In 1705, Edmond Halley (1656–1742) applied Newton's method to twenty-three cometary apparitions that had occurred between 1337 and 1698. He

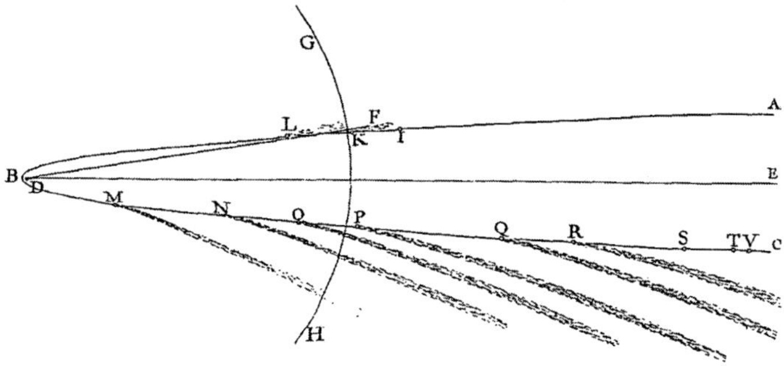

Figure 250: *The orbit of the comet of 1680, fitted to a parabola, as shown in Isaac Newton's Principia*

noted that three of these, the comets of 1531, 1607, and 1682, had very similar orbital elements, and he was further able to account for the slight differences in their orbits in terms of gravitational perturbation caused by Jupiter and Saturn. Confident that these three apparitions had been three appearances of the same comet, he predicted that it would appear again in 1758–9. Halley's predicted return date was later refined by a team of three French mathematicians: Alexis Clairaut, Joseph Lalande, and Nicole-Reine Lepaute, who predicted the date of the comet's 1759 perihelion to within one month's accuracy.[382] When the comet returned as predicted, it became known as Halley's Comet (with the latter-day designation of 1P/Halley). It will next appear in 2061.

Studies of physical characteristics

James Thomson *The Seasons* (1730; 1748)

Isaac Newton described comets as compact and durable solid bodies moving in oblique orbit and their tails as thin streams of vapor emitted by their nuclei, ignited or heated by the Sun. Newton suspected that comets were the origin of the life-supporting component of air.

As early as the 18th century, some scientists had made correct hypotheses as to comets' physical composition. In 1755, Immanuel Kant hypothesized that comets are composed of some volatile substance, whose vaporization gives rise to their brilliant displays near perihelion. In 1836, the German mathematician Friedrich Wilhelm Bessel, after observing streams of vapor during the appearance of Halley's Comet in 1835, proposed that the jet forces of evaporating material could be great enough to significantly alter a comet's orbit, and he argued that the non-gravitational movements of Encke's Comet resulted from this phenomenon.

In 1950, Fred Lawrence Whipple proposed that rather than being rocky objects containing some ice, comets were icy objects containing some dust and rock. This "dirty snowball" model soon became accepted and appeared to be supported by the observations of an armada of spacecraft (including the European Space Agency's *Giotto* probe and the Soviet Union's *Vega 1* and *Vega 2*) that flew through the coma of Halley's Comet in 1986, photographed the nucleus, and observed jets of evaporating material.

On 22 January 2014, ESA scientists reported the detection, for the first definitive time, of water vapor on the dwarf planet Ceres, the largest object in the asteroid belt. The detection was made by using the far-infrared abilities of the Herschel Space Observatory. The finding is unexpected because comets, not asteroids, are typically considered to "sprout jets and plumes". According to one of the scientists, "The lines are becoming more and more blurred between comets and asteroids." On 11 August 2014, astronomers released studies, using the Atacama Large Millimeter/Submillimeter Array (ALMA) for the first time, that detailed the distribution of HCN, HNC, H
2CO, and dust inside the comae of comets C/2012 F6 (Lemmon) and C/2012 S1 (ISON).

Spacecraft missions

- The Halley Armada describes the collection of spacecraft missions that visited and/or made observations of Halley's Comet 1980s perihelion.
- **Deep Impact**. Debate continues about how much ice is in a comet. In 2001, the *Deep Space 1* spacecraft obtained high-resolution images of the surface of Comet Borrelly. It was found that the surface of comet Borrelly is hot and dry, with a temperature of between 26 to 71 °C (79 to 160 °F), and extremely dark, suggesting that the ice has been removed by solar heating and maturation, or is hidden by the soot-like material that covers Borrelly. In July 2005, the *Deep Impact* probe blasted a crater on Comet Tempel 1 to study its interior. The mission yielded results suggesting that the majority of a comet's water ice is below the surface and that these reservoirs feed the jets of vaporised water that form the coma of Tempel 1. Renamed EPOXI, it made a flyby of Comet Hartley 2 on 4 November 2010.
- **Stardust**. Data from the Stardust mission show that materials retrieved from the tail of Wild 2 were crystalline and could only have been "born in fire," at extremely high temperatures of over 1,000 °C (1,830 °F). Although comets formed in the outer Solar System, radial mixing of material during the early formation of the Solar System is thought to

Figure 251: *Woodcut of the Great Comet of 1577*

have redistributed material throughout the proto-planetary disk. As a result, comets also contain crystalline grains that formed in the early, hot inner Solar System. This is seen in comet spectra as well as in sample return missions. More recent still, the materials retrieved demonstrate that the "comet dust resembles asteroid materials". These new results have forced scientists to rethink the nature of comets and their distinction from asteroids.

- **Rosetta**. The *Rosetta* probe orbited Comet Churyumov–Gerasimenko. On 12 November 2014, its lander Philae successfully landed on the comet's surface, the first time a spacecraft has ever landed on such an object in history.

Great comets

Approximately once a decade, a comet becomes bright enough to be noticed by a casual observer, leading such comets to be designated as great comets. Predicting whether a comet will become a great comet is notoriously difficult, as many factors may cause a comet's brightness to depart drastically from predictions. Broadly speaking, if a comet has a large and active nucleus, will pass close to the Sun, and is not obscured by the Sun as seen from the Earth when at its brightest, it has a chance of becoming a great comet. However, Comet

Kohoutek in 1973 fulfilled all the criteria and was expected to become spectacular but failed to do so. Comet West, which appeared three years later, had much lower expectations but became an extremely impressive comet.

The late 20th century saw a lengthy gap without the appearance of any great comets, followed by the arrival of two in quick succession—Comet Hyakutake in 1996, followed by Hale–Bopp, which reached maximum brightness in 1997 having been discovered two years earlier. The first great comet of the 21st century was C/2006 P1 (McNaught), which became visible to naked eye observers in January 2007. It was the brightest in over 40 years.

Sungrazing comets

A sungrazing comet is a comet that passes extremely close to the Sun at perihelion, generally within a few million kilometres. Although small sungrazers can be completely evaporated during such a close approach to the Sun, larger sungrazers can survive many perihelion passages. However, the strong tidal forces they experience often lead to their fragmentation.

About 90% of the sungrazers observed with SOHO are members of the Kreutz group, which all originate from one giant comet that broke up into many smaller comets during its first passage through the inner Solar System. The remainder contains some sporadic sungrazers, but four other related groups of comets have been identified among them: the Kracht, Kracht 2a, Marsden, and Meyer groups. The Marsden and Kracht groups both appear to be related to Comet 96P/Machholz, which is also the parent of two meteor streams, the Quadrantids and the Arietids.

Unusual comets

Of the thousands of known comets, some exhibit unusual properties. Comet Encke (2P/Encke) orbits from outside the asteroid belt to just inside the orbit of the planet Mercury whereas the Comet 29P/Schwassmann–Wachmann currently travels in a nearly circular orbit entirely between the orbits of Jupiter and Saturn. 2060 Chiron, whose unstable orbit is between Saturn and Uranus, was originally classified as an asteroid until a faint coma was noticed. Similarly, Comet Shoemaker–Levy 2 was originally designated asteroid 1990 UL_3. *(See also Fate of comets, above)*

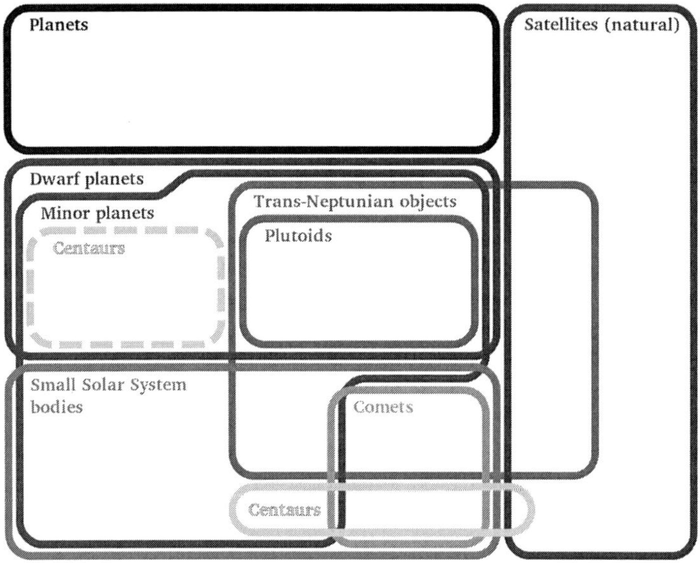

Figure 252: *Euler diagram showing the types of bodies in the Solar System.*

Centaurs

Centaurs typically behave with characteristics of both asteroids and comets. Centaurs can be classified as comets such as 60558 Echeclus, and 166P/NEAT. 166P/NEAT was discovered while it exhibited a coma, and so is classified as a comet despite its orbit, and 60558 Echeclus was discovered without a coma but later became active,[383] and was then classified as both a comet and an asteroid (174P/Echeclus). One plan for *Cassini* involved sending it to a centaur, but NASA decided to destroy it instead.

Observation

A comet may be discovered photographically using a wide-field telescope or visually with binoculars. However, even without access to optical equipment, it is still possible for the amateur astronomer to discover a sungrazing comet online by downloading images accumulated by some satellite observatories such as SOHO. SOHO's 2000th comet was discovered by Polish amateur astronomer Michał Kusiak on 26 December 2010 and both discoverers of Hale-Bopp used amateur equipment (although Hale was not an amateur).

Lost

A number of periodic comets discovered in earlier decades or previous centuries are now lost comets. Their orbits were never known well enough to predict future appearances or the comets have disintegrated. However, occasionally a "new" comet is discovered, and calculation of its orbit shows it to be an old "lost" comet. An example is Comet 11P/Tempel–Swift–LINEAR, discovered in 1869 but unobservable after 1908 because of perturbations by Jupiter. It was not found again until accidentally rediscovered by LINEAR in 2001. There are at least 18 comets that fit this category.

Gallery

Figure 253: *Comet C/2006 P1 (McNaught) taken from Victoria, Australia 2007*

Figure 254: *The Great Comet of 1882 is a member of the Kreutz group*

Figure 255: *Great Comet 1861*

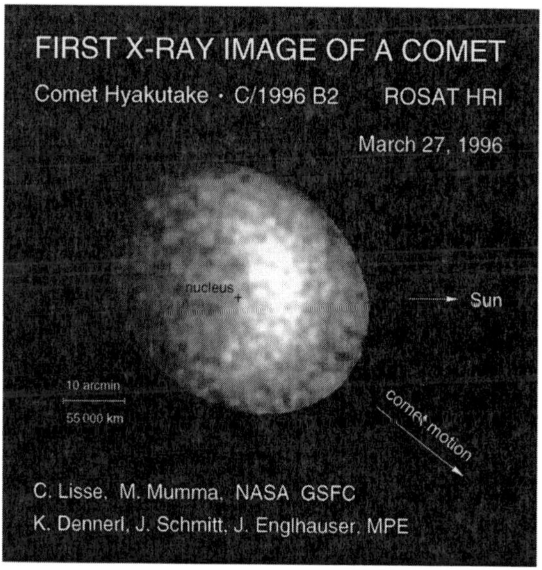

Figure 256: *Comet Hyakutake (X-ray, ROSAT satellite)*

Figure 257: *"Active asteroid" P/2013 P5 (PANSTARRS) with several tails. UNIQ-ref-0-3951097dbf03c794-QINU*

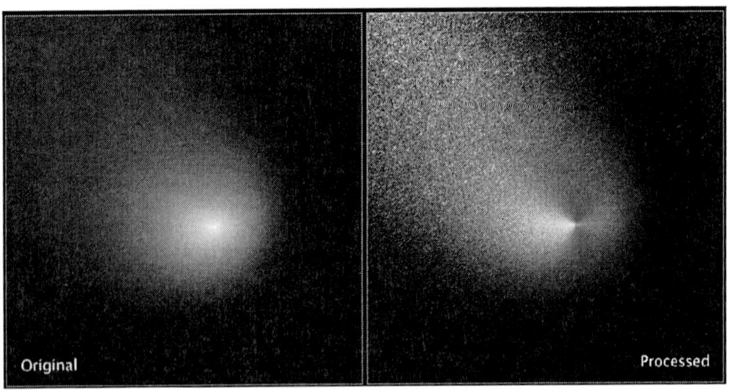

Figure 258: *Comet Siding Spring to pass near Mars on 19 October 2014 (Hubble; 11 March 2014)*

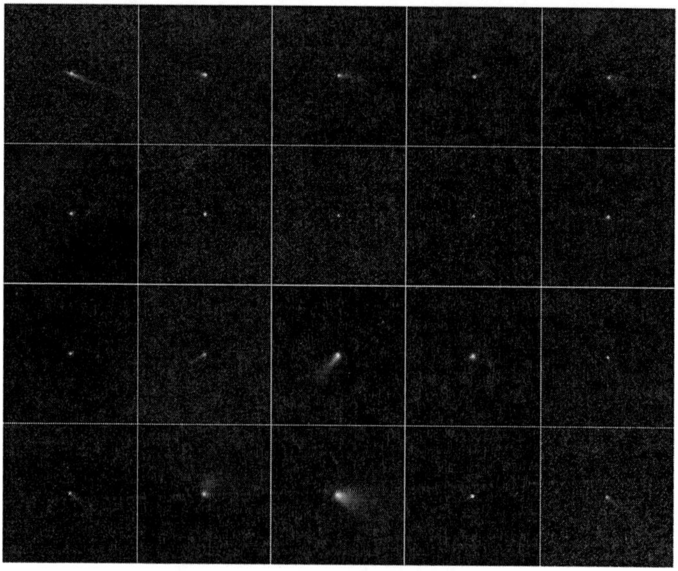

Figure 259: *Mosaic of 20 comets discovered by the WISE space telescope*

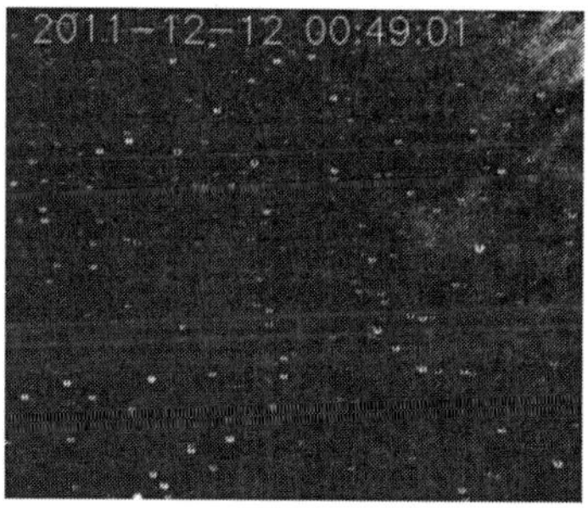

Figure 260: *C/2011 W3 (Lovejoy) heads towards the Sun*

Figure 261: *View from the impactor in its last moments before hitting the comet Tempel 1 in the Deep Impact mission*

Videos

NASA is developing a comet harpoon for returning samples to Earth.

Encke's Comet loses its tail

In popular culture

The depiction of comets in popular culture is firmly rooted in the long Western tradition of seeing comets as harbingers of doom and as omens of world-altering change. Halley's Comet alone has caused a slew of sensationalist publications of all sorts at each of its reappearances. It was especially noted that the birth and death of some notable persons coincided with separate appearances of the comet, such as with writers Mark Twain (who correctly speculated that he'd "go out with the comet" in 1910) and Eudora Welty, to whose life Mary Chapin Carpenter dedicated the song "Halley Came to Jackson".

In times past, bright comets often inspired panic and hysteria in the general population, being thought of as bad omens. More recently, during the passage of Halley's Comet in 1910, the Earth passed through the comet's tail, and erroneous newspaper reports inspired a fear that cyanogen in the tail might poison millions, whereas the appearance of Comet Hale–Bopp in 1997 triggered the mass suicide of the Heaven's Gate cult.

In science fiction, the impact of comets has been depicted as a threat overcome by technology and heroism (as in the 1998 films *Deep Impact* and *Armageddon*), or as a trigger of global apocalypse (*Lucifer's Hammer*, 1979) or zombies (*Night of the Comet*, 1984). In Jules Verne's *Off on a Comet* a group of people are stranded on a comet orbiting the Sun, while a large manned space expedition visits Halley's Comet in Sir Arthur C. Clarke's novel *2061: Odyssey Three*.

References

Bibliography

- Sagan, Carl; Druyan, Ann (1997). *Comet*[384]. London: Headline. ISBN 978-0-7472-7664-7.

Further reading

- Schechner, Sara J. (1997). *Comets, Popular Culture, and the Birth of Modern Cosmology*. Princeton University Press. ISBN 978-0-691-01150-9.
- Brandt, John C.; Chapman, Robert D. (2004). *Introduction to Comets* (2nd ed.). Cambridge University Press. ISBN 978-0-521-80863-7.

External links

- Comets[385] at DMOZ
- Comets Page[386] at NASA's Solar System Exploration[387]
- International Comet Quarterly[388]
- How to Make a Model of a Comet[389] audio slideshow – National High Magnetic Field Laboratory
- Catalogue of the Solar System Small Bodies Orbital Evolution[390]
- Information about comets and asteroids[391]

Kuiper belt

Types of distant minor planets

- Cis-Neptunian objects
  - Centaurs
  - Neptune trojans
- Trans-Neptunian objects (TNOs)‡
  - Kuiper belt objects (KBOs)
    - Classical KBOs (cubewanos)
    - Resonant KBOs
      - Plutinos (2:3 resonance)
  - Scattered disc objects (SDOs)
    - Resonant SDOs
  - Detached objects
  - Sednoids
  - Oort cloud objects (ICO/OCOs)

‡ Trans-Neptunian dwarf planets are called "plutoids"

- v
- t
- e[393]

The **Kuiper belt** (/ˈkaɪpər/ or Dutch pronunciation: [ˈkœy̯pər]),[394] occasionally called the **Edgeworth–Kuiper belt**, is a circumstellar disc in the Solar System beyond the (known) planets, extending from the orbit of Neptune (at 30 AU) to approximately 50 AU from the Sun. It is similar to the asteroid belt, but is far larger—20 times as wide and 20 to 200 times as massive. Like the asteroid belt, it consists mainly of small bodies or remnants from when the Solar System formed. While many asteroids are composed primarily of rock and metal, most Kuiper belt objects are composed largely of frozen volatiles (termed "ices"), such as methane, ammonia and water. The Kuiper belt is home to three officially recognized dwarf planets: Pluto, Haumea and Makemake. Some of the Solar System's moons, such as Neptune's Triton and Saturn's Phoebe, are thought to have originated in the region.[395]

The Kuiper belt was named after Dutch-American astronomer Gerard Kuiper, though he did not predict its existence. In 1992, 1992 QB_1 was discovered, the first Kuiper belt object (KBO) since Pluto. Since its discovery, the number of known KBOs has increased to over a thousand, and more than 100,000 KBOs over 100 km (62 mi) in diameter are thought to exist.[396] The Kuiper belt was initially thought to be the main repository for periodic comets, those with orbits lasting less than 200 years. Studies since the mid-1990s have shown that the belt is dynamically stable and that comets' true place of origin is the scattered disc, a dynamically active zone created by the outward motion of Neptune 4.5 billion years ago; scattered disc objects such as Eris have extremely eccentric orbits that take them as far as 100 AU from the Sun.[397] Because the

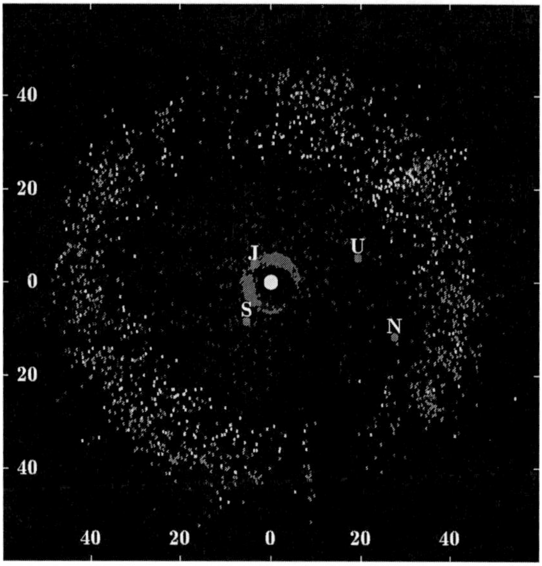

Figure 262:

Known objects in the Kuiper belt beyond the orbit of Neptune. (Scale in AU; epoch as of January 2015.)

Distances but not sizes are to scale

Source: Minor Planet Center, www<wbr/>.cfeps<wbr/>.net[392] *and others*

International Astronomical Union's Minor Planet Center, the body responsible for cataloguing minor planets in the Solar System, makes the distinction, the editorial choice for Wikipedia articles on the trans-Neptunian region is to make this distinction as well. On Wikipedia, Eris, the most-massive known trans-Neptunian object, is not part of the Kuiper belt and this makes Pluto the most-massive Kuiper belt object.</ref>

The Kuiper belt is distinct from the theoretical Oort cloud, which is a thousand times more distant and is mostly spherical. The objects within the Kuiper belt, together with the members of the scattered disc and any potential Hills cloud or Oort cloud objects, are collectively referred to as trans-Neptunian objects (TNOs). Pluto is the largest and most massive member of the Kuiper belt, and the largest and the second-most-massive known TNO, surpassed only by Eris in the scattered disc. Originally considered a planet, Pluto's status as part of the Kuiper belt caused it to be reclassified as a dwarf planet in 2006. It is compositionally similar to many other objects of the Kuiper belt and its orbital period is characteristic of a class of KBOs, known as "plutinos", that share the same 2:3 resonance with Neptune.

Figure 263: *Astronomer Gerard Kuiper, after whom the Kuiper belt is named*

History

After the discovery of Pluto in 1930, many speculated that it might not be alone. The region now called the Kuiper belt was hypothesized in various forms for decades. It was only in 1992 that the first direct evidence for its existence was found. The number and variety of prior speculations on the nature of the Kuiper belt have led to continued uncertainty as to who deserves credit for first proposing it.[398]

Hypotheses

The first astronomer to suggest the existence of a trans-Neptunian population was Frederick C. Leonard. Soon after Pluto's discovery by Clyde Tombaugh in 1930, Leonard pondered whether it was "not likely that in Pluto there has come to light the *first* of a *series* of ultra-Neptunian bodies, the remaining members of which still await discovery but which are destined eventually to be detected". That same year, astronomer Armin O. Leuschner suggested that Pluto "may be one of many long-period planetary objects yet to be discovered."

In 1943, in the *Journal of the British Astronomical Association*, Kenneth Edgeworth hypothesized that, in the region beyond Neptune, the material within the primordial solar nebula was too widely spaced to condense into planets, and so rather condensed into a myriad of smaller bodies. From this he concluded that "the outer region of the solar system, beyond the orbits of the planets, is occupied by a very large number of comparatively small bodies" and that, from time to time, one of their number "wanders from its own sphere and appears as an occasional visitor to the inner solar system",[399] becoming a comet.

In 1951, in a paper in *Astrophysics: A Topical Symposium*, Gerard Kuiper speculated on a similar disc having formed early in the Solar System's evolution, but he did not think that such a belt still existed today. Kuiper was operating on the assumption, common in his time, that Pluto was the size of Earth and had therefore scattered these bodies out toward the Oort cloud or out of the Solar System. Were Kuiper's hypothesis correct, there would not be a Kuiper belt today.

The hypothesis took many other forms in the following decades. In 1962, physicist Al G.W. Cameron postulated the existence of "a tremendous mass of small material on the outskirts of the solar system".[400] In 1964, Fred Whipple, who popularised the famous "dirty snowball" hypothesis for cometary structure, thought that a "comet belt" might be massive enough to cause the purported discrepancies in the orbit of Uranus that had sparked the search for Planet X, or, at the very least, massive enough to affect the orbits of known comets. Observation ruled out this hypothesis.

In 1977, Charles Kowal discovered 2060 Chiron, an icy planetoid with an orbit between Saturn and Uranus. He used a blink comparator, the same device that had allowed Clyde Tombaugh to discover Pluto nearly 50 years before. In 1992, another object, 5145 Pholus, was discovered in a similar orbit. Today, an entire population of comet-like bodies, called the centaurs, is known to exist in the region between Jupiter and Neptune. The centaurs' orbits are unstable and have dynamical lifetimes of a few million years. From the time of Chiron's discovery in 1977, astronomers have speculated that the centaurs therefore must be frequently replenished by some outer reservoir.[401]

Further evidence for the existence of the Kuiper belt later emerged from the study of comets. That comets have finite lifespans has been known for some time. As they approach the Sun, its heat causes their volatile surfaces to sublimate into space, gradually dispersing them. In order for comets to continue to be visible over the age of the Solar System, they must be replenished frequently. One such area of replenishment is the Oort cloud, a spherical swarm of comets extending beyond 50,000 AU from the Sun first hypothesised by Dutch astronomer Jan Oort in 1950. The Oort cloud is thought to be the point of origin of long-period comets, which are those, like Hale–Bopp, with orbits lasting thousands of years.[402]

Figure 264: *The array of telescopes atop Mauna Kea, with which the Kuiper belt was discovered*

There is another comet population, known as short-period or periodic comets, consisting of those comets that, like Halley's Comet, have orbital periods of less than 200 years. By the 1970s, the rate at which short-period comets were being discovered was becoming increasingly inconsistent with their having emerged solely from the Oort cloud.[403] For an Oort cloud object to become a short-period comet, it would first have to be captured by the giant planets. In a paper published in *Monthly Notices of the Royal Astronomical Society* in 1980, Uruguayan astronomer Julio Fernández stated that for every short-period comet to be sent into the inner Solar System from the Oort cloud, 600 would have to be ejected into interstellar space. He speculated that a comet belt from between 35 and 50 AU would be required to account for the observed number of comets. Following up on Fernández's work, in 1988 the Canadian team of Martin Duncan, Tom Quinn and Scott Tremaine ran a number of computer simulations to determine if all observed comets could have arrived from the Oort cloud. They found that the Oort cloud could not account for all short-period comets, particularly as short-period comets are clustered near the plane of the Solar System, whereas Oort-cloud comets tend to arrive from any point in the sky. With a "belt", as Fernández described it, added to the formulations, the simulations matched observations. Reportedly because the words "Kuiper" and "comet belt" appeared in the opening sentence of Fernández's paper, Tremaine named this hypothetical region the "Kuiper belt".[404]

Discovery

In 1987, astronomer David Jewitt, then at MIT, became increasingly puzzled by "the apparent emptiness of the outer Solar System". He encouraged then-graduate student Jane Luu to aid him in his endeavour to locate another object beyond Pluto's orbit, because, as he told her, "If we don't, nobody will."[405]

Using telescopes at the Kitt Peak National Observatory in Arizona and the Cerro Tololo Inter-American Observatory in Chile, Jewitt and Luu conducted their search in much the same way as Clyde Tombaugh and Charles Kowal had, with a blink comparator. Initially, examination of each pair of plates took about eight hours,[406] but the process was sped up with the arrival of electronic charge-coupled devices or CCDs, which, though their field of view was narrower, were not only more efficient at collecting light (they retained 90% of the light that hit them, rather than the 10% achieved by photographs) but allowed the blinking process to be done virtually, on a computer screen. Today, CCDs form the basis for most astronomical detectors.[407] In 1988, Jewitt moved to the Institute of Astronomy at the University of Hawaii. Luu later joined him to work at the University of Hawaii's 2.24 m telescope at Mauna Kea.[408] Eventually, the field of view for CCDs had increased to 1024 by 1024 pixels, which allowed searches to be conducted far more rapidly.[409] Finally, after five years of searching, Jewitt and Luu announced on August 30, 1992 the "Discovery of the candidate Kuiper belt object" (15760) 1992 QB1. Six months later, they discovered a second object in the region, (181708) 1993 FW.

Studies conducted since the trans-Neptunian region was first charted have shown that the region now called the Kuiper belt is not the point of origin of short-period comets, but that they instead derive from a linked population called the scattered disc. The scattered disc was created when Neptune migrated outward into the proto-Kuiper belt, which at the time was much closer to the Sun, and left in its wake a population of dynamically stable objects that could never be affected by its orbit (the Kuiper belt proper), and a population whose perihelia are close enough that Neptune can still disturb them as it travels around the Sun (the scattered disc). Because the scattered disc is dynamically active and the Kuiper belt relatively dynamically stable, the scattered disc is now seen as the most likely point of origin for periodic comets.

Name

Astronomers sometimes use the alternative name Edgeworth–Kuiper belt to credit Edgeworth, and KBOs are occasionally referred to as EKOs. Brian G. Marsden claims that neither deserves true credit: "Neither Edgeworth nor Kuiper wrote about anything remotely like what we are now seeing, but Fred Whipple did".[410] David Jewitt comments: "If anything... Fernández most nearly deserves the credit for predicting the Kuiper Belt."

KBOs are sometimes called "kuiperoids", a name suggested by Clyde Tombaugh.[411] The term "trans-Neptunian object" (TNO) is recommended for objects in the belt by several scientific groups because the term is less controversial than all others—it is not an exact synonym though, as TNOs include all objects orbiting the Sun past the orbit of Neptune, not just those in the Kuiper belt.

Figure 265: *Dust in the Kuiper belt creates a faint infrared disc.*

Structure

At its fullest extent (but excluding the scattered disc), including its outlying regions, the Kuiper belt stretches from roughly 30 to 55 AU. The main body of the belt is generally accepted to extend from the 2:3 mean-motion resonance (see below) at 39.5 AU to the 1:2 resonance at roughly 48 AU. The Kuiper belt is quite thick, with the main concentration extending as much as ten degrees outside the ecliptic plane and a more diffuse distribution of objects extending several times farther. Overall it more resembles a torus or doughnut than a belt. Its mean position is inclined to the ecliptic by 1.86 degrees.

The presence of Neptune has a profound effect on the Kuiper belt's structure due to orbital resonances. Over a timescale comparable to the age of the Solar System, Neptune's gravity destabilises the orbits of any objects that happen to lie in certain regions, and either sends them into the inner Solar System or out into the scattered disc or interstellar space. This causes the Kuiper belt to have pronounced gaps in its current layout, similar to the Kirkwood gaps in the asteroid belt. In the region between 40 and 42 AU, for instance, no objects can retain a stable orbit over such times, and any observed in that region must have migrated there relatively recently.

Classical belt

Between the 2:3 and 1:2 resonances with Neptune, at approximately 42–48 AU, the gravitational interactions with Neptune occur over an extended timescale, and objects can exist with their orbits essentially unaltered. This region is known as the classical Kuiper belt, and its members comprise roughly two thirds of KBOs observed to date. Because the first modern KBO discovered, (15760) 1992 QB1, is considered the prototype of this group, classical KBOs are often referred to as cubewanos ("Q-B-1-os"). The guidelines established by the IAU demand that classical KBOs be given names of mythological beings associated with creation.

The classical Kuiper belt appears to be a composite of two separate populations. The first, known as the "dynamically cold" population, has orbits much like the planets; nearly circular, with an orbital eccentricity of less than 0.1, and with relatively low inclinations up to about 10° (they lie close to the plane of the Solar System rather than at an angle). The cold population also contain a concentration of objects, referred to as the kernel, with semi-major axes at 44–44.5 AU. The second, the "dynamically hot" population, has orbits much more inclined to the ecliptic, by up to 30°. The two populations have been named this way not because of any major difference in temperature, but from analogy to particles in a gas, which increase their relative velocity as they become heated up. Not only are the two populations in different orbits, the cold population also differs in color and albedo, being redder and brighter, has a larger fraction of binary objects, has a different size distribution, and lacks very large objects. The difference in colors may be a reflection of different compositions, which suggests they formed in different regions. The hot population is proposed to have formed near Neptune's original orbit and to have been scattered out during the migration of the giant planets. The cold population, on the other hand, has been proposed to have formed more or less in its current position because the loose binaries would be unlikely to survive encounters with Neptune. Although the Nice model appears to be able to at least partially explain a compositional difference, it has also been suggested the color difference may reflect differences in surface evolution.

Resonances

When an object's orbital period is an exact ratio of Neptune's (a situation called a mean-motion resonance), then it can become locked in a synchronised motion with Neptune and avoid being perturbed away if their relative alignments are appropriate. If, for instance, an object orbits the Sun twice for every three Neptune orbits, and if it reaches perihelion with Neptune a quarter of an orbit away from it, then whenever it returns to perihelion, Neptune will always be in about the same relative position as it began, because it will have completed $1\ ^1/_2$ orbits in the same time. This is known as the 2:3 (or 3:2) resonance, and it corresponds to a characteristic semi-major axis of about 39.4 AU. This 2:3 resonance is populated by about 200 known objects, including Pluto together with its moons. In recognition of this, the members of this family are known

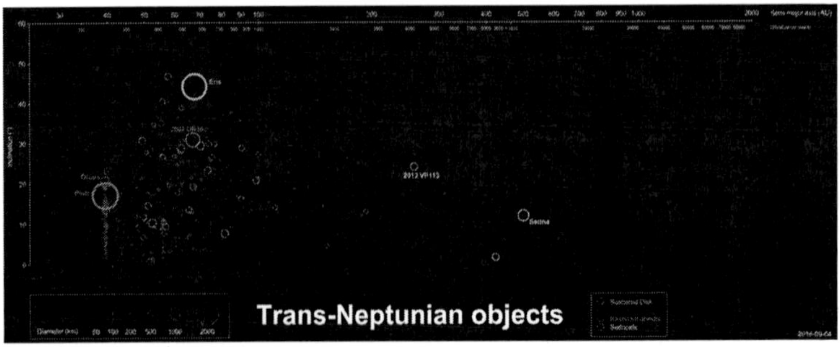

Figure 266: *Distribution of cubewanos (blue), Resonant trans-Neptunian objects (red), Sednoids (yellow) and scattered objects (grey)*

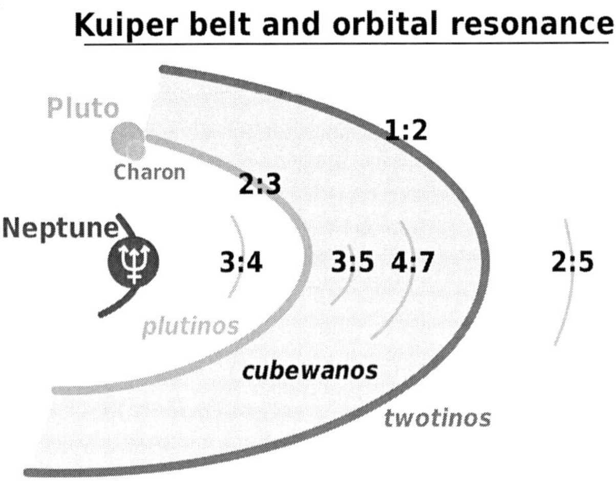

Figure 267: *Orbit classification (schematic of semi-major axes)*

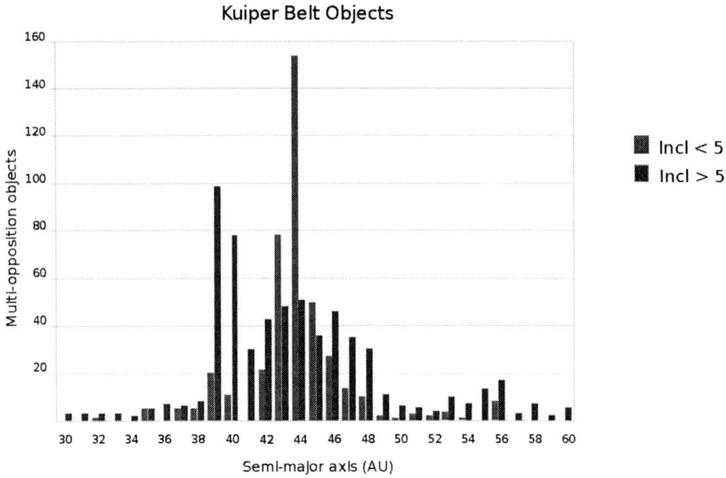

Figure 268: *Histogram of the semi-major axes of Kuiper belt objects with inclinations above and below 5 degrees. Spikes from the plutinos and the 'kernel' are visible at 39–40 AU and 44 AU.*

as plutinos. Many plutinos, including Pluto, have orbits that cross that of Neptune, though their resonance means they can never collide. Plutinos have high orbital eccentricities, suggesting that they are not native to their current positions but were instead thrown haphazardly into their orbits by the migrating Neptune. IAU guidelines dictate that all plutinos must, like Pluto, be named for underworld deities. The 1:2 resonance (whose objects complete half an orbit for each of Neptune's) corresponds to semi-major axes of ~47.7AU, and is sparsely populated. Its residents are sometimes referred to as twotinos. Other resonances also exist at 3:4, 3:5, 4:7 and 2:5.[412] Neptune has a number of trojan objects, which occupy its Lagrangian points, gravitationally stable regions leading and trailing it in its orbit. Neptune trojans are in a 1:1 mean-motion resonance with Neptune and often have very stable orbits.

Additionally, there is a relative absence of objects with semi-major axes below 39 AU that cannot apparently be explained by the present resonances. The currently accepted hypothesis for the cause of this is that as Neptune migrated outward, unstable orbital resonances moved gradually through this region, and thus any objects within it were swept up, or gravitationally ejected from it.[413]

Kuiper cliff

The 1:2 resonance appears to be an edge beyond which few objects are known. It is not clear whether it is actually the outer edge of the classical belt or just

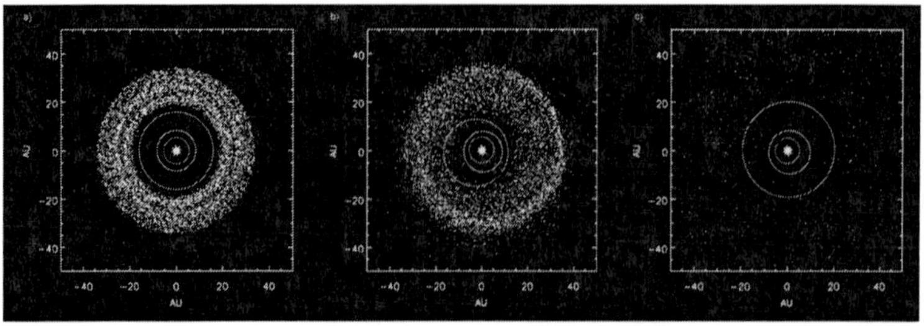

Figure 269: *Simulation showing outer planets and Kuiper belt: a) before Jupiter/Saturn 1:2 resonance, b) scattering of Kuiper belt objects into the Solar System after the orbital shift of Neptune, c) after ejection of Kuiper belt bodies by Jupiter*

the beginning of a broad gap. Objects have been detected at the 2:5 resonance at roughly 55 AU, well outside the classical belt; predictions of a large number of bodies in classical orbits between these resonances have not been verified through observation.

Based on estimations of the primordial mass required to form Uranus and Neptune, as well as bodies as large as Pluto (see below), earlier models of the Kuiper belt had suggested that the number of large objects would increase by a factor of two beyond 50 AU, so this sudden drastic falloff, known as the *Kuiper cliff*, was unexpected, and to date its cause is unknown. In 2003, Bernstein, Trilling, et al. found evidence that the rapid decline in objects of 100 km or more in radius beyond 50 AU is real, and not due to observational bias. Possible explanations include that material at that distance was too scarce or too scattered to accrete into large objects, or that subsequent processes removed or destroyed those that did. Patryk Lykawka of Kobe University claimed that the gravitational attraction of an unseen large planetary object, perhaps the size of Earth or Mars, might be responsible.

Origin

The precise origins of the Kuiper belt and its complex structure are still unclear, and astronomers are awaiting the completion of several wide-field survey telescopes such as Pan-STARRS and the future LSST, which should reveal many currently unknown KBOs. These surveys will provide data that will help determine answers to these questions.

The Kuiper belt is thought to consist of planetesimals, fragments from the original protoplanetary disc around the Sun that failed to fully coalesce into planets and instead formed into smaller bodies, the largest less than 3,000 kilometres

(1,900 mi) in diameter. Studies of the crater counts on Pluto and Charon revealed a scarcity of small craters suggesting that such objects formed directly as sizeable objects in the range of tens of kilometers in diameter rather than being accreted from much smaller, roughly kilometer scale bodies. Hypothetical mechanisms for the formation of these larger bodies include the gravitational collapse of clouds of pebbles concentrated between eddies in a turbulent protoplanetary disk or in streaming instabilities. These collapsing clouds may fragment, forming binaries.

Modern computer simulations show the Kuiper belt to have been strongly influenced by Jupiter and Neptune, and also suggest that neither Uranus nor Neptune could have formed in their present positions, because too little primordial matter existed at that range to produce objects of such high mass. Instead, these planets are estimated to have formed closer to Jupiter. Scattering of planetesimals early in the Solar System's history would have led to migration of the orbits of the giant planets: Saturn, Uranus, and Neptune drifted outwards, whereas Jupiter drifted inwards. Eventually, the orbits shifted to the point where Jupiter and Saturn reached an exact 1:2 resonance; Jupiter orbited the Sun twice for every one Saturn orbit. The gravitational repercussions of such a resonance ultimately destabilized the orbits of Uranus and Neptune, causing them to be scattered outward onto high-eccentricity orbits that crossed the primordial planetesimal disc. While Neptune's orbit was highly eccentric, its mean-motion resonances overlapped and the orbits of the planetesimals evolved chaotically, allowing planetesimals to wander outward as far as Neptune's 1:2 resonance to form a dynamically cold belt of low-inclination objects. Later, after its eccentricity decreased, Neptune's orbit expanded outward toward its current position. Many planetesimals were captured into and remain in resonances during this migration, others evolved onto higher-inclination and lower-eccentricity orbits and escaped from the resonances onto stable orbits. Many more planetesimals were scattered inward, with small fractions being captured as Jupiter trojans, as irregular satellites orbiting the giant planets, and as outer belt asteroids. The remainder were scattered outward again by Jupiter and in most cases ejected from the Solar System reducing the primordial Kuiper belt population by 99% or more.

The original version of the currently most popular model, the "Nice model", reproduces many characteristics of the Kuiper belt such as the "cold" and "hot" populations, resonant objects, and a scattered disc, but it still fails to account for some of the characteristics of their distributions. The model predicts a higher average eccentricity in classical KBO orbits than is observed (0.10–0.13 versus 0.07) and its predicted inclination distribution contains too few high inclination objects. In addition, the frequency of binary objects in the cold belt, many of which are far apart and loosely bound, also poses a problem for the model. These are predicted to have been separated during encounters with Neptune, leading some to propose that the cold disc formed at its current location, representing the only truly local population of small bodies in the solar system.

A recent modification of the Nice model has the Solar System begin with five giant planets, including an additional ice giant, in a chain of mean-motion resonances. About 400 million years after the formation of the Solar System the resonance chain is broken. Instead of being scattered into the disc, the ice giants first migrate outward several AU. This divergent migration eventually leads to a resonance crossing, destabilizing the orbits of the planets. The extra ice giant encounters Saturn and is scattered inward onto a Jupiter-crossing orbit and after a series of encounters is ejected from the Solar System. The remaining planets then continue their migration until the planetesimal disc is nearly depleted with small fractions remaining in various locations.

As in the original Nice model, objects are captured into resonances with Neptune during its outward migration. Some remain in the resonances, others evolve onto higher-inclination, lower-eccentricity orbits, and are released onto stable orbits forming the dynamically hot classical belt. The hot belt's inclination distribution can be reproduced if Neptune migrated from 24 AU to 30 AU on a 30 Myr timescale. When Neptune migrates to 28 AU, it has a gravitational encounter with the extra ice giant. Objects captured from the cold belt into the 1:2 mean-motion resonance with Neptune are left behind as a local concentration at 44 AU when this encounter causes Neptune's semi-major axis to jump outward. The objects deposited in the cold belt include some loosely bound 'blue' binaries originating from closer than the cold belt's current location. If Neptune's eccentricity remains small during this encounter, the chaotic evolution of orbits of the original Nice model is avoided and a primordial cold belt is preserved. In the later phases of Neptune's migration, a slow sweeping of mean-motion resonances removes the higher-eccentricity objects from the cold belt, truncating its eccentricity distribution.

Composition

Being distant from the Sun and major planets, Kuiper belt objects are thought to be relatively unaffected by the processes that have shaped and altered other Solar System objects; thus, determining their composition would provide substantial information on the makeup of the earliest Solar System. Due to their small size and extreme distance from Earth, the chemical makeup of KBOs is very difficult to determine. The principal method by which astronomers determine the composition of a celestial object is spectroscopy. When an object's light is broken into its component colors, an image akin to a rainbow is formed.

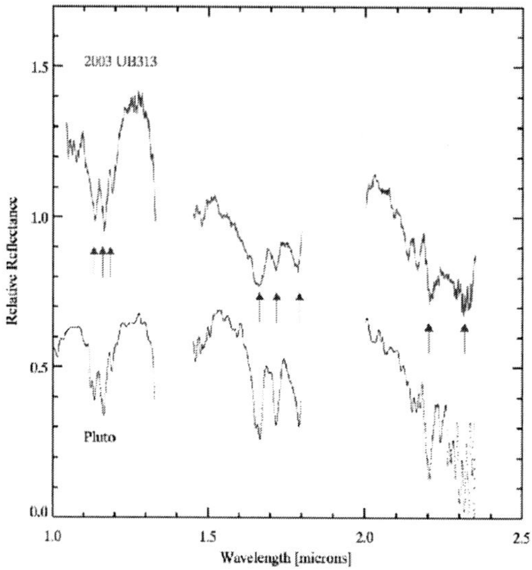

Figure 270: *The infrared spectra of both Eris and Pluto, highlighting their common methane absorption lines*

This image is called a spectrum. Different substances absorb light at different wavelengths, and when the spectrum for a specific object is unravelled, dark lines (called absorption lines) appear where the substances within it have absorbed that particular wavelength of light. Every element or compound has its own unique spectroscopic signature, and by reading an object's full spectral "fingerprint", astronomers can determine its composition.

Analysis indicates that Kuiper belt objects are composed of a mixture of rock and a variety of ices such as water, methane, and ammonia. The temperature of the belt is only about 50 K, so many compounds that would be gaseous closer to the Sun remain solid. The densities and rock–ice fractions are known for only a small number of objects for which the diameters and the masses have been determined. The diameter can be determined by imaging with a high-resolution telescope such as the Hubble Space Telescope, by the timing of an occultation when an object passes in front of a star or, most commonly, by using the albedo of an object calculated from its infrared emissions. The masses are determined using the semi-major axes and periods of satellites, which are therefore known only for a few binary objects. The densities range from less than 0.4 to 2.6 g/cm^3. The least dense objects are thought to be largely composed of ice and have significant porosity. The densest objects are likely composed of rock with a thin crust of ice. There is a trend of low densities for small objects and high densities for the largest objects. One possible explanation for this trend is that ice was lost from the surface layers when differentiated objects collided to form the largest objects.

Initially, detailed analysis of KBOs was impossible, and so astronomers were only able to determine the most basic facts about their makeup, primarily their color. These first data showed a broad range of colors among KBOs, ranging from neutral grey to deep red. This suggested that their surfaces were composed of a wide range of compounds, from dirty ices to hydrocarbons. This diversity was startling, as astronomers had expected KBOs to be uniformly dark, having lost most of the volatile ices from their surfaces to the effects of cosmic rays.[414] Various solutions were suggested for this discrepancy, including resurfacing by impacts or outgassing. Jewitt and Luu's spectral analysis of the known Kuiper belt objects in 2001 found that the variation in color was too extreme to be easily explained by random impacts. The radiation from the Sun is thought to have chemically altered methane on the surface of KBOs, producing products such as tholins. Makemake has been shown to possess a number of hydrocarbons derived from the radiation-processing of methane, including ethane, ethylene and acetylene.

Although to date most KBOs still appear spectrally featureless due to their faintness, there have been a number of successes in determining their composition. In 1996, Robert H. Brown et al. acquired spectroscopic data on the KBO 1993 SC, which revealed that its surface composition is markedly similar to that of Pluto, as well as Neptune's moon Triton, with large amounts of methane ice. For the smaller objects, only colors and in some cases the albedos have been determined. These objects largely fall into two classes: gray with low albedos, or very red with higher albedos. The difference in colors and albedos is hypothesized to be due to the retention or the loss of hydrogen sulfide (H_2S) on the surface of these objects, with the surfaces of those that formed far enough from the Sun to retain H_2S being reddened due to irradiation.

The largest KBOs, such as Pluto and Quaoar, have surfaces rich in volatile compounds such as methane, nitrogen and carbon monoxide; the presence of these molecules is likely due to their moderate vapor pressure in the 30–50 K temperature range of the Kuiper belt. This allows them to occasionally boil off their surfaces and then fall again as snow, whereas compounds with higher boiling points would remain solid. The relative abundances of these three compounds in the largest KBOs is directly related to their surface gravity and ambient temperature, which determines which they can retain. Water ice has been detected in several KBOs, including members of the Haumea family such as 1996 TO_{66}, mid-sized objects such as 38628 Huya and 20000 Varuna, and

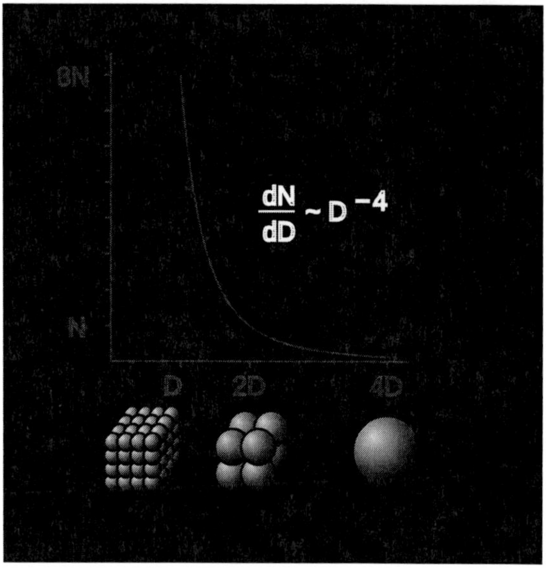

Figure 271: *Illustration of the power law*

also on some small objects. The presence of crystalline ice on large and mid-sized objects, including 50000 Quaoar where ammonia hydrate has also been detected, may indicate past tectonic activity aided by melting point lowering due to the presence of ammonia.

Mass and size distribution

Despite its vast extent, the collective mass of the Kuiper belt is relatively low. The total mass is estimated to range between 1/25 and 1/10 the mass of the Earth. Conversely, models of the Solar System's formation predict a collective mass for the Kuiper belt of 30 Earth masses. This missing >99% of the mass can hardly be dismissed, because it is required for the accretion of any KBOs larger than 100 km (62 mi) in diameter. If the Kuiper belt had always had its current low density, these large objects simply could not have formed by the collision and mergers of smaller planetesimals. Moreover, the eccentricity and inclination of current orbits makes the encounters quite "violent" resulting in destruction rather than accretion. It appears that either the current residents of the Kuiper belt have been created closer to the Sun, or some mechanism dispersed the original mass. Neptune's current influence is too weak to explain such a massive "vacuuming", though the Nice model proposes that it could have been the cause of mass removal in the past. Although the question remains open, the conjectures vary from a passing star scenario to grinding of smaller objects, via collisions, into dust small enough to be affected by solar radiation. The extent of mass loss by collisional grinding is limited by the presence of loosely bound binaries in the cold disk, which are likely to be disrupted in collisions.

Bright objects are rare compared with the dominant dim population, as expected from accretion models of origin, given that only some objects of a given size would have grown further. This relationship between $N(D)$ (the number of objects of diameter greater than D) and D, referred to as brightness slope, has been confirmed by observations. The slope is inversely proportional to some power of the diameter D:

$\frac{dN}{dD} \propto D^{-q}$ where the current measures give q = 4 ±0.5.

This implies (assuming q is not 1) that

$$N \propto D^{1-q} + \text{a constant} .$$

(The constant may be non-zero only if the power law doesn't apply at high values of D.)

Less formally, if q is 4, for example, there are 8 (=2^3) times more objects in the 100–200 km range than in the 200–400 km range, and for every object with a diameter between 1000 and 1010 km there should be around 1000 (=10^3) objects with diameter of 100 to 101 km.

If q was 1 or less, the law would imply an infinite number and mass of large objects in the Kuiper belt. If $1<q\leq 4$ there will be a finite number of objects greater than a given size, but the expected value of their combined mass would be infinite. If q is 4 or more, the law would imply an infinite mass of small objects. More accurate models find that the "slope" parameter q is in effect greater at large diameters and lesser at small diameters. It seems that Pluto is somewhat unexpectedly large, having several percent of the total mass of the Kuiper belt. It is not expected that anything larger than Pluto exists in the Kuiper belt, and in fact most of the brightest (largest) objects at inclinations less than 5° have probably been found.

For most TNOs, only the absolute magnitude is actually known, the size is inferred assuming a given albedo (not a safe assumption for larger objects).

Recent research has revealed that the size distributions of the hot classical and cold classical objects have differing slopes. The slope for the hot objects is q = 5.3 at large diameters and q = 2.0 at small diameters with the change in slope at 110 km. The slope for the cold objects is q = 8.2 at large diameters and q = 2.9 at small diameters with a change in slope at 140 km. The size distributions of the scattering objects, the plutinos, and the Neptune trojans

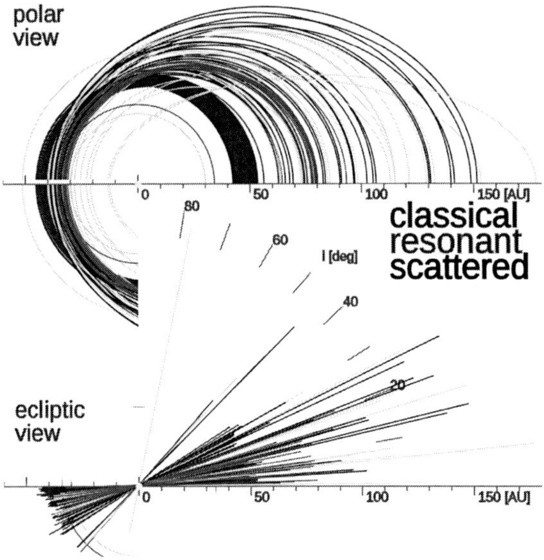

Figure 272: *Comparison of the orbits of scattered disc objects (black), classical KBOs (blue), and 2:5 resonant objects (green). Orbits of other KBOs are gray. (Orbital axes have been aligned for comparison.)*

have slopes similar to the other dynamically hot populations, but may instead have a divot, a sharp decrease in the number of objects below a specific size. This divot is hypothesized to be due to either the collisional evolution of the population, or to be due to the population having formed with no objects below this size, with the smaller objects being fragments of the original objects.

As of December 2009, the smallest Kuiper belt object detected is 980 m across. It is too dim (magnitude 35) to be seen by *Hubble* directly, but it was detected by *Hubble*'s star tracking system when it occulted a star.

Scattered objects

The scattered disc is a sparsely populated region, overlapping with the Kuiper belt but extending to beyond 100 AU. Scattered disc objects (SDOs) have very elliptical orbits, often also very inclined to the ecliptic. Most models of Solar System formation show both KBOs and SDOs first forming in a primordial belt, with later gravitational interactions, particularly with Neptune, sending the objects outward, some into stable orbits (the KBOs) and some into unstable orbits, the scattered disc. Due to its unstable nature, the scattered disc is suspected to be the point of origin of many of the Solar System's short-period

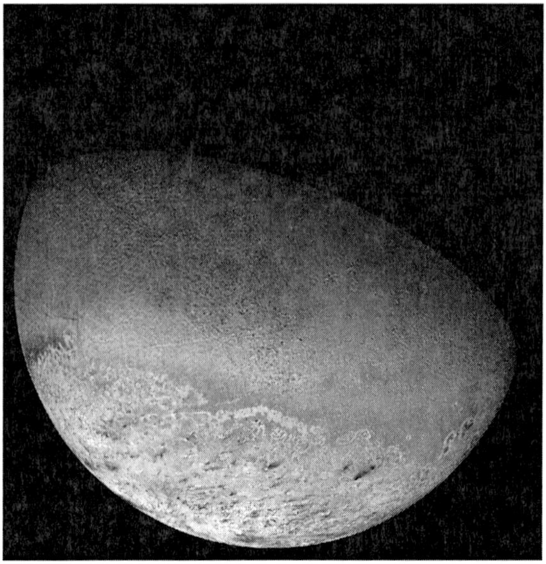

Figure 273: *Neptune's moon Triton*

comets. Their dynamic orbits occasionally force them into the inner Solar System, first becoming centaurs, and then short-period comets.

According to the Minor Planet Center, which officially catalogues all trans-Neptunian objects, a KBO, strictly speaking, is any object that orbits exclusively within the defined Kuiper belt region regardless of origin or composition. Objects found outside the belt are classed as scattered objects. In some scientific circles the term "Kuiper belt object" has become synonymous with any icy minor planet native to the outer Solar System assumed to have been part of that initial class, even if its orbit during the bulk of Solar System history has been beyond the Kuiper belt (e.g. in the scattered-disc region). They often describe scattered disc objects as "scattered Kuiper belt objects". Eris, which is known to be more massive than Pluto, is often referred to as a KBO, but is technically an SDO. A consensus among astronomers as to the precise definition of the Kuiper belt has yet to be reached, and this issue remains unresolved.

The centaurs, which are not normally considered part of the Kuiper belt, are also thought to be scattered objects, the only difference being that they were scattered inward, rather than outward. The Minor Planet Center groups the centaurs and the SDOs together as scattered objects.

Figure 274: *style=float:bottom}}*

Triton

During its period of migration, Neptune is thought to have captured a large KBO, Triton, which is the only large moon in the Solar System with a retrograde orbit (it orbits opposite to Neptune's rotation). This suggests that, unlike the large moons of Jupiter, Saturn and Uranus, which are thought to have coalesced from rotating discs of material around their young parent planets, Triton was a fully formed body that was captured from surrounding space. Gravitational capture of an object is not easy: it requires some mechanism to slow down the object enough to be caught by the larger object's gravity. A possible explanation is that Triton was part of a binary when it encountered Neptune. (Many KBOs are members of binaries. See below.) Ejection of the other member of the binary by Neptune could then explain Triton's capture. Triton is only 14% larger than Pluto, and spectral analysis of both worlds shows that their surfaces are largely composed of similar materials, such as methane and carbon monoxide. All this points to the conclusion that Triton was once a KBO that was captured by Neptune during its outward migration.

Largest KBOs

Since 2000, a number of KBOs with diameters of between 500 and 1,500 km (932 mi), more than half that of Pluto (diameter 2370 km), have been discovered. 50000 Quaoar, a classical KBO discovered in 2002, is over 1,200 km across. Makemake and Haumea, both announced on July 29, 2005, are larger still. Other objects, such as 28978 Ixion (discovered in 2001) and 20000 Varuna (discovered in 2000), measure roughly 500 km (311 mi) across.

Pluto

The discovery of these large KBOs in orbits similar to Pluto's led many to conclude that, aside from its relative size, Pluto was not particularly different from other members of the Kuiper belt. Not only are these objects similar to Pluto in size, but many also have satellites, and are of similar composition (methane and carbon monoxide have been found both on Pluto and on the largest KBOs). Thus, just as Ceres was considered a planet before the discovery of its fellow asteroids, some began to suggest that Pluto might also be reclassified.

The issue was brought to a head by the discovery of Eris, an object in the scattered disc far beyond the Kuiper belt, that is now known to be 27% more massive than Pluto. (Eris was originally thought to be larger than Pluto by volume, but the *New Horizons* mission found this not to be the case.) In response, the International Astronomical Union (IAU) was forced to define what a planet is for the first time, and in so doing included in their definition that a planet must have "cleared the neighbourhood around its orbit". As Pluto shares its orbit with many other sizable objects, it was deemed not to have cleared its orbit, and was thus reclassified from a planet to a dwarf planet, making it a member of the Kuiper belt.

Although Pluto is currently the largest known KBO, there is at least one known larger object currently outside the Kuiper belt that probably originated in it: Neptune's moon Triton (which, as explained above, is probably a captured KBO).

As of 2008, only five objects in the Solar System (Ceres, Eris, and the KBOs Pluto, Makemake and Haumea) are listed as dwarf planets by the IAU. 90482 Orcus, 28978 Ixion and many other Kuiper-belt objects are large enough to be in hydrostatic equilibrium; most of them will probably qualify when more is known about them.

Satellites

The six largest TNOs (Eris, Pluto, 2007 OR_{10}, Makemake, Haumea and Quaoar) are all known to have satellites, and two have more than one. A higher percentage of the larger KBOs have satellites than the smaller objects in the Kuiper belt, suggesting that a different formation mechanism was responsible. There are also a high number of binaries (two objects close enough in mass to be orbiting "each other") in the Kuiper belt. The most notable example is

Figure 275: *Kuiper belt object—possible target of New Horizons spacecraft (artist's concept)*

the Pluto–Charon binary, but it is estimated that around 11% of KBOs exist in binarics.

Exploration

On January 19, 2006, the first spacecraft to explore the Kuiper belt, *New Horizons*, was launched, which flew by Pluto on July 14, 2015. Beyond the Pluto flyby, the mission's goal was to locate and investigate other, farther objects in the Kuiper belt.

On October 15, 2014, it was revealed that *Hubble* had uncovered three potential targets, provisionally designated PT1 ("potential target 1"), PT2 and PT3 by the *New Horizons* team. The objects' diameters were estimated to be in the 30–55 km range; too small to be seen by ground telescopes, at distances from the Sun of 43–44 AU, which would put the encounters in the 2018–2019 period. The initial estimated probabilities that these objects were reachable within *New Horizons*' fuel budget were 100%, 7%, and 97%, respectively. All were members of the "cold" (low-inclination, low-eccentricity) classical Kuiper belt, and thus very different from Pluto. PT1 (given the temporary designation "1110113Y" on the HST web site), the most favorably situated object, was magnitude 26.8, 30–45 km in diameter, and will be encountered

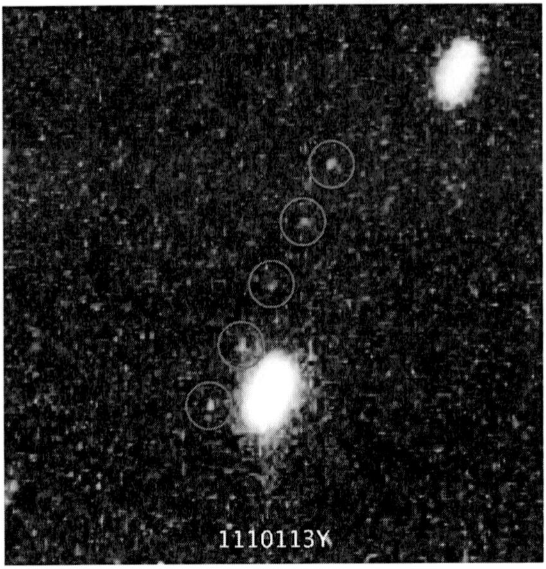

Figure 276: *The KBO 2014 MU69 (green circles), the selected target for the New Horizons Kuiper belt object mission*

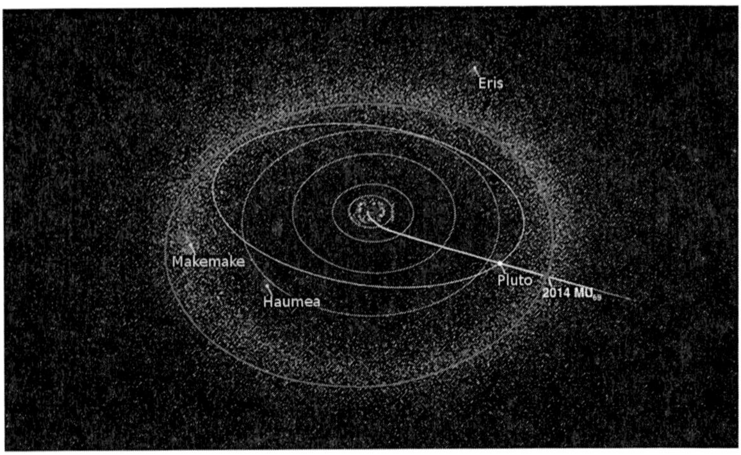

Figure 277: *Diagram showing the location of 2014 MU69 and trajectory for rendezvous*

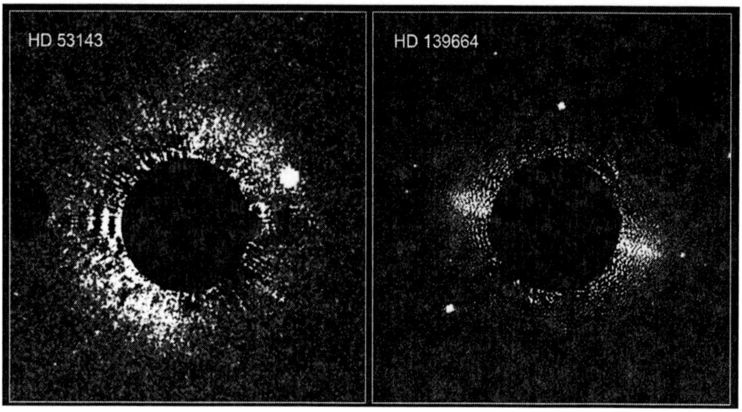

Figure 278: *Debris discs around the stars HD 139664 and HD 53143 – black circle from camera hides star to display discs.*

around January 2019. Once sufficient orbital information was provided, the Minor Planet Center gave official designations to the three target KBOs: 2014 MU_{69} (PT1), 2014 OS_{393} (PT2), and 2014 PN_{70} (PT3). By the fall of 2014, a possible fourth target, 2014 MT_{69}, had been eliminated by follow-up observations. PT2 was out of the running before the Pluto flyby.

On August 26, 2015, the first target, 2014 MU_{69}, was chosen. Course adjustment took place in late October and early November 2015, leading to a flyby in January 2019. On July 1, 2016, NASA approved additional funding for *New Horizons* to visit the object.

On December 2, 2015, *New Horizons* detected 1994 JR_1 from 270 million kilometres (170×10^6 mi) away, and the photographs show the shape of the object and one or two details.[415]

Extrasolar Kuiper belts

By 2006, astronomers had resolved dust discs thought to be Kuiper belt-like structures around nine stars other than the Sun. They appear to fall into two categories: wide belts, with radii of over 50 AU, and narrow belts (tentatively like that of the Solar System) with radii of between 20 and 30 AU and relatively sharp boundaries. Beyond this, 15–20% of solar-type stars have an observed infrared excess that is suggestive of massive Kuiper-belt-like structures. Most known debris discs around other stars are fairly young, but the two images on the right, taken by the *Hubble Space Telescope* in January 2006, are old enough (roughly 300 million years) to have settled into stable configurations. The left image is a "top view" of a wide belt, and the right image is an "edge view" of a narrow belt. Computer simulations of dust in the Kuiper belt suggest that when it was younger, it may have resembled the narrow rings seen around younger stars.

References

Bibliography

- Randall, Lisa (2015). *Dark Matter and the Dinosaurs*. New York: Ecco/HarperCollins Publishers. ISBN 978-0-06-232847-2.

External links

Wikimedia Commons has media related to ***Kuiper belt***.

- Dave Jewitt's page @ UCLA[416]
  - The belt's name[417]
- List of short period comets by family[418]
- Kuiper Belt Profile[419] by NASA's Solar System Exploration[420]
- The Kuiper Belt Electronic Newsletter[421]
- Wm. Robert Johnston's TNO page[422]
- Minor Planet Center: Plot of the Outer Solar System[423], illustrating Kuiper gap
- Website of the International Astronomical Union[424] (debating the status of TNOs)
- XXVIth General Assembly 2006[425]
- nature.com article: diagram displaying inner solar system, Kuiper Belt, and Oort Cloud[426], taken from Alan Stern, S. (2003). "The evolution of comets in the Oort cloud and Kuiper belt". *Nature*. **424** (6949): 639–42. PMID 12904784[427]. doi: 10.1038/nature01725[428].
- SPACE.com: Discovery Hints at a Quadrillion Space Rocks Beyond Neptune[429] (Sara Goudarzi) August 15, 2006 06:13 am ET
- The Outer Solar System[430] Astronomy Cast episode No. 64, includes full transcript.
- The Kuiper belt[431] at 365daysofastronomy.org
- Nine Planets' webpage on the Edgeworth-Kuiper Belt and Oort Cloud[432]
- List of TNOS[433]

<indicator name="featured-star"> ★ </indicator>

Pluto

<indicator name="pp-default"> </indicator>

Pluto

Full-disc view of Pluto in near-true color, imaged by *New Horizons*[434]

Discovery	
Discovered by	Clyde W. Tombaugh
Discovery date	February 18, 1930
Designations	
MPC designation	**(134340) Pluto**
Pronunciation	/ˈpluːtoʊ/ Media:en-us-Pluto.ogg
Named after	Pluto
Minor planet category	• Dwarf planet • Trans-Neptunian object • Plutoid • Kuiper belt object • Plutino
Adjectives	Plutonian
Orbital characteristics[435]	
Epoch J2000	
Aphelion	• 49.305 AU • (7,375.93 Gm) • February 2114
Perihelion	• 29.658 AU • (4,436.82 Gm) • (September 5, 1989)
Semi-major axis	• 39.48 AU • (5,906.38 Gm)
Eccentricity	0.2488

Orbital period	• 248.00 years • 90,560 d
Synodic period	366.73 days
Average orbital speed	4.67 km/s
Mean anomaly	14.53 deg
Inclination	• 17.16° • (11.88° to Sun's equator)
Longitude of ascending node	110.299°
Argument of perihelion	113.834°
Known satellites	5
Physical characteristics	
Mean radius	• 1,189.9±0.2 km • 0.1868 Earths
Flattening	<1%
Surface area	• 1.779×10^7 km^2 • 0.035 Earths
Volume	• $(7.057 \pm 0.004) \times 10^9$ km^3 • 0.00651 Earths
Mass	• $(1.303 \pm 0.003) \times 10^{22}$ kg • 0.00218 Earths • 0.177 Moons
Mean density	1.860±0.013 g/cm^3
Surface gravity	• 0.620 m/s^2 • 0.063 g
Escape velocity	1.212 km/s
Sidereal rotation period	• 6.387230 d • 6 d, 9 h, 17 m, 36 s
Equatorial rotation velocity	47.18 km/h
Axial tilt	122.53° (to orbit)
North pole right ascension	132.993°
North pole declination	–6.163°
Albedo	0.49 to 0.66 (geometric, varies by 35%)

Surface temp.	**min**	**mean**	**max**
Kelvin	33 K	44 K (–229 °C)	55 K

Apparent magnitude	13.65 to 16.3 (mean is 15.1)
Absolute magnitude *(H)*	–0.7

Angular diameter	0.06” to 0.11”
	Atmosphere
Surface pressure	1.0 Pa (2015)
Composition by volume	Nitrogen, methane, carbon monoxide

Pluto (minor-planet designation: **134340 Pluto**) is a dwarf planet in the Kuiper belt, a ring of bodies beyond Neptune. It was the first Kuiper belt object to be discovered.

Pluto was discovered by Clyde Tombaugh in 1930 and was originally considered to be the ninth planet from the Sun. After 1992, its status as a planet was questioned following the discovery of several objects of similar size in the Kuiper belt. In 2005, Eris, a dwarf planet in the scattered disc which is 27% more massive than Pluto, was discovered. This led the International Astronomical Union (IAU) to define the term "planet" formally in 2006, during their 26th General Assembly. That definition excluded Pluto and reclassified it as a dwarf planet.

Pluto is the largest and second-most-massive known dwarf planet in the Solar System, and the ninth-largest and tenth-most-massive known object directly orbiting the Sun. It is the largest known trans-Neptunian object by volume but is less massive than Eris. Like other Kuiper belt objects, Pluto is primarily made of ice and rock and is relatively small—about one-sixth the mass of the Moon and one-third its volume. It has a moderately eccentric and inclined orbit during which it ranges from 30 to 49 astronomical units or AU (4.4–7.4 billion km) from the Sun. This means that Pluto periodically comes closer to the Sun than Neptune, but a stable orbital resonance with Neptune prevents them from colliding. Light from the Sun takes about 5.5 hours to reach Pluto at its average distance (39.5 AU).

Pluto has five known moons: Charon (the largest, with a diameter just over half that of Pluto), Styx, Nix, Kerberos, and Hydra. Pluto and Charon are sometimes considered a binary system because the barycenter of their orbits does not lie within either body.

On July 14, 2015, the *New Horizons* spacecraft became the first spacecraft to fly by Pluto. During its brief flyby, *New Horizons* made detailed measurements and observations of Pluto and its moons. In September 2016, astronomers announced that the reddish-brown cap of the north pole of Charon is composed of tholins, organic macromolecules that may be ingredients for the emergence of life, and produced from methane, nitrogen and other gases released from the atmosphere of Pluto and transferred about 19,000 km (12,000 mi) to the orbiting moon.

Figure 279: *Mosaic of best-resolution images of Pluto from different angles*

History

Discovery

In the 1840s, Urbain Le Verrier used Newtonian mechanics to predict the position of the then-undiscovered planet Neptune after analyzing perturbations in the orbit of Uranus. Subsequent observations of Neptune in the late 19th century led astronomers to speculate that Uranus's orbit was being disturbed by another planet besides Neptune.

In 1906, Percival Lowell—a wealthy Bostonian who had founded Lowell Observatory in Flagstaff, Arizona, in 1894—started an extensive project in search of a possible ninth planet, which he termed "Planet X". By 1909, Lowell and William H. Pickering had suggested several possible celestial coordinates for such a planet. Lowell and his observatory conducted his search until his death in 1916, but to no avail. Unknown to Lowell, his surveys had captured two faint images of Pluto on March 19 and April 7, 1915, but they were not recognized for what they were. There are fourteen other known prediscovery observations, with the oldest made by the Yerkes Observatory on August 20, 1909.

Figure 280: *Clyde Tombaugh, in Kansas*

Percival's widow, Constance Lowell, entered into a ten-year legal battle with the Lowell Observatory over her husband's legacy, and the search for Planet X did not resume until 1929.[436] Vesto Melvin Slipher, the observatory director, gave the job of locating Planet X to 23-year-old Clyde Tombaugh, who had just arrived at the observatory after Slipher had been impressed by a sample of his astronomical drawings.[436]

Tombaugh's task was to systematically image the night sky in pairs of photographs, then examine each pair and determine whether any objects had shifted position. Using a blink comparator, he rapidly shifted back and forth between views of each of the plates to create the illusion of movement of any objects that had changed position or appearance between photographs. On February 18, 1930, after nearly a year of searching, Tombaugh discovered a possible moving object on photographic plates taken on January 23 and 29. A lesser-quality photograph taken on January 21 helped confirm the movement.[437] After the observatory obtained further confirmatory photographs, news of the discovery was telegraphed to the Harvard College Observatory on March 13, 1930.

Name

The discovery made headlines around the globe. Lowell Observatory, which had the right to name the new object, received more than 1,000 suggestions from all over the world, ranging from Atlas to Zymal. Tombaugh urged Slipher to suggest a name for the new object quickly before someone else did. Constance Lowell proposed *Zeus*, then *Percival* and finally *Constance*. These suggestions were disregarded.

The name Pluto, after the god of the underworld, was proposed by Venetia Burney (1918–2009), an eleven-year-old schoolgirl in Oxford, England, who was interested in classical mythology. She suggested it in a conversation with her grandfather Falconer Madan, a former librarian at the University of Oxford's Bodleian Library, who passed the name to astronomy professor Herbert Hall Turner, who cabled it to colleagues in the United States.

Each member of the Lowell Observatory was allowed to vote on a short-list of three potential names: Minerva (which was already the name for an asteroid), Cronus (which had lost reputation through being proposed by the unpopular astronomer Thomas Jefferson Jackson See), and Pluto. Pluto received every vote.[438] The name was announced on May 1, 1930. Upon the announcement, Madan gave Venetia £5 (equivalent to 300 GBP, or 450 USD in 2014)[439] as a reward.

The final choice of name was helped in part by the fact that the first two letters of *Pluto* are the initials of Percival Lowell. Pluto's astronomical symbol (♇ , Unicode U+2647, ♇) was then created as a monogram constructed from the letters "PL". Pluto's astrological symbol resembles that of Neptune (♆), but has a circle in place of the middle prong of the trident (⯓).

The name was soon embraced by wider culture. In 1930, Walt Disney was apparently inspired by it when he introduced for Mickey Mouse a canine companion named Pluto, although Disney animator Ben Sharpsteen could not confirm why the name was given. In 1941, Glenn T. Seaborg named the newly created element plutonium after Pluto, in keeping with the tradition of naming elements after newly discovered planets, following uranium, which was named after Uranus, and neptunium, which was named after Neptune.

Most languages use the name "Pluto" in various transliterations.[440] In Japanese, Houei Nojiri suggested the translation *Meiōsei* (冥王星 , "Star of the King (God) of the Underworld"), and this was borrowed into Chinese, Korean, and Vietnamese (which instead uses "Sao Diêm Vương", which was derived from the Chinese term 閻王 (Yánwáng), as "minh" is a homophone for the Sino-Vietnamese words for "dark" (冥) and "bright" (明)). Some Indian languages use the name Pluto, but others, such as Hindi, use the name of *Yama*, the God of Death in Hindu and Buddhist mythology. Polynesian languages also tend to use the indigenous god of the underworld, as in Maori *Whiro*.

Planet X disproved

Once Pluto was found, its faintness and lack of a resolvable disc cast doubt on the idea that it was Lowell's Planet X. Estimates of Pluto's mass were revised downward throughout the 20th century.

Mass estimates for Pluto

Year	Mass	Estimate by
1915	7 Earth	Lowell (prediction for Planet X)
1931	1 Earth	Nicholson & Mayall
1948	0.1 (1/10) Earth	Kuiper
1976	0.01 (1/100) Earth	Cruikshank, Pilcher, & Morrison[441]
1978	0.0015 (1/650) Earth	Christy & Harrington
2006	0.00218 (1/459) Earth	Buie et al.

Astronomers initially calculated its mass based on its presumed effect on Neptune and Uranus. In 1931, Pluto was calculated to be roughly the mass of Earth, with further calculations in 1948 bringing the mass down to roughly that of Mars. In 1976, Dale Cruikshank, Carl Pilcher and David Morrison of the University of Hawaii calculated Pluto's albedo for the first time, finding that it matched that for methane ice; this meant Pluto had to be exceptionally luminous for its size and therefore could not be more than 1 percent the mass of Earth.[441] (Pluto's albedo is 1.4–1.9 times that of Earth.)

In 1978, the discovery of Pluto's moon Charon allowed the measurement of Pluto's mass for the first time: roughly 0.2% that of Earth, and far too small to account for the discrepancies in the orbit of Uranus. Subsequent searches for an alternative Planet X, notably by Robert Sutton Harrington, failed. In 1992, Myles Standish used data from *Voyager 2*'s flyby of Neptune in 1989, which had revised the estimates of Neptune's mass downward by 0.5%—an amount comparable to the mass of Mars—to recalculate its gravitational effect on Uranus. With the new figures added in, the discrepancies, and with them the need for a Planet X, vanished. Today, the majority of scientists agree that Planet X, as Lowell defined it, does not exist. Lowell had made a prediction of Planet X's orbit and position in 1915 that was fairly close to Pluto's actual orbit

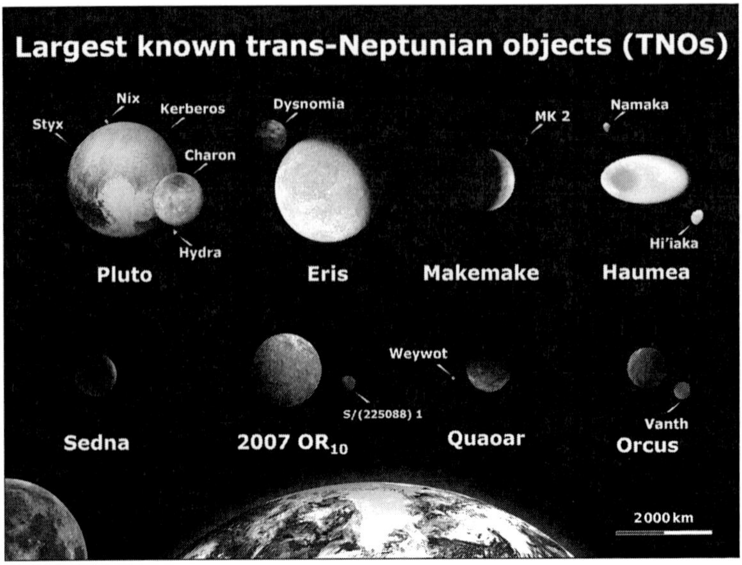

Figure 281: *style=float:bottom}}*

and its position at that time; Ernest W. Brown concluded soon after Pluto's discovery that this was a coincidence, a view still held today.

Classification

From 1992 onward, many bodies were discovered orbiting in the same volume as Pluto, showing that Pluto is part of a population of objects called the Kuiper belt. This made its official status as a planet controversial, with many questioning whether Pluto should be considered together with or separately from its surrounding population. Museum and planetarium directors occasionally created controversy by omitting Pluto from planetary models of the Solar System. The Hayden Planetarium reopened—in February 2000, after renovation—with a model of only eight planets, which made headlines almost a year later.

As objects increasingly closer in size to Pluto were discovered in the region, it was argued that Pluto should be reclassified as one of the Kuiper belt objects, just as Ceres, Pallas, Juno and Vesta lost their planet status after the discovery of many other asteroids. On July 29, 2005, astronomers at Caltech announced the discovery of a new trans-Neptunian object, Eris, which was substantially more massive than Pluto and the most massive object discovered in the Solar System since Triton in 1846. Its discoverers and the press initially called it the tenth planet, although there was no official consensus at the time on whether to call it a planet. Others in the astronomical community considered the discovery the strongest argument for reclassifying Pluto as a minor planet.

IAU classification

The debate came to a head on August 24, 2006, with an IAU resolution that created an official definition for the term "planet". According to this resolution, there are three conditions for an object in the Solar System to be considered a planet:

1. The object must be in orbit around the Sun.
2. The object must be massive enough to be rounded by its own gravity. More specifically, its own gravity should pull it into a shape defined by hydrostatic equilibrium.
3. It must have cleared the neighborhood around its orbit.

Pluto fails to meet the third condition, because its mass is only 0.07 times that of the mass of the other objects in its orbit (Earth's mass, by contrast, is 1.7 million times the remaining mass in its own orbit). The IAU further decided that bodies that, like Pluto, meet criteria 1 and 2 but do not meet criterion 3 would be called dwarf planets. On September 13, 2006, the IAU included Pluto, and Eris and its moon Dysnomia, in their Minor Planet Catalogue, giving them the official minor planet designations "(134340) Pluto", "(136199) Eris", and "(136199) Eris I Dysnomia". Had Pluto been included upon its discovery in 1930, it would have likely been designated 1164, following 1163 Saga, which was discovered a month earlier.

There has been some resistance within the astronomical community toward the reclassification. Alan Stern, principal investigator with NASA's *New Horizons* mission to Pluto, derided the IAU resolution, stating that "the definition stinks, for technical reasons". Stern contended that, by the terms of the new definition, Earth, Mars, Jupiter, and Neptune, all of which share their orbits with asteroids, would be excluded. He argued that all big spherical moons, including the Moon, should likewise be considered planets. He also stated that because less than five percent of astronomers voted for it, the decision was not representative of the entire astronomical community. Marc W. Buie, then at the Lowell Observatory petitioned against the definition. Others have supported the IAU. Mike Brown, the astronomer who discovered Eris, said "through this whole crazy circus-like procedure, somehow the right answer was stumbled on. It's been a long time coming. Science is self-correcting eventually, even when strong emotions are involved."

Public reception to the IAU decision was mixed. Many accepted the reclassification, but some sought to overturn the decision with online petitions urging the IAU to consider reinstatement. A resolution introduced by some members of the California State Assembly facetiously called the IAU decision a "scientific heresy". The New Mexico House of Representatives passed a resolution in honor of Tombaugh, a longtime resident of that state, that declared that Pluto will always be considered a planet while in New Mexican skies and that March 13, 2007, was Pluto Planet Day. The Illinois Senate passed a similar resolution in 2009, on the basis that Clyde Tombaugh, the discoverer of Pluto, was born in Illinois. The resolution asserted that Pluto was "unfairly downgraded to a 'dwarf' planet" by the IAU." Some members of the public have also rejected the change, citing the disagreement within the scientific community on the issue, or for sentimental reasons, maintaining that they have always known Pluto as a planet and will continue to do so regardless of the IAU decision.

In 2006, in its 17th annual words-of-the-year vote, the American Dialect Society voted *plutoed* as the word of the year. To "pluto" is to "demote or devalue someone or something".

Researchers on both sides of the debate gathered on August 14–16, 2008, at the Johns Hopkins University Applied Physics Laboratory for a conference that included back-to-back talks on the current IAU definition of a planet. Entitled "The Great Planet Debate", the conference published a post-conference press release indicating that scientists could not come to a consensus about the definition of planet. On June 11, 2008, the IAU had announced in a press release that the term "plutoid" would henceforth be used to refer to Pluto and other objects that have an orbital semi-major axis greater than that of Neptune and enough mass to be of near-spherical shape.

Orbit

Pluto's orbital period is presently about 248 years. Its orbital characteristics are substantially different from those of the planets, which follow nearly circular orbits around the Sun close to a flat reference plane called the ecliptic. In contrast, Pluto's orbit is moderately inclined relative to the ecliptic (over 17°) and moderately eccentric (elliptical). This eccentricity means a small region of Pluto's orbit lies nearer the Sun than Neptune's. The Pluto–Charon barycenter came to perihelion on September 5, 1989, and was last closer to the Sun than Neptune between February 7, 1979, and February 11, 1999.

In the long term, Pluto's orbit is chaotic. Computer simulations can be used to predict its position for several million years (both forward and backward in time), but after intervals longer than the Lyapunov time of 10–20 million years, calculations become speculative: Pluto is sensitive to immeasurably small details of the Solar System, hard-to-predict factors that will gradually change Pluto's position in its orbit.

The semi-major axis of Pluto's orbit varies between about 39.3 and 39.6 au with a period of about 19,951 years, corresponding to an orbital period varying between 246 and 249 years. The semi-major axis and period are presently getting longer.

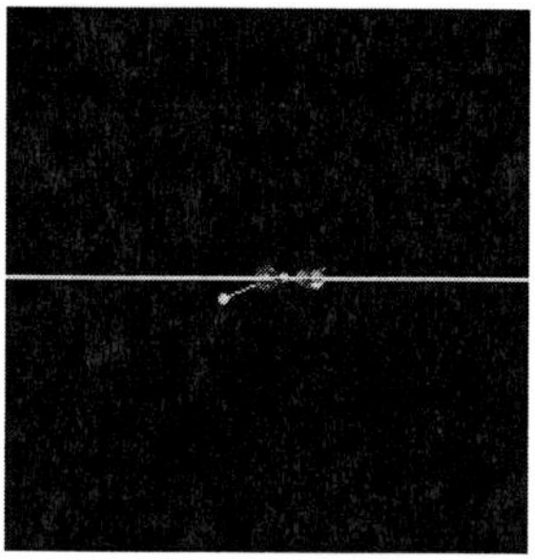

Orbit of Pluto—ecliptic view. This "side view" of Pluto's orbit (in red) shows its large inclination to the ecliptic.

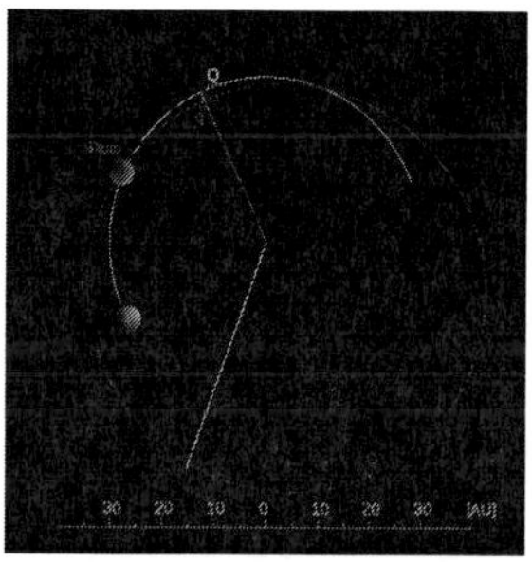

Orbit of Pluto—polar view. This "view from above" shows how Pluto's orbit (in red) is less circular than Neptune's (in blue), and how Pluto is sometimes closer to the Sun than Neptune. The darker halves of both orbits show where they pass below the plane of the ecliptic.

Relationship with Neptune

Despite Pluto's orbit appearing to cross that of Neptune when viewed from directly above, the two objects' orbits are aligned so that they can never collide or even approach closely.

The two orbits do not intersect. When Pluto is closest to the Sun, and hence closest to Neptune's orbit as viewed from above, it is also the farthest above Neptune's path. Pluto's orbit passes about 8 AU above that of Neptune, preventing a collision.

This alone is not enough to protect Pluto; perturbations from the planets (especially Neptune) could alter Pluto's orbit (such as its orbital precession) over millions of years so that a collision could be possible. Pluto is also protected by its 2:3 orbital resonance with Neptune: for every two orbits that Pluto makes around the Sun, Neptune makes three. Each cycle lasts about 495 years. This pattern is such that, in each 495-year cycle, the first time Pluto is near perihelion, Neptune is over 50° behind Pluto. By Pluto's second perihelion, Neptune will have completed a further one and a half of its own orbits, and so will be nearly 130° ahead of Pluto. Pluto and Neptune's minimum separation is over 17 AU, which is greater than Pluto's minimum separation from Uranus (11 AU). The minimum separation between Pluto and Neptune actually occurs near the time of Pluto's aphelion.

The 2:3 resonance between the two bodies is highly stable, and is preserved over millions of years. This prevents their orbits from changing relative to one another, and so the two bodies can never pass near each other. Even if Pluto's orbit were not inclined, the two bodies could never collide.

Other factors

Numerical studies have shown that over millions of years, the general nature of the alignment between the orbits of Pluto and Neptune does not change. There are several other resonances and interactions that enhance Pluto's stability. These arise principally from two additional mechanisms (besides the 2:3 mean-motion resonance).

First, Pluto's argument of perihelion, the angle between the point where it crosses the ecliptic and the point where it is closest to the Sun, librates around 90°. This means that when Pluto is closest to the Sun, it is at its farthest above the plane of the Solar System, preventing encounters with Neptune. This is a consequence of the Kozai mechanism, which relates the eccentricity of an orbit to its inclination to a larger perturbing body—in this case Neptune. Relative to Neptune, the amplitude of libration is 38°, and so the angular separation of Pluto's perihelion to the orbit of Neptune is always greater than 52° (90°–38°). The closest such angular separation occurs every 10,000 years.

Second, the longitudes of ascending nodes of the two bodies—the points where they cross the ecliptic—are in near-resonance with the above libration. When the two longitudes are the same—that is, when one could draw a straight line through both nodes and the Sun—Pluto's perihelion lies exactly at 90°, and hence it comes closest to the Sun when it is highest above Neptune's orbit. This is known as the *1:1 superresonance*. All the Jovian planets, particularly Jupiter, play a role in the creation of the superresonance.

After passing the ascending node, Pluto is interior to Neptune's orbit and moving faster, approaching Neptune from behind. The strong gravitational pull between the two causes angular momentum to be transferred to Pluto, at Neptune's expense. This moves Pluto into a slightly larger orbit, where it travels slightly more slowly, according to Kepler's third law. As its orbit changes, this has the gradual effect of changing the perihelion and longitude of its orbit (and, to a lesser degree, that of Neptune). After many such repetitions, Pluto is sufficiently slowed, and Neptune sufficiently speeded up, that Neptune begins to catch up with Pluto at the opposite side of its orbit (near the opposing node to where we began). The process is then reversed, and Pluto loses angular momentum to Neptune, until Pluto is sufficiently sped up that it begins to catch Neptune again at the original node. The whole process takes about 20,000 years to complete.

Quasi-satellite

In 2012, it was hypothesized that 15810 Arawn could be a quasi-satellite of Pluto, a specific type of co-orbital configuration. According to the hypothesis, the object would be a quasi-satellite of Pluto for about 350,000 years out of every two-million-year period. This hypothesis was disproven in 2016, when more-accurate observations of the position of Arawn were made by *New Horizons*.

Rotation

Pluto's rotation period, its day, is equal to 6.39 Earth days. Like Uranus, Pluto rotates on its "side" in its orbital plane, with an axial tilt of 120°, and so its seasonal variation is extreme; at its solstices, one-fourth of its surface is in continuous daylight, whereas another fourth is in continuous darkness. The reason for this unusual orientation has been debated. Research from University of Arizona has suggested that it may be due to the way that a body's spin will always adjust to minimise energy. This could mean a body reorienting itself to put extraneous mass near the equator and regions lacking mass tend towards the poles. This is called *polar wander*. According to a paper released from the University of Arizona, this could be caused by masses of frozen nitrogen building up in shadowed areas of the dwarf planet. These masses would cause the body to reorientate itself, leading to its unusual axial tilt of 120°. The buildup of nitrogen is due to Pluto's vast distance from the Sun. At the equator, temperatures can drop to −240 °C (−400.0 °F; 33.1 K), causing nitrogen to freeze as water would freeze on Earth. The same effect seen on Pluto would be observed on Earth if the Antarctic ice sheet was several times larger.

Figure 282: *High-resolution MVIC image of Pluto in enhanced color to bring out differences in surface composition*

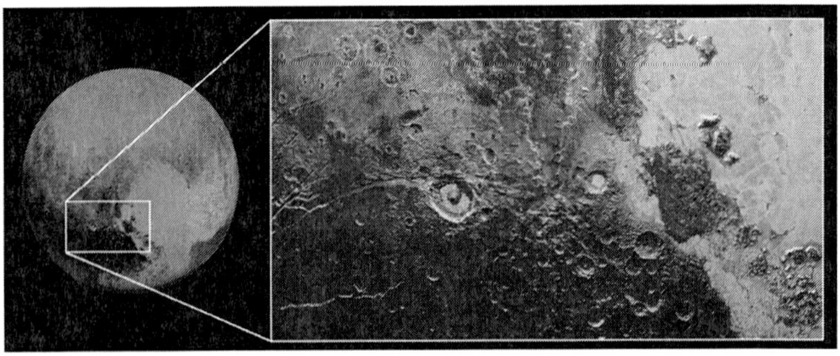

Figure 283: *Regions where water ice has been detected (blue regions)*

Geology

Surface

The plains on Pluto's surface are composed of more than 98 percent nitrogen ice, with traces of methane and carbon monoxide. Nitrogen and carbon monoxide are most abundant on the anti-Charon face of Pluto (around 180° longitude, where Tombaugh Regio's western lobe, Sputnik Planitia, is located), whereas methane is most abundant near 300° east. The mountains are made of water ice. Pluto's surface is quite varied, with large differences in both brightness and color. Pluto is one of the most contrastive bodies in the Solar System, with as much contrast as Saturn's moon Iapetus. The color varies from charcoal black, to dark orange and white. Pluto's color is more similar to that of Io with slightly more orange and significantly less red than Mars. Notable geographical features include Tombaugh Regio, or the "Heart" (a large bright area on the side opposite Charon), Cthulhu Regio, or the "Whale" (a large dark area on the trailing hemisphere), and the "Brass Knuckles" (a series of equatorial dark areas on the leading hemisphere). Sputnik Planitia, the western lobe of the "Heart", is a 1000 km-wide basin of frozen nitrogen and carbon monoxide ices, divided into polygonal cells which are interpreted as convection cells that carry floating blocks of water ice crust and sublimation pits towards their margins; there are obvious signs of glacial flows both into and out of the basin. It has no craters that were visible to *New Horizons*, indicating that its surface is less than 10 million years old. Latest studies have shown that the surface has an age of 180000^{+90000}_{-40000} years. The New Horizons science team summarized initial findings as "Pluto displays a surprisingly wide variety of geological landforms, including those resulting from glaciological and surface–atmosphere interactions as well as impact, tectonic, possible cryovolcanic, and mass-wasting processes."

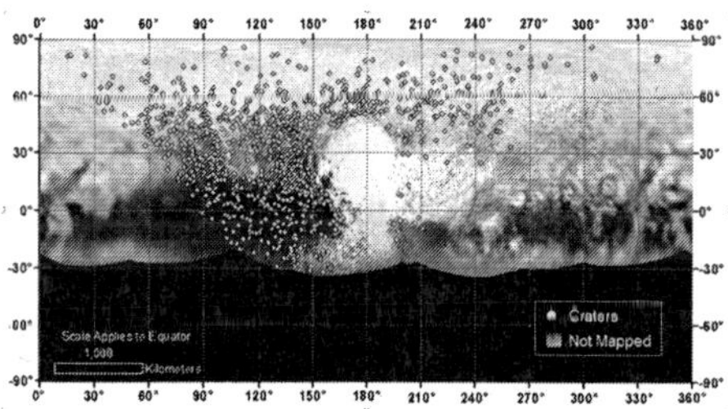

Distribution of over 1000 craters of all ages on Pluto. The variation in density (with none found in Sputnik Planitia) indicates a long history of varying geological activity.

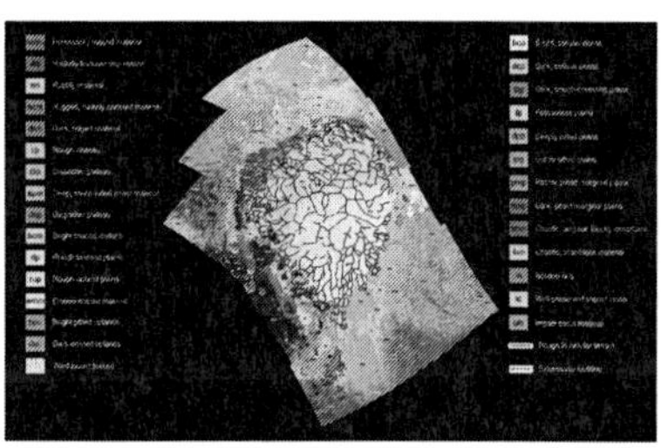

Geologic map of Sputnik Planitia and surroundings (context), with convection cell margins outlined in black

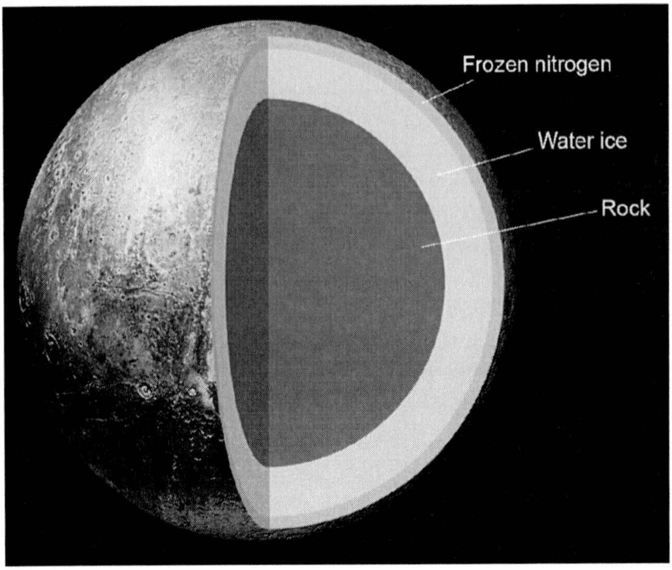

Figure 284:
Internal structure of Pluto

- *1. Frozen nitrogen*
- *2. Water ice*
- *3. Rock*

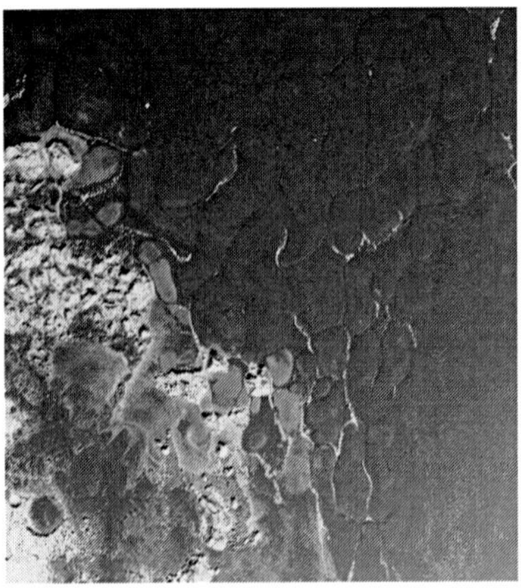

Sputnik Planitia is covered with churning nitrogen ice "cells" that are geologically young and turning over due to convection.

Internal structure

Pluto's density is 1.860±0.013 g/cm^3. Because the decay of radioactive elements would eventually heat the ices enough for the rock to separate from them, scientists expect that Pluto's internal structure is differentiated, with the rocky material having settled into a dense core surrounded by a mantle of water ice. The diameter of the core is hypothesized to be approximately 1700 km, 70% of Pluto's diameter. It is possible that such heating continues today, creating a subsurface ocean of liquid water 100 to 180 km thick at the core–mantle boundary. In September 2016, scientists at Brown University simulated the impact believed to have formed Sputnik Planitia, and showed that it might have been the result of liquid water upwelling from below after the collision, implying the existence of a subsurface ocean at least 100 km deep. Pluto has no magnetic field.

Mass and size

Selected size estimates for Pluto

Year	Radius	Notes
1993	1195 km	Millis, et al. (if no haze)
1993	1180 km	Millis, et al. (surface & haze)
1994	1164 km	Young & Binzel
2006	1153 km	Buie, et al.
2007	1161 km	Young, Young, & Buie
2011	1180 km	Zalucha, et al.
2014	1184 km	Lellouch, et al.
2015	1187 km	*New Horizons* measurement (from optical data)
2017	1189.9 km	*New Horizons* measurement (from radio occultation data)

Pluto's diameter is 2379.8±0.4 km and its mass is $(1.303\pm0.003)\times10^{22}$ kg, 17.7% that of the Moon (0.22% that of Earth). Its surface area is 1.779×10^{7} km^2, or roughly the same surface area as Russia. Its surface gravity is 0.063 *g* (compared to 1 *g* for Earth).

Figure 285: *Size comparisons: Earth, the Moon, and Pluto*

The discovery of Pluto's satellite Charon in 1978 enabled a determination of the mass of the Pluto–Charon system by application of Newton's formulation of Kepler's third law. Observations of Pluto in occultation with Charon allowed scientists to establish Pluto's diameter more accurately, whereas the invention of adaptive optics allowed them to determine its shape more accurately.

With less than 0.2 lunar masses, Pluto is much less massive than the terrestrial planets, and also less massive than seven moons: Ganymede, Titan, Callisto, Io, the Moon, Europa, and Triton. The mass is much less than thought before Charon was discovered.

Pluto is more than twice the diameter and a dozen times the mass of the dwarf planet Ceres, the largest object in the asteroid belt. It is less massive than the dwarf planet Eris, a trans-Neptunian object discovered in 2005, though Pluto has a larger diameter of 2379.8 km compared to Eris's approximate diameter of 2326 km.

Determinations of Pluto's size had been complicated by its atmosphere, and hydrocarbon haze. In March 2014, Lellouch, de Bergh et al. published findings regarding methane mixing ratios in Pluto's atmosphere consistent with a Plutonian diameter greater than 2360 km, with a "best guess" of 2368 km. On July 13, 2015, images from NASA's *New Horizons* mission Long Range Reconnaissance Imager (LORRI), along with data from the other instruments,

Figure 286: *A near-true-color image taken by New Horizons after its flyby. Numerous layers of blue haze float in Pluto's atmosphere. Along and near the limb, mountains and their shadows are visible.*

determined Pluto's diameter to be 2,370 km (1,470 mi), which was later revised to be 2,372 km (1,474 mi) on July 24, and later to 2374±8 km. Using radio occultation data from the *New Horizons* Radio Science Experiment (REX), the diameter was found to be 2379.8±0.4 km.

Atmosphere

Pluto has a tenuous atmosphere consisting of nitrogen (N_2), methane (CH_4), and carbon monoxide (CO), which are in equilibrium with their ices on Pluto's surface. According to the measurements by *New Horizons*, the surface pressure is about 1 Pa (10 μbar), roughly one million to 100,000 times less than Earth's atmospheric pressure. It was initially thought that, as Pluto moves away from the Sun, its atmosphere should gradually freeze onto the surface; studies of *New Horizons* data and ground-based occultations show that Pluto's atmospheric density increases, and that it likely remains gaseous throughout Pluto's orbit. *New Horizons* observations showed that atmospheric escape of nitrogen to be 10,000 times less than expected. Alan Stern has contended that even a small increase in Pluto's surface temperature can lead to exponential increases in Pluto's atmospheric density; from 18 hPa to as much as 280 hPa (three times that of Mars to a quarter that of the Earth). At such densities, nitrogen could flow across the surface as liquid. Just like sweat cools the body as it evaporates from the skin, the sublimation of Pluto's atmosphere cools its

Figure 287: *Image of Pluto in X-rays by Chandra X-ray Observatory (blue spot). The X-rays are probably created by interaction of the gases surrounding Pluto with solar wind, although details of their origin are not clear.*

surface. The presence of atmospheric gases was traced up to 1670 kilometers high; the atmosphere does not have a sharp upper boundary.

The presence of methane, a powerful greenhouse gas, in Pluto's atmosphere creates a temperature inversion, with the average temperature of its atmosphere tens of degrees warmer than its surface, though observations by *New Horizons* have revealed Pluto's upper atmosphere to be far colder than expected (70 K, as opposed to about 100 K). Pluto's atmosphere is divided into roughly 20 regularly spaced haze layers up to 150 km high, thought to be the result of pressure waves created by airflow across Pluto's mountains.

Satellites

Pluto has five known natural satellites: Charon, first identified in 1978 by astronomer James Christy; Nix and Hydra, both discovered in 2005; Kerberos, discovered in 2011; and Styx, discovered in 2012. The satellites' orbits are circular (eccentricity < 0.006) and coplanar with Pluto's equator (inclination < 1°), and therefore tilted approximately 120° relative to Pluto's orbit. The Plutonian system is highly compact: the five known satellites orbit within the inner 3% of the region where prograde orbits would be stable. Closest to Pluto is Charon, which is large enough to be in hydrostatic equilibrium and to cause the

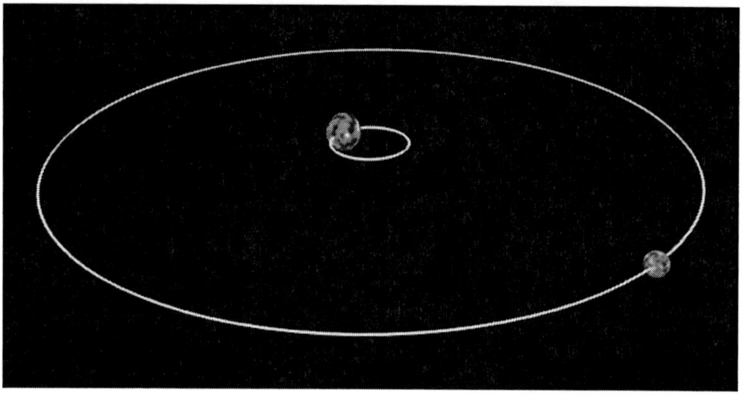

Figure 288: *An oblique view of the Pluto–Charon system showing that Pluto orbits a point outside itself. The two bodies are mutually tidally locked.*

barycenter of the Pluto–Charon system to be outside Pluto. Beyond Charon there are four much smaller circumbinary moons, Styx, Nix, Kerberos, and Hydra.

The orbital periods of all Pluto's moons are linked in a system of orbital resonances and near resonances. When precession is accounted for, the orbital periods of Styx, Nix, and Hydra are in an exact 18:22:33 ratio. There is a sequence of approximate ratios, 3:4:5:6, between the periods of Styx, Nix, Kerberos, and Hydra with that of Charon; the ratios become closer to being exact the further out the moons are.

The Pluto–Charon system is one of the few in the Solar System whose barycenter lies outside the primary body; the Patroclus–Menoetius system is a smaller example, and the Sun–Jupiter system is the only larger one. The similarity in size of Charon and Pluto has prompted some astronomers to call it a double dwarf planet. The system is also unusual among planetary systems in that each is tidally locked to the other, which means that Pluto and Charon always have the same hemisphere facing each other. From any position on either body, the other is always at the same position in the sky, or always obscured. This also means that the rotation period of each is equal to the time it takes the entire system to rotate around its barycenter.

In 2007, observations by the Gemini Observatory of patches of ammonia hydrates and water crystals on the surface of Charon suggested the presence of active cryo-geysers.

Pluto's moons are hypothesized to have been formed by a collision between Pluto and a similar-sized body, early in the history of the Solar System. The collision released material that consolidated into the moons around Pluto.

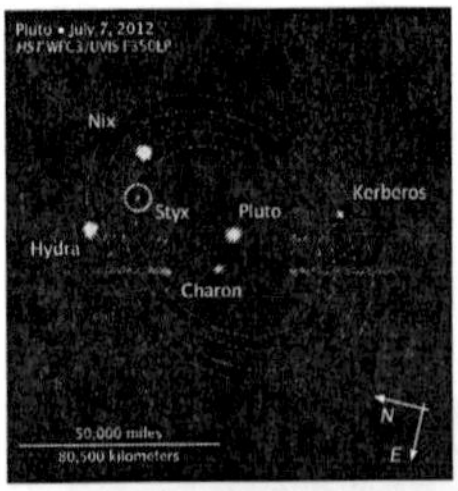

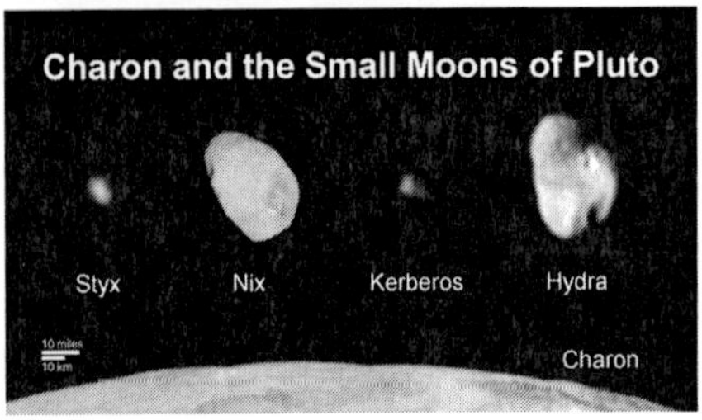

1. The Pluto system: Pluto, Charon, Styx, Nix, Kerberos, and Hydra, imaged by the Hubble Space Telescope in July 2012. **2.** Pluto and Charon, to scale. Image acquired by *New Horizons* on July 8, 2015. **3.** Family portrait of the five moons of Pluto, to scale. **4.** Pluto's moon Charon as viewed by *New Horizons* on July 13, 2015

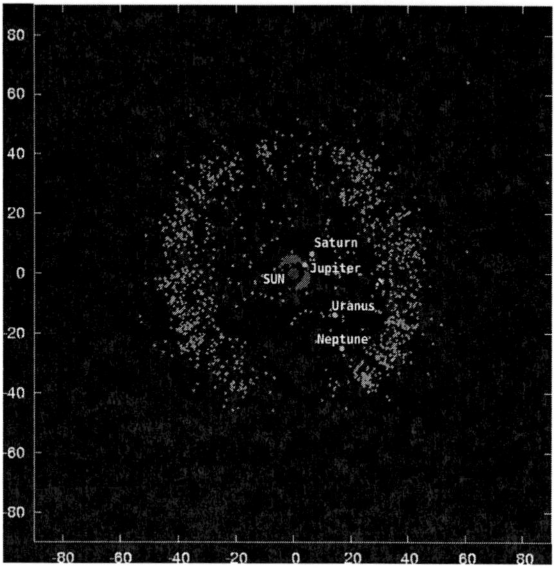

Figure 289: *Plot of the known Kuiper belt objects, set against the four giant planets*

Origin

Pluto's origin and identity had long puzzled astronomers. One early hypothesis was that Pluto was an escaped moon of Neptune, knocked out of orbit by its largest current moon, Triton. This idea was eventually rejected after dynamical studies showed it to be impossible because Pluto never approaches Neptune in its orbit.

Pluto's true place in the Solar System began to reveal itself only in 1992, when astronomers began to find small icy objects beyond Neptune that were similar to Pluto not only in orbit but also in size and composition. This trans-Neptunian population is thought to be the source of many short-period comets. Pluto is now known to be the largest member of the Kuiper belt, a stable belt of objects located between 30 and 50 AU from the Sun. As of 2011, surveys of the Kuiper belt to magnitude 21 were nearly complete and any remaining Pluto-sized objects are expected to be beyond 100 AU from the Sun. Like other Kuiper-belt objects (KBOs), Pluto shares features with comets; for example, the solar wind is gradually blowing Pluto's surface into space. It has been claimed that if Pluto were placed as near to the Sun as Earth, it would develop a tail, as comets do. This claim has been disputed with the argument that Pluto's escape velocity is too high for this to happen.[442]

Though Pluto is the largest Kuiper belt object discovered, Neptune's moon Triton, which is slightly larger than Pluto, is similar to it both geologically and atmospherically, and is thought to be a captured Kuiper belt object. Eris (see above) is about the same size as Pluto (though more massive) but is not strictly considered a member of the Kuiper belt population. Rather, it is considered a member of a linked population called the scattered disc.

A large number of Kuiper belt objects, like Pluto, are in a 2:3 orbital resonance with Neptune. KBOs with this orbital resonance are called "plutinos", after Pluto.

Like other members of the Kuiper belt, Pluto is thought to be a residual planetesimal; a component of the original protoplanetary disc around the Sun that failed to fully coalesce into a full-fledged planet. Most astronomers agree that Pluto owes its current position to a sudden migration undergone by Neptune early in the Solar System's formation. As Neptune migrated outward, it approached the objects in the proto-Kuiper belt, setting one in orbit around itself (Triton), locking others into resonances, and knocking others into chaotic orbits. The objects in the scattered disc, a dynamically unstable region overlapping the Kuiper belt, are thought to have been placed in their current positions by interactions with Neptune's migrating resonances. A computer model created in 2004 by Alessandro Morbidelli of the Observatoire de la Côte d'Azur in Nice suggested that the migration of Neptune into the Kuiper belt may have been triggered by the formation of a 1:2 resonance between Jupiter and Saturn, which created a gravitational push that propelled both Uranus and Neptune into higher orbits and caused them to switch places, ultimately doubling Neptune's distance from the Sun. The resultant expulsion of objects from the proto-Kuiper belt could also explain the Late Heavy Bombardment 600 million years after the Solar System's formation and the origin of the Jupiter trojans. It is possible that Pluto had a near-circular orbit about 33 AU from the Sun before Neptune's migration perturbed it into a resonant capture. The Nice model requires that there were about a thousand Pluto-sized bodies in the original planetesimal disk, which included Triton and Eris.

Observation and exploration

Pluto's distance from Earth makes its in-depth study and exploration difficult. On July 14, 2015, NASA's *New Horizons* space probe flew through the Pluto system, providing much information about it.

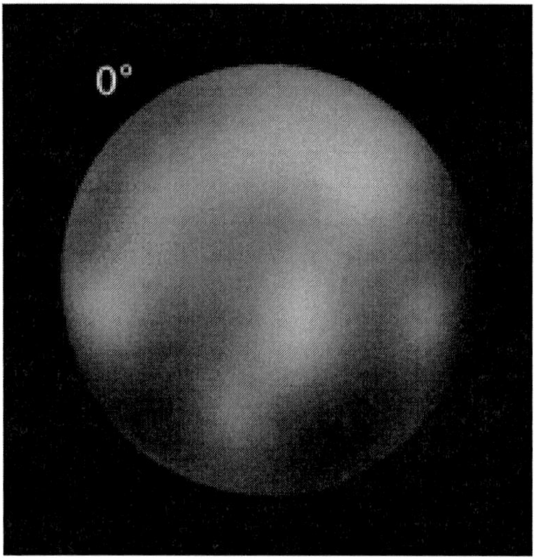

Figure 290: *Computer-generated rotating image of Pluto based on observations by the Hubble Space Telescope in 2002–2003*

Observation

Pluto's visual apparent magnitude averages 15.1, brightening to 13.65 at perihelion. To see it, a telescope is required; around 30 cm (12 in) aperture being desirable. It looks star-like and without a visible disk even in large telescopes, because its angular diameter is only 0.11".

The earliest maps of Pluto, made in the late 1980s, were brightness maps created from close observations of eclipses by its largest moon, Charon. Observations were made of the change in the total average brightness of the Pluto–Charon system during the eclipses. For example, eclipsing a bright spot on Pluto makes a bigger total brightness change than eclipsing a dark spot. Computer processing of many such observations can be used to create a brightness map. This method can also track changes in brightness over time.

Better maps were produced from images taken by the Hubble Space Telescope (HST), which offered higher resolution, and showed considerably more detail, resolving variations several hundred kilometers across, including polar regions and large bright spots. These maps were produced by complex computer processing, which finds the best-fit projected maps for the few pixels of the Hubble images. These remained the most detailed maps of Pluto until the flyby of *New Horizons* in July 2015, because the two cameras on the HST used for these maps were no longer in service.

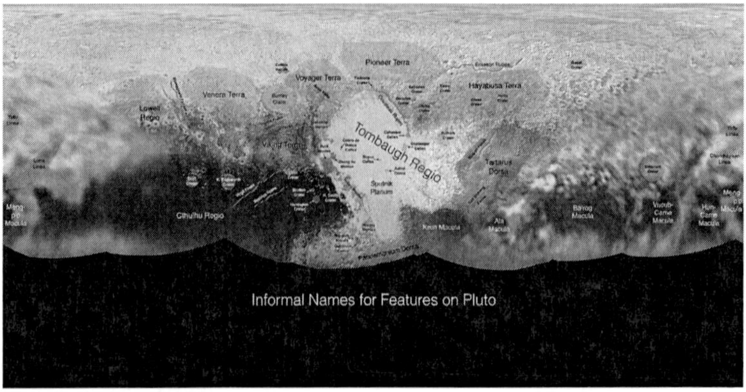

Figure 291: *The portions of Pluto's surface mapped by New Horizons (annotated)*

Exploration

The *New Horizons* spacecraft, which flew by Pluto in July 2015, is the first and so far only attempt to explore Pluto directly. Launched in 2006, it captured its first (distant) images of Pluto in late September 2006 during a test of the Long Range Reconnaissance Imager. The images, taken from a distance of approximately 4.2 billion kilometers, confirmed the spacecraft's ability to track distant targets, critical for maneuvering toward Pluto and other Kuiper belt objects. In early 2007 the craft made use of a gravity assist from Jupiter.

New Horizons made its closest approach to Pluto on July 14, 2015, after a 3,462-day journey across the Solar System. Scientific observations of Pluto began five months before the closest approach and continued for at least a month after the encounter. Observations were conducted using a remote sensing package that included imaging instruments and a radio science investigation tool, as well as spectroscopic and other experiments. The scientific goals of *New Horizons* were to characterize the global geology and morphology of Pluto and its moon Charon, map their surface composition, and analyze Pluto's neutral atmosphere and its escape rate. On October 25, 2016, at 05:48 pm ET, the last bit of data (of a total of 50 billion bits of data; or 6.25 gigabytes) was received from *New Horizons* from its close encounter with Pluto.

Videos

Pluto flyover animated (July 14, 2015)

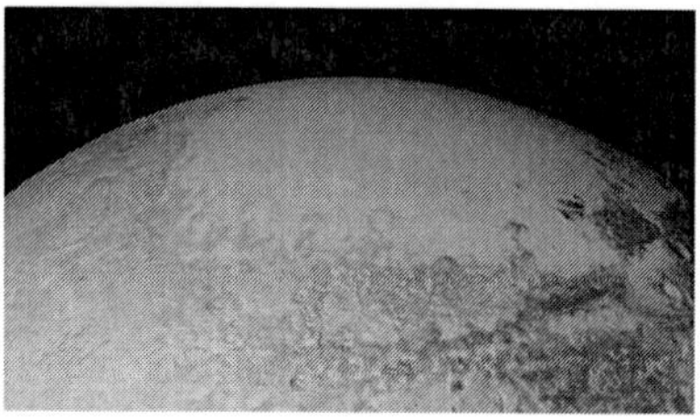

(00:30; released September 18, 2015)

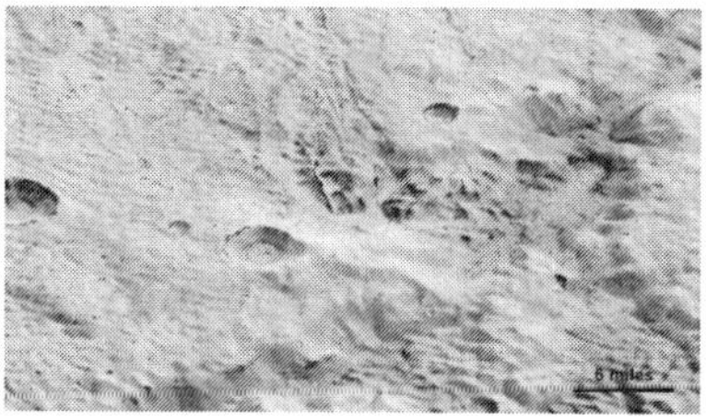

(00:50; released December 5, 2015)

This mosaic strip – extending across the hemisphere that faced the New Horizons spacecraft as it flew past Pluto. (No Audio – 1080p 60fps)

Further reading

- Stern, S A and Tholen, D J (1997), *Pluto and Charon*, University of Arizona Press ISBN 978-0816518401

External links

- *New Horizons* homepage[443]
- Pluto Profile[444] at NASA's Solar System Exploration site[445]
- NASA Pluto factsheet[446]
- Website of the observatory that discovered Pluto[447]
- Earth telescope image of Pluto system[448]
- Keck infrared with AO of Pluto system[449]
- Gray, Meghan (2009). "Pluto"[450]. *Sixty Symbols*. Brady Haran for the University of Nottingham.
- Video – Pluto – viewed through the years (GIF)[451] (NASA; animation; July 15, 2015).
- Video – Pluto – "FlyThrough" (00:22; MP4)[452] (YouTube)[453] (NASA; animation; August 31, 2015).
- "A Day on Pluto Video made from July 2015 New Horizon Images"[454] Scientific American

- NASA CGI video[455] of Pluto flyover (July 14, 2017)
- CGI video[456] simulation of rotating Pluto by Seán Doran (see album[457] for more)

<indicator name="featured-star"> </indicator>

Charon (moon)

Charon

Charon in approximate true color as viewed by *New Horizons* on 13 July 2015

Discovery	
Discovered by	James W. Christy
Discovery date	22 June 1978
Designations	
Pronunciation	/ˈʃærən/ ***SHARR**-ən* or /ˈkɛərən/ ***KAIR**-ən*[458]
Named after	Charon
Alternative names	(134340) Pluto I
Adjectives	Charonian
Orbital characteristics	
Epoch 2452600.5	
Semi-major axis	17536±4 km to system barycenter, 19571±4 km to the center of Pluto
Eccentricity	0.00
Orbital period	6.3872304±0.0000011 d (6 d, 9 h, 17 m, 36.7 ± 0.1 s)
Average orbital speed	0.21 km/s[459]
Inclination	0.001° (to Pluto's equator) 119.591°±0.014° (to Pluto's orbit) 112.783°±0.014° (to the ecliptic)

Longitude of ascending node	223.046°±0.014° (to vernal equinox)
Satellite of	Pluto
Physical characteristics	
Mean radius	606±3 km[460] (0.095 Earths, 0.51 Plutos)
Flattening	<1%[461]
Surface area	$4. \times 10^6$ km^2 (0.0090 Earths)
Volume	$(9.32 \pm 0.14) \times 10^8$ km^3 (0.00086 Earths)
Mass	$(1.586 \pm 0.015) \times 10^{21}$ kg $(2.66 \times 10^{-4}$ Earths) (12.2% of Pluto)
Mean density	1.707±0.013 g/cm^3
Surface gravity	0.288 m/s^2
Escape velocity	0.59 km/s 0.37 mi/s
Rotation period	synchronous
Albedo	0.2 to 0.5 at a solar phase angle of 15°
Temperature	−220 °C (53 K)
Apparent magnitude	16.8
Absolute magnitude *(H)*	1
Angular diameter	55 milli-arcsec

Charon, also known as **(134340) Pluto I**, is the largest of the five known natural satellites of the dwarf planet Pluto. It was discovered in 1978 at the United States Naval Observatory in Washington, D.C., using photographic plates taken at the United States Naval Observatory Flagstaff Station (NOFS). With half the diameter and one eighth the mass of Pluto, it is a very large moon in comparison to its parent body. Its gravitational influence is such that the barycenter of the Pluto–Charon system lies outside Pluto. In September 2016, astronomers announced that the reddish-brown cap of the north pole of Charon is composed of tholins, organic macromolecules that may be essential ingredients of life, and produced from methane, nitrogen and related gases released from the atmosphere of Pluto and transferred over about 19,000 km (12,000 mi) distance to the orbiting moon. The *New Horizons* spacecraft is the only probe that has visited the Pluto system. It approached Charon to within 27,000 km (17,000 mi) in 2015.

Figure 292: *Mosaic of best-resolution images of Charon from different angles*

History

Discovery

Charon was discovered by United States Naval Observatory astronomer James Christy, using the 1.55-meter telescope at NOFS, and was formally announced to the world via the International Astronomical Union on July 7, 1978 On June 22, 1978, he had been examining highly magnified images of Pluto on photographic plates taken at the 1.55-meter (61 in) Flagstaff telescope two months prior. Christy noticed that a slight elongation appeared periodically. Later, the bulge was confirmed on plates dating back to April 29, 1965.

Subsequent observations of Pluto determined that the bulge was due to a smaller accompanying body. The periodicity of the bulge corresponded to Pluto's rotation period, which was previously known from Pluto's light curve. This indicated a synchronous orbit, which strongly suggested that the bulge effect was real and not spurious. This resulted in reassessments of Pluto's size, mass, and other physical characteristics, because the calculated mass and albedo of the Pluto–Charon system had previously been attributed to Pluto alone.

Doubts about Charon's existence were erased when it and Pluto entered a five-year period of mutual eclipses and transits between 1985 and 1990. This

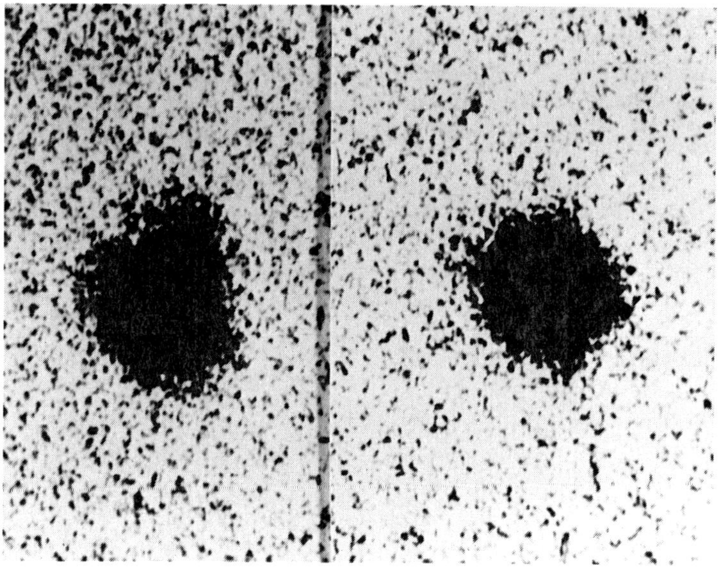

Figure 293: *Charon's discovery at the Naval Observatory Flagstaff Station as a time-varying bulge on the image of Pluto (seen near the top at left, but absent on the right). Negative image.*

occurs when the Pluto–Charon orbital plane is edge-on as seen from Earth, which only happens at two intervals in Pluto's 248-year orbital period. It was fortuitous that one of these intervals happened to occur soon after Charon's discovery.

Name

Charon was originally known by the temporary designation **S/1978 P 1**, according to the then recently instituted convention. On June 24, 1978, Christy first suggested the name *Charon* as a scientific-sounding version of his wife Charlene's nickname, "Char". Although colleagues at the Naval Observatory proposed *Persephone,* Christy stuck with *Charon* after discovering that it coincidentally refers to a Greek mythological figure:[462] Charon (/ˈkɛərɒn/ or /ˈkɛərən/; Greek Χάρων) is the ferryman of the dead, closely associated in myth with the god Hades, whom the Romans identified with their god Pluto. The IAU officially adopted the name in late 1985 and it was announced on January 3, 1986.

There is minor debate over the preferred pronunciation of the name. The practice of following the classical pronunciation established for the mythological ferryman Charon (IPA [ˈkɛːrən]) is used by major English-language dictionaries, such as the Merriam-Webster and Oxford English Dictionary.[463] These indicate only one pronunciation of "Charon" when referring specifically to Pluto's moon: with an initial "k" sound. Speakers of many languages other than English, and many English-speaking astronomers as well, follow this pronunciation.

However, Christy himself pronounced the *ch* as *sh* (IPA [ʃ]), after his wife Charlene. Because of this, as an acknowledgement of Christy and sometimes as an in-joke or shibboleth, the initial *sh* pronunciation is common among astronomers when speaking English,[464,465,466,467] and this is the prescribed pronunciation at NASA and of the *New Horizons* team.[468,469]

Formation

Simulation work published in 2005 by Robin Canup suggested that Charon could have been formed by a collision around 4.5 billion years ago, much like Earth and the Moon. In this model, a large Kuiper belt object struck Pluto at high velocity, destroying itself and blasting off much of Pluto's outer mantle, and Charon coalesced from the debris. However, such an impact should result in an icier Charon and rockier Pluto than scientists have found. It is now thought that Pluto and Charon might have been two bodies that collided before going into orbit about each other. The collision would have been violent enough to boil off volatile ices like methane (CH
4) but not violent enough to have destroyed either body. The very similar density of Pluto and Charon implies that the parent bodies were not fully differentiated when the impact occurred.

Orbit

Charon and Pluto orbit each other every 6.387 days. The two objects are gravitationally locked to one another, so each keeps the same face towards the other. This is a case of mutual tidal locking, as compared to that of the Earth and the Moon, where the Moon always shows the same face to Earth, but not vice versa. The average distance between Charon and Pluto is 19,570 kilometres (12,160 mi). The discovery of Charon allowed astronomers to calculate accurately the mass of the Plutonian system, and mutual occultations revealed their sizes. However, neither indicated the two bodies' individual masses, which could only be estimated, until the discovery of Pluto's outer moons in late 2005. Details in the orbits of the outer moons reveal that Charon has approximately 12.18% of the mass of Pluto. This shows it to have a density of 1.702 ± 0.021 g/cm^3, suggesting a composition of 55 ± 5% rock to 45% ice, whereas Pluto is somewhat denser and about 70% rock.

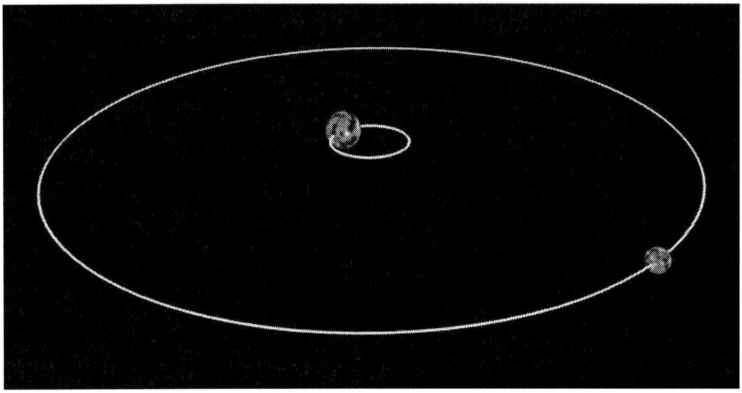

Figure 294: *An oblique view of the Pluto–Charon system showing that Pluto orbits a point outside itself. Also visible is the mutual tidal locking between the two bodies.*

Physical characteristics

Charon's diameter is 1,212 kilometres (753 mi), just over half that of Pluto, and larger than the dwarf planet Ceres, and the twelfth largest natural satellite in the solar system. Charon is sufficiently massive to have collapsed into a spheroid under its own gravity. Charon's slow rotation means that there is almost no flattening. Its equatorial and polar radii differ by less than 1%.

Interior

Charon's volume and mass allow calculation of its density, 1.702±0.021 g/cm^3, from which it can be determined that Charon is slightly less dense than Pluto and thus contains a larger proportion of ice relative to rock in its interior. The difference is considerably lower than that of most suspected collisional satellites. Before *New Horizons*' flyby, there were two conflicting theories about Charon's internal structure: some scientists thought Charon to be a differentiated body like Pluto, with a rocky core and an icy mantle, whereas others thought it would be uniform throughout. Evidence in support of the former position was found in 2007, when observations by the Gemini Observatory of patches of ammonia hydrates and water crystals on the surface of Charon suggested the presence of active cryogeysers. The fact that the ice was still in crystalline form suggested it had been deposited recently, because solar radiation would have degraded it to an amorphous state after roughly thirty thousand years.

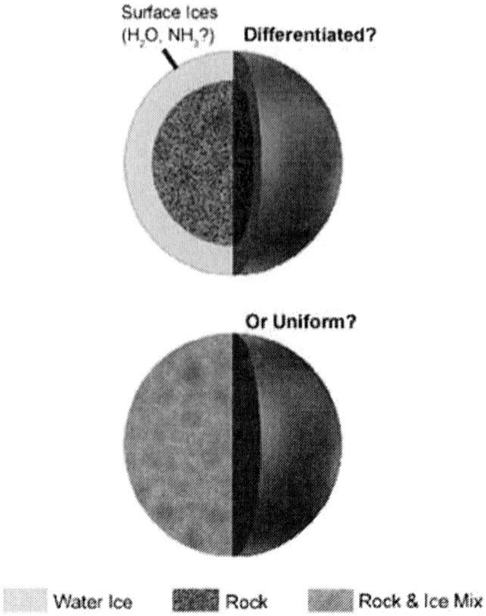

Figure 295: *The two conflicting theories about Charon's internal structure*

Figure 296: *Size comparisons: Earth, the Moon, and Charon*

Figure 297: *Charon in enhanced color to bring out differences in surface composition.*

Surface

Unlike Pluto's surface, which is composed of nitrogen and methane ices, Charon's surface appears to be dominated by the less volatile water ice. In 2007, observations by the Gemini Observatory of patches of ammonia hydrates and water crystals on the surface of Charon suggested the presence of active cryogeysers and cryovolcanoes.

Photometric mapping of Charon's surface shows a latitudinal trend in albedo, with a bright equatorial band and darker poles. The north polar region is dominated by a very large dark area informally dubbed "Mordor" by the *New Horizons* team. The favored explanation for this phenomenon is that they are formed by condensation of gases that escaped from Pluto's atmosphere. In winter, the temperature is –258 °C, and these gases, which include nitrogen, carbon monoxide, and methane, condense into their solid forms; when these ices are subjected to solar radiation, they chemically react to form various reddish tholins. Later, when the area is again heated by the Sun as Charon's seasons change, the temperature at the pole rises to –213 °C, resulting in the volatiles sublimating and escaping Charon, leaving only the tholins behind. Over millions of years, the residual tholin builds up thick layers, obscuring

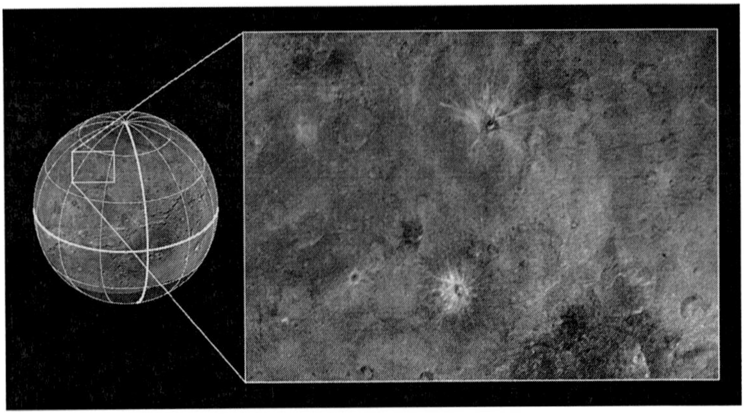

Figure 298: *The youngest crater of Charon.*

the icy crust. In addition to Mordor, *New Horizons* found evidence of extensive past geology that suggests that Charon is probably differentiated; in particular, the southern hemisphere has fewer craters than the northern and is considerably less rugged, suggesting that a massive resurfacing event—perhaps prompted by the partial or complete freezing of an internal ocean—occurred at some point in the past and removed many of the earlier craters.

Charon has a series of extensive grabens or canyons, such as Serenity Chasma, which extend as an equatorial belt for at least 1,000 km. Argo Chasma potentially reaches as deep as 9 km, with steep cliffs that may rival Verona Rupes on Miranda for the title of tallest cliff in the solar system.

Mountain in a moat

In a released photo by *New Horizons*, an unusual surface feature has captivated and baffled the scientist team of the mission. The image reveals a mountain rising out of a depression, in a zoomed in view. It's "a large mountain sitting in a moat", said Jeff Moore, of NASA's Ames Research Center, in a statement. "This is a feature that has geologists stunned and stumped", he added. *New Horizons* captured the photo from a distance of 49,000 miles (79,000 km).

Observation and exploration

Since the first blurred images of the moon (1), images showing Pluto and Charon resolved into separate disks were taken for the first time by the Hubble Space Telescope in the 1990s (2). The telescope was responsible for the best, yet low quality images of the moon. In 1994, the clearest picture of the Pluto-Charon system showed two distinct and well defined circles (3). The image was taken by Hubble's Faint Object Camera (FOC) when the system was 2.6 billion miles (4.4 billion kilometers) away from Earth Later, the development of adaptive optics made it possible to resolve Pluto and Charon into separate disks using ground-based telescopes.

In June 2015, the *New Horizons* spacecraft captured consecutive images of the Pluto–Charon system as it approached it. The images were put together in an animation. It was the best image of Charon to that date (4). In July 2015, the *New Horizons* spacecraft made its closest approach to the Pluto system. It is the only spacecraft to date to have visited and studied Charon.

Timeline of Charon observations

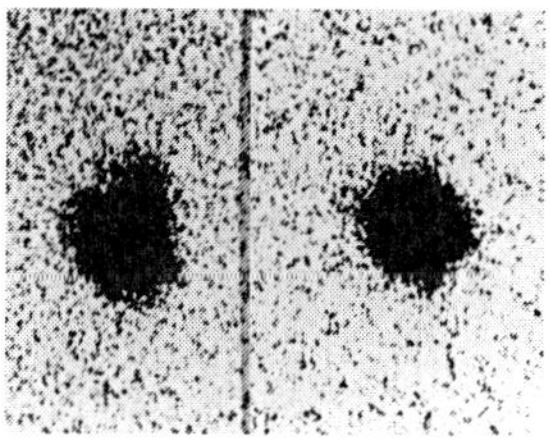

(1) Discovery;

1978

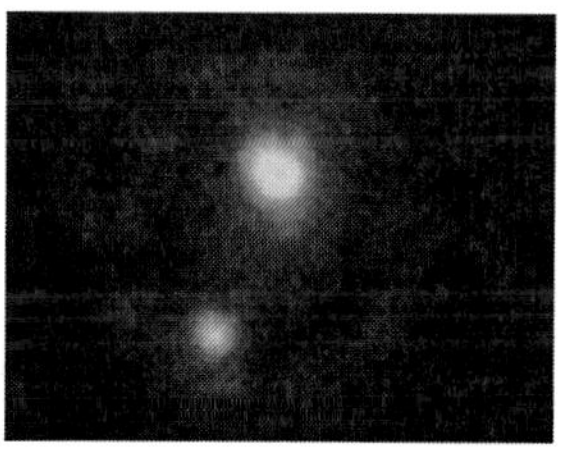

(2) HST – before correction;

1990

(3) HST – after correction;

1994

Figure 299: *Pluto and Charon, to scale.*
Photo taken by New Horizons on approach

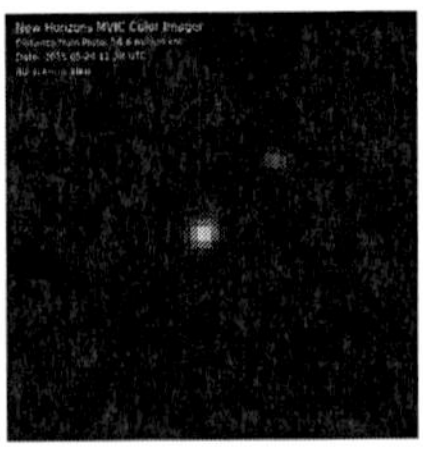

(4) 1st color animated view;

2015

Classification

The center of mass (barycenter) of the Pluto–Charon system lies outside either body. Because neither object truly orbits the other, and Charon has 12.2% the mass of Pluto, it has been argued that Charon should be considered to be part of a binary system with Pluto. The International Astronomical Union (IAU) states that Charon is considered to be just a satellite of Pluto, but the idea that Charon might be classified a dwarf planet in its own right may be considered at a later date.

In a draft proposal for the 2006 redefinition of the term, the IAU proposed that a planet be defined as a body that orbits the Sun that is large enough for gravitational forces to render the object (nearly) spherical. Under this proposal, Charon would have been classified as a planet, because the draft explicitly defined a planetary satellite as one in which the barycenter lies within the major body. In the final definition, Pluto was reclassified as a dwarf planet, but the formal definition of a planetary satellite was not decided upon. Charon is not in the list of dwarf planets currently recognized by the IAU. Had the draft proposal been accepted, even the Moon would be classified as a planet in billions of years when the tidal acceleration that is gradually moving the Moon away from Earth takes it far enough away that the center of mass of the system no longer lies within Earth.

The other moons of Pluto, Nix, Hydra, Kerberos and Styx, orbit the same barycenter, but they are not large enough to be spherical, and they are simply considered to be satellites of Pluto (or of Pluto–Charon).

Gallery

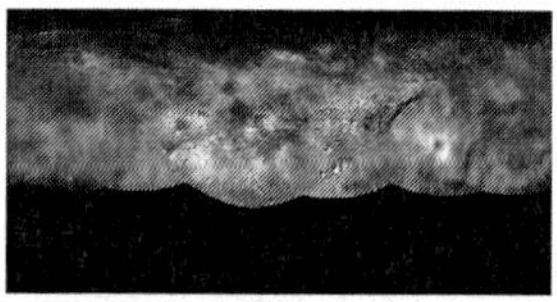

Global map of Charon

Identically processed enhanced-color views of Pluto and Charon

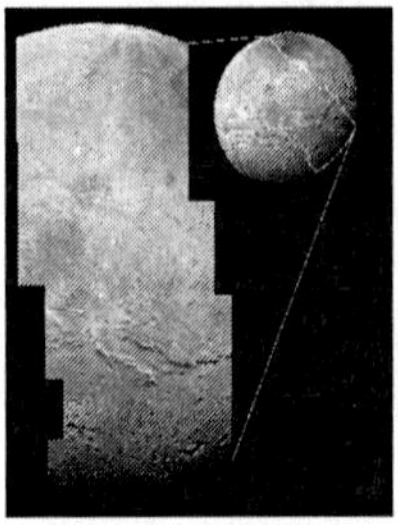

High resolution enhanced-color mosaic of Charon

Pluto and Charon as viewed by *New Horizons*(color; July 11, 2015).

Pluto and Charon as viewed by *New Horizons*(false-color; July 13, 2015).

Charon – night-side as viewed by *New Horizons*(July 17, 2015).

Videos

External links

Wikimedia Commons has media related to ***Charon (moon)***.

- Charon Profile[470] at NASA's Solar System Exploration site[471]
- James W. Christy and Robert S. Harrington, "The satellite of Pluto", *The Astronomical Journal* **83** (1978) 1005[472]
- Marc W. Buie, Phases of Charon as seen from Pluto[473], Lowell Observatory
- Buie, Surface of Charon and Pluto[474]
- Hubble reveals new map of Pluto[475], BBC News, September 12, 2005
- IAU Circular No. 3241[476] describing the discovery
- Measuring the Size of a Small, Frost World[477] (ESO press release January 2006)
- M. J. Person et al.: Charon's Radius and Density from the Combined Data Sets of the 2005 July 11 Occultation[478] (submitted to the Astronomical Journal, February 3, 2006)
- Cryovolcanism on Charon and other Kuiper Belt Objects[479]
- New Horizons Camera Spots Pluto's Largest Moon – July 10, 2013[480]
- New Horizons in the PHSF clean room at KSC, Nov 4, 2005[481]
- NASA CGI video[482] of Charon flyover (14 July 2017)
- CGI video[483] simulation of rotating Charon by Seán Doran (see album[484] for more)

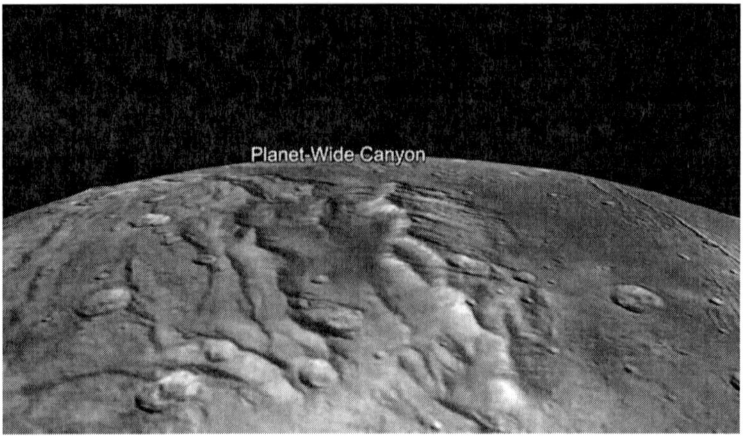

Figure 300:
Charon – Flyover video animation (00:20)
(released October 1, 2015).

Makemake

Makemake

Makemake and its moon (indicated by the arrow), as seen by the Hubble Space Telescope

Discovery	
Discovered by	• Michael E. Brown • Chad Trujillo • David Rabinowitz
Discovery date	March 31, 2005
Designations	
MPC designation	(136472) Makemake

Pronunciation	/ˌmækiˈmæki/, /ˌmɑːkiˈmɑːki/ or /ˌmɑːkeɪˈmɑːkeɪ/ 🔊 Media:en-us-Makemake.ogg[485]</ref>
Named after	Makemake
Alternative designations	2005 FY_9
Minor planet category	Dwarf planet cubewano scattered-near
Adjectives	Makemakean
Orbital characteristics	
Epoch JD 2457000.5 (9 December 2014)	
Aphelion	52.840 AU
Perihelion	38.590 AU
Semi-major axis	45.715 AU
Eccentricity	0.15586
Orbital period	309.09 yr (112,897 d)
Average orbital speed	4.419 km/s
Mean anomaly	15
Inclination	29.00685°
Longitude of ascending node	79.3659°
Argument of perihelion	297.240°
Known satellites	1
Physical characteristics	
Dimensions	• $(1434^{+48}_{-18}) \times (1420^{+18}_{-24}$ km) • (1502±45) × (1430±9) km
Mean radius	• 715^{+19}_{-11} km • 739±17 km
Flattening	0.05[486]
Surface area	$(6.9\pm0.3)\times10^6\ \text{km}^2$[487]
Volume	$(1.7\pm0.1)\times10^9\ \text{km}^3$
Mass	$< 4.4 \times 10^{21}$ kg
Mean density	1.4–3.2 g/cm^3 $<3.05\ g/cm^3$
Sidereal rotation period	7.771±0.003 h
Geometric albedo	$0.81^{+0.03}_{-0.05}$
Temperature	32–36 K (single-terrain model) 40–44 K (two-terrain model)

Spectral type	B–V=0.83, V–R=0.5
Apparent magnitude	17.0 (opposition)
Absolute magnitude (*H*)	-0.3

Makemake (minor-planet designation **136472 Makemake**) is a dwarf planet and perhaps the largest Kuiper belt[488] object in the classical population,[489] with a diameter approximately two thirds that of Pluto. Makemake has one known satellite, S/2015 (136472) 1. Makemake's extremely low average temperature, about 30 K (–243.2 °C), means its surface is covered with methane, ethane, and possibly nitrogen ices.

Makemake was discovered on March 31, 2005, by a team led by Michael E. Brown, and announced on July 29, 2005. Initially, it was known as **2005 FY$_9$** and later given the minor-planet number 136472. Makemake was recognized as a dwarf planet by the International Astronomical Union (IAU) in July 2008. Its name derives from Makemake in the mythology of the Rapa Nui people of Easter Island.

History

Discovery

Makemake was discovered on March 31, 2005, by a team at the Palomar Observatory, led by Michael E. Brown, and was announced to the public on July 29, 2005. The team had planned to delay announcing their discoveries of the bright objects Makemake and Eris until further observations and calculations were complete, but announced them both on July 29 when the discovery of another large object they had been tracking, Haumea, was controversially announced on July 27 by a different team in Spain.

Despite its relative brightness (it is about a fifth as bright as Pluto),[490] Makemake was not discovered until after many much fainter Kuiper belt objects. Most searches for minor planets are conducted relatively close to the ecliptic (the region of the sky that the Sun, Moon and planets appear to lie in, as seen from Earth), due to the greater likelihood of finding objects there. It probably escaped detection during the earlier surveys due to its relatively high orbital inclination, and the fact that it was at its farthest distance from the ecliptic at the time of its discovery, in the northern constellation of Coma Berenices.

Besides Pluto, Makemake is the only other dwarf planet that was bright enough that Clyde Tombaugh could have detected it during his search for trans-Neptunian planets around 1930. At the time of Tombaugh's survey, Makemake was only a few degrees from the ecliptic, near the border of Taurus and

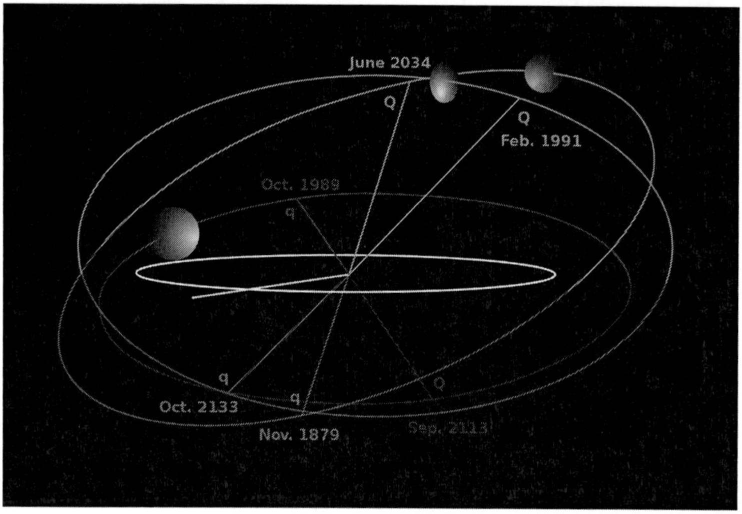

Figure 301: *Orbits of Makemake (blue), Haumea (green), contrasted with the orbit of Pluto (red) and the ecliptic (grey). The perihelia (q) and the aphelia (Q) are marked with the dates of passage. The positions on April 2006 are marked with the spheres illustrating relative sizes and differences in albedo and colour.*

Auriga,[491] at an apparent magnitude of 16.0. This position, however, was also very near the Milky Way, and Makemake would have been almost impossible to find against the dense background of stars. Tombaugh continued searching for some years after the discovery of Pluto, but he did not find Makemake or any other trans-Neptunian objects.

Name

The provisional designation 2005 FY_9 was given to Makemake when the discovery was made public. Before that, the discovery team used the codename "Easterbunny" for the object, because of its discovery shortly after Easter.

In July 2008, in accordance with IAU rules for classical Kuiper belt objects, 2005 FY_9 was given the name of a creator deity. The name of Makemake, the creator of humanity and god of fertility in the myths of the Rapa Nui, the native people of Easter Island, was chosen in part to preserve the object's connection with Easter.

Orbit and classification

As of December 2015[492], Makemake is 52.4 AU (7.84×10^9 km) from the Sun, almost as far from the Sun as it ever reaches on its orbit. Makemake follows an orbit very similar to that of Haumea: highly inclined at 29° and a moderate eccentricity of about 0.16. Nevertheless, Makemake's orbit is slightly farther from the Sun in terms of both the semi-major axis and perihelion. Its orbital period is nearly 310 years, more than Pluto's 248 years and Haumea's 283 years. Both Makemake and Haumea are currently far from the ecliptic—the angular distance is almost 29°. Makemake is approaching its 2033 aphelion, whereas Haumea passed its aphelion in early 1992.

Makemake is a classical Kuiper belt object (KBO), which means its orbit lies far enough from Neptune to remain stable over the age of the Solar System. Unlike plutinos, which can cross Neptune's orbit due to their 2:3 resonance with the planet, the classical objects have perihelia further from the Sun, free from Neptune's perturbation. Such objects have relatively low eccentricities (e below 0.2) and orbit the Sun in much the same way the planets do. Makemake, however, is a member of the "dynamically hot" class of classical KBOs, meaning that it has a high inclination compared to others in its population. Makemake is, probably coincidentally, near the 11:6 resonance with Neptune.[493]

Physical characteristics

Brightness, size, and rotation

Makemake is currently visually the second-brightest Kuiper belt object after Pluto, having a March opposition apparent magnitude of 17.0 in the constellation Coma Berenices. This is bright enough to be visible using a high-end amateur telescope.

Combining the detection in infrared by the *Spitzer Space Telescope* and *Herschel Space Telescope* with the similarities of spectrum with Pluto yielded an estimated diameter from 1,360 to 1,480 km. From the 2011 stellar occultation by Makemake, its dimensions have been initially measured to be $(1502 \pm 45) \times (1430 \pm 9)$ km. However, this analysis of the occultation data was later reanalyzed, which led to the dimension estimate of (1434+48
-18) × (1420+18
-24 km) without a pole-orientation constraint. This means that Makemake is slightly larger than that of Haumea, making it likely the fourth-largest known trans-Neptunian object after Pluto, Eris, and 2007 OR_{10}, though the error bars with the latter overlap. Makemake was the fourth dwarf planet recognized, because it has a bright V-band absolute magnitude of –0.44. Makemake has a

Figure 302: *Makemake (apparent magnitude 16.9)*

high geometrical albedo of 0.81+0.01
-0.02.

The rotation period of Makemake is estimated at 7.77 hours. Its lightcurve amplitude is small, only 0.03 *mag*. This was thought to be due to Makemake currently being viewed pole on from Earth; however, S/2015 (136472) 1's orbital plane (which is probably orbiting with little inclination relative to Makemake's equator due to tides resulting from its rapid rotation) is edge-on from Earth, implying that Makemake is really being viewed equator-on.

Spectra and surface

Like Pluto, Makemake appears red in the visible spectrum, and significantly redder than the surface of Eris (see colour comparison of TNOs). The near-infrared spectrum is marked by the presence of the broad methane (CH_4) absorption bands. Methane is observed also on Pluto and Eris, but its spectral signature is much weaker.

Spectral analysis of Makemake's surface revealed that methane must be present in the form of large grains at least one centimetre in size. In addition to methane, large amounts of ethane and tholins as well as smaller amounts of ethylene, acetylene and high-mass alkanes (like propane) may be present, most likely created by photolysis of methane by solar radiation. The tholins are

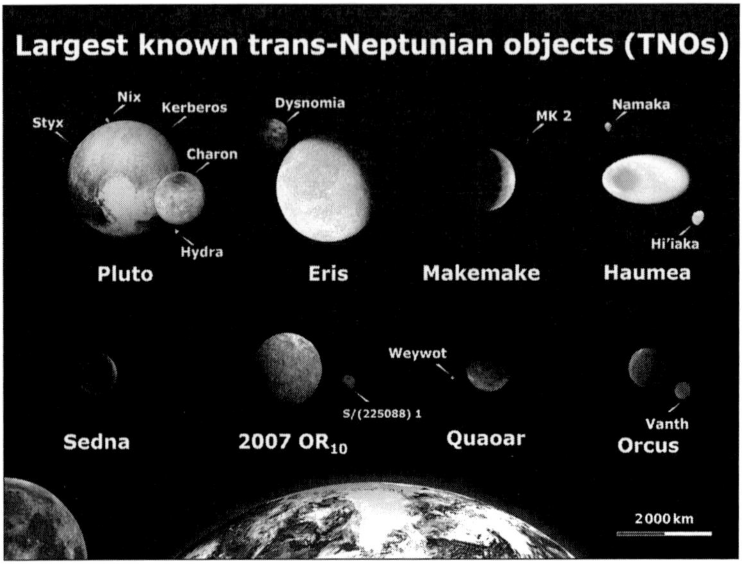

Figure 303: *style=float:bottom}}*

probably responsible for the red color of the visible spectrum. Although evidence exists for the presence of nitrogen ice on its surface, at least mixed with other ices, there is nowhere near the same level of nitrogen as on Pluto and Triton, where it composes more than 98 percent of the crust. The relative lack of nitrogen ice suggests that its supply of nitrogen has somehow been depleted over the age of the Solar System.

The far-infrared (24–70 μm) and submillimeter (70–500 μm) photometry performed by Spitzer and Herschel telescopes revealed that the surface of Makemake is not homogeneous. Although the majority of it is covered by nitrogen and methane ices, where the albedo ranges from 78 to 90%, there are small patches of dark terrain whose albedo is only 2 to 12%, and that make up 3–7% of the surface. These studies were made before S/2015 (136472) 1 was discovered; thus, these small dark patches may actually have been the dark surface of the satellite rather than any actual surface features on Makemake. However, another experiment has refuted these studies. Between 2006 and 2013, the Telescopio Nazionale Galileo acquired new visible and near infra-red spectra for Makemake that covered nearly 80% of its surface; this study found that the variation in the spectra were negligible, suggesting that Makemake's surface may indeed be homogenous.

Atmosphere

Makemake was expected to have an atmosphere similar to that of Pluto but with a lower surface pressure. However, on 23 April 2011 Makemake passed in front of an 18th-magnitude star and abruptly blocked its light. The results showed that Makemake presently lacks a substantial atmosphere and placed an upper limit of 4–12 nanobar on the pressure at its surface.

The presence of methane and possibly nitrogen suggests that Makemake could have a transient atmosphere similar to that of Pluto near its perihelion. Nitrogen, if present, will be the dominant component of it. The existence of an atmosphere also provides a natural explanation for the nitrogen depletion: because the gravity of Makemake is weaker than that of Pluto, Eris and Triton, a large amount of nitrogen was probably lost via atmospheric escape; methane is lighter than nitrogen, but has significantly lower vapor pressure at temperatures prevalent at the surface of Makemake (32–36 K), which hinders its escape; the result of this process is a higher relative abundance of methane. However, studies of Pluto's atmosphere by *New Horizons* suggest that methane, not nitrogen, is the dominant escaping gas, suggesting that the reasons for Makemake's absence of nitrogen may be more complicated.

Satellite

S/2015 (136472) 1

Dwarf planet Makemake and its moon

Discovery	
Discovered by	• HST/Wide Field Camera 3 • Alex Parker • Marc Buie • Will Grundy • Keith Noll
Discovery date	April 2015
Designations	

Pronunciation	//
Alternative designations	MK 2 (unofficial)
Orbital characteristics	
Semi-major axis	21,000 km to 300,000 km
Orbital period	12.4 days to 660 days
Satellite of	Makemake
Physical characteristics	
Mean radius	~87.5km
Geometric albedo	~0.1
Apparent magnitude	25.1

S/2015 (136472) 1, nicknamed MK 2 by the discovery team, is the only known moon of Makemake. It is estimated to be 175 km (110 mi) km in diameter (for an assumed albedo of 4%) and has a semi-major axis at least 21,000 km (13,000 mi) from Makemake. Its orbital period is $\geq$ 12 days (the minimum values are those for a circular orbit; the actual orbital eccentricity is unknown). Observations leading to its discovery occurred in April 2015, using the *Hubble Space Telescope*'s Wide Field Camera 3 and its discovery was announced on 26 April 2016.

Most other large trans-Neptunian objects have at least one satellite: Eris has one, Haumea has two, Pluto has five, and 2007 OR_{10} has one satellite. 10% to 20% of all trans-Neptunian objects are expected to have one or more satellites. Because satellites offer a simple method to measure an object's mass, Makemake's satellite should lead to better estimates of its mass.

Observations

A preliminary examination of the imagery suggests that S/2015 (136472) 1 has a reflectivity similar to charcoal, making it an extremely dark object. This is somewhat surprising because Makemake is the second-brightest known object in the Kuiper belt. One hypothesis to explain this is that its gravity is not strong enough to prevent bright but volatile ices from being lost to space when it is heated by the distant Sun.

Further observations will be needed in order to determine S/2015 (136472) 1's orbit. If it is circular, it would suggest that S/2015 (136472) 1 was formed by an ancient impact event, but if it is elliptical, it suggests that it may have been captured.

Alex Parker, the leader of the team that performed the analysis of the images at the Southwest Research Institute, said that S/2015 (136472) 1's orbit appears to be aligned edge-on to Earth-based observatories. This would make it much more difficult to detect because it would be lost in Makemake's glare much of the time and, along with its dark surface, would contribute to previous surveys failing to observe it.

Exploration

It was calculated that a flyby mission to Makemake could take just over 16 years using a Jupiter gravity assist, based on a launch date of 21 August 2024 or 24 August 2036. Makemake would be approximately 52 AU from the Sun when the spacecraft arrives.

External links

Wikimedia Commons has media related to ***Makemake***.

- MPEC listing for Makemake[494]
- AstDys[495] orbital elements
- Orbital simulation[496] from JPL (Java) / Ephemeris[497]
- Press release[498] from WHT and TNG on Makemake's similarity to Pluto.
- Makemake Sky Charts and Coordinates[499]
- Precovery image[500] with the 1.06 m Kleť Observatory telescope on April 20, 2003
- Makemake as seen on 2010-02-18 UT[501] with the Keck 1
- Makemake of the Outer Solar System[502] APOD July 15, 2008
- Simulation of Makemake (2005 FY_9)'s orbit[503]

<indicator name="featured-star"> </indicator>

Haumea

Haumea

Keck image of Haumea and its two moons. Hiʻiaka is above Haumea (center), and Namaka is directly below.

Discovery	
Discovered by	Brown et al.; Ortiz et al. (neither official)
Discovery date	2004 December 28 (Brown et al.); 2005 July (Ortiz et al.)
Designations	
MPC designation	(136108) Haumea
Pronunciation	/haʊˈmeɪ.ə/ or /ˌhɑːuːˈmeɪ.ə/[504] or ***HAH**-oo-**MAY**-ə* with four syllables according to Brown's students.[505]</ref>
Named after	Haumea
Alternative designations	2003 EL_{61}
Minor planet category	Dwarf planet, plutoid, TNO, 7:12 resonance, (delisted cubewano) Haumea family, and trinary
Adjectives	Haumean
Orbital characteristics	
Epoch 2014-12-09 (JD 2457000.5)	
Aphelion	51.483 AU (7.702 Tm)
Perihelion	34.952 AU (5.229 Tm)
Semi-major axis	43.218 AU (6.465 Tm)
Eccentricity	0.19126
Orbital period	284.12 yr (103774 d)
Average orbital speed	4.531 km/s
Mean anomaly	209.07°
Inclination	28.19°
Longitude of ascending node	121.79°
Argument of perihelion	240.20°
Known satellites	2
Physical characteristics	
Dimensions	2,322 × 1,701 × 1,138 km 1,920 × 1,540 × 990 km

Mean radius	816 km 620+34 -29 km (Herschel) 690 km
Surface area	$\approx 8\times10^6$ km^2
Volume	$\approx 3.5\times10^9$ km^3
Mass	$(4.006\pm0.040)\times10^{21}$ kg 0.00066 Earths
Mean density	2.6 g/cm^3 1.885–1.757 g/cm^3
Equatorial surface gravity	≈ 0.401 m/s^2
Equatorial escape velocity	≈ 0.809 km/s
Sidereal rotation period	0.163146±0.000004 d (3.9155±0.0001 h)
Geometric albedo	0.804+0.062 -0.095 0.70–0.75 0.51 ± 0.02
Temperature	< 50 K
Spectral type	(Neutral) B–V = 0.64, V–R = 0.33 B_0–V_0 = 0.646
Apparent magnitude	17.3 (opposition)
Absolute magnitude *(H)*	0.2

Haumea, minor-planet designation **136108 Haumea**, is a dwarf planet located beyond Neptune's orbit. It was discovered in 2004 by a team headed by Mike Brown of Caltech at the Palomar Observatory in the United States and independently in 2005, by a team headed by José Luis Ortiz Moreno at the Sierra Nevada Observatory in Spain, though the latter claim has been contested. On September 17, 2008, it was recognized as a dwarf planet by the International Astronomical Union (IAU) and named after Haumea, the Hawaiian goddess of childbirth.

Haumea's mass is about one-third that of Pluto, and 1/1400 that of Earth. Although its shape has not been directly observed, calculations from its light curve indicate that it is a triaxial ellipsoid, with its major axis twice as long as its minor. Its gravity is thought to be sufficient for it to have relaxed into hydrostatic equilibrium, making it a dwarf planet. Haumea's elongated shape together with its rapid rotation, high density, and high albedo (from a surface of crystalline water ice), are thought to be the consequences of a giant collision, which left Haumea the largest member of a collisional family that includes several large trans-Neptunian objects (TNOs) and Haumea's two known moons, Hi'iaka and Namaka.

History

Discovery

Two teams claim credit for the discovery of Haumea. Mike Brown and his team at Caltech discovered Haumea in December 2004 on images they had taken on May 6, 2004. On July 20, 2005, they published an online abstract of a report intended to announce the discovery at a conference in September 2005. At around this time, José Luis Ortiz Moreno and his team at the Instituto de Astrofísica de Andalucía at Sierra Nevada Observatory in Spain found Haumea on images taken on March 7–10, 2003. Ortiz emailed the Minor Planet Center with their discovery on the night of July 27, 2005.

Brown initially conceded discovery credit to Ortiz,[506] but came to suspect the Spanish team of fraud upon learning that his observation logs were accessed from the Spanish observatory the day before the discovery announcement. These logs included enough information to allow the Ortiz team to precover Haumea in their 2003 images, and they were accessed again just before Ortiz scheduled telescope time to obtain confirmation images for a second announcement to the MPC on July 29. Ortiz later admitted he had accessed the Caltech observation logs but denied any wrongdoing, stating he was merely verifying whether they had discovered a new object.

IAU protocol is that discovery credit for a minor planet goes to whoever first submits a report to the MPC (Minor Planet Center) with enough positional data for a decent determination of its orbit, and that the credited discoverer has priority in choosing a name. However, the IAU announcement on September 17, 2008, that Haumea had been accepted as a dwarf planet, did not mention a discoverer. The location of discovery was listed as the Sierra Nevada Observatory of the Spanish team, but the chosen name, Haumea, was the Caltech proposal; Ortiz's team had proposed "Ataecina", named for the ancient Iberian goddess of Spring.

Name

Until it was given a permanent name, the Caltech discovery team used the nickname "Santa" among themselves, because they had discovered Haumea on December 28, 2004, just after Christmas. The Spanish team were the first to file a claim for discovery to the Minor Planet Center, in July 2005. On July 29, 2005, Haumea was given the provisional designation **2003 EL$_{61}$**, based on the date of the Spanish discovery image. On September 7, 2006, it was numbered and admitted into the official minor planet catalogue as (136108) 2003 EL_{61}.

Following guidelines established by the IAU that classical Kuiper belt objects be given names of mythological beings associated with creation, in September 2006 the Caltech team submitted formal names from Hawaiian mythology to the IAU for both (136108) 2003 EL_{61} and its moons, in order "to pay homage to the place where the satellites were discovered". The names were proposed by David Rabinowitz of the Caltech team. *Haumea* is the matron goddess of the island of Hawai‘i, where the Mauna Kea Observatory is located. In addition, she is identified with Papa, the goddess of the earth and wife of Wākea (space), which is appropriate because Haumea is thought to be composed almost entirely of solid rock, without the thick ice mantle over a small rocky core typical of other known Kuiper belt objects. Lastly, Haumea is the goddess of fertility and childbirth, with many children who sprang from different parts of her body; this corresponds to the swarm of icy bodies thought to have broken off the dwarf planet during an ancient collision. The two known moons, also believed to have formed in this manner, are thus named after two of Haumea's daughters, Hi‘iaka and Nāmaka.

The proposal by the Ortiz team, Ataecina, did not meet IAU naming requirements, because Ataecina is not a creation deity. *(See Ataecina § Dwarf planet.)*

Classification

Haumea is a plutoid, a dwarf planet beyond Neptune's orbit. It is classified as a dwarf planet because it is presumed to be massive enough to have been rounded by its own gravity into a shape in hydrostatic equilibrium, but not massive enough to have cleared its neighbourhood of similarly sized objects. Haumea was initially listed as a classical Kuiper belt object in 2006 by the Minor Planet Center, but no longer. The nominal trajectory suggests that Haumea is in a weak 7:12 orbital resonance with Neptune, which would make it a resonant object instead. Although there are precovery images of Haumea dating back to March 22, 1955 from the Palomar Mountain Digitized Sky Survey, further observations of the orbit will be required to verify its dynamic status.

Orbit and rotation

Haumea has an orbital period of 284 Earth years, a perihelion of 35 AU, and an orbital inclination of 28°. It passed aphelion in early 1992, and is currently more than 50 AU from the Sun. Haumea's orbit has a slightly greater eccentricity than that of the other members of its collisional family. This is thought

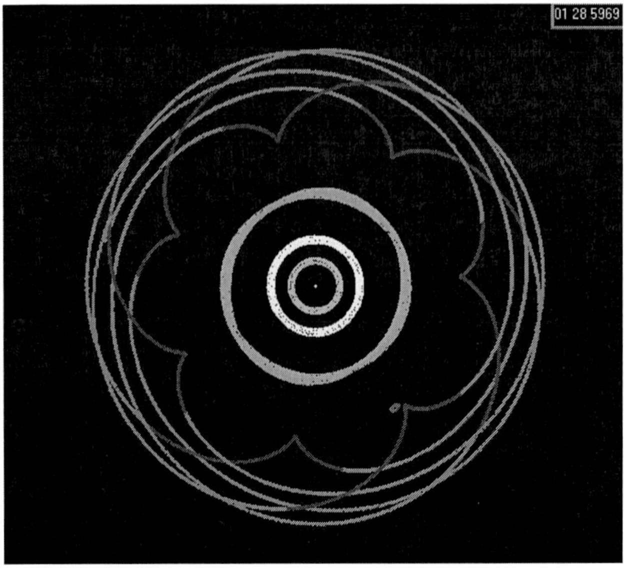

Figure 304: *The nominal libration of Haumea in a rotating frame, with Neptune stationary (see 2 Pallas for an example of non-librating)*

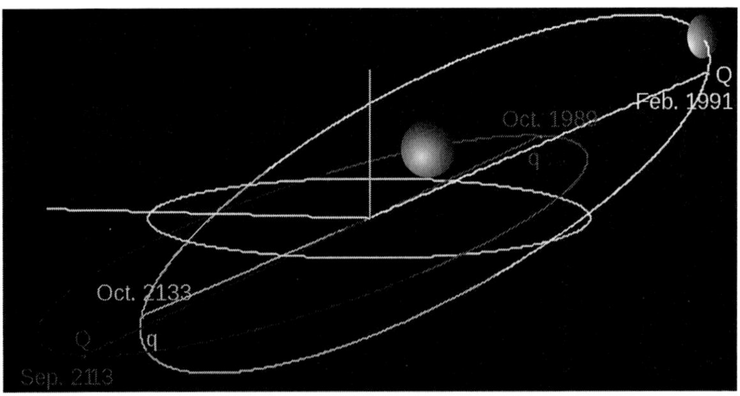

Figure 305: *Orbits of Haumea (yellow) and Pluto (red), relative to that of Neptune (grey), as of May 2009*[507]

to be due to Haumea's weak 7:12 orbital resonance with Neptune gradually modifying its initial orbit over the course of a billion years, through the Kozai effect, which allows the exchange of an orbit's inclination for increased eccentricity.

With a visual magnitude of 17.3, Haumea is the third-brightest object in the Kuiper belt after Pluto and Makemake, and easily observable with a large amateur telescope. However, because the planets and most small Solar System bodies share a common orbital alignment from their formation in the primordial disk of the Solar System, most early surveys for distant objects focused on the projection on the sky of this common plane, called the ecliptic. As the region of sky close to the ecliptic became well explored, later sky surveys began looking for objects that had been dynamically excited into orbits with higher inclinations, as well as more distant objects, with slower mean motions across the sky. These surveys eventually covered the location of Haumea, with its high orbital inclination and current position far from the ecliptic.

Haumea displays large fluctuations in brightness over a period of 3.9 hours, which can only be explained by a rotational period of this length. This is faster than any other known equilibrium body in the Solar System, and indeed faster than any other known body larger than 100 km in diameter. While most rotating bodies in equilibrium are flattened into oblate spheroids, Haumea rotates so quickly that it is distorted into a triaxial ellipsoid. If Haumea were to rotate much more rapidly, it would distort itself into a dumbbell shape and split in two. This rapid rotation is thought to have been caused by the impact that created its satellites and collisional family.

Physical characteristics

Because Haumea has moons, the mass of the system can be calculated from their orbits using Kepler's third law. The result is 4.2×10^{21} kg, 28% the mass of the Plutonian system and 6% that of the Moon. Nearly all of this mass is in Haumea.

Size, shape, and composition

The size of a Solar System object can be deduced from its optical magnitude, its distance, and its albedo. Objects appear bright to Earth observers either because they are large or because they are highly reflective. If their reflectivity (albedo) can be ascertained, then a rough estimate can be made of their size. For most distant objects, the albedo is unknown, but Haumea is large and bright enough for its thermal emission to be measured, which has given an approximate value for its albedo and thus its size. However, the calculation of its dimensions is complicated by its rapid rotation. The rotational physics

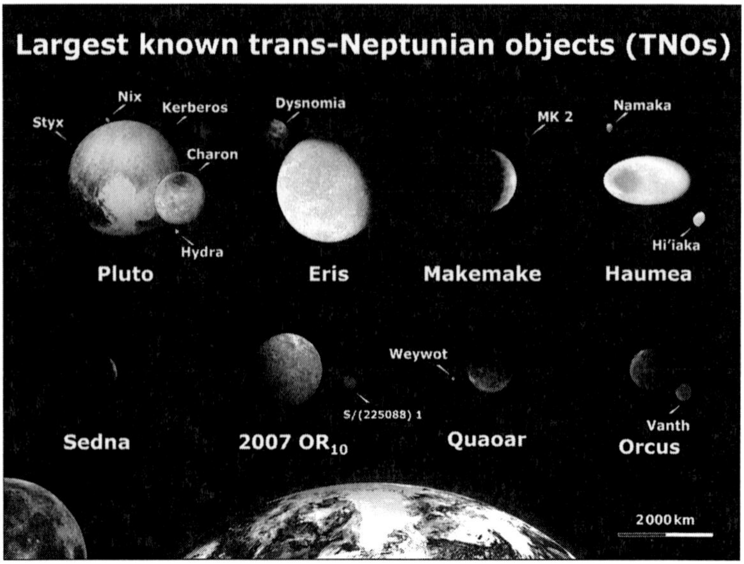

Figure 306: *style=float:bottom}}*

of deformable bodies predicts that over as little as a hundred days, a body rotating as rapidly as Haumea will have been distorted into the equilibrium form of a triaxial ellipsoid. It is thought that most of the fluctuation in Haumea's brightness is caused not by local differences in albedo but by the alternation of the side view and end view as seen from Earth.

The rotation and amplitude of Haumea's light curve were argued to place strong constraints on its composition. If Haumea were in hydrostatic equilibrium and had a low density like Pluto, with a thick mantle of ice over a small rocky core, its rapid rotation would have elongated it to a greater extent than the fluctuations in its brightness allow. Such considerations constrained its density to a range of 2.6–3.3 g/cm^3. By comparison, the Moon, which is rocky, has a density of 3.3 g/cm^3, whereas Pluto, which is typical of icy objects in the Kuiper belt, has a density of 1.86 g/cm^3. Haumea's possible high density covered the values for silicate minerals such as olivine and pyroxene, which make up many of the rocky objects in the Solar System. This also suggested that the bulk of Haumea was rock covered with a relatively thin layer of ice. A thick ice mantle more typical of Kuiper belt objects may have been blasted off during the impact that formed the Haumean collisional family.

Several ellipsoid-model calculations of Haumea's dimensions have been made. The first model produced after Haumea's discovery was calculated from ground-based observations of Haumea's light curve at optical wavelengths: it

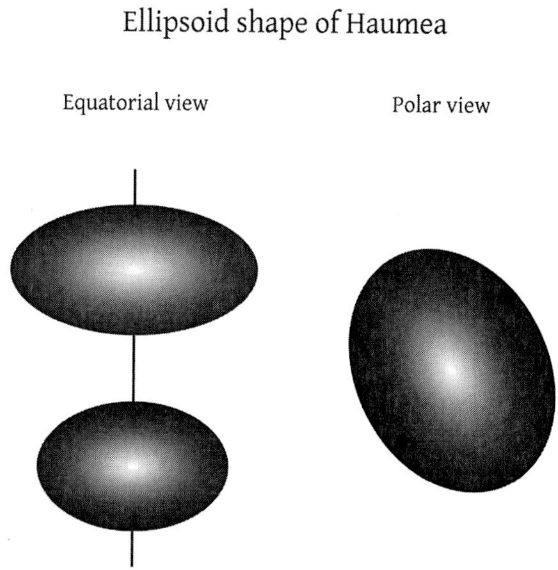

Figure 307: *The calculated ellipsoid shape of Haumea, 1,960×1,518×996 km (assuming an albedo of 0.73). At the left are the minimum and maximum equatorial silhouettes (1,960×996 and 1,518×996 km); at the right is the view from the pole (1,960×1,518 km).*

provided a total length of 1,960 to 2,500 km and a visual albedo (p_V) greater than 0.6. The most likely shape is a triaxial ellipsoid with approximate dimensions of 2,000 × 1,500 × 1,000 km, with an albedo of 0.71. Observations by the Spitzer Space Telescope give a diameter of 1,150+250
−100 km and an albedo of 0.84+0.1
−0.2, from photometry at infrared wavelengths of 70 μm. Subsequent light-curve analyses have suggested an equivalent circular diameter of 1,450 km. In 2010 an analysis of measurements taken by Herschel Space Telescope together with the older Spitzer Telescope measurements yielded a new estimate of the equivalent diameter of Haumea—about 1300 km. These independent size estimates overlap at an average geometric mean diameter of roughly 1,400 km.

However the observations of a stellar occultation in January 2017 cast a doubt on all those conclusions. The measured shape of Haumea, while elongated as presumed before, appeared to have significantly larger dimensions – according to the data obtained from the occultation Haumea is approximately the diameter of Pluto along its longest axis and about half that at its poles. So, the resulting density of about 1.8 g/cm^3 is more in line with densities of other large TNOs. This resulting shape is inconsistent with a homogenous body in

Figure 308: *Haumea is believed to rotate in just under 4 hours. This rapid rotation causes it to be elongated.*

hydrostatic equilibrium; although it is expected to be differentiated, it is not known whether that would make its shape consistent with hydrostatic equilibrium. Nevertheless Haumea appears to be one of the largest trans-Neptunian objects discovered, smaller than Eris, Pluto, similar to Makemake, and possibly 2007 OR10, and larger than Sedna, Quaoar, and Orcus.

Surface

In 2005, the Gemini and Keck telescopes obtained spectra of Haumea which showed strong crystalline water ice features similar to the surface of Pluto's moon Charon. This is peculiar, because crystalline ice forms at temperatures above 110 K, whereas Haumea's surface temperature is below 50 K, a temperature at which amorphous ice is formed. In addition, the structure of crystalline ice is unstable under the constant rain of cosmic rays and energetic particles from the Sun that strike trans-Neptunian objects. The timescale for the crystalline ice to revert to amorphous ice under this bombardment is on the order of ten million years, yet trans-Neptunian objects have been in their present cold-temperature locations for timescales of billions of years. Radiation damage should also redden and darken the surface of trans-Neptunian objects where the common surface materials of organic ices and tholin-like compounds are present, as is the case with Pluto. Therefore, the spectra and colour suggest Haumea and its family members have undergone recent resurfacing that produced fresh ice. However, no plausible resurfacing mechanism has been suggested.

Haumea is as bright as snow, with an albedo in the range of 0.6–0.8, consistent with crystalline ice. Other large TNOs such as Eris appear to have albedos as high or higher. Best-fit modeling of the surface spectra suggested that 66% to 80% of the Haumean surface appears to be pure crystalline water ice, with one contributor to the high albedo possibly hydrogen cyanide or phyllosilicate clays. Inorganic cyanide salts such as copper potassium cyanide may also be present.

However, further studies of the visible and near infrared spectra suggest a homogeneous surface covered by an intimate 1:1 mixture of amorphous and crystalline ice, together with no more than 8% organics. The absence of ammonia hydrate excludes cryovolcanism and the observations confirm that the collisional event must have happened more than 100 million years ago, in agreement with the dynamic studies. The absence of measurable methane in the spectra of Haumea is consistent with a warm collisional history that would have removed such volatiles, in contrast to Makemake.

In addition to the large fluctuations in Haumea's light curve due to the body's shape, which affect all colours equally, smaller independent colour variations seen in both visible and near-infrared wavelengths show a region on the surface that differs both in colour and in albedo. More specifically, a large dark red area on Haumea's bright white surface was seen in September 2009, possibly an impact feature, which indicates an area rich in minerals and organic (carbon-rich) compounds, or possibly a higher proportion of crystalline ice. Thus Haumea may have a mottled surface reminiscent of Pluto, if not as extreme.

Ring

A stellar occultation observed on 21 January 2017 and described in an 11 October 2017 *Nature* paper indicated the presence of a ring around Haumea. This represents the first ring system discovered for a TNO. The ring has a radius of about 2,287 km, a width of ~70 km and an opacity of 0.5. It is well within Haumea's Roche limit, which would be at a radius of about 4,400 km if it were spherical (being nonspherical pushes the limit out farther). The ring plane approximately coincides with Haumea's equatorial plane and the orbital plane of its larger, outer moon Hi'iaka. The ring is also close to the 3:1 resonance with Haumea's rotation (which is at a radius of 2,285 ± 8 km). The ring contributes around 5% to the total brightness of Haumea.

Figure 309: *Artist's conception of Haumea with its moons Hi'iaka and Namaka. The moons are much more distant than depicted here.*

Moons

Two small satellites have been discovered orbiting Haumea, (136108) Haumea I Hi'iaka and (136108) Haumea II Namaka. Brown's team discovered both in 2005, through observations of Haumea using the W.M. Keck Observatory.

Hi'iaka, at first nicknamed "Rudolph" by the Caltech team, was discovered January 26, 2005. It is the outer and, at roughly 310 km in diameter, the larger and brighter of the two, and orbits Haumea in a nearly circular path every 49 days. Strong absorption features at 1.5 and 2 micrometres in the infrared spectrum are consistent with nearly pure crystalline water ice covering much of the surface. The unusual spectrum, along with similar absorption lines on Haumea, led Brown and colleagues to conclude that capture was an unlikely model for the system's formation, and that the Haumean moons must be fragments of Haumea itself.

Namaka, the smaller, inner satellite of Haumea, was discovered on June 30, 2005, and nicknamed "Blitzen". It is a tenth the mass of Hi'iaka, orbits Haumea in 18 days in a highly elliptical, non-Keplerian orbit, and as of 2008[507] is inclined 13° from the larger moon, which perturbs its orbit. The relatively large eccentricities together with the mutual inclination of the orbits of the satellites are unexpected as they should have been damped by the tidal effects.

A relatively recent passage by a 3:1 resonance with Hi'iaka might explain the current excited orbits of the Haumean moons.

At present, the orbits of the Haumean moons appear almost exactly edge-on from Earth, with Namaka periodically occulting Haumea. Observation of such transits would provide precise information on the size and shape of Haumea and its moons, as happened in the late 1980s with Pluto and Charon. The tiny change in brightness of the system during these occultations will require at least a medium-aperture professional telescope for detection. Hi'iaka last occulted Haumea in 1999, a few years before discovery, and will not do so again for some 130 years. However, in a situation unique among regular satellites, Namaka's orbit is being greatly torqued by Hi'iaka, which preserved the viewing angle of Namaka–Haumea transits for several more years.

Collisional family

Haumea is the largest member of its collisional family, a group of astronomical objects with similar physical and orbital characteristics thought to have formed when a larger progenitor was shattered by an impact. This family is the first to be identified among TNOs and includes—beside Haumea and its moons—(55636) 2002 TX300 ($\approx$364 km), (24835) 1995 SM55 ($\approx$174 km), (19308) 1996 TO66 ($\approx$200 km), (120178) 2003 OP32 ($\approx$230 km), and (145453) 2005 RR43 ($\approx$252 km). Brown and colleagues proposed that the family were a direct product of the impact that removed Haumea's ice mantle, but a second proposal suggests a more complicated origin: that the material ejected in the initial collision instead coalesced into a large moon of Haumea, which was later shattered in a second collision, dispersing its shards outwards. This second scenario appears to produce a dispersion of velocities for the fragments that is more closely matched to the measured velocity dispersion of the family members.

The presence of the collisional family could imply that Haumea and its "offspring" might have originated in the scattered disc. In today's sparsely populated Kuiper belt, the chance of such a collision occurring over the age of the Solar System is less than 0.1 percent. The family could not have formed in the denser primordial Kuiper belt because such a close-knit group would have been disrupted by Neptune's migration into the belt—the believed cause of the belt's current low density. Therefore, it appears likely that the dynamic scattered disc region, in which the possibility of such a collision is far higher, is the place of origin for the object that generated Haumea and its kin.

Because it would have taken at least a billion years for the group to have diffused as far as it has, the collision which created the Haumea family is believed to have occurred very early in the Solar System's history.

Exploration

It was calculated that a flyby mission to Haumea could take 14.25 years using a Jupiter gravity assist, based on a launch date of 25 September 2025. Haumea would be 48.18 AU from the Sun when the spacecraft arrives. A flight time of 16.45 years can be achieved with launch dates on 1 November 2026, 23 September 2037 and 29 October 2038. Haumea could become a target for an exploration mission, and an example of this work is a preliminary study on a probe to Haumea and its moons (at 35–51 AU).[508] Probe mass, power source, and propulsion systems are key technology areas for this type of mission.

External links

Wikimedia Commons has media related to ***Haumea***.

- Visualization of Haumea's orbit[509] by NASA
- (136108) Haumea, Hi'iaka, and Namaka[510] at Johnston's Archive.com (updated September 17, 2008)
- International Year of Astronomy 2009 podcast: Dwarf Planet Haumea (Darin Ragozzine)[511]
- Haumea as seen on June 10, 2011[512] by Mike Brown using the 4.20 m (165 in) WHT / ∼0:30–3:30 dip in the brightness of Haumea+Namaka comes when Namaka crosses Haumea[513] (Hi'iaka, the outer moon, is blended in the images, but it rotates every 4.5 hr and adds a little variation)

<indicator name="featured-star"> </indicator>

Scattered disc

Types of distant minor planets

- Cis-Neptunian objects
  - Centaurs
  - Neptune trojans
- Trans-Neptunian objects (TNOs)‡
  - Kuiper belt objects (KBOs)
    - Classical KBOs (cubewanos)
    - Resonant KBOs
      - Plutinos (2:3 resonance)
  - Scattered disc objects (SDOs)
    - Resonant SDOs
  - Detached objects
  - Sednoids
  - Oort cloud objects (ICO/OCOs)

‡ Trans-Neptunian dwarf planets are called "plutoids"

- v
- t
- e[514]

The **scattered disc** (or **scattered disk**) is a distant circumstellar disc in the Solar System that is sparsely populated by icy small solar system bodies, and are a subset of the broader family of trans-Neptunian objects. The scattered-disc objects (SDOs) have orbital eccentricities ranging as high as 0.8, inclinations as high as 40°, and perihelia greater than 30 astronomical units (4.5×10^9 km; 2.8×10^9 mi). These extreme orbits are thought to be the result of gravitational "scattering" by the gas giants, and the objects continue to be subject to perturbation by the planet Neptune.

Although the closest scattered-disc objects approach the Sun at about 30–35 AU, their orbits can extend well beyond 100 AU. This makes scattered objects among the most distant and coldest objects in the Solar System.[515] The innermost portion of the scattered disc overlaps with a torus-shaped region of orbiting objects traditionally called the Kuiper belt, but its outer limits reach much further away from the Sun and further above and below the ecliptic than the Kuiper belt proper.[516] In this article, the scattered disc will be considered a separate population from the Kuiper belt.</ref>

Because of its unstable nature, astronomers now consider the scattered disc to be the place of origin for most periodic comets in the Solar System, with the centaurs, a population of icy bodies between Jupiter and Neptune, being the intermediate stage in an object's migration from the disc to the inner Solar System. Eventually, perturbations from the giant planets send such objects towards the Sun, transforming them into periodic comets. Many Oort cloud objects are also thought to have originated in the scattered disc. Detached

Figure 310: *Eris, the largest known scattered-disc object (center), and its moon Dysnomia (left of object)*

objects are not sharply distinct from scattered disc objects, and some such as Sedna have sometimes been considered to be included in this group.

Discovery

Traditionally, devices like a blink comparator were used in astronomy to detect objects in the Solar System, because these objects would move between two exposures—this involved time-consuming steps like exposing and developing photographic plates or films, and people then using a blink comparator to manually detect prospective objects. During the 1980s, the use of CCD-based cameras in telescopes made it possible to directly produce electronic images that could then be readily digitized and transferred to digital images. Because the CCD captured more light than film (about 90% versus 10% of incoming light) and the blinking could now be done at an adjustable computer screen, the surveys allowed for higher throughput. A flood of new discoveries was the result: over a thousand trans-Neptunian objects were detected between 1992 and 2006.

The first scattered-disc object (SDO) to be recognised as such was 1996 TL_{66}, originally identified in 1996 by astronomers based at Mauna Kea in Hawaii. Three more were identified by the same survey in 1999: 1999 CV_{118}, 1999

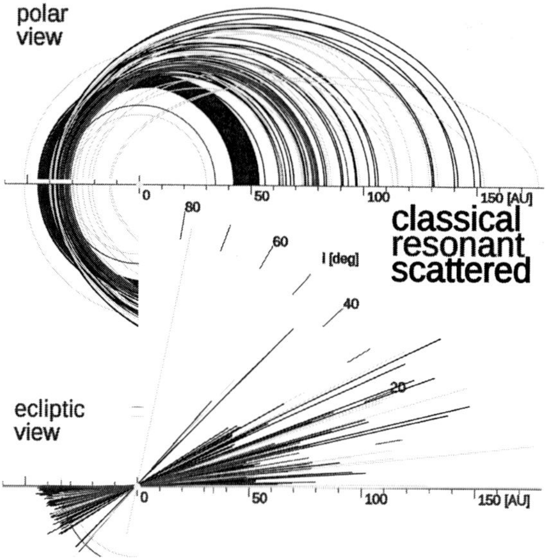

Figure 311: *The eccentricity and inclination of the scattered-disc population compared to the classical and 5:2 resonant Kuiper-belt objects*

CY$_{118}$, and 1999 CF$_{119}$. The first object presently classified as an SDO to be discovered was 1995 TL$_8$, found in 1995 by Spacewatch.[517]

As of 2011, over 200 SDOs have been identified, including 2007 UK$_{126}$ (discovered by Schwamb, Brown, and Rabinowitz), 2002 TC$_{302}$ (NEAT), Eris (Brown, Trujillo, and Rabinowitz), Sedna (Brown, Trujillo, and Rabinowitz) and 2004 VN112 (Deep Ecliptic Survey). Although the numbers of objects in the Kuiper belt and the scattered disc are hypothesized to be roughly equal, observational bias due to their greater distance means that far fewer SDOs have been observed to date.

Subdivisions of trans-Neptunian space

Known trans-Neptunian objects are often divided into two subpopulations: the Kuiper belt and the scattered disc. A third reservoir of trans-Neptunian objects, the Oort cloud, has been hypothesized, although no confirmed direct observations of the Oort cloud have been made. Some researchers further suggest a transitional space between the scattered disc and the inner Oort cloud, populated with "detached objects".

Scattered disc versus Kuiper belt

The Kuiper belt is a relatively thick torus (or "doughnut") of space, extending from about 30 to 50 AU comprising two main populations of Kuiper belt objects (KBOs): the classical Kuiper-belt objects (or "cubewanos"), which lie in orbits untouched by Neptune, and the resonant Kuiper-belt objects; those which Neptune has locked into a precise orbital ratio such as 3:2 (the object goes around twice for every three Neptune orbits) and 2:1 (the object goes around once for every two Neptune orbits). These ratios, called orbital resonances, allow KBOs to persist in regions which Neptune's gravitational influence would otherwise have cleared out over the age of the Solar System, since the objects are never close enough to Neptune to be scattered by its gravity. Those in 3:2 resonances are known as "plutinos", because Pluto is the largest member of their group, whereas those in 2:1 resonances are known as "twotinos".

In contrast to the Kuiper belt, the scattered-disc population can be disturbed by Neptune. Scattered-disc objects come within gravitational range of Neptune at their closest approaches ($\sim$30 AU) but their farthest distances reach many times that. Ongoing research suggests that the centaurs, a class of icy planetoids that orbit between Jupiter and Neptune, may simply be SDOs thrown into the inner reaches of the Solar System by Neptune, making them "cis-Neptunian" rather than trans-Neptunian scattered objects.[518] Some objects, like (29981) 1999 TD_{10}, blur the distinction and the Minor Planet Center (MPC), which officially catalogues all trans-Neptunian objects, now lists centaurs and SDOs together.

The MPC also makes a clear distinction between the Kuiper belt and the scattered disc, separating those objects in stable orbits (the Kuiper belt) from those in scattered orbits (the scattered disc and the centaurs). However, the difference between the Kuiper belt and the scattered disc is not clear-cut, and many astronomers see the scattered disc not as a separate population but as an outward region of the Kuiper belt. Another term used is "scattered Kuiper-belt object" (or SKBO) for bodies of the scattered disc.

Morbidelli and Brown propose that the difference between objects in the Kuiper belt and scattered-disc objects is that the latter bodies "are transported in semi-major axis by close and distant encounters with Neptune," but the former experienced no such close encounters. This delineation is inadequate (as they note) over the age of the Solar System, since bodies "trapped in resonances" could "pass from a scattering phase to a non-scattering phase (and vice versa) numerous times." That is, trans-Neptunian objects could travel back and forth between the Kuiper belt and the scattered disc over time. Therefore, they chose instead to define the regions, rather than the objects, defining the scattered disc as "the region of orbital space that can be visited by bodies that have encountered Neptune" within the radius of a Hill sphere, and the Kuiper belt as its "complement ... in the $a > 30$ AU region"; the region of the Solar System populated by objects with semi-major axes greater than 30 AU.

Detached objects

The Minor Planet Center classifies the trans-Neptunian object 90377 Sedna as a scattered-disc object. Its discoverer Michael E. Brown has suggested instead that it should be considered an inner Oort-cloud object rather than a member of the scattered disc, because, with a perihelion distance of 76 AU, it is too remote to be affected by the gravitational attraction of the outer planets. Under this definition, an object with a perihelion greater than 40 AU could be classified as outside the scattered disc.

Sedna is not the only such object: (148209) 2000 CR105 (discovered before Sedna) and 2004 VN112 have a perihelion too far away from Neptune to be influenced by it. This led to a discussion among astronomers about a new minor planet set, called the *extended scattered disc* (**E-SDO**). 2000 CR105 may also be an inner Oort-cloud object or (more likely) a transitional object between the scattered disc and the inner Oort cloud. More recently, these objects have been referred to as *"detached"*,[519] or *distant detached objects* (**DDO**).

There are no clear boundaries between the scattered and detached regions. Gomes et al. define SDOs as having "highly eccentric orbits, perihelia beyond Neptune, and semi-major axes beyond the 1:2 resonance." By this definition, all distant detached objects are SDOs. Since detached objects' orbits cannot be produced by Neptune scattering, alternative scattering mechanisms have been put forward, including a passing star or a distant, planet-sized object.

A scheme introduced by a 2005 report from the Deep Ecliptic Survey by J. L. Elliott et al. distinguishes between two categories: *scattered-near* (i.e. typical SDOs) and *scattered-extended* (i.e. detached objects). Scattered-near objects are those whose orbits are non-resonant, non-planetary-orbit-crossing and have a Tisserand parameter (relative to Neptune) less than 3. Scattered-extended objects have a Tisserand parameter (relative to Neptune) greater than 3 and have a time-averaged eccentricity greater than 0.2.

An alternative classification, introduced by B. J. Gladman, B. G. Marsden and C. Van Laerhoven in 2007, uses 10-million-year orbit integration instead of the Tisserand parameter. An object qualifies as an SDO if its orbit is not resonant, has a semi-major axis no greater than 2000 AU, and, during the integration, its semi-major axis shows an excursion of 1.5 AU or more. Gladman et al. suggest the term *scattering disk object* to emphasize this present mobility. If the object is not an SDO as per the above definition, but the eccentricity of

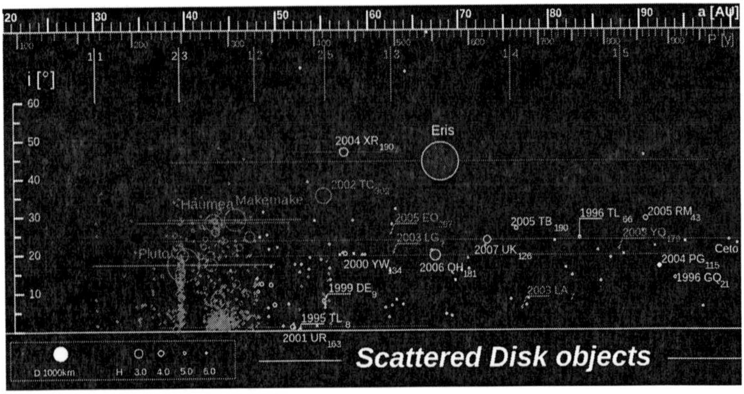

Figure 312: *The semi-major axes and inclinations of all known scattered-disc objects (in blue) up to 100 AU together with Kuiper-belt objects (in grey) and resonant objects (in green). The eccentricity of the orbits is represented by segments (extending from the perihelion to the aphelion) with the inclination represented on Y axis.*

its orbit is greater than 0.240, it is classified as a *detached TNO*. (Objects with smaller eccentricity are considered classical.) In this scheme, the disc extends from the orbit of Neptune to 2000 AU, the region referred to as the inner Oort cloud.

Orbits

The scattered disc is a very dynamic environment. Because they are still capable of being perturbed by Neptune, SDOs' orbits are always in danger of disruption; either of being sent outward to the Oort cloud or inward into the centaur population and ultimately the Jupiter family of comets. For this reason Gladman et al. prefer to refer to the region as the scattering disc, rather than scattered. Unlike Kuiper-belt objects (KBOs), the orbits of scattered-disc objects can be inclined as much as 40° from the ecliptic.

SDOs are typically characterized by orbits with medium and high eccentricities with a semi-major axis greater than 50 AU, but their perihelia bring them within influence of Neptune. Having a perihelion of roughly 30 AU is one of the defining characteristics of scattered objects, as it allows Neptune to exert its gravitational influence.

The classical objects (cubewanos) are very different from the scattered objects: more than 30% of all cubewanos are on low-inclination, near-circular

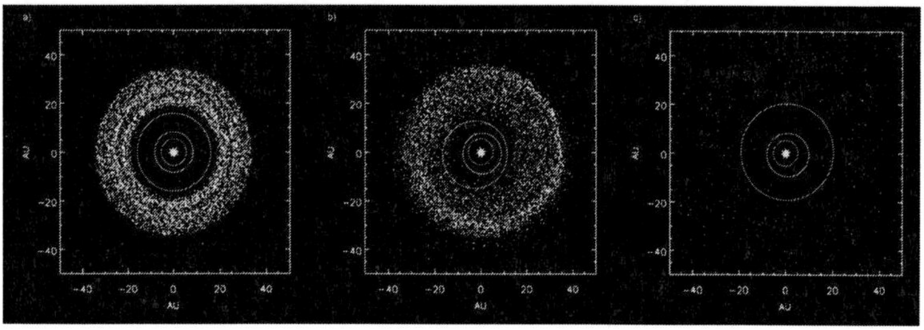

Figure 313: *Simulation showing Outer Planets and Kuiper Belt: a) Before Jupiter/ Saturn 2:1 resonance b) Scattering of Kuiper-belt objects into the Solar System after the orbital shift of Neptune c) After ejection of Kuiper-belt bodies by Jupiter*

orbits whose eccentricities peak at 0.25. Classical objects possess eccentricities ranging from 0.2 to 0.8. Though the inclinations of scattered objects are similar to the more extreme KBOs, very few scattered objects have orbits as close to the ecliptic as much of the KBO population.

Although motions in the scattered disc are random, they do tend to follow similar directions, which means that SDOs can become trapped in temporary resonances with Neptune. Examples of resonant orbits within the scattered disc include 1:3, 2:7, 3:11, 5:22 and 4:79.

Formation

The scattered disc is still poorly understood: no model of the formation of the Kuiper belt and the scattered disc has yet been proposed that explains all their observed properties.

According to contemporary models, the scattered disc formed when Kuiper belt objects (KBOs) were "scattered" into eccentric and inclined orbits by gravitational interaction with Neptune and the other outer planets. The amount of time for this process to occur remains uncertain. One hypothesis estimates a period equal to the entire age of the Solar System; a second posits that the scattering took place relatively quickly, during Neptune's early migration epoch.

Models for a continuous formation throughout the age of the Solar System illustrate that at weak resonances within the Kuiper belt (such as 5:7 or 8:1), or at the boundaries of stronger resonances, objects can develop weak orbital instabilities over millions of years. The 4:7 resonance in particular has large instability. KBOs can also be shifted into unstable orbits by close passage of massive objects, or through collisions. Over time, the scattered disc would gradually form from these isolated events.

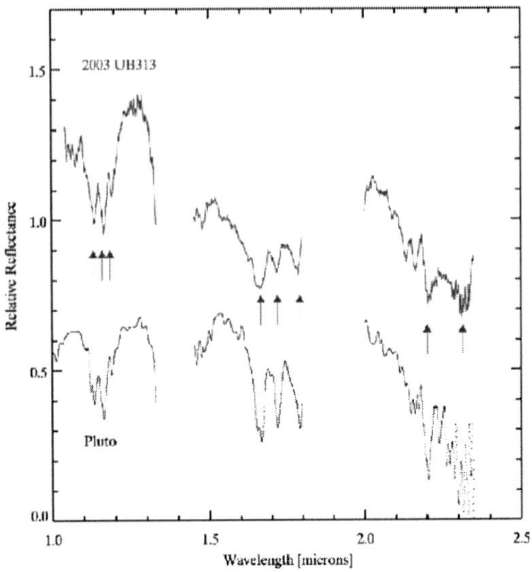

Figure 314: *The infrared spectra of both Eris and Pluto, highlighting their common methane absorption lines*

Computer simulations have also suggested a more rapid and earlier formation for the scattered disc. Modern theories indicate that neither Uranus nor Neptune could have formed *in situ* beyond Saturn, as too little primordial matter existed at that range to produce objects of such high mass. Instead, these planets, and Saturn, may have formed closer to Jupiter, but were flung outwards during the early evolution of the Solar System, perhaps through exchanges of angular momentum with scattered objects. Once the orbits of Jupiter and Saturn shifted to a 2:1 resonance (two Jupiter orbits for each orbit of Saturn), their combined gravitational pull disrupted the orbits of Uranus and Neptune, sending Neptune into the temporary "chaos" of the proto-Kuiper belt. As Neptune traveled outward, it scattered many trans-Neptunian objects into higher and more eccentric orbits. This model states that 90% or more of the objects in the scattered disc may have been "promoted into these eccentric orbits by Neptune's resonances during the migration epoch...[therefore] the scattered disc might not be so scattered."

Composition

Scattered objects, like other trans-Neptunian objects, have low densities and are composed largely of frozen volatiles such as water and methane. Spectral analysis of selected Kuiper belt and scattered objects has revealed signatures of similar compounds. Both Pluto and Eris, for instance, show signatures for methane.

Astronomers originally supposed that the entire trans-Neptunian population would show a similar red surface colour, as they were thought to have originated in the same region and subjected to the same physical processes. Specifically, SDOs were expected to have large amounts of surface methane, chemically altered into complex organic molecules by energy from the Sun. This would absorb blue light, creating a reddish hue. Most classical objects display this colour, but scattered objects do not; instead, they present a white or greyish appearance.

One explanation is the exposure of whiter subsurface layers by impacts; another is that the scattered objects' greater distance from the Sun creates a composition gradient, analogous to the composition gradient of the terrestrial and gas giant planets. Mike Brown, discoverer of the scattered object Eris, suggests that its paler colour could be because, at its current distance from the Sun, its atmosphere of methane is frozen over its entire surface, creating an inches-thick layer of bright white ice. Pluto, conversely, being closer to the Sun, would be warm enough that methane would freeze only onto cooler, high-albedo regions, leaving low-albedo tholin-covered regions bare of ice.

Comets

The Kuiper belt was initially thought to be the source of the Solar System's ecliptic comets. However, studies of the region since 1992 have shown that the orbits within the Kuiper belt are relatively stable, and that ecliptic comets originate from the scattered disc, where orbits are generally less stable.

Comets can loosely be divided into two categories: short-period and long-period—the latter being thought to originate in the Oort cloud. The two major categories of short-period comets are Jupiter-family comets (JFCs) and Halley-type comets. Halley-type comets, which are named after their prototype, Halley's Comet, are thought to have originated in the Oort cloud but to have been drawn into the inner Solar System by the gravity of the giant planets, whereas the JFCs are thought to have originated in the scattered disc. The centaurs are thought to be a dynamically intermediate stage between the scattered disc and the Jupiter family.

There are many differences between SDOs and JFCs, even though many of the Jupiter-family comets may have originated in the scattered disc. Although the centaurs share a reddish or neutral coloration with many SDOs, their nuclei are bluer, indicating a fundamental chemical or physical difference. One hypothesis is that comet nuclei are resurfaced as they approach the Sun by subsurface materials which subsequently bury the older material.

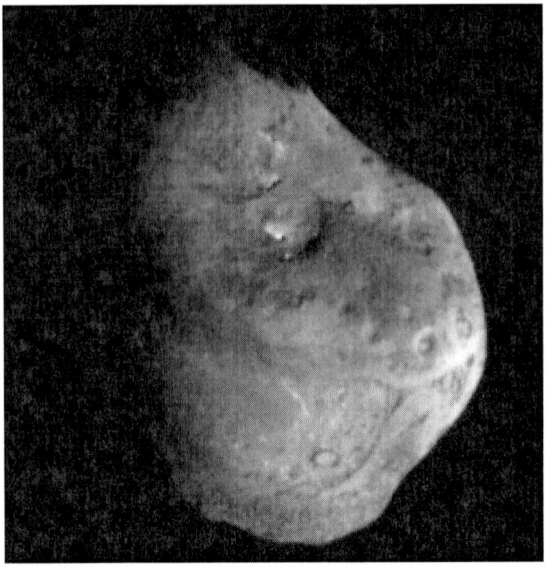

Figure 315: *Tempel 1, a Jupiter-family comet*

References

<indicator name="featured-star"> ★ </indicator>

Eris (dwarf planet)

Eris

Eris (center) and Dysnomia (left of center), taken by the Hubble Space Telescope

Discovery	
Discovered by	• M. E. Brown • C. A. Trujillo • D. L. Rabinowitz
Discovery date	January 5, 2005[520]
Designations	
MPC designation	(136199) Eris
Pronunciation	/ˈɪərɪs/ or /ˈɛrɪs/[521]
Named after	Eris
Alternative designations	2003 UB_{313}
Minor planet category	• Dwarf planet • TNO • Plutoid • SDO • Binary
Adjectives	Eridian
Orbital characteristics	
Epoch December 9, 2014 (JD 2457000.5)	
Aphelion	• 97.651 AU • 14.602×10^9 km
Perihelion	• 37.911 AU • 5.723×10^9 km
Semi-major axis	• 67.781 AU • 10.166×10^9 km
Eccentricity	0.44068
Orbital period	• 203,830 d • 558.04 yr

<table>
<tr><th>Average orbital speed</th><td>3.4338 km/s</td></tr>
<tr><th>Mean anomaly</th><td>204.16°</td></tr>
<tr><th>Inclination</th><td>44.0445°</td></tr>
<tr><th>Longitude of ascending node</th><td>35.9531°</td></tr>
<tr><th>Argument of perihelion</th><td>150.977°</td></tr>
<tr><th>Known satellites</th><td>Dysnomia</td></tr>
<tr><th colspan="2">Physical characteristics</th></tr>
<tr><th>Mean radius</th><td>1163±6 km</td></tr>
<tr><th>Surface area</th><td>$(1.70 \pm 0.02) \times 10^7$ km^2[522]</td></tr>
<tr><th>Volume</th><td>$(6.59 \pm 0.10) \times 10^9$ km^3</td></tr>
<tr><th>Mass</th><td>• $(1.66 \pm 0.02) \times 10^{22}$ kg
• 0.0028 Earths
• 0.23 Moons</td></tr>
<tr><th>Mean density</th><td>2.52±0.07 g/cm^3[523]</td></tr>
<tr><th>Equatorial surface gravity</th><td>0.82±0.02 m/s^2
0.083±0.002 *g*[524]</td></tr>
<tr><th>Equatorial escape velocity</th><td>1.38±0.01 km/s</td></tr>
<tr><th>Sidereal rotation period</th><td>25.9±0.5 hr</td></tr>
<tr><th>Geometric albedo</th><td>0.96+0.09
−0.04</td></tr>
<tr><th></th><td><table><tr><th>Surface temp.</th><th>min</th><th>mean</th><th>max</th></tr><tr><th>(approx)</th><td>30 K</td><td>42.5 K</td><td>55 K</td></tr></table></td></tr>
<tr><th>Spectral type</th><td>B−V=0.78, V−R=0.45</td></tr>
<tr><th>Apparent magnitude</th><td>18.7</td></tr>
<tr><th>Absolute magnitude (*H*)</th><td>−1.17+0.06
−0.11[525]</ref></td></tr>
<tr><th>Angular diameter</th><td>40 milli-arcsec</td></tr>
</table>

Eris (minor-planet designation **136199 Eris**) is the most massive and second-largest dwarf planet known in the Solar System. Eris was discovered in January 2005 by a Palomar Observatory-based team led by Mike Brown, and its identity was verified later that year. In September 2006 it was named after Eris, the Greek goddess of strife and discord.

Among known bodies of the Solar System, Eris is the ninth most massive directly orbiting the Sun, and the 16th most massive overall, because seven moons are more massive than all known dwarf planets. It is also the largest which has not yet been visited by a spacecraft. Eris was measured to be 2,326 ± 12 kilometers (1,445.3 ± 7.5 mi) in diameter. Eris' mass is about

0.27% of the Earth mass, about 27% more than dwarf planet Pluto, although Pluto is slightly larger by volume.

Eris is a trans-Neptunian object (TNO) and a member of a high-eccentricity population known as the scattered disk. It has one known moon, Dysnomia. As of February 2016[526], its distance from the Sun was 96.3 astronomical units (1.441×10^{10} km; 8.95×10^9 mi), roughly three times that of Pluto. With the exception of some long-period comets, Eris and Dysnomia are currently the second-most-distant known natural objects in the Solar System,[527] the farthest object being V774104 discovered in November 2015 at 103 AU.

Because Eris appeared to be larger than Pluto, NASA initially described it as the Solar System's tenth planet. This, along with the prospect of other objects of similar size being discovered in the future, motivated the International Astronomical Union (IAU) to define the term *planet* for the first time. Under the IAU definition approved on August 24, 2006, Eris is a "dwarf planet", along with objects such as Pluto, Ceres, Haumea and Makemake, thereby reducing the number of known planets in the Solar System to eight, the same as before Pluto's discovery in 1930. Observations of a stellar occultation by Eris in 2010 showed that its diameter was 2,326 ± 12 kilometers (1,445.3 ± 7.5 mi), very slightly less than Pluto, which was measured by *New Horizons* as 2,372 ± 4 kilometers (1,473.9 ± 2.5 mi) in July 2015.

History

Discovery

Eris was discovered by the team of Mike Brown, Chad Trujillo, and David Rabinowitz on January 5, 2005, from images taken on October 21, 2003. The discovery was announced on July 29, 2005, the same day as Makemake and two days after Haumea, due in part to events that would later lead to controversy about Haumea. The search team had been systematically scanning for large outer Solar System bodies for several years, and had been involved in the discovery of several other large TNOs, including 50000 Quaoar, 90482 Orcus, and 90377 Sedna.

Routine observations were taken by the team on October 21, 2003, using the 1.2 m Samuel Oschin Schmidt telescope at Palomar Observatory, California, but the image of Eris was not discovered at that point due to its very slow motion across the sky: The team's automatic image-searching software excluded all objects moving at less than 1.5 arcseconds per hour to reduce the number of false positives returned. When Sedna was discovered, it was moving at 1.75 arcsec/h, and in light of that the team reanalyzed their old data with a lower limit on the angular motion, sorting through the previously excluded images by eye. In January 2005, the re-analysis revealed Eris's slow motion against the background stars.

Follow-up observations were then carried out to make a preliminary determination of Eris's orbit, which allowed the object's distance to be estimated. The team had planned to delay announcing their discoveries of the bright objects Eris and Makemake until further observations and calculations were complete, but announced them both on July 29 when the discovery of another large TNO they had been tracking, Haumea, was controversially announced on July 27 by a different team in Spain.

More observations released in October 2005 revealed that Eris has a moon, later named Dysnomia. Observations of Dysnomia's orbit permitted scientists to determine the mass of Eris, which in June 2007 they calculated to be $(1.66 \pm 0.02) \times 10^{22}$ kg, 27%±2% greater than Pluto's.

Name

Eris is named after the Greek goddess Eris (Greek *"Ερις)*, a personification of strife and discord. The name was assigned on September 13, 2006, following an unusually long period in which the object was known by the provisional designation **2003 UB$_{313}$**, which was granted automatically by the IAU under their naming protocols for minor planets. The regular adjectival form of *Eris* is *Eridian.*

Xena

Due to uncertainty over whether the object would be classified as a planet or a minor planet, because different nomenclature procedures apply to these different classes of objects, the decision on what to name the object had to wait until after the August 24, 2006, IAU ruling. As a result, for a time the object became known to the wider public as *Xena*.

"Xena" was an informal name used internally by the discovery team. It was inspired by the title character of the television series *Xena: Warrior Princess*. The discovery team had reportedly saved the nickname "Xena" for the first body they discovered that was larger than Pluto. According to Brown,

> *We chose it since it started with an X (planet "X"), it sounds mythological (OK, so it's TV mythology, but Pluto is named after a cartoon, right?),*[528] *and (this part is actually true) we've been working to get more female deities out there (e.g. Sedna). Also, at the time, the TV show was still on TV, which shows you how long we've been searching!*

"We assumed [that] a real name would come out fairly quickly, [but] the process got stalled", Mike Brown said in interview,

> *One reporter [Ken Chang] called me up from the New York Times who happened to have been a friend of mine from college, [and] I was a little less guarded with him than I am with the normal press. He asked me, "What's the name you guys proposed?" and I said, "Well, I'm not going to tell." And he said, "Well, what do you guys call it when you're just talking amongst yourselves?"... As far as I remember this was the only time I told anybody this in the press, and then it got everywhere, which I only sorta felt bad about—I kinda like the name.*

Choosing an official name

According to science writer Govert Schilling, Brown initially wanted to call the object "Lila", after a concept in Hindu mythology that described the cosmos as the outcome of a game played by Brahma. The name was very similar to "Lilah", the name of Brown's newborn daughter. Brown was mindful of not making his name public before it had been officially accepted. He had done so with Sedna a year previously, and had been heavily criticized. However, no objection was raised to the Sedna name other than the breach of protocol, and no competing names were suggested for Sedna.

He listed the address of his personal web page announcing the discovery as */~mbrown/planetlila* and in the chaos following the controversy over the discovery of Haumea, forgot to change it. Rather than needlessly anger more of his fellow astronomers, he simply said that the webpage had been named for his daughter and dropped "Lila" from consideration.

Brown had also speculated that *Persephone*, the wife of the god Pluto, would be a good name for the object. The name had been used several times in science fiction, and was popular with the public, having handily won a poll conducted by *New Scientist* magazine ("Xena", despite only being a nickname, came fourth). This was not possible once the object was classified as a dwarf planet, because there is already an asteroid with that name, 399 Persephone.

With the dispute resolved, the discovery team proposed *Eris* on September 6, 2006. On September 13, 2006 this name was accepted as the official name by the IAU. Brown decided that, because the object had been considered a planet for so long, it deserved a name from Greek or Roman mythology, like the other planets. The asteroids had taken the vast majority of Graeco-Roman names. *Eris*, whom Brown described as his favorite goddess, had fortunately escaped inclusion. The name in part reflects the discord in the astronomical community caused by the debate over the object's (and Pluto's) classification.

Classification

Eris is a trans-Neptunian dwarf planet (plutoid). Its orbital characteristics more specifically categorize it as a scattered-disk object (SDO), or a TNO that has been "scattered" from the Kuiper belt into more-distant and unusual orbits following gravitational interactions with Neptune as the Solar System was forming. Although its high orbital inclination is unusual among the known SDOs, theoretical models suggest that objects that were originally near the inner edge of the Kuiper belt were scattered into orbits with higher inclinations than objects from the outer belt. Inner-belt objects are expected to be generally more massive than outer-belt objects, and so astronomers expect to discover more large objects like Eris in high-inclination orbits, which have traditionally been neglected.

Because Eris was initially thought to be larger than Pluto, it was described as the "tenth planet" by NASA and in media reports of its discovery. In response to the uncertainty over its status, and because of ongoing debate over whether Pluto should be classified as a planet, the IAU delegated a group of astronomers to develop a sufficiently precise definition of the term *planet* to decide the issue. This was announced as the IAU's *Definition of a Planet in the Solar System,* adopted on August 24, 2006. At this time, both Eris and Pluto were classified as *dwarf planets,* a category distinct from the new definition of *planet.* Brown has since stated his approval of this classification. The IAU subsequently added Eris to its Minor Planet Catalogue, designating it *(136199) Eris.*

Orbit

Eris has an orbital period of 558 years. Its maximum possible distance from the Sun (aphelion) is 97.65 AU, and its closest (perihelion) is 37.91 AU. It came to perihelion between 1698 and 1699, to aphelion around 1977, and will return to perihelion around 2256 to 2258. Eris and its moon are currently the second most distant known objects in the Solar System, after V774104, apart from long-period comets and space probes. Approximately forty known TNOs, most notably 2006 SQ372, 2000 OO_{67} and Sedna, though currently closer to the Sun than Eris, have greater average orbital distances than Eris's semimajor axis of 67.7 AU.

Eris's orbit is highly eccentric, and brings Eris to within 37.9 AU of the Sun, a typical perihelion for scattered objects. This is within the orbit of Pluto, but still safe from direct interaction with Neptune (29.8–30.4 AU). Pluto, on the other hand, like other plutinos, follows a less inclined and less eccentric orbit and, protected by orbital resonance, can cross Neptune's orbit. It is possible that Eris is in a 17:5 resonance with Neptune, though further observations will be required to check that hypothesis. Unlike the eight planets, whose orbits all lie roughly in the same plane as the Earth's, Eris's orbit is highly inclined: It is tilted at an angle of about 44 degrees to the ecliptic. In about 800 years, Eris will be closer to the Sun than Pluto for some time (see the graph at the right).

As of February 2016[526], Eris has an apparent magnitude of 18.7, making it bright enough to be detectable to some amateur telescopes. A 200-millimeter (7.9 in) telescope with a CCD can detect Eris under favorable conditions.[529] The reason it had not been noticed until now is its steep orbital inclination; searches for large outer Solar System objects tend to concentrate on the ecliptic plane, where most bodies are found.

Because of the high inclination of its orbit, Eris only passes through a few constellations of the traditional Zodiac; it is now in the constellation Cetus. It was in Sculptor from 1876 until 1929 and Phoenix from roughly 1840 until 1875. In 2036 it will enter Pisces and stay there until 2065, when it will enter Aries. It will then move into the northern sky, entering Perseus in 2128 and Camelopardalis (where it will reach its northernmost declination) in 2173.

Most-distant known objects in the Solar System as of 25 August 2017[526]

Object name	Distance from the Sun (AU)			Apparent magnitude	Absolute magnitude (H)
	Current	Perihelion	Aphelion		
V774104	103	Unknown	Unknown	24	4
Eris	96.19	37.8	97.6	18.7	−1.1
2014 UZ224	91.22	38.0	179.8	23.2	3.5
2007 OR10	87.81	33.0	100.8	21.7	1.8
Sedna	85.33	76.0	939	21.0	1.5
2013 FS28	85.27	34.6	347.6	24.5	4.9
2014 FC69	84.56	40.3	106.9	24.1	4.6
2006 QH181	83.72	37.8	96.7	23.6	4.3
2012 VP113	83.54	80.5	438	23.4	4.0
2013 FY27	80.13	36.1	81.8	22.1	3.0
2010 GB174	71.53	48.7	693	25.1	6.5
2014 FJ72	70.87	38.7	152.2	24.2	5.6
2012 FH84	68.68	45.8	80.6	25.7	7.3
2015 GR50	68.33	35.6	78.6	25.1	6.7
2015 GP50	67.99	35.9	89.1	24.8	6.5

2013 FQ$_{28}$	67.63	48.7	80.6	24.4	6.0
2013 UJ$_{15}$	64.93	36.3	69.2	25.2	7.0
2014 FL$_{72}$	64.10	38.2	170.4	25.0	6.8
2011 GM$_{89}$	63.83	37.2	68.8	25.6	7.1
2015 RR$_{245}$	63.58	33.7	129.2	22.1	3.9
2014 SU$_{349}$	62.55	30.8	109.8	25.0	7.0
2013 AT$_{183}$	62.39	36.0	88.1	22.0	4.7
2014 SV$_{349}$	61.97	34.2	89.0	23.0	5.0
2014 FE$_{72}$	61.58	36.3	4274.0	24.1	6.1
2015 RX$_{245}$	61.38	45.5	768.9	24.2	6.2
2000 CR$_{105}$	61.24	44.3	412	23.9	6.3
2008 ST$_{291}$	60.44	42.4	154.5	22.2	4.2
2014 FM$_{72}$	60.20	34.4	76.6	24.1	6.2
2014 FF$_{72}$	60.09	37.1	63.3	24.8	6.9
2003 QX$_{113}$	60.04	36.7	62.1	22.5	4.7
2015 KH$_{162}$	59.45	41.5	82.8	21.6	3.9
2013 UT$_{15}$	59.30	43.9	347.9	24.2	6.3
2014 FL$_{70}$	59.24	33.1	77.0	24.0	6.5
2013 SY$_{99}$	59.13	49.9	1301.4	24.5	6.8
2014 FH$_{72}$	59.06	37.3	77.3	25.1	7.2

This table includes all known objects currently located at least twice as far as Neptune.
See List of trans-Neptunian objects for more.

- view
- talk
- edit[530]

Size, mass and density

Size estimates

Year	Radius (diameter)	Source
2005	1,199 (2,397) km	*Hubble*
2007	1,300 (2,600) km	Spitzer
2011	1,163 (2,326) km	Occultation

In 2005, the diameter of Eris was measured to be 2397±100 km, using images from the *Hubble Space Telescope* (HST). The size of an object is determined from its absolute magnitude (H) and the albedo (the amount of light it reflects).

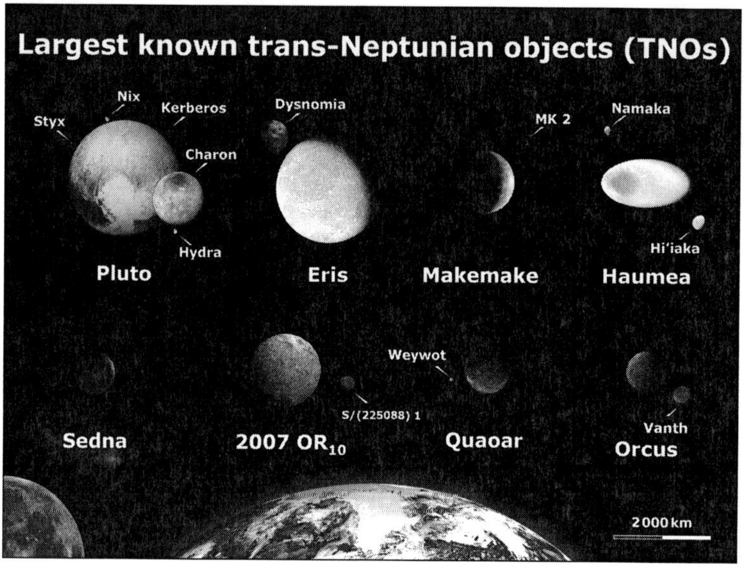

Figure 316: *style=float:bottom}}*

At a distance of 97 AU, an object with a diameter of 3,000 km would have an angular size of 40 milliarcseconds, which is directly measurable with the *Hubble Space Telescope*. Although resolving such small objects is at the very limit of its capabilities,[531] sophisticated image processing techniques such as deconvolution can be used to measure such angular sizes fairly accurately.[532]

This makes Eris around the same size as Pluto, which is 2372+4 km across. It also indicates an albedo of 0.96, higher than that of any other large body in the Solar System except Enceladus. It is speculated that the high albedo is due to the surface ices being replenished because of temperature fluctuations as Eris's eccentric orbit takes it closer and farther from the Sun.

In 2007, a series of observations of the largest trans-Neptunian objects with the Spitzer Space Telescope gave an estimate of Eris's diameter of 2600+400 -200 km. The Spitzer and Hubble estimates overlap in the range of 2,400–2,500 km, 4–8% larger than Pluto. Astronomers now suspect that Eris's spin axis is currently pointing toward the Sun, which would make the sunlit hemisphere warmer than average and skew any infrared measurements toward higher values. So the outcome from the 2010 Chile occultation is actually more in line with the Hubble result from 2005.

In November 2010, Eris was the subject of one of the most distant stellar occultations yet from Earth. Preliminary data from this event cast doubt on previous

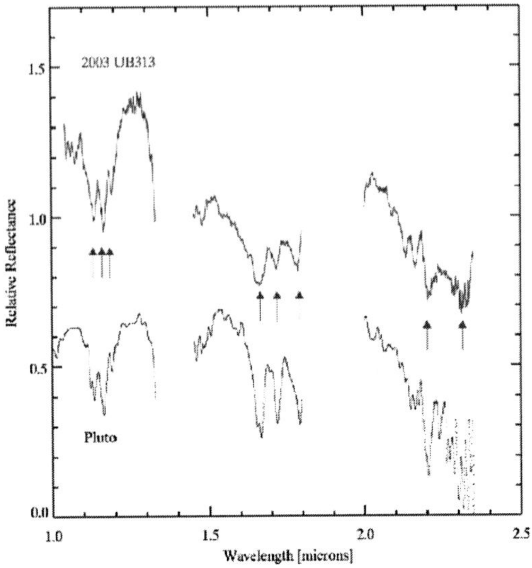

Figure 317: *The infrared spectrum of Eris, compared to that of Pluto, shows the marked similarities between the two bodies. Arrows denote methane absorption lines.*

size estimates. The teams announced their final results from the occultation in October 2011, with an estimated diameter of 2326+6
−6 km. The mass of Eris can be calculated with much greater precision. Based on the currently accepted value for Dysnomia's period—15.774 days— Eris is 27 percent more massive than Pluto. If the 2011 occultation results are used, then Eris has a density of 2.52±0.07 g/cm^3, substantially denser than Pluto, and thus must be composed largely of rocky materials.

Models of internal heating via radioactive decay suggest that Eris could have an internal ocean of liquid water at the mantle–core boundary.

In July 2015, after nearly ten years of Eris being considered the ninth-largest object known to directly orbit the sun, close-up imagery from the New Horizons mission more accurately determined Pluto's volume to be slightly larger than Eris's, rather than slightly smaller as previously thought. Eris is now the tenth-largest object known to directly orbit the sun by volume, though not by mass.

Surface and atmosphere

The discovery team followed up their initial identification of Eris with spectroscopic observations made at the 8 m Gemini North Telescope in Hawaii on January 25, 2005. Infrared light from the object revealed the presence of methane ice, indicating that the surface may be similar to that of Pluto, which at the time was the only TNO known to have surface methane, and of Neptune's moon Triton, which also has methane on its surface. No surface details can be resolved from Earth or its orbit with any instrument currently available.

Due to Eris's distant eccentric orbit, its surface temperature is estimated to vary between about 30 and 56 K (–243.2 and –217.2 °C).

Unlike the somewhat reddish Pluto and Triton, Eris appears almost white. Pluto's reddish color is thought to be due to deposits of tholins on its surface, and where these deposits darken the surface, the lower albedo leads to higher temperatures and the evaporation of methane deposits. In contrast, Eris is far enough from the Sun that methane can condense onto its surface even where the albedo is low. The condensation of methane uniformly over the surface reduces any albedo contrasts and would cover up any deposits of red tholins.

Even though Eris can be up to three times farther from the Sun than Pluto, it approaches close enough that some of the ices on the surface might warm enough to sublime. Because methane is highly volatile, its presence shows either that Eris has always resided in the distant reaches of the Solar System, where it is cold enough for methane ice to persist, or that the celestial body has an internal source of methane to replenish gas that escapes from its atmosphere. This contrasts with observations of another discovered TNO, Haumea, which reveal the presence of water ice but not methane.

Satellite

In 2005, the adaptive optics team at the Keck telescopes in Hawaii carried out observations of the four brightest TNOs (Pluto, Makemake, Haumea, and Eris), using the newly commissioned laser guide star adaptive optics system. Images taken on September 10 revealed a moon in orbit around Eris. In keeping with the "Xena" nickname already in use for Eris, Brown's team nicknamed the moon "Gabrielle", after the television warrior princess' sidekick. When Eris received its official name from the IAU, the moon received the name *Dysnomia,* after the Greek goddess of lawlessness who was Eris's daughter. Brown says he picked it for similarity to his wife's name, Diane. The name also retains an oblique reference to Eris's old informal name *Xena*, portrayed on TV by Lucy Lawless.

Exploration

It was calculated that a flyby mission to Eris could take 24.66 years using a Jupiter gravity assist, based on launch dates of 3 April 2032 or 7 April 2044. Eris would be 92.03 or 90.19 AU from the Sun when the spacecraft arrives.

References

External links

Wikimedia Commons has media related to ***Eris***.

- MPC Database entry for (136199) Eris[533]
- Michael Brown's webpage about Eris[534]
- Brown's webpage about Dysnomia[535]
- 2007 KCET interview of Mike Brown about Eris and Haumea with Julia Sweeney[536]
- compiled list of data[537]
- MPEC listing for 2003 UB$_{313}$[538]
- Java 3D orbit visualization[539]
- Spaceflight Now article about 2003 UB$_{313}$ (Eris), 2003 EL$_{61}$, and 2005 FY$_9$ (Makcmakc)[540]
- Slacker Astronomy Interview With Co-Discoverer Trujillo[541]
- Trans-Neptunian Object 2003 UB$_{313}$[542]—IAU statement regarding the planetary status of Eris
- Simulation of 2003 UB$_{313}$'s orbit[543]
- Keck observatory page about the discovery of Dysnomia[544]
- Caltech Press Release, 7/29/2005[545] "Planetary Scientists Discover Tenth Planet".
- Press release on the Spitzer Space Telescope trying to image 2003 UB$_{313}$ again[546]

<indicator name="featured-star"> </indicator>

Heliosphere

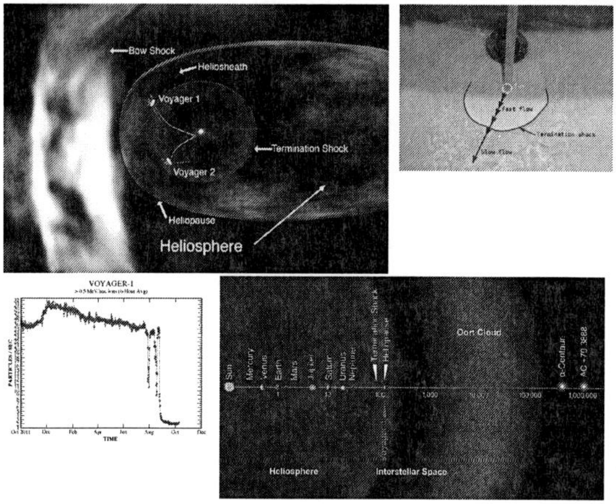

- **Top**: Diagram of the heliosphere as it travels through the interstellar medium:
  a) *Termination shock*: the solar wind collides for the first time with the interstellar medium, slows down and changes direction.
  b) *Heliosheath*: the outer region of the heliosphere; the solar wind is compressed and turbulent
  c) *Heliopause*: the boundary between solar wind and interstellar wind where they are in equilibrium.
  d) *Bow shock*: the shock wave caused by the heliosphere in the direction it travels. (Might not exist.)
- **Middle**: water running into a sink as an analogy for the heliosphere and its different zones (left) and Voyager measuring a drop of the solar wind's high-energy particles at the termination shock (right)
- **Bottom**: Logarithmic scale of the Solar System and Voyager's position

The **heliosphere** is the bubble-like region of space dominated by the Sun, which extends far beyond the orbit of Pluto. Plasma "blown" out from the Sun, known as the solar wind, creates and maintains this bubble against the outside pressure of the interstellar medium, the hydrogen and helium gas that permeates the Milky Way Galaxy. The solar wind flows outward from the Sun until encountering the **termination shock**, where motion slows abruptly. The Voyager spacecraft have explored the outer reaches of the heliosphere, passing through the shock and entering the **heliosheath**, a transitional region which is in turn bounded by the outermost edge of the heliosphere, called the **heliopause**. The shape of the heliosphere is controlled by the interstellar medium through which it is traveling, as well as the Sun and is not perfectly spherical. The limited data available and unexplored nature of these structures have resulted in many theories.

On September 12, 2013, NASA announced that Voyager 1 left the heliosphere on August 25, 2012, when it measured a sudden increase in plasma density

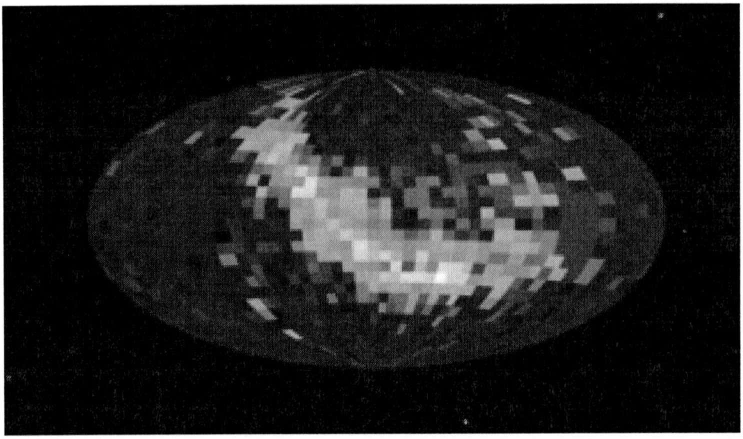

Figure 318: *Energetic neutral atoms map by IBEX. Credit: NASA / Goddard Space Flight Center Scientific Visualization Studio.*

of about forty times. Because the heliopause marks one boundary between the Sun's solar wind and the rest of the galaxy, a spacecraft such as Voyager 1 which has departed the heliosphere, can be said to have reached interstellar space.

Summary

Except for regions near obstacles such as planets or comets, the heliosphere is dominated by material emanating from the Sun, although cosmic rays, fast-moving neutral atoms, and cosmic dust can penetrate the heliosphere from the outside. Originating at the extremely hot surface of the corona, solar wind particles reach escape velocity, streaming outwards at 300 to 800 km/s (671 thousand to 1.79 million mph or 1 to 2.9 million km/h).[547] As it begins to interact with the interstellar medium, its velocity slows to a stop. The point where the solar wind becomes slower than the speed of sound is called the termination shock; the solar wind continues to slow as it passes through the heliosheath leading to a boundary called the heliopause, where the interstellar medium and solar wind pressures balance. The termination shock was traversed by Voyager 1 in 2004, and Voyager 2 in 2007.

It was thought that beyond the heliopause there was a bow shock, but data from Interstellar Boundary Explorer suggested the velocity of the Sun through the interstellar medium is too low for it to form. It may be a more gentle "bow wave". Voyager data led to a new theory that the heliosheath has "magnetic bubbles" and a stagnation zone.

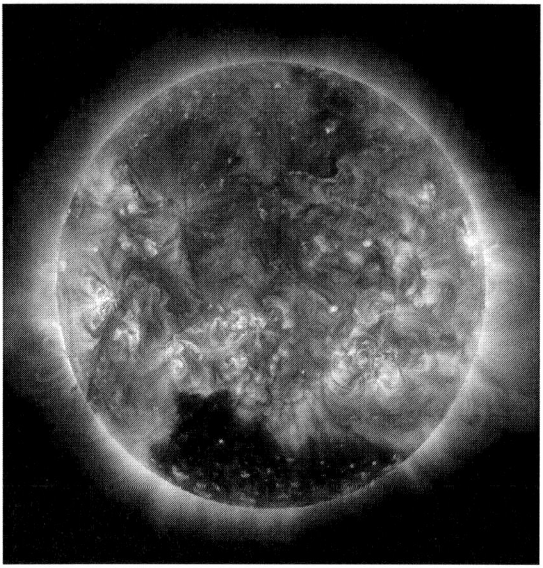

Figure 319: *Earth's Sun as seen at a wavelength of 19.3 nanometers (ultraviolet)*

The 'stagnation region' within the heliosheath, starting around 113 au (1.69×10^{10} km; 1.05×10^{10} mi), was detected by Voyager 1 in 2010. There the solar wind velocity drops to zero, the magnetic field intensity doubles and high-energy electrons from the galaxy increase 100-fold. Starting in May 2012 at 120 au (1.8×10^{10} km; 1.1×10^{10} mi), Voyager 1 detected a sudden increase in cosmic rays, an apparent signature of approach to the heliopause. In December 2012 NASA announced that in late August 2012 Voyager 1, at about 122 au (1.83×10^{10} km; 1.13×10^{10} mi) from the Sun, entered a new region they called the "magnetic highway", an area still under the influence of the Sun but with some dramatic differences. In the summer of 2013, NASA announced that Voyager 1 had reached interstellar space as of August 25, 2012.

Cassini and IBEX data challenged the "heliotail" theory in 2009. In July 2013, IBEX results revealed a 4-lobed tail on the Solar System's heliosphere.

Solar wind

The solar wind consists of particles (ionized atoms from the solar corona) and fields (in particular, magnetic fields).Wikipedia:Please clarify Because the Sun rotates once approximately every 25 days, the magnetic field transported by the solar wind gets wrapped into a spiral. Variations in the Sun's magnetic field

Figure 320: *The heliospheric current sheet out to the orbit of Jupiter*

are carried outward by the solar wind and can produce magnetic storms in the Earth's own magnetosphere.Wikipedia:Please clarify

Structure

Heliospheric current sheet

The heliospheric current sheet is a ripple in the heliosphere created by the rotating magnetic field of the Sun. Extending throughout the heliosphere, it is considered the largest structure in the Solar System and is said to resemble a "ballerina's skirt".

Outer structure

The outer structure of the heliosphere is determined by the interactions between the solar wind and the winds of interstellar space. The solar wind streams away from the Sun in all directions at speeds of several hundred km/s in the Earth's vicinity. At some distance from the Sun, well beyond the orbit of Neptune, this supersonic wind must slow down to meet the gases in the interstellar medium. This takes place in several stages:

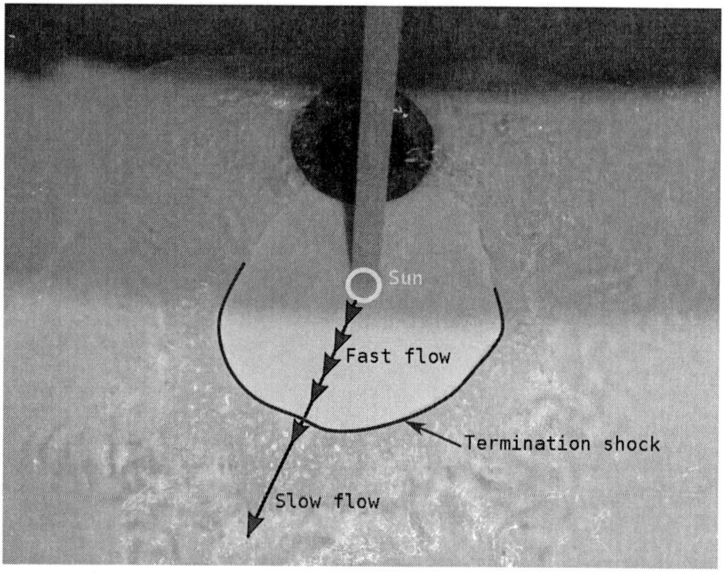

Figure 321: *A termination shock in a sink basin*

- The solar wind is traveling at supersonic speeds within the Solar System. At the termination shock, a standing shock wave, the solar wind falls below the speed of sound and becomes subsonic.
- It was previously thought that, once subsonic, the solar wind would be shaped by the ambient flow of the interstellar medium, forming blunt nose on one side and comet-like heliotail behind, a region called the heliosheath. However, observations in 2009 showed that this model is incorrect. As of 2011, it is thought to be filled with a magnetic bubble "foam".
- The outer surface of the heliosheath, where the heliosphere meets the interstellar medium, is called the **heliopause**. This is the edge of the entire heliosphere. Observations in 2009 led to changes to this model.
- In theory, the heliopause causes turbulence in the interstellar medium as the sun orbits the Galactic Center. Outside the heliopause, would be a turbulent region caused by the pressure of the advancing heliopause against the interstellar medium. However, the velocity of Solar wind relative to the interstellar medium is probably too low for a bow shock.

Termination shock

The termination shock is the point in the heliosphere where the solar wind slows down to subsonic speed (relative to the Sun) because of interactions with the local interstellar medium. This causes compression, heating, and a change in the magnetic field. In the Solar System the termination shock is believed to be 75 to 90 astronomical units from the Sun. In 2004, Voyager 1 crossed the Sun's termination shock followed by Voyager 2 in 2007.

The shock arises because solar wind particles are emitted from the Sun at about 400 km/s, while the speed of sound (in the interstellar medium) is about 100 km/s. (The exact speed depends on the density, which fluctuates considerably.) The interstellar medium, although very low in density, nonetheless has a constant pressure associated with it; the pressure from the solar wind decreases with the square of the distance from the Sun. As one moves far enough away from the Sun, the pressure of the solar wind drops to where it can no longer maintain supersonic flow against the pressure of the interstellar medium, at which point the solar wind slows to below its speed of sound, causing a shock wave. Further from the Sun, the termination shock is followed by the heliopause, where the two pressures become equal and solar wind particles are stopped by the interstellar medium.

Other termination shocks can be seen in terrestrial systems; perhaps the easiest may be seen by simply running a water tap into a sink creating a hydraulic jump. Upon hitting the floor of the sink, the flowing water spreads out at a speed that is higher than the local wave speed, forming a disk of shallow, rapidly diverging flow (analogous to the tenuous, supersonic solar wind). Around the periphery of the disk, a shock front or wall of water forms; outside the shock front, the water moves slower than the local wave speed (analogous to the subsonic interstellar medium).

Evidence presented at a meeting of the American Geophysical Union in May 2005 by Ed Stone suggests that the *Voyager 1* spacecraft passed the termination shock in December 2004, when it was about 94 AU from the Sun, by virtue of the change in magnetic readings taken from the craft. In contrast, *Voyager 2* began detecting returning particles when it was only 76 AU from the Sun, in May 2006. This implies that the heliosphere may be irregularly shaped, bulging outwards in the Sun's northern hemisphere and pushed inward in the south.

Heliosheath

The heliosheath is the region of the heliosphere beyond the termination shock. Here the wind is slowed, compressed and made turbulent by its interaction with the interstellar medium. Its distance from the Sun is approximately 80 to 100 astronomical units (AU) at its closest point.

A proposed model hypothesizes that the heliosheath is shaped like the coma of a comet, and trails several times that distance in the direction opposite to the

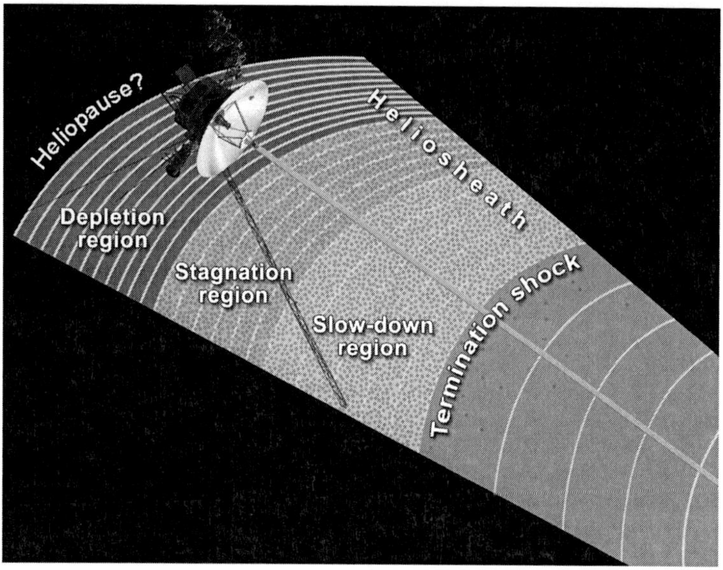

Figure 322: *This picture about the heliosphere was released on June 28, 2013 and incorporates results from the Voyager spacecraft.*

Sun's path through space. At its windward side, its thickness is estimated to be between 10 and 100 AU. However, observations in 2009 showed that model may be incorrect.

The *Voyager 1* and *Voyager 2* spacecraft are currently studying the heliosheath. In late 2010, *Voyager 1* reached a region of the heliosheath where the solar wind's velocity had dropped to zero. In 2011, astronomers announced that the *Voyagers* had determined that the heliosheath is not smooth, but is filled with 100 million-mile-wide bubbles created by the impact of the solar wind and the interstellar medium. *Voyager 1* and *2* began detecting evidence for the bubbles in 2007 and 2008, respectively. The probably sausage-shaped bubbles are formed by magnetic reconnection between oppositely oriented sectors of the solar magnetic field as the solar wind slows down. They probably represent self-contained structures that have detached from the interplanetary magnetic field.

Heliopause

The heliopause is the theoretical boundary where the Sun's solar wind is stopped by the interstellar medium; where the solar wind's strength is no longer great enough to push back the stellar winds of the surrounding stars. This is the boundary where the interstellar medium and solar wind pressures balance. The crossing of the heliopause should be signaled by a sharp drop in the temperature of charged particles, a change in the direction of the magnetic field, and an increase in the amount of galactic cosmic rays. In May 2012, *Voyager 1* detected a rapid increase in such cosmic rays (a 9% increase in a month, following a more gradual increase of 25% from Jan. 2009 to Jan. 2012), suggesting it was approaching the heliopause. In the fall of 2013, NASA announced that Voyager 1 had crossed the heliopause as of August 25, 2012. This was at a distance of 121 AU (18 billion km) from the Sun. Contrary to predictions, data from Voyager 1 indicates the magnetic field of the galaxy is aligned with the solar magnetic field.

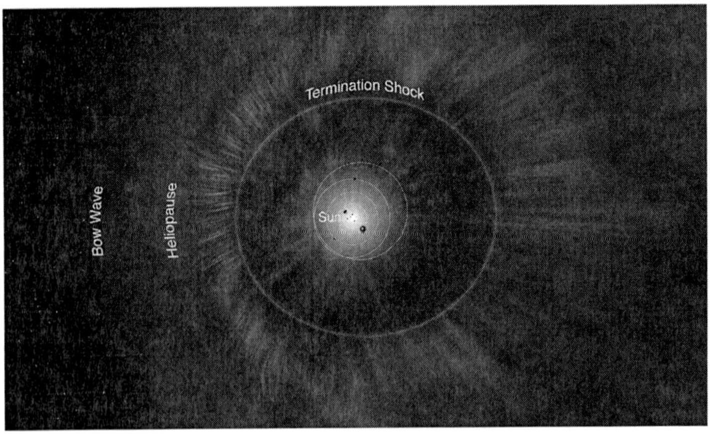

File:Ibexheliosphererevised.jpg

"..solar wind and the interstellar medium interact to create a region known as the inner heliosheath, bounded on the inside by the termination shock, and on the outside by the heliopause."-NASA

Heliotail

The heliotail is the tail of the heliosphere, and thus the solar system's tail. It can be compared to the tail of a comet (however, a comet's tail does not stretch behind it as it moves; it is always pointing away from the Sun). The tail is a region where the Sun's solar wind slows down and ultimately escapes the heliosphere, slowly evaporating because of charge exchange.[548] The shape of the

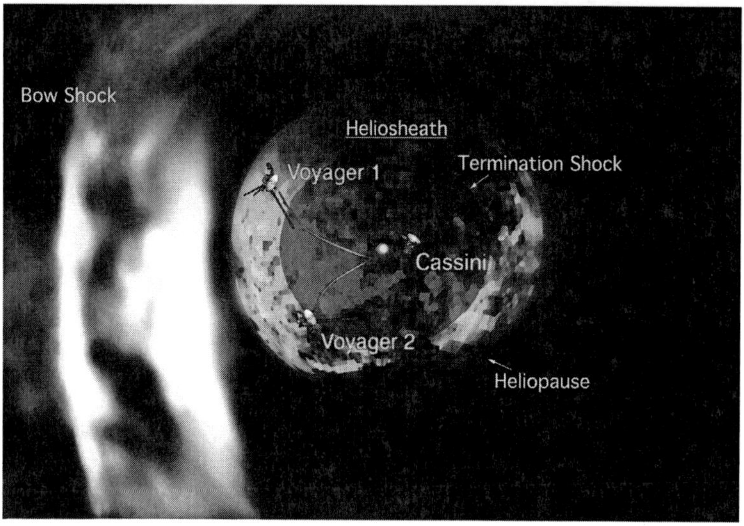

Figure 323: *The bubble-like heliosphere moving through the interstellar medium*

heliotail (newly found by NASA's Interstellar Boundary Explorer - IBEX), is that of a four-leaf clover.[549] The particles in the tail do not shine, therefore it cannot be seen with conventional optical instruments. IBEX made the first observations of the heliotail by measuring the energy of "energetic neutral atoms", neutral particles created by collisions in the solar system's boundary zone.

The tail has been shown to contain fast and slow particles; the slow particles are on the side and the fast particles are encompassed in the center. The shape of the tail can be linked to the sun sending out fast solar winds near its poles and slow solar wind near its equator more recently. The clover-shaped tail moves further away from the sun, which makes the charged particles begin to morph into a new orientation.

Space beyond the heliosphere

The heliopause is the final known boundary between the heliosphere and the interstellar space that is filled with material, especially plasma, not from our own star, the Sun, but from other stars. Even so, just outside the heliosphere (i.e. the "solar bubble") there is a transitional region, as detected by Voyager 1. Just as some interstellar pressure was detected as early as 2004, some of the Sun's material seeps into the interstellar medium. The heliosphere is thought

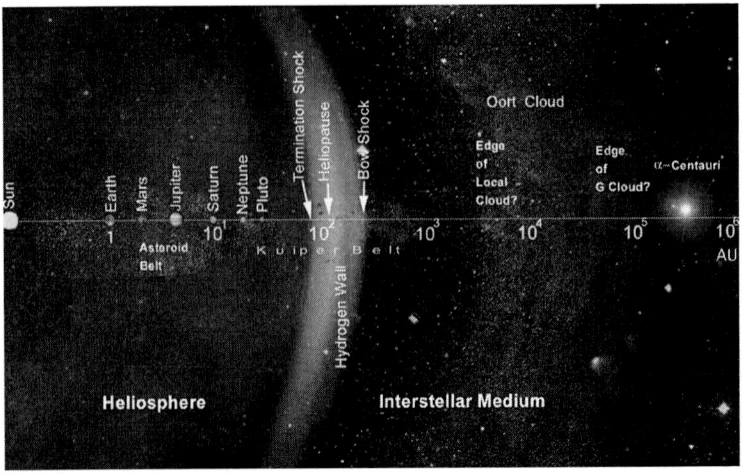

Figure 324: *Solar map with the location of the hypothetical hydrogen wall and bow shock*

to reside in the Local Interstellar Cloud inside the Local Bubble, which is a region in the Orion Arm of the Milky Way Galaxy.

Outside the heliosphere there is a forty-fold increase in plasma density. There is also a radical reduction in the detection of certain types of particles from the Sun, and a large increase in Galactic cosmic rays.

The flow of the interstellar medium (ISM) into the heliosphere has been measured by at least 11 different spacecraft as of 2013. By 2013, it was suspected that the direction of the flow had changed over time. The flow, coming from Earth's perspective from the constellation Scorpius, has probably changed direction by several degrees since the 1970s.

Hydrogen wall

According to one hypothesis, there exists a region of hot hydrogen known as the **hydrogen wall** between the bow shock and the heliopause. The wall is composed of interstellar material interacting with the edge of the heliosphere. One paper released in 2013 studied the concept of a bow wave and hydrogen wall.

Another hypothesis suggests that the heliopause could be smaller on the side of the Solar System facing the Sun's orbital motion through the galaxy. It may also vary depending on the current velocity of the solar wind and the local density of the interstellar medium. It is known to lie far outside the orbit of Neptune. The current mission of the *Voyager 1* and *2* spacecraft is to find and study the termination shock, heliosheath, and heliopause. Meanwhile, the Interstellar Boundary Explorer (IBEX) mission is attempting to image the heliopause from Earth orbit within two years of its 2008 launch. Initial results (October 2009) from IBEX suggest that previous assumptions are insufficiently cognisant of the true complexities of the heliopause.

When particles emitted by the sun bump into the interstellar ones, they slow down while releasing energy. Many particles accumulate in and around the heliopause, highly energised by their negative acceleration, creating a shock wave.Wikipedia:Citation needed An alternative definition is that the heliopause is the magnetopause between the Solar System's magnetosphere and the galaxy's plasma currents.Wikipedia:Citation needed

Bow shock

It was long hypothesized that the Sun produces a "shock wave" in its travels within the interstellar medium. It would occur if the interstellar medium is moving supersonically "toward" the Sun, since its solar wind moves "away" from the Sun supersonically. When the interstellar wind hits the heliosphere it slows and creates a region of turbulence. A bow shock was thought to possibly occur at about 230 AU, but in 2012 it was determined it probably does not exist. This conclusion resulted from new measurements: The velocity of the LISM (Local Interstellar Medium) relative to the Sun's was previously measured to be 26.3 km/s by Ulysses, whereas IBEX measured it at 23.2 km/s.

This phenomenon has been observed outside the Solar System, around stars other than the Sun, by NASA's now retired orbital GALEX telescope. The red giant star Mira in the constellation Cetus has been shown to have both a debris tail of ejecta from the star and a distinct shock in the direction of its movement through space (at over 130 kilometers per second).

Observational methods

Detection by spacecraft

The precise distance to, and shape of the heliopause is still uncertain. Interplanetary/interstellar spacecraft such as *Pioneer 10*, *Pioneer 11* and *Voyager 2* are traveling outward through the Solar System and will eventually pass through the heliopause.

Figure 325: *Pioneer H, on display at the National Air and Space Museum, was a canceled probe to study the Sun.*

Cassini results

Rather than a comet-like shape, the heliosphere appears to be bubble-shaped according to data from Cassini's Ion and Neutral Camera (MIMI / INCA). Rather than being dominated by the collisions between the solar wind and the interstellar medium, the INCA (ENA) maps suggest that the interaction is controlled more by particle pressure and magnetic field energy density. The new shape from the data is thought more like a spherical bubble, than a cometary shape.

IBEX results

Initial data from Interstellar Boundary Explorer (IBEX), launched in October 2008, revealed a previously unpredicted "very narrow ribbon that is two to three times brighter than anything else in the sky." Initial interpretations suggest that "the interstellar environment has far more influence on structuring the heliosphere than anyone previously believed"[550] "No one knows what is creating the ENA (energetic neutral atoms) ribbon, ..."

"The IBEX results are truly remarkable! What we are seeing in these maps does not match with any of the previous theoretical models of this region. It will be exciting for scientists to review these (ENA) maps and revise the way we understand our heliosphere and how it interacts with the galaxy."[551] In October 2010, significant changes were detected in the ribbon after 6 months,

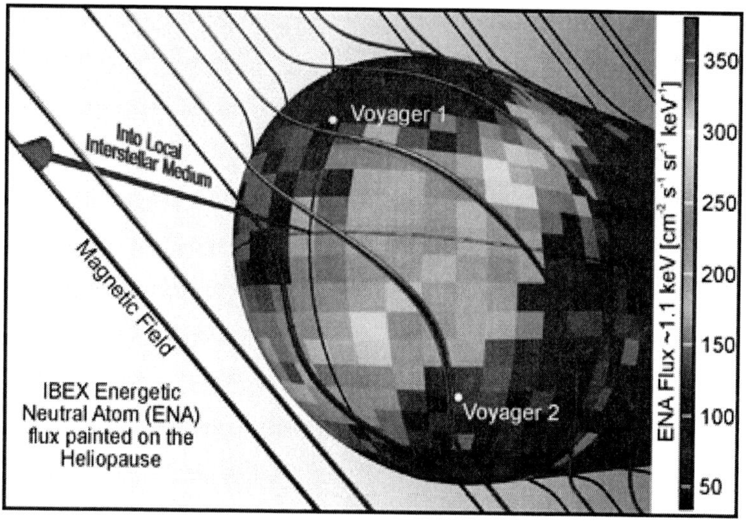

based on the second set of IBEX observations. IBEX data did not support the existence of a bow shock, but there might be a 'bow wave' according to one study.

Locally

Of particular interest is the Earth's interaction with the heliosphere, but its extent and interaction with other bodies in the solar system have also been studied. Some examples of missions that have or continue to collect data related to the heliosphere include *(see also List of heliophysics missions)*:

- Solar Anomalous and Magnetospheric Particle Explorer
- Solar and Heliospheric Observatory
- Solar Dynamics Observatory
- STEREO
- Ulysses (spacecraft)

During a total eclipse the high-temperature corona can be more readily observed from Earth solar observatories. During the Apollo program the Solar wind was measured on the Moon via the Solar Wind Composition Experiment. Some examples of Earth surface based Solar observatories include the McMath–Pierce solar telescope or the newer GREGOR Solar Telescope, and the refurbished Big Bear Solar Observatory.

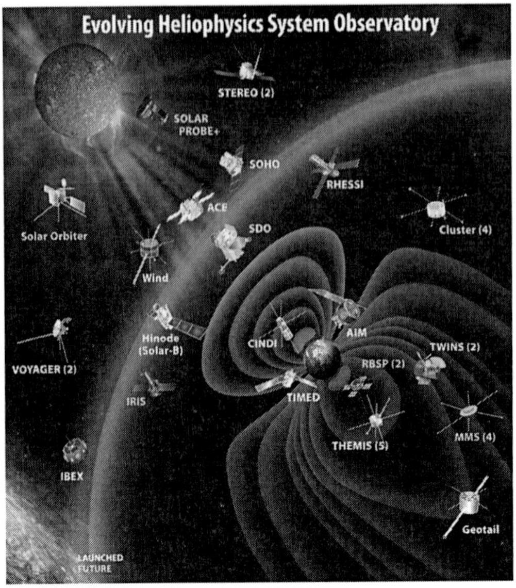

Figure 326: *Overview of Heliophysics spacecraft circa 2011*

Timeline

- As of March 2005, it was reported that measurements by the Solar Wind Anisotropies (SWAN) instrument on board the Solar and Heliospheric Observatory (SOHO) have shown that the heliosphere, the solar wind-filled volume which prevents the Solar System from becoming embedded in the local (ambient) interstellar medium, is not axisymmetrical, but is distorted, very likely under the effect of the local galactic magnetic field.
- As of 2008, there is a previously unpredicted narrow ribbon of ENAs.
- As of October 2009, the heliosphere may be bubble, not comet shaped.
- As of October 2010, significant changes were detected in the ribbon after 6 months, based on the second set of IBEX observations.
- As of June 2011, the heliosheath area is thought to be filled with magnetic bubbles (each about 1 AU wide), creating a "foamy zone". The theory helps explain in situ heliosphere measurements by the two Voyager probes.
- As of May 2012, IBEX data implies there is probably not a bow "shock".
- As of June 2012 at 119 AU Voyager 1 detected an increase in cosmic rays.
- Between late August and early September 2012, Voyager 1 witnessed a sharp drop in protons from the sun, from 25 particles per sec in late

August, to about 2 particles per second by early October. It was later determined that it entered interstellar space on August 25, 2012.

Gallery

These depictions include features that may not reflect most recent models.

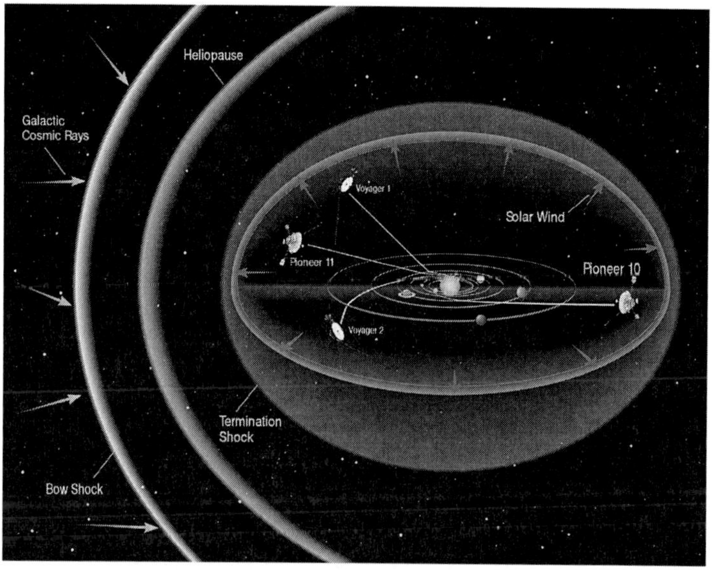

Figure 327: *Heliosphere: classical model*

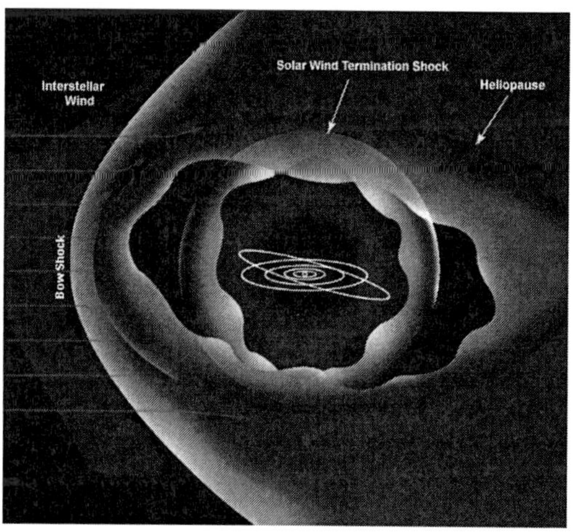

Figure 328: *Heliosphere interacting with the interstellar wind*

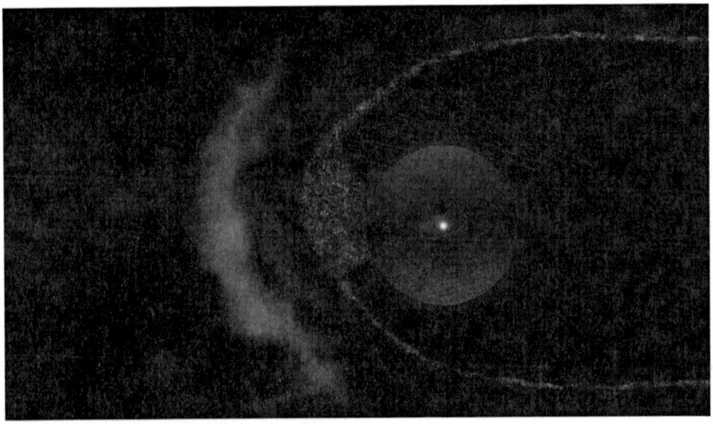

Figure 329: *Moving heliosphere, showing a heliosheath filled with magnetic bubble "foam" (red)*

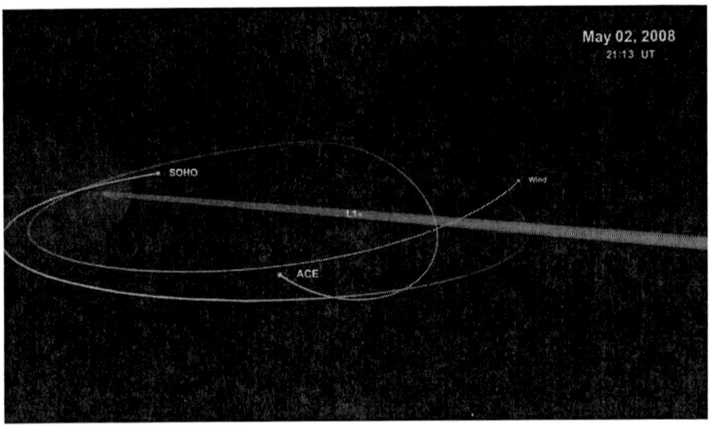

Figure 330: *Fragment of the film Sentinels of the Heliosphere, tracking some researching satellites deployed to analyse the Sun*

References

Notes

- "Heliopause Seems to Be 23 Billion Kilometres"[552]. Universe Today. December 9, 2003. Retrieved 2007-08-08.
- "Space probes reveal Solar System's bullet shape"[553]. COSMOS magazine. May 11, 2007. Retrieved 2007-05-12.

External links

Wikimedia Commons has media related to ***Heliosphere***.

- Moving into Interstellar Space (Artist Concept)[554]
- "Cassini Data Helps Redraw Shape of Our Solar System" 2010[555]
- Publications in Refereed Journals[556]
- Voyager Interstellar Mission Objectives[557]
- The Heliosphere (Cosmicopia)[558]
- NASA GALEX (Galaxy evolution Explorer) homepage at Caltech[559]
- The Solar and Heliospheric Research Group at the University of Michigan[560]
- Ribbon at Edge of Our Solar System: Will the Sun Enter a Million-Degree Cloud of Interstellar Gas this century ?[561]
- A Big Surprise from the Edge of the Solar System[562] (NASA 06.09.11)
- Schwadron, N. A.; et al. (6 September 2011). "Does the Space Environment Affect the Ecosphere?"[563] (PDF). *Eos*. American Geophysical Union. **92** (36): 297–301. Bibcode: 2011EOSTr..92..297S[564]. doi: 10.1029/2011eo360001[565].

Old theories

Warning, content may be outdated

- The Heliosphere[566], MIT Space Plasma Group
- UI's Don Gurnett Says Voyager 1 Is Approaching Edge Of Solar System[567] December 8, 2003 Univ. of Iowa Press release
- NASA's Interstellar Probe[568] (2000)
- CNN: NASA: Voyager I enters solar system's final frontier[569] – May 25, 2005
- *New Scientist*: Voyager 1 reaches the edge of the solar system[570] – May 25, 2005
- Surprises from the Edge of the Solar System[571] – Voyager 1 Newest Findings as of September 2006
- The heliospheric hydrogen wall and astrospheres[572]
- Heliosphere[573], has a diagram.
- Heliosphere[574] Astronomy Cast episode #65, includes full transcript.

Detached object

Detached objects are a dynamical class of minor planets in the outer reaches of the Solar System and belong to the broader family of trans-Neptunian objects (TNOs). These objects have orbits whose points of closest approach to the Sun (perihelion) are sufficiently distant from the gravitational influence of Neptune that they are only moderately affected by Neptune and the other known planets: this makes them appear to be "detached" from the Solar System.[575]

In this way, detached objects differ substantially from most other known TNOs, which form a loosely defined set of populations that have been perturbed to varying degrees onto their current orbit by gravitational encounters with the giant planets, predominantly Neptune. Detached objects have larger perihelia than these other TNO populations, including the objects in orbital resonance with Neptune, such as Pluto, the classical Kuiper belt objects in non-resonant orbits such as Makemake, and the scattered disk objects like Eris.

Detached objects have also been referred to in the scientific literature as **extended scattered disc objects** (E-SDO), **distant detached objects** (DDO), or **scattered–extended**, as in the formal classification by the Deep Ecliptic Survey. This reflects the dynamical gradation that can exist between the orbital parameters of the scattered disk and the detached population.

At least nine such bodies have been securely identified, of which the largest, most distant, and best known is Sedna. Those with perihelia greater than 75 AU are termed sednoids. As of 2016, there are two known sednoids, Sedna and 2012 VP113.

Orbits

Detached objects have perihelia much larger than Neptune's aphelion. They often have highly elliptical, very large orbits with semi-major axes of up to a few hundred astronomical units (AU, the radius of Earth's orbit). Such orbits cannot have been created by gravitational scattering by the giant planets (in particular, Neptune). Instead, a number of explanations have been put forward, including an *encounter with a passing star* or a *distant planet-sized object*, or *Neptune itself* (which may once have had a much more eccentric orbit, from which it could have tugged the objects to their current orbit) or *ejected planets* (present in the early Solar System that were ejected).

The classification suggested by the Deep Ecliptic Survey team introduces a formal distinction between *scattered-near* objects (which could be scattered by Neptune) and *scattered-extended* objects (e.g. 90377 Sedna) using a Tisserand's parameter value of 3.

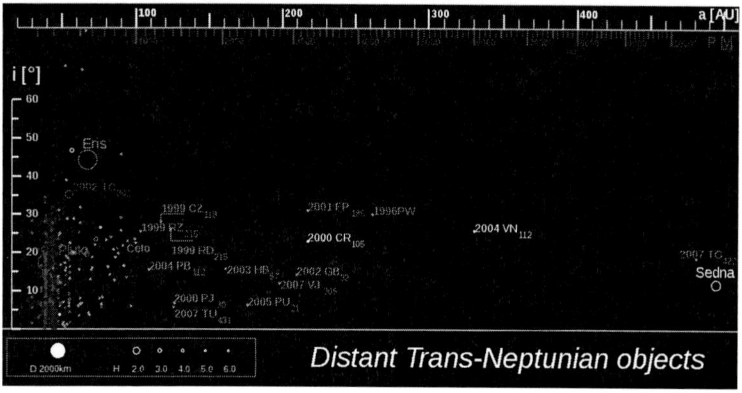

Figure 331:
Trans-Neptunian objects plotted by their distance and inclination. Objects beyond a distance of 100 AU display their designation.

The Planet Nine hypothesis suggests that the orbits of several detached objects can be explained by the gravitational influence of a large, unobserved planet between 200 AU and 1200 AU from the Sun and/or the influence of Neptune.

Classification

Types of distant minor planets

- Cis-Neptunian objects
  - Centaurs
  - Neptune trojans
- Trans-Neptunian objects (TNOs)‡
  - Kuiper belt objects (KBOs)
    - Classical KBOs (cubewanos)
    - Resonant KBOs
      - Plutinos (2:3 resonance)
  - Scattered disc objects (SDOs)
    - Resonant SDOs
  - Detached objects
  - Sednoids
  - Oort cloud objects (ICO/OCOs)

‡ Trans-Neptunian dwarf planets are called "plutoids"

- v
- t
- e[576]

Detached objects are one of five distinct dynamical classes of TNO; the other four classes are classical Kuiper-belt objects, resonant objects, scattered-disc objects (SDO), and sednoids. Detached objects generally have a perihelion distance greater than 40 AU, deterring strong interactions with Neptune, which has an approximately circular orbit about 30 AU from the Sun. However, there are no clear boundaries between the scattered and detached regions, since both can coexist as TNOs in an intermediate region with perihelion distance between 37 and 40 AU. One such intermediate body with a well determined orbit is (120132) 2003 FY_{128}.

The discovery of 90377 Sedna in 2003, together with a few other objects discovered around that time such as (148209) 2000 CR_{105} and 2004 XR_{190}, has motivated discussion of a category of distant objects that may also be inner Oort cloud objects or (more likely) transitional objects between the scattered disc and the inner Oort cloud.

Although Sedna is officially considered a scattered-disc object by the MPC, its discoverer Michael E. Brown has suggested that because its perihelion distance of 76 AU is too distant to be affected by the gravitational attraction of the outer planets it should be considered an inner-Oort-cloud object rather than a member of the scattered disc. This classification of Sedna as a detached object is accepted in recent publications.[577]

This line of thinking suggests that the lack of a significant gravitational interaction with the outer planets creates an extended–outer group starting somewhere between Sedna (perihelion 76 AU) and more conventional SDOs like 1996 TL_{66} (perihelion 35 AU), which is listed as a scattered–near object by the Deep Ecliptic Survey.

Influence of Neptune

One of the problems with defining this extended category is that weak resonances may exist and would be difficult to prove due to chaotic planetary perturbations and the current lack of knowledge of the orbits of these distant objects. They have orbital periods of more than 300 years and most have only been observed over a short observation arc of a couple years. Due to their great distance and slow movement against background stars, it may be decades before most of these distant orbits are determined well enough to confidently confirm or rule out a resonance. Further improvement in the orbit and potential resonance of these objects will help to understand the migration of the giant planets and the formation of the Solar System. For example, simulations by Emel'yanenko and Kiseleva in 2007 show that many distant objects could be in resonance with Neptune. They show a 10% likelihood that 2000 CR_{105} is in a 20:1 resonance, a 38% likelihood that 2003 QK_{91} is in a 10:3 resonance, and an 84% likelihood that (82075) 2000 YW_{134} is in an 8:3 resonance.[578] The likely dwarf planet (145480) 2005 TB_{190} appears to have less than a 1% likelihood of being in a 4:1 resonance.

Influence of hypothetical planet(s) beyond Neptune

Mike Brown—who made the Planet Nine hypothesis—makes an observation that "all of the known distant objects which are pulled even a little bit away from the Kuiper seem to be clustered under the influence of this hypothetical planet (specifically, objects with semimajor axis > 100 AU and perihelion > 42 AU)." Carlos de la Fuente Marcos and Ralph de la Fuente Marcos have calculated that some of the statistically significant commensurabilities are compatible with the Planet Nine hypothesis; in particular, a number of objects[579,580] which are called Extreme trans Neptunian objects (**ETNOs**).</ref> may be trapped in the 5:3 and 3:1 mean-motion resonances with a putative Planet Nine with a semimajor axis ~700 AU.

Possible detached objects

This is a list of known objects by decreasing perihelion, that could not be easily scattered by Neptune's current orbit and therefore are likely to be detached objects, but that lie inside the perihelion gap of ≈50–75 AU that defines the sednoids:

Designation number[581]	Name	Diameter (km)	H	Perihelion (AU)	Semi-major axis (AU)	Aphelion (AU)	Argument of perihelion (°)	Dis-covery Year	Discoverer	Method[582]	Refs
48639	1995 TL_8	≈350	5.2	40.0	52.5	64.5		1995	A. Gleason	assumed	
148209	2000 CR_{105}	≈250	6.1	44.0	224	403	316.5	2000	Lowell Observatory	assumed	
	2004 VN_{112}	130–300	6.4	47.3	329	610	327.2	2004	CTIO	assumed	
	2004 XR_{190}	335–850	4.4	51.5	57.7	64		2004	Lynne Jones et al.	assumed	
145480	2005 TB_{190}	≈500	4.7	46.2	76.4	106.5		2005	Becker, A. C. et al.	assumed	
	2008 ST_{291}	≈612	4.2	42.5	98.6	154.8		2008	Meg Schwamb et al.		
	2010 ER65		5.4	40	99.77	159.5	324	2010	European Southern Observatory, La Silla		
	2010 GB_{174}	242	6.6	48.5	361	673	347.3	2010	Canada–France–Hawaii Tele-scope		
	2013 FT28		6.7	43.6	310	576.6	40.2	2013	Cerro Tololo Obser-vatory, La Serena		
	2013 GP_{136}		6.6	41.1	151.75	262.38	42	2013	Mauna Kea		
	2013 JD64		8.0	42.6	72.6	102.6	177.5	2013	Mauna Kea		

	2013 YJ151		5.4	40.9	72.9	104.9	142	2013	Pan-STARRS 1, Haleakala		
	2014 FC69	300–700	4.6	40.2	73.57	106.9	191.3	2014	Cerro Tololo Observatory, La Serena		
	2014 FC72		5.4	51.3	76.9	101	32.4	2014	Pan-STARRS 1, Haleakala		
	2014 SR349		6.6	47.6	289	539.4	341.4	2014	Cerro Tololo Observatory, La Serena		
	2014 SS349		7.5	45.5	142.3	239.2	147.8	2014	Cerro Tololo Observatory, La Serena		
	2014 WT69		5.7	44.5	76.7	108.8	139.56	2014	Pan-STARRS 1, Haleakala		

As of July 2017, there are 46 known objects with perihelions over 40 AU and semimajor axies over 50 AU.[583]

Objects with a>50, q>40

Name	H	q (AU)	a (AU)	Q (AU)	Period (years)
2001 FL193	8.7	40.25	50.27	60.29	356
2014 WK509	4	40.17	50.97	61.77	364
(48639) 1995 TL8	5	40.04	52.29	64.54	378
2014 EZ51	3.7	40.61	52.32	64.04	378
2002 CP154	6.5	42.1	52.57	63.04	381
2014 OJ394	5	40.67	52.76	64.86	383
2006 AO101	7.1	41.92	52.92	63.93	385
2007 LE38	7	41.78	54.29	66.79	400
2005 CG81	6.1	41.08	54.3	67.52	400
2013 VD24	7.7	41.26	54.51	67.76	402
2015 GP50	6.5	40.46	55.34	70.23	412
(495603) 2015 AM281	4.8	41.4	55.48	69.56	413
2010 DN93	4.7	45.08	55.67	66.25	415
2012 FH84	7.2	42.66	56.43	70.19	424
2004 XR190	4.1	51.17	57.52	63.88	436
(82075) 2000 YW134	4.7	41.24	58.14	75.03	443
2015 KH162	3.9	41.46	62.19	82.91	490
2014 JM80	5.5	46.01	62.69	79.37	496
2013 FQ28	6	45.84	63.21	80.58	503
2015 FJ345	7.9	50.7	63.21	75.72	503
(385607) 2005 EO297	7.2	41.2	63.32	85.44	504
2004 PD112	6.1	43.58	64.29	85.01	516
2014 QR441	6.8	42.6	67.49	92.37	554
2009 KX36	8	44.72	67.81	90.9	558
2013 JD64	8	42.61	72.85	103.1	622
2013 YJ151	5.4	40.88	72.88	104.88	622
2014 FC69	4.6	40.59	73	105.4	624
(145480) 2005 TB190	4.5	46.2	75.2	104.2	652
2014 FC72	4.5	51.68	76.32	100.96	667
2014 FZ71	6.9	55.93	76.38	96.83	668
2008 ST291	4.2	42.31	98.81	155.3	982

2010 ER65	5	40.02	100.13	160.24	1002
2014 SS349	7.6	45.46	142.37	239.28	1699
(496315) 2013 GP136	6.6	41.07	154.81	268.55	1926
2015 KH163	7.9	40.07	156.54	273.02	1959
2013 UT15	6.3	43.88	195.26	346.64	2728
(148209) 2000 CR105	6.3	44.29	226.14	407.99	3401
2012 VP113	4	80.49	256.64	432.78	4111
2014 SR349	6.6	47.57	289	530.42	4913
2013 FT28	6.7	43.6	310.07	576.55	5460
(474640) 2004 VN112	6.5	47.32	317.65	587.98	5661
2010 GB174	6.5	48.76	369.73	690.71	7109
2015 RX245	6.2	45.48	407.19	768.89	8217
(90377) Sedna	1.5	76.04	499.43	922.82	11161
2013 SY99	6.7	49.91	678.96	1308.01	17691
2015 KG163	8.2	40.51	810.11	1579.71	23058

References

<indicator name=”good-star”> ⊕ </indicator>

Sednoid

A **sednoid** is a trans-Neptunian object with a perihelion greater than 50 AU and a semi-major axis greater than 150 AU. Only two objects are known from this population, 90377 Sedna and 2012 VP113, both of which have perihelia greater than 75 AU, but it is suspected that there are many more. These objects lie outside an apparently nearly empty gap in the Solar System starting at about 50 AU, and have no significant interaction with the planets. They are usually grouped with the detached objects. Some astronomers, such as Scott Shepard, consider the sednoids to be inner Oort cloud objects (OCOs), though the inner Oort cloud, or Hills cloud, was originally predicted to lie beyond 2,000 AU, several times as far as the aphelia of the two known sednoids.

Figure 332: *Sedna, the eponymous and first known sednoid*

Unexplained orbits

The sednoids' orbits cannot be explained by perturbations from the giant planets, nor by interaction with the galactic tides. If they formed in their current locations, their orbits must originally have been circular; otherwise accretion (the coalescence of smaller bodies into larger ones) would not have been possible because the large relative velocities between planetesimals would have been too disruptive. Their present elliptical orbits can be explained by several hypotheses:

1. These objects could have had their orbits and perihelion distances "lifted" by the passage of a nearby star when the Sun was still embedded in its birth star cluster.
2. Their orbits could have been disrupted by an as-yet-unknown planet-sized body beyond the Kuiper belt such as the hypothesized Planet Nine.
3. They could have been captured from around passing stars, most likely in the Sun's birth cluster.

Known members

Sednoids

Number	Name	Diameter (km)	Perihelion (AU)	Semimajor axis (AU)	Aphelion (AU)	Argument of perihelion (°)	Year discovered
90377	Sedna	995 ± 80	76.06	515.79	955.51	311.38	2003
	2012 VP_{113}	600	80.50	261.00	441.49	293.78	2012

The two sednoids, like all of the more extreme detached objects (objects with semi-major axes > 150 AU and perihelia > 30 AU; the orbit of Neptune), have a similar orientation (argument of perihelion) of ≈ 0° (338°±38°). This is not due to an observational bias and is unexpected, because interaction with the giant planets should have randomized their arguments of perihelion (ω), with precession periods between 40 Myr and 650 Myr and 1.5 Gyr for Sedna. This suggests that one or more undiscovered massive perturbers may exist in the outer Solar System. A super-Earth at 250 AU would cause these objects to librate around ω = 0°±60° for billions of years. There are multiple possible configurations and a low-albedo super-Earth at that distance would have an apparent magnitude below the current all-sky-survey detection limits. This hypothetical super-Earth has been dubbed Telisto and Planet Nine. Larger, more-distant perturbers would also be too faint to be detected.

Fourteen known objects have a semi-major axis greater than 150 AU, a perihelion beyond Neptune, an argument of perihelion of 340°±55°, and an observation arc of more than 1 year.

On 10 November 2015, a third candidate sednoid, V774104, was reported at a distance of 103 AU from the Sun, but its observation arc was too short to know whether its perihelion was outside Neptune's influence.

Sednoids might constitute a proper dynamical class, but they may have a heterogeneous origin; the spectral slopes of (474640) 2004 VN_{112}, 2013 RF_{98}, 2012 VP_{113}, 2002 GB_{32} and 2003 HB_{57} are very different from that of 90377 Sedna.

Theoretical population

Each of the proposed mechanisms for Sedna's extreme orbit would leave a distinct mark on the structure and dynamics of any wider population. If a trans-Neptunian planet was responsible, all such objects would share roughly the same perihelion (≈80 AU). If Sedna were captured from another planetary system that rotated in the same direction as the Solar System, then all of its population would have orbits on relatively low inclinations and have semi-major axes ranging from 100–500 AU. If it rotated in the opposite direction, then two populations would form, one with low and one with high inclinations. The perturbations from passing stars would produce a wide variety of perihelia and inclinations, each dependent on the number and angle of such encounters.

Acquiring a larger sample of such objects would therefore help in determining which scenario is most likely. "I call Sedna a fossil record of the earliest Solar System", said Brown in 2006. "Eventually, when other fossil records are found, Sedna will help tell us how the Sun formed and the number of stars that were close to the Sun when it formed." A 2007–2008 survey by Brown, Rabinowitz and Schwamb attempted to locate another member of Sedna's hypothetical population. Although the survey was sensitive to movement out to 1,000 AU and discovered the likely dwarf planet 2007 OR_{10}, it detected no new sednoids. Subsequent simulations incorporating the new data suggested about 40 Sedna-sized objects probably exist in this region, with the brightest being about Eris's magnitude (–1.0).

External links

- New icy body hints at planet lurking beyond Pluto[584]

Oort cloud

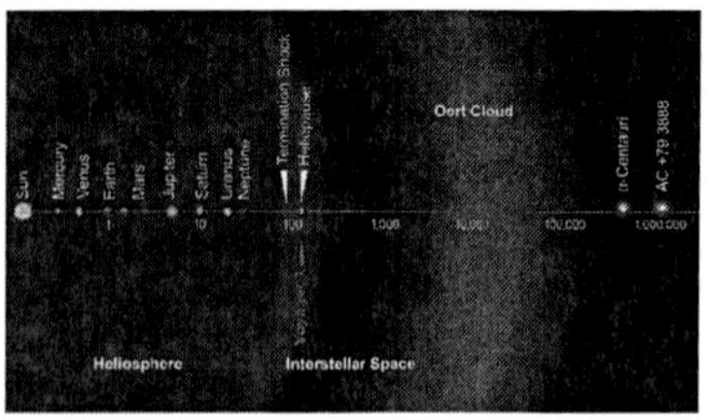

This graphic shows the distance from the Oort cloud to the rest of the Solar System and two of the nearest stars measured in astronomical units. The scale is logarithmic, with each specified distance ten times further out than the previous one. Red arrow indicates location of Voyager 1, a space probe that will reach the Oort cloud in 300 years.

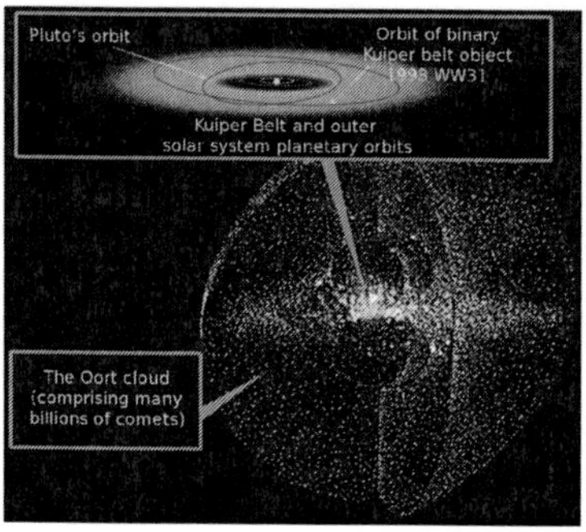

An artist's impression of the Oort cloud and the Kuiper belt (inset). Sizes of individual objects have been exaggerated for visibility.

Types of distant minor planets

- Cis-Neptunian objects
  - Centaurs
  - Neptune trojans
- Trans-Neptunian objects (TNOs)‡
  - Kuiper belt objects (KBOs)
    - Classical KBOs (cubewanos)
    - Resonant KBOs
      - Plutinos (2:3 resonance)
  - Scattered disc objects (SDOs)
    - Resonant SDOs
  - Detached objects
  - Sednoids
  - Oort cloud objects (ICO/OCOs)

‡ Trans-Neptunian dwarf planets are called "plutoids"

- v
- t
- e[585]

The **Oort cloud** (/ˈɔːrt/ or /ˈʊərt/, named after the Dutch astronomer Jan Oort), sometimes called the **Öpik–Oort cloud**, is a theoretical cloud of predominantly icy planetesimals believed to surround the Sun to as far as somewhere between 50,000 and 200,000 AU (0.8 and 3.2 ly).[586] It is divided into two regions: a disc-shaped inner Oort cloud (or Hills cloud) and a spherical outer Oort cloud. Both regions lie beyond the heliosphere and in interstellar space. The Kuiper belt and the scattered disc, the other two reservoirs of trans-Neptunian objects, are less than one thousandth as far from the Sun as the Oort cloud.

The outer limit of the Oort cloud defines the cosmographical boundary of the Solar System and the extent of the Sun's Hill sphere. The outer Oort cloud is only loosely bound to the Solar System, and thus is easily affected by the gravitational pull both of passing stars and of the Milky Way itself. These forces occasionally dislodge comets from their orbits within the cloud and send them toward the inner Solar System. Based on their orbits, most of the short-period comets may come from the scattered disc, but some may still have originated from the Oort cloud.

Astronomers conjecture that the matter composing the Oort cloud formed closer to the Sun and was scattered far into space by the gravitational effects of the giant planets early in the Solar System's evolution. Although no confirmed direct observations of the Oort cloud have been made, it may be the source of all long-period and Halley-type comets entering the inner Solar System, and many of the centaurs and Jupiter-family comets as well.

Hypothesis

In 1932 the Estonian astronomer Ernst Öpik postulated that long-period comets originated in an orbiting cloud at the outermost edge of the Solar System. The Dutch astronomer Jan Oort independently revived the idea in 1950 as a means to resolve a paradox:

- Over the course of the Solar System's existence the orbits of comets are unstable and eventually dynamics dictate that a comet must either collide with the Sun or a planet or else be ejected from the Solar System by planetary perturbations.
- Moreover, their volatile composition means that as they repeatedly approach the Sun, radiation gradually boils the volatiles off until the comet splits or develops an insulating crust that prevents further outgassing.

Thus, Oort reasoned, a comet could not have formed while in its current orbit and must have been held in an outer reservoir for almost all of its existence.

There are two main classes of comet: short-period comets (also called ecliptic comets) and long-period comets (also called nearly isotropic comets). Ecliptic comets have relatively small orbits, below 10 AU, and follow the ecliptic plane, the same plane in which the planets lie. All long-period comets have very large orbits, on the order of thousands of AU, and appear from every direction in the sky. Oort noted that there was a peak in numbers of long-period comets with aphelia (their farthest distance from the Sun) of roughly 20,000 AU, which suggested a reservoir at that distance with a spherical, isotropic distribution. Those relatively rare comets with orbits of about 10,000 AU have probably gone through one or more orbits through the Solar System and have had their orbits drawn inward by the gravity of the planets.

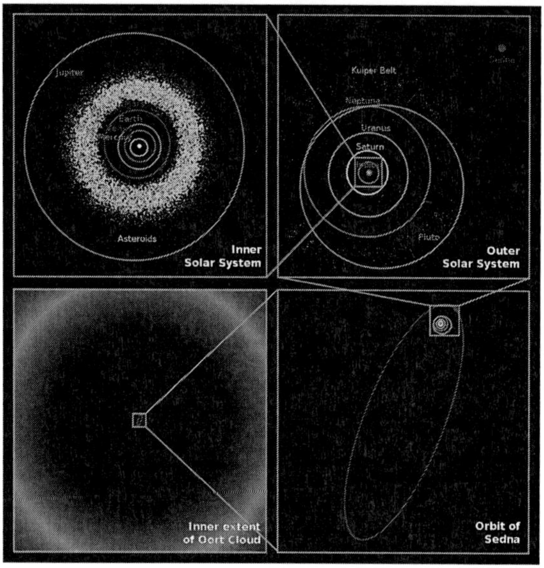

Figure 333: *The presumed distance of the Oort cloud compared to the rest of the Solar System*

Structure and composition

The Oort cloud is thought to occupy a vast space from somewhere between 2,000 and 5,000 AU (0.03 and 0.08 ly) to as far as 50,000 AU (0.79 ly) from the Sun. Some estimates place the outer edge at between 100,000 and 200,000 AU (1.58 and 3.16 ly). The region can be subdivided into a spherical outer Oort cloud of 20,000–50,000 AU (0.32–0.79 ly), and a torus-shaped inner Oort cloud of 2,000–20,000 AU (0.0–0.3 ly). The outer cloud is only weakly bound to the Sun and supplies the long-period (and possibly Halley-type) comets to inside the orbit of Neptune. The inner Oort cloud is also known as the Hills cloud, named after Jack G. Hills, who proposed its existence in 1981. Models predict that the inner cloud should have tens or hundreds of times as many cometary nuclei as the outer halo; it is seen as a possible source of new comets to resupply the tenuous outer cloud as the latter's numbers are gradually depleted. The Hills cloud explains the continued existence of the Oort cloud after billions of years.

The outer Oort cloud may have trillions of objects larger than 1 km (0.62 mi), and billions with absolute magnitudes[587] brighter than 11 (corresponding to approximately 20-kilometre (12 mi) diameter), with neighboring objects tens of millions of kilometres apart. Its total mass is not known, but, assuming that Halley's Comet is a suitable prototype for comets within the outer Oort cloud, roughly the combined mass is 3×10^{25} kilograms (6.6×10^{25} lb), or five times that of Earth. Earlier it was thought to be more massive (up to 380 Earth masses), but improved knowledge of the size distribution of long-period comets led to lower estimates. The mass of the inner Oort cloud has not been characterized.

If analyses of comets are representative of the whole, the vast majority of Oort-cloud objects consist of ices such as water, methane, ethane, carbon monoxide and hydrogen cyanide. However, the discovery of the object 1996 PW, an object whose appearance was consistent with a D-type asteroid in an orbit typical of a long-period comet, prompted theoretical research that suggests that the Oort cloud population consists of roughly one to two percent asteroids. Analysis of the carbon and nitrogen isotope ratios in both the long-period and Jupiter-family comets shows little difference between the two, despite their presumably vastly separate regions of origin. This suggests that both originated from the original protosolar cloud, a conclusion also supported by studies of granular size in Oort-cloud comets and by the recent impact study of Jupiter-family comet Tempel 1.

Origin

The Oort cloud is thought to be a remnant of the original protoplanetary disc that formed around the Sun approximately 4.6 billion years ago. The most widely accepted hypothesis is that the Oort cloud's objects initially coalesced much closer to the Sun as part of the same process that formed the planets and minor planets, but that gravitational interaction with young gas giants such as Jupiter ejected the objects into extremely long elliptic or parabolic orbits. Recent research has been cited by NASA hypothesizing that a large number of Oort cloud objects are the product of an exchange of materials between the Sun and its sibling stars as they formed and drifted apart, and it is suggested that many—possibly the majority of—Oort cloud objects did not form in close proximity to the Sun. Simulations of the evolution of the Oort cloud from the beginnings of the Solar System to the present suggest that the cloud's mass peaked around 800 million years after formation, as the pace of accretion and collision slowed and depletion began to overtake supply.

Models by Julio Ángel Fernández suggest that the scattered disc, which is the main source for periodic comets in the Solar System, might also be the primary source for Oort cloud objects. According to the models, about half of the objects scattered travel outward toward the Oort cloud, whereas a quarter are shifted inward to Jupiter's orbit, and a quarter are ejected on hyperbolic orbits. The scattered disc might still be supplying the Oort cloud with material. A third of the scattered disc's population is likely to end up in the Oort cloud after 2.5 billion years.

Computer models suggest that collisions of cometary debris during the formation period play a far greater role than was previously thought. According to these models, the number of collisions early in the Solar System's history was so great that most comets were destroyed before they reached the Oort cloud. Therefore, the current cumulative mass of the Oort cloud is far less than was once suspected. The estimated mass of the cloud is only a small part of the 50–100 Earth masses of ejected material.

Gravitational interaction with nearby stars and galactic tides modified cometary orbits to make them more circular. This explains the nearly spherical shape of the outer Oort cloud. On the other hand, the Hills cloud, which is bound more strongly to the Sun, has not acquired a spherical shape. Recent studies have shown that the formation of the Oort cloud is broadly compatible with the hypothesis that the Solar System formed as part of an embedded cluster of 200–400 stars. These early stars likely played a role in the cloud's formation, since the number of close stellar passages within the cluster was much higher than today, leading to far more frequent perturbations.

In June 2010 Harold F. Levison and others suggested on the basis of enhanced computer simulations that the Sun "captured comets from other stars while it was in its birth cluster". Their results imply that "a substantial fraction of the Oort cloud comets, perhaps exceeding 90%, are from the protoplanetary discs of other stars".

Comets

Comets are thought to have two separate points of origin in the Solar System. Short-period comets (those with orbits of up to 200 years) are generally accepted to have emerged from either the Kuiper belt or the scattered disc, which are two linked flat discs of icy debris beyond Neptune's orbit at 30 AU and jointly extending out beyond 100 AU from the Sun. Long-period comets, such as comet Hale–Bopp, whose orbits last for thousands of years, are thought to originate in the Oort cloud. The orbits within the Kuiper belt are relatively stable, and so very few comets are thought to originate there. The scattered disc, however, is dynamically active, and is far more likely to be the place of origin for comets. Comets pass from the scattered disc into the realm of the outer planets, becoming what are known as centaurs. These centaurs are then sent farther inward to become the short-period comets.

There are two main varieties of short-period comet: Jupiter-family comets (those with semi-major axes of less than 5 AU) and Halley-family comets.

Figure 334: *Comet Hale–Bopp, an archetypical Oort-cloud comet*

Halley-family comets, named for their prototype, Halley's Comet, are unusual in that although they are short-period comets, it is hypothesized that their ultimate origin lies in the Oort cloud, not in the scattered disc. Based on their orbits, it is suggested they were long-period comets that were captured by the gravity of the giant planets and sent into the inner Solar System. This process may have also created the present orbits of a significant fraction of the Jupiter-family comets, although the majority of such comets are thought to have originated in the scattered disc.

Oort noted that the number of returning comets was far less than his model predicted, and this issue, known as "cometary fading", has yet to be resolved. No known dynamical process can explain this undercount of observed comets. Hypotheses for this discrepancy include the destruction of comets due to tidal stresses, impact or heating; the loss of all volatiles, rendering some comets invisible, or the formation of a non-volatile crust on the surface. Dynamical studies of Oort cloud comets have shown that their occurrence in the outer-planet region is several times higher than in the inner-planet region. This discrepancy may be due to the gravitational attraction of Jupiter, which acts as a kind of barrier, trapping incoming comets and causing them to collide with it, just as it did with Comet Shoemaker–Levy 9 in 1994.

Tidal effects

Most of the comets seen close to the Sun seem to have reached their current positions through gravitational perturbation of the Oort cloud by the tidal force exerted by the Milky Way. Just as the Moon's tidal force deforms Earth's oceans, causing the tides to rise and fall, the galactic tide also distorts the orbits of bodies in the outer Solar System. In the charted regions of the Solar System, these effects are negligible compared to the gravity of the Sun, but in the outer reaches of the system, the Sun's gravity is weaker and the gradient of the Milky Way's gravitational field has substantial effects. Galactic tidal forces stretch the cloud along an axis directed toward the galactic centre and compress it along the other two axes; these small perturbations can shift orbits in the Oort cloud to bring objects close to the Sun. The point at which the Sun's gravity concedes its influence to the galactic tide is called the tidal truncation radius. It lies at a radius of 100,000 to 200,000 AU, and marks the outer boundary of the Oort cloud.

Some scholars theorise that the galactic tide may have contributed to the formation of the Oort cloud by increasing the perihelia (smallest distances to the Sun) of planetesimals with large aphelia (largest distances to the Sun). The effects of the galactic tide are quite complex, and depend heavily on the behaviour of individual objects within a planetary system. Cumulatively, however, the effect can be quite significant: up to 90% of all comets originating from the Oort cloud may be the result of the galactic tide. Statistical models of the observed orbits of long-period comets argue that the galactic tide is the principal means by which their orbits are perturbed toward the inner Solar System.

Stellar perturbations and stellar companion hypotheses

Besides the galactic tide, the main trigger for sending comets into the inner Solar System is thought to be interaction between the Sun's Oort cloud and the gravitational fields of nearby stars or giant molecular clouds. The orbit of the Sun through the plane of the Milky Way sometimes brings it in relatively close proximity to other stellar systems. For example, 70 thousand years ago, Scholz's star passed through the outer Oort cloud (although its low mass and high relative velocity limited its effect). During the next 10 million years the known star with the greatest possibility of perturbing the Oort cloud is Gliese 710. This process also scatters Oort cloud objects out of the ecliptic plane, potentially also explaining its spherical distribution.

In 1984, Physicist Richard A. Muller postulated that the Sun has a heretofore undetected companion, either a brown dwarf or a red dwarf, in an elliptical orbit within the Oort cloud. This object, known as Nemesis, was hypothesized to pass through a portion of the Oort cloud approximately every 26 million years, bombarding the inner Solar System with comets. However, to date no evidence of Nemesis has been found, and many lines of evidence (such as crater counts), have thrown its existence into doubt. Recent scientific analysis no longer supports the idea that extinctions on Earth happen at regular, repeating intervals. Thus, the Nemesis hypothesis is no longer needed.

A somewhat similar hypothesis was advanced by astronomer John J. Matese of the University of Louisiana at Lafayette in 2002. He contends that more comets are arriving in the inner Solar System from a particular region of the Oort cloud than can be explained by the galactic tide or stellar perturbations alone, and that the most likely cause is a Jupiter-mass object in a distant orbit. This hypothetical gas giant was nicknamed Tyche. The WISE mission, an all-sky survey using parallax measurements in order to clarify local star distances, was capable of proving or disproving the Tyche hypothesis. In 2014, NASA announced that the WISE survey had ruled out any object as they had defined it.

Future exploration

Space probes have yet to reach the area of the Oort cloud. *Voyager 1*, the fastest and farthest of the interplanetary space probes currently leaving the Solar System, will reach the Oort cloud in about 300 years and would take about 30,000 years to pass through it. However, around 2025, the radioisotope thermoelectric generators on *Voyager 1* will no longer supply enough power to operate any of its scientific instruments, preventing any exploration by *Voyager 1*. The other four probes currently escaping the Solar System either are already or are predicted to be non-functional when they reach the Oort cloud; however, it may be possible to find an object from the cloud that has been knocked into the inner Solar System.

In the 1980s there was a concept for a probe to reach 1,000 AU in 50 years called *TAU*; among its missions would be to look for the Oort cloud.

In the 2014 Announcement of Opportunity for the Discovery program, an observatory to detect the objects in the Oort cloud (and Kuiper belt) called the "Whipple Mission" was proposed. It would monitor distant stars with a photometer, looking for transits up to 10 thousand AU away. The observatory was proposed for halo orbiting around L2 with a suggested 5-year mission. It has been suggested that the Kepler observatory may also be able to detect objects in the Oort cloud.[588]

Figure 335: *Artist's impression of the TAU spacecraft*

External links

<indicator name="spoken-icon"> </indicator>

Wikimedia Commons has media related to ***Oort cloud***.

- Oort Cloud Profile[589] by NASA's Solar System Exploration[590]
- The Kuiper Belt and The Oort Cloud[591]
- The effect of perturbations by the Alpha Cen A/B system on the Oort Cloud[592]
- Reassessing the formation of the Inner Oort cloud in an embedded star cluster II: Probing the inner edge[593] (Brasser; Schwamb : 7 Nov 2014 : arXiv:1411.1844)

<indicator name="featured-star"> ★ </indicator>

Vulcanoid

The **vulcanoids** are a hypothetical population of asteroids that orbit the Sun in a dynamically stable zone inside the orbit of the planet Mercury. They are named after the hypothetical planet Vulcan, whose existence was disproven in 1915. So far, no vulcanoids have been discovered, and it is not yet clear if any exist.

If they do exist, the vulcanoids could easily evade detection because they would be very small and near the bright glare of the Sun. Due to their proximity to the Sun, searches from the ground can only be carried out during twilight or solar eclipses. Any vulcanoids must be between about 100 metres (330 ft) and 6 kilometres (3.7 mi) in diameter and are probably located in nearly circular orbits near the outer edge of the gravitationally stable zone.

The vulcanoids, should they be found, may provide scientists with material from the first period of planet formation, as well as insights into the conditions prevalent in the early Solar System. Although every other gravitationally stable region in the Solar System has been found to contain objects, non-gravitational forces (such as the Yarkovsky effect) or the influence of a migrating planet in the early stages of the Solar System's development may have depleted this area of any asteroids that may have been there.

History and observation

Celestial bodies interior to the orbit of Mercury have been hypothesized, and searched for, for centuries. The German astronomer Christoph Scheiner thought he had seen small bodies passing in front of the Sun in 1611, but these were later shown to be sunspots. In the 1850s, Urbain Le Verrier made detailed calculations of Mercury's orbit and found a small discrepancy in the planet's perihelion precession from predicted values. He postulated that the gravitational influence of a small planet or ring of asteroids within the orbit of Mercury would explain the deviation. Shortly afterward, an amateur astronomer named

Figure 336: *Artist's impression of a vulcanoid asteroid*

Edmond Lescarbault claimed to have seen Le Verrier's proposed planet transit the Sun. The new planet was quickly named Vulcan but was never seen again, and the anomalous behaviour of Mercury's orbit was explained by Einstein's general theory of relativity in 1915. The vulcanoids take their name from this hypothetical planet. What Lescarbault saw was probably another sunspot.

Vulcanoids, should they exist, would be difficult to detect due to the strong glare of the nearby Sun, and ground-based searches can only be carried out during twilight or during solar eclipses. Several searches during eclipses were conducted in the early 1900s, which did not reveal any vulcanoids, and observations during eclipses remain a common search method. Conventional telescopes cannot be used to search for them because the nearby Sun could damage their optics.

In 1998, astronomers analysed data from the SOHO spacecraft's LASCO instrument, which is a set of three coronagraphs. The data taken between January and May of that year did not show any vulcanoids brighter than magnitude 7. This corresponds to a diameter of about 60 kilometres (37 mi), assuming the asteroids have an albedo similar to that of Mercury. In particular, a large planetoid at a distance of 0.18 AU, predicted by the theory of scale relativity, was ruled out.

Figure 337: *A total solar eclipse. These events provide an opportunity to search for vulcanoids from the ground.*

Later attempts to detect the vulcanoids involved taking astronomical equipment above the interference of Earth's atmosphere, to heights where the twilight sky is darker and clearer than on the ground. In 2000, planetary scientist Alan Stern performed surveys of the vulcanoid zone using a Lockheed U-2 spy plane. The flights were conducted at a height of 21,300 metres (69,900 ft) during twilight. In 2002, he and Dan Durda performed similar observations on an F-18 fighter jet. They made three flights over the Mojave desert at an altitude of 15,000 metres (49,000 ft) and made observations with the Southwest Universal Imaging System—Airborne (SWUIS-A).

Even at these heights the atmosphere is still present and can interfere with searches for vulcanoids. In 2004, a sub-orbital spaceflight was attempted in order to get a camera above Earth's atmosphere. A Black Brant rocket was launched from White Sands, New Mexico, on January 16, carrying a powerful camera named VulCam, on a ten-minute flight. This flight reached an altitude of 274,000 metres (899,000 ft) and took over 50,000 images. None of the images revealed any vulcanoids, but there were technical problems.

Searches of NASA's two STEREO spacecraft data have failed to detect any vulcanoid asteroids. It is doubtful that there are any vulcanoids larger than 5.7 kilometres (3.5 mi) in diameter.

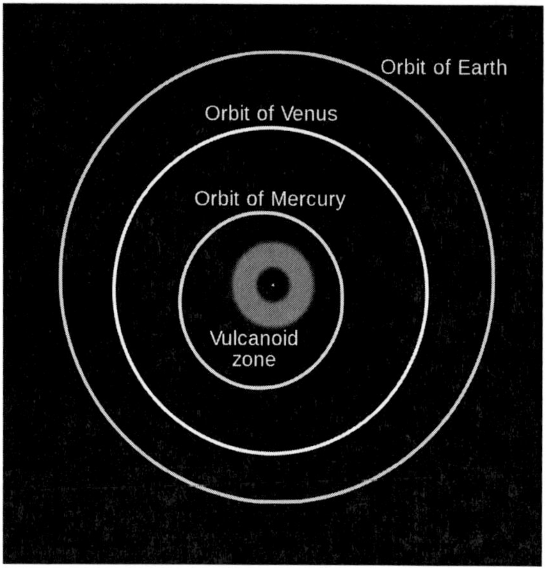

Figure 338: *The zone, represented by the orange region, in which vulcanoids may exist, compared with the orbits of Mercury, Venus and Earth*

The MESSENGER space probe may provide evidence regarding vulcanoids. Its opportunities will be limited because its instruments need to be pointed away from the Sun at all times to avoid damage. The spacecraft has already taken a few of a planned series of images of the outer regions of the vulcanoid zone.Wikipedia:Manual of Style/Dates and numbers#Chronological items

Orbit

A vulcanoid is an asteroid in a stable orbit with a semi-major axis less than that of Mercury (i.e. 0.387 AU). This does not include objects like sungrazing comets, which, although they have perihelia inside the orbit of Mercury, have far greater semi-major axes.

The vulcanoids are thought to exist in a gravitationally stable band inside the orbit of Mercury, at distances of 0.06–0.21 AU from the Sun. All other similarly stable regions in the Solar System have been found to contain objects, although non-gravitational forces such as radiation pressure, Poynting–Robertson drag and the Yarkovsky effect may have depleted the vulcanoid area of its original contents. There may be no more than 300–900 vulcanoids larger than 1 kilometre (0.62 mi) in radius remaining, if any. The gravitational stability of the vulcanoid zone is due in part to the fact that there is only one neighbouring planet. In that respect it can be compared to the Kuiper belt.

The outer edge of the vulcanoid zone is approximately 0.21 AU from the Sun. More distant objects are unstable due to the gravitational influence of Mercury and would be perturbed into Mercury-crossing orbits on timescales of the order of 100 million years. The inner edge is not sharply defined: objects closer than 0.06 AU are highly susceptible to Poynting–Robertson drag and the Yarkovsky effect, and even out to 0.09 AU vulcanoids would have temperatures of 1,000 K or more, which is hot enough for evaporation of rocks to be the limiting factor in their lifetime.

The volume of the vulcanoid zone is very small compared to that of the asteroid belt. Collisions between objects in the vulcanoid zone would be frequent and highly energetic, tending to lead to the destruction of the objects. The most favourable location for vulcanoids is probably in circular orbits near the outer edge of the vulcanoid zone. Vulcanoids are unlikely to have inclinations of more than about 10° to the ecliptic. Mercury trojans, asteroids trapped in Mercury's Lagrange points, are also possible.

Physical characteristics

Any vulcanoids that exist must be relatively small. Previous searches, particularly from the STEREO spacecrafts, rule out asteroids larger than 6 kilometres (3.7 mi) in diameter. The minimum size is about 100 metres (330 ft); particles smaller than 0.2 μm are strongly repulsed by radiation pressure, and objects smaller than 70 m would be drawn into the Sun by Poynting–Robertson drag. Between these upper and lower limits, a population of asteroids between 1 kilometre (0.62 mi) and 6 kilometres (3.7 mi) in diameter is thought to be possible. They would be almost hot enough to glow red hot.

It is thought that the vulcanoids would be very rich in elements with a high melting point, such as iron and nickel. They are unlikely to possess a regolith because such fragmented material heats and cools more rapidly, and is affected more strongly by the Yarkovsky effect, than solid rock. Vulcanoids are probably similar to Mercury in colour and albedo, and may contain material left over from the earliest stages of the Solar System's formation.

There is evidence that Mercury was struck by a large object relatively late in its development, a collision which stripped away much of Mercury's crust and mantle, and explaining the thinness of Mercury's mantle compared to the mantles of the other terrestrial planets. If such an impact occurred, much of the resulting debris might still be orbiting the Sun in the vulcanoid zone.

Significance

Vulcanoids, being an entirely new class of celestial bodies, would be interesting in their own right, but discovering whether or not they exist would yield insights into the formation and evolution of the Solar System. If they exist they might contain material left over from the earliest period of planet formation, and help determine the conditions under which the terrestrial planets, particularly Mercury, formed. In particular, if vulcanoids exist or did exist in the past, they would represent an additional population of impactors that have affected no other planet but Mercury, making that planet's surface appear older than it actually is. If vulcanoids are found not to exist, this would place different constraints on planet formation and suggest that other processes have been at work in the inner Solar System, such as planetary migration clearing out the area.

References

<indicator name="good-star"> ⊕ </indicator>

Planets beyond Neptune

Following the discovery of the planet Neptune in 1846, there was considerable speculation that another planet might exist beyond its orbit. The search began in the mid-19th century and culminated at the start of the 20th with Percival Lowell's quest for Planet X. Lowell proposed the Planet X hypothesis to explain apparent discrepancies in the orbits of the giant planets, particularly Uranus and Neptune, speculating that the gravity of a large unseen ninth planet could have perturbed Uranus enough to account for the irregularities.[594]

Clyde Tombaugh's discovery of Pluto in 1930 appeared to validate Lowell's hypothesis, and Pluto was officially named the ninth planet. In 1978, Pluto was conclusively determined to be too small for its gravity to affect the giant planets, resulting in a brief search for a tenth planet. The search was largely abandoned in the early 1990s, when a study of measurements made by the *Voyager 2* spacecraft found that the irregularities observed in Uranus's orbit were due to a slight overestimation of Neptune's mass. After 1992, the discovery of numerous small icy objects with similar or even wider orbits than Pluto led to a debate over whether Pluto should remain a planet, or whether it and its neighbours should, like the asteroids, be given their own separate classification. Although a number of the larger members of this group were initially described as planets, in 2006 the International Astronomical Union reclassified Pluto and

Figure 339: *Percival Lowell, originator of the Planet X hypothesis*

its largest neighbours as dwarf planets, leaving Neptune the farthest known planet in the Solar System.

While today[595] the astronomical community widely agrees that Planet X, as originally envisioned, does not exist, the concept of an as-yet-unobserved planet has been revived by a number of astronomers to explain other anomalies observed in the outer Solar System. As of March 2014, observations with the WISE telescope have ruled out the possibility of a Saturn-sized object (95 Earth mass) out to 10,000 AU, and a Jupiter-sized ($\approx$318 Earth mass) or larger object out to 26,000 AU.

In 2014, based on similarities of the orbits of a group of recently discovered extreme trans-Neptunian objects, astronomers hypothesized the existence of a super-Earth planet, 2 to 15 times the mass of the Earth and beyond 200 AU with possibly a high inclined orbit at some 1500 AU. In 2016, further work showed this unknown distant planet is likely on an inclined, eccentric orbit that goes no closer than about 200 AU and no farther than about 1200 AU from the Sun. The orbit is predicted to be anti-aligned to the clustered extreme trans-Neptunian objects. Because Pluto is no longer considered a planet by the International Astronomical Union, this new hypothetical object has become known as Planet Nine.

Figure 340: *Jacques Babinet, an early proponent of a trans-Neptunian planet*

Early speculation

In the 1840s, the French mathematician Urbain Le Verrier used Newtonian mechanics to analyse perturbations in the orbit of Uranus, and hypothesised that they were caused by the gravitational pull of a yet-undiscovered planet. Le Verrier predicted the position of this new planet and sent his calculations to German astronomer Johann Gottfried Galle. On 23 September 1846, the night following his receipt of the letter, Galle and his student Heinrich d'Arrest discovered Neptune, exactly where Le Verrier had predicted.[596] There remained some slight discrepancies in the giant planets' orbits. These were taken to indicate the existence of yet another planet orbiting beyond Neptune.

Even before Neptune's discovery, some speculated that one planet alone was not enough to explain the discrepancy. On 17 November 1834, the British amateur astronomer the Reverend Thomas John Hussey reported a conversation he had had with French astronomer Alexis Bouvard to George Biddell Airy, the British Astronomer Royal. Hussey reported that when he suggested to Bouvard that the unusual motion of Uranus might be due to the gravitational influence of an undiscovered planet, Bouvard replied that the idea had occurred to him, and that he had corresponded with Peter Andreas Hansen, director of the Seeberg Observatory in Gotha, about the subject. Hansen's opinion was that a single body could not adequately explain the motion of Uranus, and postulated that two planets lay beyond Uranus.

In 1848, Jacques Babinet raised an objection to Le Verrier's calculations, claiming that Neptune's observed mass was smaller and its orbit larger than Le Verrier had initially predicted. He postulated, based largely on simple subtraction from Le Verrier's calculations, that another planet of roughly 12 Earth masses, which he named "Hyperion", must exist beyond Neptune. Le Verrier denounced Babinet's hypothesis, saying, "[There is] absolutely nothing by which one could determine the position of another planet, barring hypotheses in which imagination played too large a part."

In 1850 James Ferguson, Assistant Astronomer at the United States Naval Observatory, noted that he had "lost" a star he had observed, GR1719k, which Lt. Matthew Maury, the superintendent of the Observatory, claimed was evidence that it must be a new planet. Subsequent searches failed to recover the "planet" in a different position, and in 1878, CHF Peters, director of the Hamilton College Observatory in New York, showed that the star had not in fact vanished, and that the previous results had been due to human error.

In 1879, Camille Flammarion noted that the comets 1862 III and 1889 III had aphelia of 47 and 49 AU, respectively, suggesting that they might mark the orbital radius of an unknown planet that had dragged them into an elliptical orbit. Astronomer George Forbes concluded on the basis of this evidence that two planets must exist beyond Neptune. He calculated, based on the fact that four comets possessed aphelia at around 100 AU and a further six with aphelia clustered at around 300 AU, the orbital elements of a pair of hypothetical trans-Neptunian planets. These elements concorded suggestively with those made independently by another astronomer named David Peck Todd, suggesting to many that they might be valid. However, sceptics argued that the orbits of the comets involved were still too uncertain to produce meaningful results. George Forbes is today considered to be the first describing Planet Nine.

In 1900 and 1901, Harvard College Observatory director William Henry Pickering led two searches for trans-Neptunian planets. The first was begun by Danish astronomer Hans Emil Lau who, after studying the data on the orbit of Uranus from 1690 to 1895, concluded that one trans-Neptunian planet alone could not account for the discrepancies in its orbit, and postulated the position of two planets he believed were responsible. The second was launched when Gabriel Dallet suggested that a single trans-Neptunian planet lying at 47 AU could account for the motion of Uranus. Pickering agreed to examine plates for any suspected planets. In neither case were any found.

In 1909, Thomas Jefferson Jackson See, an astronomer with a reputation as an egocentric contrarian, opined "that there is certainly one, most likely two and possibly three planets beyond Neptune". Tentatively naming the first planet "Oceanus", he placed their respective distances at 42, 56 and 72 AU from the Sun. He gave no indication as to how he determined their existence, and no known searches were mounted to locate them.

In 1911, Indian astronomer Venkatesh P. Ketakar suggested the existence of two trans-Neptunian planets, which he named Brahma and Vishnu, by reworking the patterns observed by Pierre-Simon Laplace in the planetary satellites of Jupiter and applying them to the outer planets. The three inner Galilean moons of Jupiter, Io, Europa and Ganymede, are locked in a complicated 1:2:4 resonance called a Laplace resonance. Ketakar suggested that Uranus, Neptune and his hypothetical trans-Neptunian planets were locked in Laplace-like resonances. His calculations predicted a mean distance for Brahma of 38.95 AU and an orbital period of 242.28 Earth years (3:4 resonance with Neptune). When Pluto was discovered 19 years later, its mean distance of 39.48 AU and orbital period of 248 Earth years were close to Ketakar's prediction (Pluto in fact has a 2:3 resonance with Neptune). Ketakar made no predictions for the orbital elements other than mean distance and period. It is not clear how Ketakar arrived at these figures, and his second planet, Vishnu, was never located.

Planet X

In 1894, with the help of William Pickering, Percival Lowell, a wealthy Bostonian, founded the Lowell Observatory in Flagstaff, Arizona. In 1906, convinced he could resolve the conundrum of Uranus's orbit, he began an extensive project to search for a trans-Neptunian planet, which he named *Planet X*, a name previously used by Gabriel Dallet. The *X* in the name represents an unknown and is pronounced as the letter, as opposed to the Roman numeral for 10 (at the time, Planet X would have been the ninth planet). Lowell's hope in tracking down Planet X was to establish his scientific credibility, which had eluded him due to his widely derided belief that channel-like features visible on the surface of Mars were canals constructed by an intelligent civilization.[597]

Lowell's first search focused on the ecliptic, the plane encompassed by the zodiac where the other planets in the Solar System lie. Using a 5-inch photographic camera, he manually examined over 200 three-hour exposures with a magnifying glass, and found no planets. At that time Pluto was too far above the ecliptic to be imaged by the survey.[598] After revising his predicted possible locations, Lowell conducted a second search from 1914 to 1916. In 1915, he published his *Memoir of a Trans-Neptunian Planet*, in which he concluded that Planet X had a mass roughly seven times that of Earth—about half that of Neptune—and a mean distance from the Sun of 43 AU. He assumed Planet X would be a large, low-density object with a high albedo, like the giant planets.

Figure 341: *Clyde William Tombaugh*

As a result, it would show a disc with diameter of about one arcsecond and an apparent magnitude of between 12 and 13—bright enough to be spotted.[599]

Separately, in 1908, Pickering announced that, by analysing irregularities in Uranus's orbit, he had found evidence for a ninth planet. His hypothetical planet, which he termed "Planet O" (because it came after "N", i.e. Neptune), possessed a mean orbital radius of 51.9 AU and an orbital period of 373.5 years. Plates taken at his observatory in Arequipa, Peru, showed no evidence for the predicted planet, and British astronomer P. H. Cowell showed that the irregularities observed in Uranus's orbit virtually disappeared once the planet's displacement of longitude was taken into account. Lowell himself, despite his close association with Pickering, dismissed Planet O out of hand, saying, "This planet is very properly designated "O", [for it] is nothing at all."[600] Unbeknownst to Pickering, four of the photographic plates taken in the search for "Planet O" by astronomers at the Mount Wilson Observatory in 1919 captured images of Pluto, though this was only recognised years later. Pickering went on to suggest many other possible trans-Neptunian planets up to the year 1932, which he named *P*, *Q*, *R*, *S*, *T* and *U*; none were ever detected.

Discovery of Pluto

Lowell's sudden death in 1916 temporarily halted the search for Planet X. Failing to find the planet, according to one friend, "virtually killed him".[601]

Lowell's widow, Constance, engaged in a legal battle with the observatory over Lowell's legacy which halted the search for Planet X for several years.[602] In 1925, the observatory obtained glass discs for a new 13 in (33 cm) wide-field telescope to continue the search, constructed with funds from Abbott Lawrence Lowell,[603] Percival's brother. In 1929 the observatory's director, Vesto Melvin Slipher, summarily handed the job of locating the planet to Clyde Tombaugh, a 22-year-old Kansas farm boy who had only just arrived at the Lowell Observatory after Slipher had been impressed by a sample of his astronomical drawings.

Tombaugh's task was to systematically capture sections of the night sky in pairs of images. Each image in a pair was taken two weeks apart. He then placed both images of each section in a machine called a blink comparator, which by exchanging images quickly created a time lapse illusion of the movement of any planetary body. To reduce the chances that a faster-moving (and thus closer) object be mistaken for the new planet, Tombaugh imaged each region near its opposition point, 180 degrees from the Sun, where the apparent retrograde motion for objects beyond Earth's orbit is at its strongest. He also took a third image as a control to eliminate any false results caused by defects in an individual plate. Tombaugh decided to image the entire zodiac, rather than focus on those regions suggested by Lowell.

By the beginning of 1930, Tombaugh's search had reached the constellation of Gemini. On 18 February 1930, after searching for nearly a year and examining nearly 2 million stars, Tombaugh discovered a moving object on photographic plates taken on 23 January and 29 January of that year.[604] A lesser-quality photograph taken on January 21 confirmed the movement. Upon confirmation, Tombaugh walked into Slipher's office and declared, "Doctor Slipher, I have found your Planet X." The object lay just six degrees from one of two locations for Planet X Lowell had suggested; thus it seemed he had at last been vindicated. After the observatory obtained further confirmatory photographs, news of the discovery was telegraphed to the Harvard College Observatory on March 13, 1930. The new object was later precovered on photographs dating back to 19 March 1915.[605] The decision to name the object *Pluto* was intended in part to honour Percival Lowell, as his initials made up the word's first two letters. After discovering Pluto, Tombaugh continued to search the ecliptic for other distant objects. He found hundreds of variable stars and asteroids, as well as two comets, but no further planets.

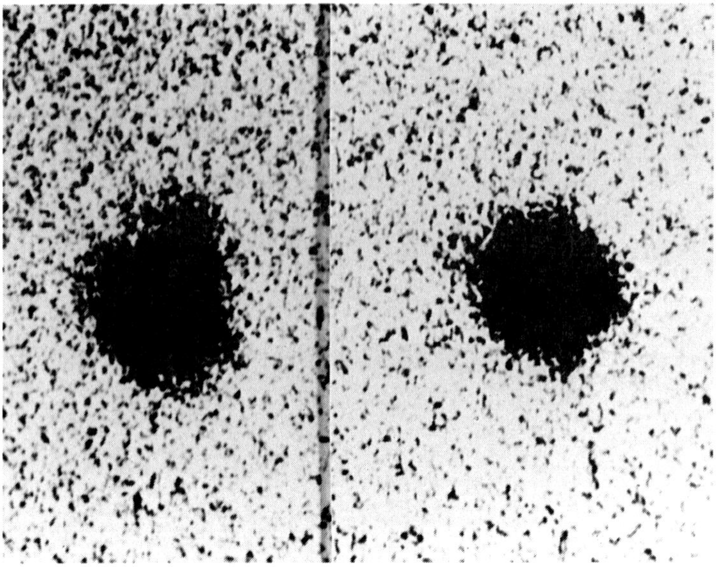

Figure 342: *Discovery image of Charon*

Pluto loses Planet X title

To the observatory's disappointment and surprise, Pluto showed no visible disc; it appeared as a point, no different from a star, and, at only 15th magnitude, was six times dimmer than Lowell had predicted, which meant it was either very small, or very dark. Because Lowell astronomers thought Pluto was massive enough to perturb planets, they assumed that its albedo could be no less than 0.07 (meaning that it reflected only 7% of the light that hit it); about as dark as asphalt and similar to that of Mercury, the least reflective planet known. This would give Pluto an estimated mass of no more than 70% that of Earth. Observations also revealed that Pluto's orbit was very elliptical, far more than that of any other planet.

Almost immediately, some astronomers questioned Pluto's status as a planet. Barely a month after its discovery was announced, on 14 April 1930, in an article in the *New York Times*, Armin O. Leuschner suggested that Pluto's dimness and high orbital eccentricity made it more similar to an asteroid or comet: "The Lowell result confirms the possible high eccentricity announced by us on April 5. Among the possibilities are a large asteroid greatly disturbed in its orbit by close approach to a major planet such as Jupiter, or it may be one of many long-period planetary objects yet to be discovered, or a bright cometary object." In that same article, Harvard Observatory director Harlow

Shapley wrote that Pluto was a "member of the Solar System not comparable with known asteroids and comets, and perhaps of greater importance to cosmogony than would be another major planet beyond Neptune." In 1931, using a mathematical formula, Ernest W. Brown asserted (in agreement with E. C. Bower), that the presumed irregularities in the orbit of Uranus could not be due to the gravitational effect of a more distant planet, and thus that Lowell's supposed prediction was "purely accidental".

Throughout the mid-20th century, estimates of Pluto's mass were revised downward. In 1931, Nicholson and Mayall calculated its mass, based on its supposed effect on the giant planets, as roughly that of Earth; a value somewhat in accord with the 0.91 Earth mass calculated in 1942 by Lloyd R. Wylie at the US Naval Observatory, using the same assumptions. In 1949, Gerard Kuiper's measurements of Pluto's diameter with the 200 inch telescope at Mount Palomar Observatory led him to the conclusion that it was midway in size between Mercury and Mars and that its mass was most probably about 0.1 Earth mass.

In 1973, Dennis Rawlins conjectured, based on the similarities in the periodicity and amplitude of brightness variation between Pluto and Neptune's moon Triton, that Pluto's mass must be similar to Triton's. This is, in fact, true, and had been argued by astronomers Walter Baade and E. C. Bower as early as 1934. However, because Triton's mass was then believed to be roughly 2.5% that of the Earth–Moon system (more than ten times its actual value), Rawlins's determination for Pluto's mass was similarly incorrect. It was nonetheless a meagre enough value for him to conclude that Pluto was not Planet X.[606] In 1976, Dale Cruikshank, Carl Pilcher and David Morrison of the University of Hawaii analysed spectra from Pluto's surface and determined that it must contain methane ice, which is highly reflective. This meant that Pluto, far from being dark, was in fact exceptionally bright, and thus was probably no more than 0.01 Earth mass.[607]

Mass estimates for Pluto:

Year	Mass	Notes
1931	1 Earth	Nicholson & Mayall
1942	0.91 Earth	Wylie
1948	0.1 (1/10 Earth)	Kuiper
1973	0.025 (1/40 Earth)	Rawlins
1976	0.01 (1/100 Earth)	Cruikshank, Pilcher, & Morrison
1978	0.002 (1/500 Earth)	Christy & Harrington

2006	0.00218 (1/459 Earth)	Buie et al.

Pluto's size was finally determined conclusively in 1978, when American astronomer James W. Christy discovered its moon Charon. This enabled him, together with Robert Sutton Harrington of the U.S. Naval Observatory, to measure the mass of the Pluto–Charon system directly by observing the moon's orbital motion around Pluto. They determined Pluto's mass to be 1.31×10^{22} kg; roughly one five-hundredth that of Earth or one-sixth that of the Moon, and far too small to account for the observed discrepancies in the orbits of the outer planets. Lowell's "prediction" had been a coincidence: If there was a Planet X, it was not Pluto.[608]

Further searches for Planet X

After 1978, a number of astronomers kept up the search for Lowell's Planet X, convinced that, because Pluto was no longer a viable candidate, an unseen tenth planet must have been perturbing the outer planets.

In the 1980s and 1990s, Robert Harrington led a search to determine the real cause of the apparent irregularities.[609] He calculated that any Planet X would be at roughly three times the distance of Neptune from the Sun; its orbit would be highly eccentric, and strongly inclined to the ecliptic—the planet's orbit would be at roughly a 32-degree angle from the orbital plane of the other known planets. This hypothesis was met with a mixed reception. Noted Planet X sceptic Brian G. Marsden of the Minor Planet Center pointed out that these discrepancies were a hundredth the size of those noticed by Le Verrier, and could easily be due to observational error.[610]

In 1972, Joseph Brady of the Lawrence Livermore National Laboratory studied irregularities in the motion of Halley's Comet. Brady claimed that they could have been caused by a Jupiter-sized planet beyond Neptune at 59 AU that is in a retrograde orbit around the Sun. However, both Marsden and Planet X proponent P. Kenneth Seidelmann attacked the hypothesis, showing that Halley's Comet randomly and irregularly ejects jets of material, causing changes to its own orbital trajectory, and that such a massive object as Brady's Planet X would have severely affected the orbits of known outer planets.[611]

Although its mission did not involve a search for Planet X, the IRAS space observatory made headlines briefly in 1983 due to an "unknown object" that was at first described as "possibly as large as the giant planet Jupiter and possibly so close to Earth that it would be part of this Solar System". Further analysis revealed that of several unidentified objects, nine were distant galaxies and the tenth was "interstellar cirrus"; none were found to be Solar System bodies.

In 1988, A. A. Jackson and R. M. Killen studied the stability of Pluto's resonance with Neptune by placing test "Planet X-es" with various masses and at various distances from Pluto. Pluto and Neptune's orbits are in a 3:2 resonance, which prevents their collision or even any close approaches, regardless of their separation in the z axis. It was found that the hypothetical object's mass had to exceed 5 Earth masses to break the resonance, and the parameter space is quite large and a large variety of objects could have existed beyond Pluto without disturbing the resonance. Four test orbits of a trans-Plutonian planet have been integrated forward for four million years in order to determine the effects of such a body on the stability of the Neptune–Pluto 3:2 resonance. Planets beyond Pluto with masses of 0.1 and 1.0 Earth masses in orbits at 48.3 and 75.5 AU, respectively, do not disturb the 3:2 resonance. Test planets of 5 Earth masses with semi-major axes of 52.5 and 62.5 AU disrupt the four-million-year libration of Pluto's argument of perihelion.

Planet X disproved

Harrington died in January 1993, without having found Planet X.[612] Six months before, E. Myles Standish had used data from *Voyager 2*'s 1989 flyby of Neptune, which had revised the planet's total mass downward by 0.5%—an amount comparable to the mass of Mars—to recalculate its gravitational effect on Uranus. When Neptune's newly determined mass was used in the Jet Propulsion Laboratory Developmental Ephemeris (JPL DE), the supposed discrepancies in the Uranian orbit, and with them the need for a Planet X, vanished. There are no discrepancies in the trajectories of any space probes such as *Pioneer 10*, *Pioneer 11*, *Voyager 1*, and *Voyager 2* that can be attributed to the gravitational pull of a large undiscovered object in the outer Solar System.[613] Today, most astronomers agree that Planet X, as Lowell defined it, does not exist.

Discovery of further trans-Neptunian objects

After the discovery of Pluto and Charon, no more trans-Neptunian objects (TNOs) were found until (15760) 1992 QB_1 in 1992. Since then, thousands of such objects have been discovered. Most are now recognized as part of the Kuiper belt, a swarm of icy bodies left over from the Solar System's formation that orbit near the ecliptic plane just beyond Neptune. Though none were as large as Pluto, some of these distant trans-Neptunian objects, such as Sedna, were initially described in the media as "new planets".

In 2005, astronomer Mike Brown and his team announced the discovery of 2003 UB_{313} (later named Eris after the Greek goddess of discord and strife), a trans-Neptunian object then thought to be just barely larger than Pluto. Soon

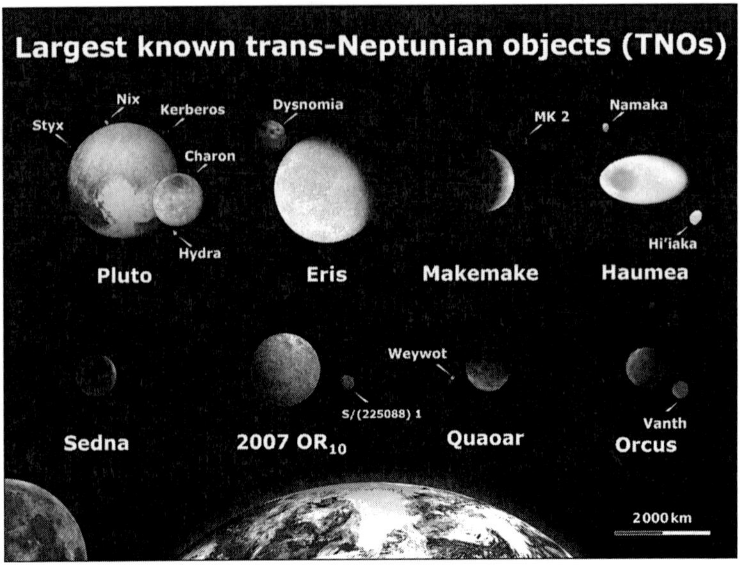

Figure 343: *style=float:bottom}}*

afterwards, a NASA Jet Propulsion Laboratory press release described the object as the "tenth planet".

Eris was never officially classified as a planet, and the 2006 definition of planet defined both Eris and Pluto not as planets but as dwarf planets because they have not cleared their neighbourhoods. They do not orbit the Sun alone, but as part of a population of similarly sized objects. Pluto itself is now recognized as being a member of the Kuiper belt and the largest dwarf planet, larger than the more-massive Eris.

A number of astronomers, most notably Alan Stern, the head of NASA's New Horizons mission to Pluto, contend that the IAU's definition is flawed, and that Pluto and Eris, and all large trans-Neptunian objects, such as Makemake, Sedna, Quaoar, Varuna and Haumea, should be considered planets in their own right. However, the discovery of Eris did not rehabilitate the Planet X theory because it is far too small to have significant effects on the outer planets' orbits.

Subsequently proposed trans-Neptunian planets

Although most astronomers accept that Lowell's Planet X does not exist, a number have revived the idea that a large unseen planet could create observable gravitational effects in the outer Solar System. These hypothetical objects are

Figure 344: *The orbit of Sedna (red) set against the orbits of Jupiter (orange), Saturn (yellow), Uranus (green), Neptune (blue), and Pluto (purple)*

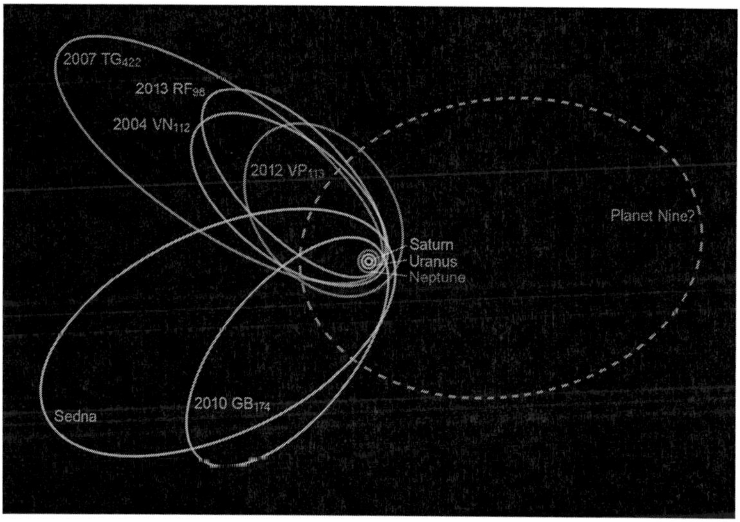

Figure 345: *Prediction of hypothetical Planet Nine's orbit based on unique clustering*

often referred to as "Planet X", although the conception of these objects may differ considerably from that proposed by Lowell.

Orbits of distant objects

Sedna's orbit

When Sedna was discovered, its extreme orbit raised questions about its origin. Its perihelion is so distant (approximately 75 AU) that no currently observed mechanism can explain Sedna's eccentric distant orbit. It is too far from the planets to have been affected by the gravity of Neptune or the other giant planets and too bound to the Sun to be affected by outside forces such as the galactic tides. Hypotheses to explain its orbit include that it was affected by a passing star, that it was captured from another planetary system, or that it was tugged into its current position by a trans-Neptunian planet. The most obvious solution to determining Sedna's peculiar orbit would be to locate a number of objects in a similar region, whose various orbital configurations would provide an indication as to their history. If Sedna had been pulled into its orbit by a trans-Neptunian planet, any other objects found in its region would have a similar perihelion to Sedna (around 80 AU).

Elongated orbits of group of Kuiper belt objects

In 2012, Rodney Gomes modelled the orbits of 92 Kuiper belt objects and found that six of those orbits were far more elongated than the model predicted. He concluded that the simplest explanation was the gravitational pull of a distant planetary companion, such as a Neptune-sized object at 1500 AU or a Mars-sized object at around 53 AU.

Discovery of 2012 VP_{113} and the orbital clustering of Kuiper belt objects

In 2014, astronomers announced the discovery of 2012 VP113, a large object with a Sedna-like 4200-year orbit and a perihelion of roughly 80 AU, which led them to suggest that it offered evidence of a potential trans-Neptunian planet. Trujillo and Sheppard argued that the orbital clustering of arguments of perihelia for VP113 and other extremely distant TNOs suggests the existence of a "super-Earth" of between 2 and 15 Earth masses beyond 200 AU and possibly on an inclined orbit at 1500 AU.

In 2014 astronomers at the Universidad Complutense in Madrid suggested that the available data actually indicate more than one trans-Neptunian planet.

Further analysis & Planet Nine hypothesis

On January 20, 2016, Brown and Konstantin Batygin published an article corroborating Trujillo and Sheppard's initial findings; proposing a super-Earth (dubbed Planet Nine) based on a statistical clustering of the arguments of perihelia (noted before) near zero and also ascending nodes near 113° of six distant trans-Neptunian objects. They estimated it to be ten times the mass of Earth (about 60% the mass of Neptune) with a semimajor axis of approximately 400–1500 AU.

Probability

Even without gravitational evidence, Mike Brown, the discoverer of Sedna, has argued that Sedna's 12,000-year orbit means that probability alone suggests that an Earth-sized object exists beyond Neptune. Sedna's orbit is so eccentric that it spends only a small fraction of its orbital period near the Sun, where it can be easily observed. This means that unless its discovery was a freak accident, there is probably a substantial population of objects roughly Sedna's diameter yet to be observed in its orbital region. Mike Brown noted that "Sedna is about three-quarters the size of Pluto. If there are sixty objects three-quarters the size of Pluto [out there] then there are probably forty objects the size of Pluto ... If there are forty objects the size of Pluto, then there are probably ten that are twice the size of Pluto. There are probably three or four that are three times the size of Pluto, and the biggest of these objects ... is probably the size of Mars or the size of the Earth."[614] However, he notes that, should such an object be found, even though it might approach Earth in size, it would still be a dwarf planet by the current definition, because it would not have cleared its neighbourhood sufficiently.

Kuiper cliff

Additionally, speculation of a possible trans-Neptunian planet has revolved around the so-called "Kuiper cliff". The Kuiper belt terminates suddenly at a distance of 48 AU from the Sun. Brunini and Melita have speculated that this sudden drop-off may be attributed to the presence of an object with a mass between that of Mars and Earth located beyond 48 AU. The presence of an object with a mass similar to that of Mars in a circular orbit at 60 AU leads to a trans-Neptunian object population incompatible with observations. For instance, it would severely deplete the plutino population. Astronomers have not excluded the possibility of an object with a mass similar to that of Earth located farther than 100 AU with an eccentric and inclined orbit. Computer simulations by Patryk Lykawka of Kobe University have suggested that an object with a mass between 0.3 and 0.7 Earth masses, ejected outward by Neptune early in the Solar System's formation and currently in an elongated orbit between 101 and 200 AU from the Sun, could explain the Kuiper cliff and the peculiar detached objects such as Sedna and 2012 VP113. Although some astronomers, such as Renu Malhotra and David Jewitt, have cautiously supported these claims, others, such as Alessandro Morbidelli, have dismissed them as "contrived". In 2017, Malhotra and Kat Volk argued that an unexpected variance in inclination for KBOs farther than 50 AU provided evidence of a possible Mars-sized planet residing at the edge of the Solar System.

Other proposed planets

Tyche was a hypothetical gas giant proposed to be located in the Solar System's Oort cloud. It was first proposed in 1999 by astrophysicists John Matese, Patrick Whitman and Daniel Whitmire of the University of Louisiana at Lafayette. They argued that evidence of Tyche's existence could be seen in a supposed bias in the points of origin for long-period comets. In 2013, Matese and Whitmire re-evaluated the comet data and noted that Tyche, if it existed, would be detectable in the archive of data that was collected by NASA's Wide-field Infrared Survey Explorer (WISE) telescope. In 2014, NASA announced that the WISE survey had ruled out any object with Tyche's characteristics, indicating that Tyche as hypothesized by Matese, Whitman, and Whitmire does not exist.

The oligarch theory of planet formation states that there were hundreds of planet-sized objects, known as oligarchs, in the early stages of the Solar System's evolution. In 2005, astronomer Eugene Chiang speculated that although some of these oligarchs became the planets we know today, most would have been flung outward by gravitational interactions. Some may have escaped the Solar System altogether to become free-floating planets, whereas others would be orbiting in a halo around the Solar System, with orbital periods of millions of years. This halo would lie at between 1,000 and 10,000 AU from the Sun, or between a third and a thirtieth the distance to the Oort cloud.[615]

In December 2015, astronomers at the Atacama Large Millimeter Array (ALMA) detected a brief series of 350 GHz pulses that they concluded must either be a series of independent sources, or a single, fast moving source. Deciding that the latter was the most likely, they calculated based on its speed that, were it bound to the Sun, the object, which they named "Gna" after a fast-moving messenger goddess in Norse mythology, would be about 12–25 AU distant and have a dwarf planet-sized diameter of 220 to 880 km. However, if it were a rogue planet not gravitationally bound to the Sun, and as far away as 4000 AU, it could be much larger. The paper was never formally accepted, and has been withdrawn until the detection is confirmed. Scientists' reactions to the notice were largely sceptical; Mike Brown commented that, "If it is true that ALMA accidentally discovered a massive outer solar system object in its tiny, tiny, tiny, field of view, that would suggest that there are something like 200,000 Earth-sized planets in the outer solar system ... Even better, I just realized that this many Earth-sized planets existing would destabilize the entire solar system and we would all die."

Constraints on additional planets

As of 2016 the following observations severely constrain the mass and distance of any possible additional Solar System planet:

- An analysis of mid-infrared observations with the WISE telescope have ruled out the possibility of a Saturn-sized object (95 Earth masses) out to 10,000 AU, and a Jupiter-sized or larger object out to 26,000 AU.
- Using modern data on the anomalous precession of the perihelia of Saturn, Earth, and Mars, Lorenzo Iorio concluded that any unknown planet with a mass of 0.7 times that of Earth must be farther than 350–400 AU; one with a mass of 2 times that of Earth, farther than 496–570 AU; and finally one with a mass of 15 times that of Earth, farther than 970–1111 AU. Moreover, Iorio stated that the modern ephemerides of the Solar System outer planets has provided even tighter constraints: no celestial body with a mass of 15 times that of Earth can exist closer than 1100–1300 AU. However, work by another group of astronomers using a more comprehensive model of the Solar System found that Iorio's conclusion was only partially correct. Their analysis of *Cassini* data on Saturn's orbital residuals found that observations were inconsistent with a planetary body with the orbit and mass similar to those of Batygin and Brown's Planet Nine having a true anomaly of –130° to –110° or –65° to 85°. Furthermore, the analysis found that Saturn's orbit is slightly better explained if such a body is located at a true anomaly of $117.8^{\circ}{}^{+11^{\circ}}_{-10^{\circ}}$. At this location, it would be approximately 630 AU from the Sun.

Bibliography

- Ken Croswell (1997). *Planet Quest: The Epic Discovery of Alien Solar Systems*. New York: The Free Press. ISBN 978-0-684-83252-4.
- Mark Littman (1990). *Planets Beyond: Discovering the Outer Solar System*. New York: Wiley. ISBN 978-0-471-51053-6.
- Govert Schilling (2009). *The Hunt for Planet X: New Worlds and the Fate of Pluto*. New York: Springer. ISBN 978-0-387-77804-4.
- Clyde W. Tombaugh (1946). "The Search for the Ninth Planet, Pluto". *Astronomical Society of the Pacific Leaflets*. **5**: 73–80. Bibcode: 1946ASPL....5...73T[616].

Further reading

- "Planet X Discovered??" by PBS SpaceTime[617]
- SEDS on Planet X[618]
- E. Myles Standish, Jr. (May 1993). "Planet X: No Dynamical Evidence in the Optical Observations". *Astronomical Journal*. **105** (5): 2000–2006. Bibcode: 1993AJ....105.2000S[619]. doi: 10.1086/116575[620].
- Gerald D. Quinlan (May 6, 1993). "Planet X: A Myth Exposed". *Nature*. **363** (6424): 18–19. Bibcode: 1993Natur.363...18Q[621]. doi: 10.1038/363018b0[622].
- Daniel P. Whitmire & John J. Matese (January 3, 1985). "Periodic Comet Showers and Planet X". *Nature*. **313** (5997): 36–38. Bibcode: 1985Natur.313...36W[623]. doi: 10.1038/313036a0[624].
- Jones, R. L.; M. E. Brown; P. A. Abell; A. C. Becker; et al. (2009). "Next steps in understanding the outer solar system: A whitepaper submitted to the 2010 Decadal Survey Committee"[625] (PDF). Submission to the National Academy of Sciences Board on Physics and Astronomy. Retrieved 2009-04-10.
- Andrew Coates provides a simplified summary of the history behind the search & claims[626] for the planet x

<indicator name="featured-star"> </indicator>

Planet Nine

<indicator name=”good-star”> ⊕ </indicator>

Planet Nine

Artist’s impression of Planet Nine as an ice giant eclipsing the central Milky Way, with the Sun in the distance. Neptune’s orbit is shown as a small ellipse around the Sun. (See labeled version.)

Orbital characteristics	
Aphelion	1,200 AU (est.)
Perihelion	200 AU (est.)
Semi-major axis	700 AU (est.)
Eccentricity	0.6 (est.)
Orbital period	10,000 to 20,000 years
Inclination	30° to ecliptic (est.)
Argument of perihelion	150° (est.)
Physical characteristics	
Mean radius	13,000 to 26,000 km (8,000–16,000 mi) 2–4 $R_\oplus$ (est.)
Mass	6×10^{25} kg (est.) $\geq 10\ M_\oplus$ (est.)
Apparent magnitude	>22.5 (est.)

Planet Nine is a hypothetical large planet in the far outer Solar System, the gravitational effects of which would explain the improbable orbital configuration of a group of trans-Neptunian objects (TNOs) that orbit mostly beyond the Kuiper belt.

Recent speculation about the alignment of the eTNOs being due to a distant massive planet began with a 2014 letter to the journal *Nature*, in which astronomers Chad Trujillo and Scott S. Sheppard inferred the possible existence of a massive trans-Neptunian planet from similarities in the orbits of the distant trans-Neptunian objects Sedna and 2012 VP113. On 20 January 2016, researchers Konstantin Batygin and Michael E. Brown at Caltech explained how a massive outer planet would be the likeliest explanation for the similarities in orbits of six distant objects, and proposed specific orbital parameters. The predicted planet could be a super-Earth, with an estimated mass of 10 Earths (approximately 5,000 times the mass of Pluto), a diameter two to four times that of Earth, and a highly elliptical orbit with an orbital period of approximately 15,000 years.

In addition to the clustering of the perihelia and the orbital poles of distant objects, Planet Nine offers explanations for the high perihelion of Sedna and 2012 VP_{113}, for objects with orbits roughly perpendicular to those of the planets, for high inclination TNOs with semimajor axes less than 100 AU, and for the tilt of the Sun's rotation axis. The objects it dynamically controls would form a cloud centered on its semi-major axis with a wide range of inclinations.

Batygin and Brown suggested Planet Nine was a primordial giant planet core that was ejected from its initial orbit by an encounter with Jupiter during the nebular epoch of the Solar System and was later perturbed into a stable orbit by a distant encounter with a passing star or by the solar nebula. Others have proposed that it was captured in a similar manner from another star, or that it formed on a very distant circular orbit and was perturbed into its current eccentric orbit during a distant encounter with another star.

Naming

Planet Nine does not have an official name, and it will not receive one unless its existence is confirmed, typically through optical imaging. Once confirmed, the IAU will certify a name, with priority typically given to a name proposed by its discoverers. It will likely be a name chosen from Roman or Greek mythology.

In their original paper, Batygin and Brown simply referred to the object as "perturber", and only in later press releases did they use "Planet Nine". They have also used the names "Jehoshaphat" and "George" for Planet Nine. Brown has stated: "We actually call it Phattie[627] when we're just talking to each other."

Postulated characteristics

Orbit

Planet Nine is hypothesized to follow a highly elliptical orbit around the Sun, with an orbital period of 10,000–20,000 years. The period with which it goes around the sun is a rational multiple of the periods of all the furthest Kuiper belt objects. The planet's orbit would have a semi-major axis of approximately 700 AU, or about 20 times the distance from Neptune to the Sun, although it

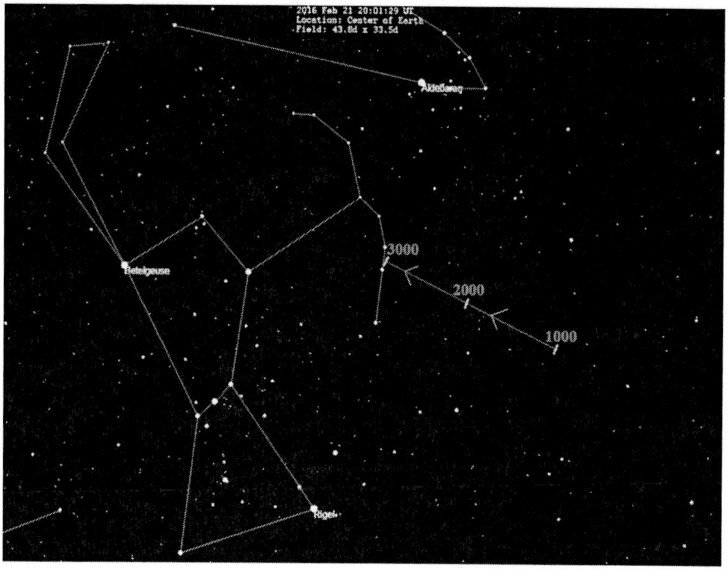

Figure 346: *One hypothetical path through the sky of Planet Nine near aphelion crossing Orion west to east with about 2,000 years of motion. It is derived from that employed in the artistic conception on Brown's blog.*

might come as close as 200 AU (30 billion km, 19 billion mi), and its inclination is estimated to be roughly 30°±10°.[628] The high eccentricity of Planet Nine's orbit could take it as far away as 1,200 AU at its aphelion.

The aphelion, or farthest point from the Sun, would be in the general direction of the constellation of Taurus, whereas the perihelion, the nearest point to the Sun, would be in the general direction of the southerly areas of Serpens (Caput), Ophiuchus, and Libra.[629,630]

Brown thinks that if Planet Nine is confirmed to exist, a probe could fly by it in as little as 20 years, with a powered slingshot around the Sun.

Size and composition

The planet is estimated to have 10 times the mass and two to four times the diameter of Earth. An object with the same diameter as Neptune has not been excluded by previous surveys. An infrared survey by the Wide-field Infrared Survey Explorer (WISE) in 2009 allowed for a Neptune-sized object beyond 700 AU. A similar study in 2014 focused on possible higher-mass bodies in the outer Solar System and ruled out Jupiter-mass objects out to 26,000 AU.

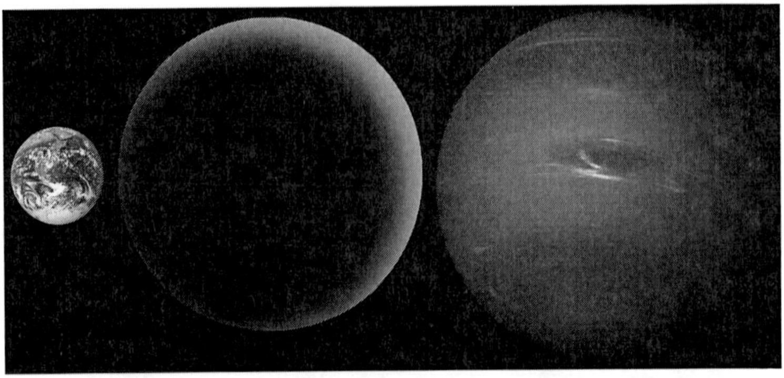

Figure 347: *Planet Nine is hypothesized to be two to four times the diameter of Earth; similar to the ice giants Uranus and Neptune.*

Brown thinks that no matter where it is speculated to be, if Planet Nine exists, then its mass is higher than what is required to clear its feeding zone in 4.6 billion years, and thus that it dominates the outer edge of the Solar System, which is sufficient to make it a planet by current definitions. Using a metric based on work by Jean-Luc Margot, Brown calculated that only at the smallest size and farthest distance was it on the border of being called a dwarf planet. Margot himself says that Planet Nine satisfies the quantitative criterion for orbit-clearing developed by him in 2015, and that according to that criterion, Planet Nine will qualify as a planet—if and when it is detected.

Brown speculates that the predicted planet is most likely an ejected ice giant, similar in composition to Uranus and Neptune: a mixture of rock and ice with a small envelope of gas. Based on its size, the considerable gravitational pull of Planet Nine could theoretically promote life in subsurface oceans with liquid water in nearby moons. Subsurface oceans have been discovered on the moons Enceladus (water), Dione (possibly water), and subsurface water is postulated for Neptune's largest moon Triton, and possibly also found on Europa.

Dynamical effects

The dynamical effects of Planet Nine would explain five aspects of the Solar System: the clustering of the orbits of extreme transneptunian objects (eTNOs); the high perihelia of objects like Sedna that are detached from Neptune's influence; the high inclinations of eTNOs with orbits roughly perpendicular to the plane of the Solar System, high inclination Kuiper belt objects with semi-major axis less than 100 AU, and the obliquity, or tilt, of the Sun's axis relative to the ecliptic. While other mechanisms have been offered for

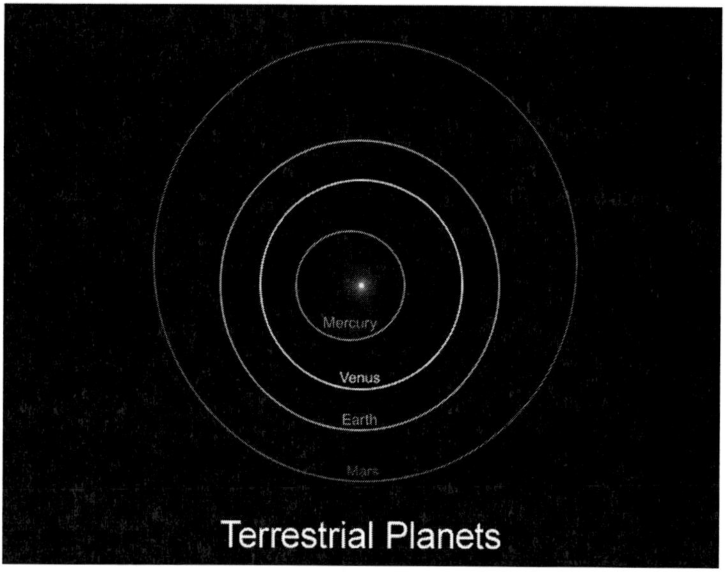

Figure 348: *Orbital correlations among six distant trans-Neptunian objects led to the hypothesis. (See: Final frame orbits)*

many of these peculiarities the gravitational influence Planet Nine is the only one that explains all five.

Orbital clustering and high perihelion objects

The clustering of the orbits of extreme transneptunian objects was first described by Chad Trujillo and Scott S. Sheppard, who noted similarities between the orbits of Sedna and 2012 VP_{113}. Upon further analysis they observed that the arguments of perihelion of 12 eTNOs with perihelia greater than 30 AU and semi-major axes greater than 150 AU were clustered near zero degrees. Trujillo and Sheppard proposed that this alignment was caused by a massive unknown planet beyond Neptune via the Kozai mechanism. (see Trujillo and Sheppard (2014) for more details.)

Caltech's Konstantin Batygin and Michael E. Brown, looking to refute the mechanism proposed by Trujillo and Sheppard, also examined the orbits the extreme transeptunian objects. After eliminating the objects in Trujillo and Sheppard's original analysis that were unstable due to close approaches to Neptune or were affected by Neptune mean-motion resonances they determined that the argument of perihelion for the remaining six objects (namely Sedna, 2012 VP_{113}, 2004 VN_{112}, 2010 GB_{174}, 2010 VZ_{98}, and 2000 CR_{105}) were

318°±8°. This was out of alignment with how the Kozai mechanism would align these orbits, at c. 0° or 180°.[631]

1. For bodies whose values of a and e are such that they could encounter the planets only near perihelion (or aphelion), such encounters may be prevented by the high inclination and the libration of ω about 90° or 270° (even when the encounters occur, they do not affect much the minor planet's orbit due to comparatively high relative velocities).
2. Another mechanism is viable when at low inclinations when ω oscillates around 0° or 180° and the minor planet's semimajor axis is close to that of the perturbing planet: in this case the °node crossing occurs always near perihelion and aphelion, far from the planet itself, provided the eccentricity is high enough and the orbit of the planet is almost circular.</ ref>

Batygin and Brown also found that the orbits of the six objects with semimajor axes greater than 250 AU and perihelia beyond 30 AU (namely Sedna, 2012 VP113, 2007 TG422, 2004 VN_{112}, 2013 RF98, and 2010 GB174) were aligned in space with their perihelia in roughly the same direction. The orbits of the six objects were also tilted with respect to that of the ecliptic and approximately co-planar. They determined that there was only a 0.007% likelihood that this combination of alignments was due to chance.

These six objects had been discovered by six different surveys on six different telescopes. That made it less likely that the clumping might be due to an observation bias such as pointing a telescope at a particular part of the sky. These six were the only minor planets known to have perihelia greater than 30 AU and a semi-major axis greater than 250 AU, as of January 2016. Most have perihelia significantly beyond Neptune, which orbits 30 AU from the Sun. Generally, TNOs with perihelia smaller than 36 AU experience strong encounters with Neptune. All six objects are relatively small, but currently relatively bright because they are near their closest distance to the Sun in their elliptical orbits.

Among the extreme transneptunian objects are two high perihelion objects: Sedna and 2012 VP113. Sedna and 2012 VP113 are distant detached objects with perihelia greater than 70 AU. Their high perihelia keep them at a sufficient distance to avoid significant gravitational perturbations from Neptune. Previous explanations for the high perihelion of Sedna include a close encounter with an unknown planet on a distant orbit and a distant encounter with random star or a member of the Sun's birth cluster that passed near the Solar System.

A companion paper by Brown and Batygin extended the cluster to include a seventh object, 2000 CR_{105}, with a semi-major axis of 227 AU; and noted the absence of objects with perihelia beyond 42 AU and semi-major axes between 100 AU and 200 AU.

Extreme trans-Neptunian objects

Since early 2016, seven more extreme trans-Neptunian objects have been discovered with orbits that have a perihelion greater than 30 AU and a semi-major axis greater than 250 AU. These are also included in the orbital diagrams and tables below.

The Extreme Trans-Neptunian object orbits

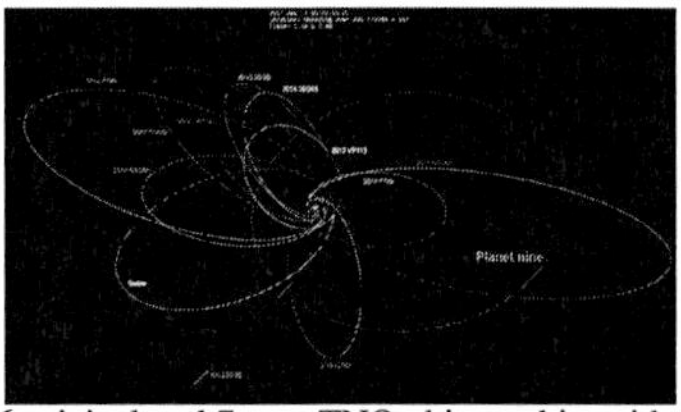

6 original and 7 new TNO object orbits with current positions near their perihelion in purple, with hypothetical Planet Nine orbit in green

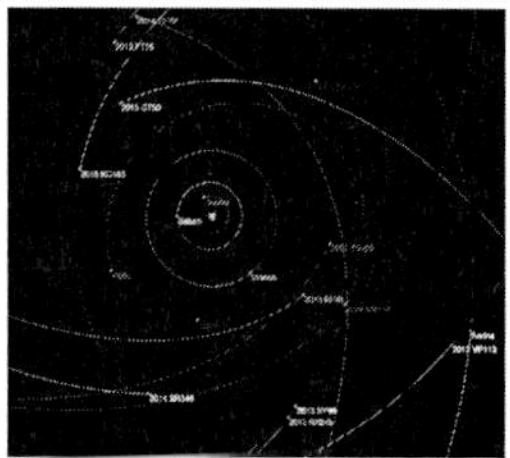

Closer up view of the 13 TNO current positions

Extreme Trans-Neptunian objects with perihelion greater than 30 AU and a semi-major axis greater than 250 AUWikipedia:Verifiability

Object	Orbit							orbital plane			Body		
	Barycentric[632] yet an epoch 2017 heliocentric period of ~10,400 years. The barycentric solution is a much more stabe ~11,300 years.</ref> Orbital period (years)	Barycentric Semimaj. (AU)	Peri. (AU)	Barycentric Aphel. (AU)	Curr. dist. from Sun (AU)	Ecc.	Arg. peri ω (°)	incl. i (°)	Long. asc □ or Ω (°)	Long. peri ϖ=ω +Ω (°)	Hv	Current mag.	Diam. (km)
2012 VP_{113}	4,300	266	80.27	441	83.5	0.69	292.8	24.1	90.8	23.6	4.0	23.3	600
2013 RF_{98}	6,900	364	36.10	690	36.8	0.90	311.8	29.6	67.6	19.4	8.7	24.4	70
2004 VN_{112}	5,900	327	47.32	607	47.7	0.85	327.1	25.6	66.0	33.1	6.5	23.3	200
2007 TG_{422}	11,300	503	35.57	970	37.3	0.93	285.7	18.6	112.9	38.6	6.2	22.0	200
90377 Sedna	11,400	507	76.04	936	85.5	0.85	311.5	11.9	144.5	96.0	1.5	20.9	1,000
2010 GB_{174}	6,600	351	48.76	654	71.2	0.87	347.8	21.5	130.6	118.4	6.5	25.1	200
2013 FT_{28}	5,050	295	43.60	546	57.0	0.86	40.2	17.3	217.8	258.0 (*)	6.7	24.4	200
2014 FE_{72}	58,000	1,500	36.31	2,960	61.5	0.98	134.4	20.6	336.8	111.2	6.1	24.0	200
2014 SR_{349}	5,160	299	47.57	549	56.3	0.84	341.4	18.0	34.8	16.2	6.6	24.2	200
2013 SY_{99}	19,700	730	49.91	1,410	60.3	0.93	32.4	4.2	29.5	61.9	6.7	24.5	250

2015 GT_{50}	5510	310	38.45	580	41.7	0.89	129.2	8.8	46.1	175.3 (*)	8.5	24.9	80
2015 RX_{245}	8,920	430	45.48	815	61.4	0.89	65.4	12.2	8.6	74.0	6.2	24.2	250
2015 KG_{163}	17,730	680	40.51	1,320	40.8	0.95	32.0	14.0	219.1	251.1 (*)	8.1	24.3	100
Ideal range for ETNOs under the hypothesis		>250	>30			>0.5		10~30		2~120			
Hypoth-esized Planet Nine	**~15,000**	**~700**	**~200**	**~1,200**	**~1,000?**	**~0.6**	**~150**	**~30**	**91±15**	**241±15**		**>22**	**~40,000**

(*) longitude of perihelion, ϖ, outside expected range

Numerical simulations of extreme TNOs

The clustering of the orbits of extreme transeptunian objects and raising of their perihelia is reproduced in simulations that include Planet Nine. In simulations conducted by Batygin and Brown swarms of planetesimals that began with large semi-major axis orbits at random orientations and initially interacting with Neptune were were sculpted into collinear groups of spatially confined orbits by a distant planet if it is substantially more massive than Earth and on a highly eccentric orbit. These surviving objects were in orbits that were oriented with their long axes anti-aligned with respect to the massive planet and were roughly co-planar with it. The objects were were also found to be in resonance with the massive planet. The resonances included high-order resonances, for example 27:17, and were interconnected, yielding an orbital evolution that was fundamentally chaotic, causing their semi-major axes to vary unpredictably on million-year timescales. The perihelia of these objects were also raised temporarily, producing Sedna-like orbits, before being returned to orbits more typical of typical trans-neptunian object after several hundred million years. Some of the eTNOs also evolved onto orbits perpendicular to the plane of the Solar System.

Batygin and Brown found that the distribution of the orbits of known extreme transeptunian objects is best reproduced in simulations using a 10 $M_{\oplus}$[633] planet in the following orbit:

- semi-major axis $a \approx 700$ AU, orbital period 18,000 years
- eccentricity $e \approx 0.6$, perihelion ≈ 280 AU, aphelion $\approx 1{,}120$ AU
- inclination $i \approx 30°$ to the ecliptic
- longitude of the ascending node $\Omega \approx 94°$.[634]
- argument of perihelion $\omega \approx 139°$ and longitude of perihelion $\varpi = 235° \pm 12°$

This orbit results in strong anti-alignment beyond 250 AU, weak alignment between 150 AU and 250 AU, and little effect inside 150 AU. Anti-alignment occurs with variable success using a semi-major axis between 400 AU and 1500 AU and an eccentricity between 0.5 and 0.8. Anti-alignment weakens as Planet Nine's inclination is increased.

Numerical simulations including a distant massive planet were also conducted by other groups. Lawler etal found that the dynamical effect of Planet Nine would generate a cloud of objects centered around Planet Nine's semi-major axis, with most at greater semi-major axes. A massive planet in a circular orbit at 250 AU increased the number of high perihelion objects threefold; one at 500 AU with and eccentrictiy of 0.5 increased these tenfold, and left a significant fraction with inclinations greater than 60°. Simulations conducted by Nesvorny etal of the giant planet migration described in the Nice model that included Planet Nine in its nominal orbit yield similar results, with roughly 0.3–0.4 Earth masses of an initial 20 Earth mass planetesimal disk remaining in what they called the Planet Nine cloud at the end of a simulation of the last 4 billion years.

Secular and resonant dynamics of extreme TNOs

The anti-alignment and the raising of the perihelia of extreme trans-neptunian objects with semi-major axes greater than 250 AU is produced by the secular effects of Planet Nine. Without a distant massive planet and beyond the influence of Neptune the perihelia of eTNOs would precess at rates that varied with their semi-major axes and eccentricities. Their longitudes of perihelion would circulate, passing through all values ranging from 0 to 360, and any initial alignment would be smeared out by their varied precession rates. The distance of their perihlia (and their eccentricities) in contrast would remain unchanged. The secular effects of a distant massive planet alters this behavior, causing variations in the perihelia with time and in some cases causes the longitudes of perihelion to librate or oscillate within a limited range of values. For an object with an initial perihelion near Neptune's orbit and anti-aligned with Planet Nine its perihelia can increase as it precesses, following a curve on a plot of longitudes of perihelion vs eccentricity. Those objects that are stable and anti-aligned follow closed curves reaching lower eccentricitics and high perihelia before returning to their initial orbits. The orbits of objects that are unstable, for example some that are initially aligned with Planet Nine, precess until parts of the orbits are tangent to that of Planet Nine leading to close encounters. Some stable orbits do exist for aligned objects with larger eccentricities. Objects in these orbits have yet to be observed, however, and an additional perturbation would have been required to be captured in these orbits.

The long term stability of anti-aligned extreme trans-neptunian objects with orbits that intersect that of Planet Nine is the result of their resonant dynamics. Objects in resonance with a massive planet are phase protected, preventing them from making close approaches to the planet. This protection is due to the perturbations by the massive planet, which slowly alter the orbit of a resonant object if it drifts out of alignment. In a system where Planet Nine and the eTNOs orbits orbit in the plane of the other planets the boundaries of the resonances are well defined, allowing eTNOs that are captured in strong resonances with Planet Nine to remain in them for the life of the Solar System. At large semi-major axes, beyond a 3:1 resonance with Planet Nine, most are in anti-aligned orbits; at smaller semi-major axes an increasing fraction of objects are in circulating orbits. For a system where the orbits of Planet Nine and the objects are inclined the resonant dynamics is more complex. The variety of possible orientations of the orbits makes the boundaries of the resonances fuzzy, replacing the well defined boundaries with chaotic bands. When the secular evolution of the ETNO's orbit causes it to enter the chaotic bands the object's orbit hops between resonances, including higher order resonances, causing its semi-major axes to diffuse until its secular evolution causes it to exit the band. As a result eTNOs undergo extended periods in stable resonances interspersed with periods of chaotic diffusion of their semi-major axes.

The bending of the Laplace plane of the Solar System toward that of Planet Nine's orbit can produce the apparent clustering of the longitude of the ascending nodes and arguments of perihelion of the extreme TNOs. The Laplace plane defines the center about which the pole of an object's orbit precesses with time. At larger semi-major the angular momentum of planet Nine causes the Laplace plane to bend toward that of its orbit. As a result, when the poles of the eTNO orbit precess about the Laplace plane, they tend to remain on one side of the ecliptic pole, producing clustering of their longitudes of ascending nodes. For objects with small inclination relative to Planet Nine, which were found to be more stable in simulations, this creates an apparent libration of their ascending nodes with respect to the ecliptic. In combination with the anti-alignment of the longitudes of perihelion this can also produce clustering of the arguments of perihelion.

Objects with perpendicular orbits

The secular effects of Planet Nine can deliver extreme trans-neptunian objects onto orbits roughly perpendicular to the plane of the Solar System. The high inclination orbits are generated by a high order secular resonance involving a linear combination of the orbits arguments and longitudes of perihelion: $\Delta\varpi$ - 2ω. Objects that reach low eccentricities during their secular evolution can enter this resonance. The resonance causes their eccentricities and inclinations to increase, delivering them onto perpendicular orbits with low perihelia where they are more readily observed. The orbits then evolve onto retrograde orbits with lower eccentricities after which they pass through a second phase of high eccentricity perpendicular orbits before returning to low eccentricity, low inclination orbits. Unlike the Kozai mechanism this resonance causes objects to reach their maximum eccentricities when in nearly perpendicular orbits. In simulations conducted by Batygin and Brown this evolution was relatively common, with 38% of stable objects undergoing it at least once. Sallinfest etal also observed this behavior in their study of the secular dynamics of eTNOs and noted that it caused the perihelia to fall below 30 AU for objects with semi-major axes greater than 300 AU. Six objects with semi-major axes greater than 250 AU and perihelia beyond Jupiter's orbit are currently known:

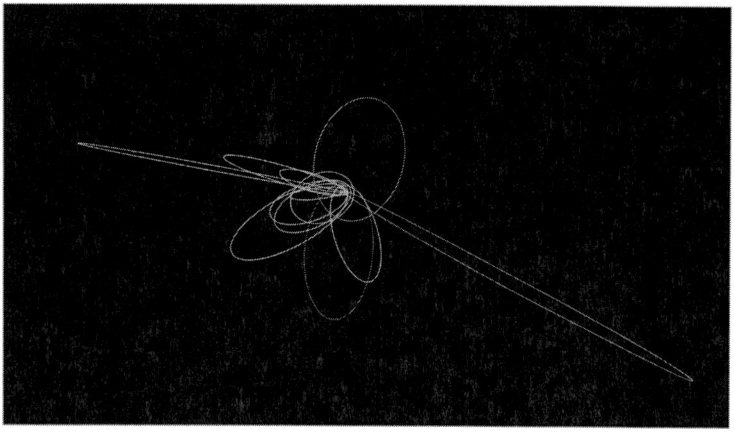

Figure 349: *The orbits of the five objects with high-inclination orbits (nearly perpendicular to the ecliptic) are shown here as cyan ellipses with the hypothetical Planet Nine in orange. Those of four are towards the left in this view, and that of one (2012 DR30) is towards the right, with an aphelion over 2,000 AU.*

High-inclination Trans-Neptunian objects with a semi-major axis greater than 250 AU

Object	Orbit						Body	
	Perihelion (AU) Figure 9	Semimaj. (AU) Figure 9	Current distance from Sun (AU)	inc (°)	Ec-cen.	Arg. peri ω (°)	Mag.	Diam. (km)
(336756) 2010 NV1	9.4	323	14	141	0.97	133	22	20–45
(418993) 2009 MS9	11.1	348	12	68	0.97	129	21	30–60
2010 BK118	6.3	484	11	144	0.99	179	21	20–50
2013 BL76	8.5	1,213	11	99	0.99	166	21.6	15–40
2012 DR30	14	1,404	17	78	0.99	195	19.6	185
2014 LM28	16.8	268		85	0.94	38	22	

High inclination TNOs

A population of high inclination trans-Neptunian objects with semi-major axes less than 100 AU may be generated by the combined effects of Planet Nine and the other giant planets. Extreme transneptunain objects that reach high inclination orbits while in the higher order secular resonance can have their perihelia lowered sufficiently for their orbits to intersect that of Neptune or the other giant planets. Encounters with one of these planets can then lower their semi-major axes to below 100 AU where their evolution would no longer be controlled by Planet Nine, leaving them on orbits like 2008 KV_{42}. The orbital distribution of the longest lived of these objects is nonuniform. Most objects have orbits with perihelia ranging from 5 AU to 35 AU and inclinations below 110 degree, beyond a gap with few objects are others with inclinations near 150 degrees and perihelia near 10 AU.

Solar obliquity

Analyses conducted contemporarily and independently by Bailey, Batygin and Brown; by Gomes, Deienno and Morbidelli; and later by Lai suggest that Planet Nine could be responsible for inducing the spin–orbit misalignment of the Solar System. The Sun's axis of rotation is tilted approximately six degrees from the orbital plane of the giant planets. The exact reason for this discrepancy remains an open question in astronomy. The analysis used computer simulations to show that both the magnitude and direction of tilt can be explained by the gravitational torques exerted by Planet Nine on the other planets over the lifetime of the Solar System. These observations are consistent with the Planet Nine hypothesis, but do not prove that Planet Nine exists, as there could be some other reason, or more than one reason, for the spin–orbit misalignment of the Solar System.

Oort cloud and comets

Numerical simulations of the migration of the giant planets show that the number of objects captured in the Oort cloud is reduced if Planet Nine was in its predicted orbit at that time. This reduction of objects captured in the Oort cloud also occurred in simulations with the giant planets on their current orbits.

The inclination distribution of Jupiter-family (or ecliptic) comets would become broader under the influence of Planet Nine. Jupiter-family comets originate primarily from the scattering objects, trans-Neptunian objects with semi-major axes that vary over time due to distant encounters with Neptune. In a model including Planet Nine, the scattering objects that reach large semi-major axes dynamically interact with Planet Nine, increasing their inclinations. As a result, the population of the scattering objects, and the population of comets derived from it, is left with a broader inclination distribution. This inclination distribution is broader than is observed, in contrast to a five-planet Nice model without a Planet Nine that can closely match the observed inclination distribution.

In a model including Planet Nine, part of the population of Halley-type comets is derived from the Planet Nine cloud. The dynamical effects of the planet drive oscillations of the perihelia of the objects in the Planet Nine cloud, delivering some of them into planet-crossing orbits. Encounters with the other planets can then alter their orbits placing them on low-perihelion orbits. The first step of this process is slow, requiring more than 100 million years, compared to comets from the Oort cloud, which can be dropped into low-perihelion orbits in one period. The Planet Nine cloud contributes roughly one-third of the total population of comets, which is similar to that without Planet Nine due to a reduced number of Oort cloud comets.

Origin

A number of possible origins for Planet Nine have been examined including its ejection from the neighborhood of the current giant planets, capture from another star, and in situ formation.

In their initial paper Batygin and Brown proposed that Planet Nine formed closer to the Sun and was ejected into a distant eccentric orbit following a close encounter with Jupiter or Saturn during the nebular epoch. Gravitational interactions with nearby stars in the Sun's birth cluster, or dynamical friction from the gaseous remnants of the solar nebula, then reduced the eccentricity of its orbit, raising its perihelion, leaving it on a very wide but stable orbit. Had it not been flung into the Solar System's farthest reaches, Planet Nine could have accreted more mass from the proto-planetary disk and developed into the core of a gas giant. Instead, its growth was halted early, leaving it with a lower mass of five times Earth's mass, similar to that of Uranus and Neptune. For Planet Nine to have been captured in a distant, stable orbit its ejection must have occurred early, between three million and ten million years after the formation of the Solar System. This timing suggests that Planet Nine is not the planet ejected in a five-planet version of the Nice model, unless that occurred too early to be the cause of the Late Heavy Bombardment, which would then require another explanation. These ejections, however, are likely to have been two events well separated in time.

Close encounters between the Sun and other stars in its birth cluster could have resulted in the capture of a planet from beyond the Solar System. Three-body interactions during these encounters can perturb the path of planets on distant orbits around another star, or free-floating planets, in a process similar to the capture of irregular satellites around the giant planets, leaving one in a stable orbit around the Sun. A planet that originated in a system with a number of Neptune-massed planets and without Jupiter-massed planets, could be scattered into a more long-lasting distant eccentric orbit, increasing its chances of capture by another star. Although this increases the odds of the Sun capturing another planet from another star, a wide variety of orbits are possible, reducing the probability of a planet being captured on an orbit like that proposed for Planet Nine to 1–2 percent. In a simulation where the capture takes place between co-planar systems a large number of other objects are also captured into orbits aligned with the planet, potentially allowing this capture scenario to be distinguished from others. The likelihood of the capture of a free-floating planet is much smaller, with only 5–10 of 10,000 simulated free-floating planets being captured on orbits similar to that proposed for Planet Nine.

A planet could also be perturbed from a distant circular orbit into an eccentric orbit by an encounter with another star. The in situ formation of a planet at this distance would require a very massive and extensive disk, or the outward drift of solids in a dissipating disk forming a narrow ring from which the planet accreted over a billion years. If a planet formed at such a great distance while the Sun was in its birth cluster, the probability of it remaining bound to the Sun in a highly eccentric orbit is roughly 10%. A previous paper found that a massive disk extending beyond 80 AU would drive Kozai oscillations of objects scattered outward by Jupiter and Saturn leaving some of them in high inclination (inc > 50°), low eccentricity orbits which have not been observed.

Ethan Siegel, who is deeply skeptical of the existence of an undiscovered new planet in the Solar System, nevertheless speculates that at least one super-Earth, which have been commonly discovered in other planetary systems but have not been discovered in the Solar System, might have been ejected from the Solar System during a dynamical instability in the early Solar System. Hal Levison thinks that the chance of an ejected object ending up in the inner Oort cloud is only about 2%, and speculates that many objects must have been thrown past the Oort cloud if one has entered a stable orbit.

Astronomers expect that the discovery of Planet Nine would aid in understanding the processes behind the formation of the Solar System and other planetary systems, as well as how unusual the Solar System is, with a lack of planets with masses between that of Earth and that of Neptune, compared to other planetary systems.

Alternate hypotheses

Temporary or coincidental nature of clustering

Cory Shankman et al. examined the consequences of a distant massive perturber on the TNOs used to infer the planet's existence. They simulated clones (objects with similar orbits) of 15 objects with semi-major axis > 150 AU and perihelion > 30 AU under the influence of a 10 Earth-massed Planet Nine in Batygin and Brown's best fit orbit.[635] While longitude of perihelion alignment of the objects with semi-major axis > 250 AU was observed in their simulations, the alignment of the arguments of perihelion was not. The simulations also revealed an increase in the inclinations of many objects, thereby predicting a larger reservoir of high-inclination TNOs.[636] These objects should have been detected in existing surveys but are still unseen so far—suggesting there is a currently missing or unseen signature of Planet Nine. A previously published paper concluded that current observations are insufficient to identify this signature, however. The perihelia of many of the objects also rose and fell smoothly, inconsistent with the current absence of extreme TNOs with perihelia between 50 AU and 70 AU. Their perihelia also reached values where the objects would not be observed and, after declining, fell low enough for the objects to enter planet-crossing orbits leading to their ejection from the Solar System. These factors would require a population of Sednas significantly larger than current estimates, and inconsistent with current models of the early Solar System, to explain current observations. Based on these challenges Shankman et al. concluded that the existence of Planet Nine is unlikely and that the currently observed alignment of the existing TNOs is a temporary phenomenon that will disappear as more objects are detected.

The results of the Outer Solar System Survey (OSSOS) suggests that the observed clustering is the result of a combination of observing bias and small number statistics. OSSOS, a well-characterized survey of the outer Solar System with known biases, observed eight trans-Neptunian objects with semi-major axis > 150 AU with orbits oriented on a wide range of directions. After accounting for the known observational biases of the survey, no evidence for the arguments of perihelion (ω) clustering identified by Trujillo and Sheppard was seen[637] and the orientation of the orbits of the objects with the largest semi-major axis was statistically consistent with random. A previously released paper by Mike Brown analyzed the discovery locations of eccentric trans-Neptunian objects. While identifying some biases he found that even with these biases the clustering of longitudes of perihelion of the known objects would be observed only 1.2% of the time if their actual distribution was uniform.

Inclination instability due to mass of undetected objects

Ann-Marie Madigan and Michael McCourt postulate that an inclination instability in a distant massive belt is responsible for the alignment of the arguments of perihelion of the ETNOs. The inclination instability occurs in a disk of particles in eccentric orbits around a massive object. The self-gravity of this disk causes its spontaneous organization, increasing the inclinations of the objects and aligning the arguments of perihelion, forming it into a cone above or below the original plane. This process requires an extended time and significant mass of the disk, on the order of a billion years for a 1–10 Earth-mass disk.[638] While an inclination instability can align the arguments of perihelion and raise perihelia, producing detached objects, it does not align the longitudes of perihelion. Mike Brown considers Planet Nine a more probable explanation, noting that current surveys do not support the existence of a scattered-disk region of sufficient mass to support this idea of "inclination instability".

Object in lower-eccentricity orbit

Renu Malhotra, Kathryn Volk, and Xianyu Wang have proposed that the four detached objects with the longest orbital periods, those with perihelia beyond 40 AU and semi-major axes greater than 250 AU, are in n:1 or n:2 mean-motion resonances with a hypothetical planet. Two other objects with semi-major axes greater than 150 AU are also potentially in resonance with this planet. Their proposed planet could be on a lower eccentricity, low inclination orbit, with eccentricity $e < 0.18$ and inclination $i \approx 11°$. The eccentricity is limited in this case by the requirement that close approaches of 2010 GB174 to the planet are avoided. If the ETNOs are in periodic orbits of the third kind, with their stability enhanced by the libration of their arguments of perihelion, the planet could be in a higher inclination orbit, with $i \approx 48°$. Unlike Batygin and Brown, Malhotra, Volk and Wang do not specify that most of the distant detached objects would have orbits anti-aligned with the massive planet.[639]

Proposed resonances of distant Trans-Neptunian objects

Body	Orbital period Heliocentric (years)	Orbital period Barycentric (years)	Semimaj. (AU)	Ratio
2013 GP136		1,830	151.8	9:1
2000 CR105		3,304	221.59±0.16	5:1
2012 VP113	4268±179	4,300	265.8±3.3	4:1
2004 VN112	5845±30	5,900	319.6±6.0	3:1
2010 GB174	7150±827	6,600	350.7±4.7	5:2
90377 Sedna		≈ 11,400	506.84±0.51	3:2

Hypothetical planet		**≈ 17,000**	**≈ 665**	**1:1**

Alignment due to the Kozai mechanism

Trujillo and Sheppard (2014)

The initial argument that the clustering of orbital elements of extreme trans-Neptunian objects such as sednoids might be caused by a massive unknown planet beyond Neptune was published in 2014 by astronomers Chad Trujillo and Scott S. Sheppard. Trujillo and Sheppard analyzed the orbits of twelve trans-Neptunian objects (TNOs) with perihelia greater than 30 AU and semi-major axes greater than 150 AU, and found they had a clustering of orbital characteristics, particularly their arguments of perihelion (which indicates the orientation of elliptical orbits within their orbital planes). In numerical simulations including only the known giant planets the arguments of perihelion of these objects circulated at varying rates which after billions of years would leave the perihelia of the twelve TNOs randomized, like in the rest of the trans-Neptunian region, unless there is something holding them in place. Trujillo and Sheppard suggested that the massive unknown planet at a few hundred astronomical units caused the arguments of perihelion of the extreme trans-Neptunian objects to librate about 0° or 180°[640,641]</ref> via the Kozai mechanism so that their orbits cross the plane of the planet's orbit near perihelion and aphelion, at the closest and farthest points from the planet. Numerical simulations with a single body of 2–15 Earth masses in a circular low-inclination orbit between 200 AU and 300 AU indicated that the arguments of perihelia of Sedna and 2012 VP_{113} would librate around 0° for billions of years (although the lower perihelion objects did not) and would undergo periods of libration with a Neptune mass object in a high inclination orbit at 1500 AU.

These simulations showed the basic idea of how a single large planet can shepherd the smaller extreme trans-Neptunian objects into similar types of orbits. It was a basic proof of concept simulation that did not obtain a unique orbit for the planet as they state there are many possible orbital configurations the planet could have. Thus they did not fully formulate a model that successfully incorporated all the clustering of the extreme objects with an orbit for the planet. But they were the first to notice there was a clustering in the orbits of extremely distant objects and that the most likely reason was from an unknown massive distant planet. Their work is very similar to how Alexis Bouvard noticed Uranus' motion was peculiar and suggested that it was likely gravitational forces from an unknown 8th planet, which led to the discovery of Neptune.Wikipedia:No original research

de la Fuente Marcos et al. (2014)

In June 2014, Raúl and Carlos de la Fuente Marcos included a thirteenth minor planet and noted that all have their argument of perihelion close to 0°. In a further analysis, Carlos and Raúl de la Fuente Marcos with Sverre J. Aarseth confirmed that the only known way that the observed alignment of the arguments of perihelion can be explained is by an undetected planet. They also theorized that a set of extreme trans-Neptunian objects (ETNOs) are kept bunched together by a Kozai mechanism similar to the one between Comet 96P/Machholz and Jupiter. They speculated that it would have a mass between that of Mars and Saturn and would orbit at some 200 AU from the Sun. However, they also struggled to explain the orbital alignment using a model with only one unknown planet.[642] They therefore suggested that this planet is itself in resonance with a more-massive world about 250 AU from the Sun, just like the one predicted in the work by Trujillo and Sheppard. They also did not rule out the possibility that the planet could have to be much farther away but much more massive in order to have the same effect and admitted the hypothesis needed more work. They also did not rule out other explanations and expected more clarity as researchers study orbits of more such distant objects. A later analysis of the distributions of the directions of perihelia and orbital poles of the ETNOs also suggests that one additional planet may not be sufficient to explain the observed clustering.

Previous models with additional planets

Attempts to detect planets beyond Neptune by indirect means such as orbital perturbation date back beyond the discovery of Pluto. A few observations were directly related to the Planet Nine hypothesis:

- George Forbes was the first to postulate the existence of trans-Neptunian planets in 1880, and his work is considered similar to more recent Planet Nine theories. In Forbes's model the planet had a semi-major axis of ~300 AU (roughly three hundred times the distance from Earth to the Sun); locations were based on clustering of the aphelion distances of periodic comets.
- The discovery of Sedna with its peculiar orbit in 2004 led to the conclusion that something beyond the known eight planets had perturbed Sedna away from the Kuiper belt. That object could have been an unknown planet on a distant orbit, a random star that passed near the Solar System, or a member of the Sun's birth cluster. The announcement in March 2014 of the discovery of a second sednoid, 2012 VP113, which shared some orbital characteristics with Sedna and other extreme trans-Neptunian objects, further raised the possibility of an unseen super-Earth in a large orbit.

- In 2008 Tadashi Mukai and Patryk Sofia Lykawka suggested that a distant Mars- or Earth-sized minor planet currently in a highly eccentric orbit between 100 and 200 AU and orbital period of 1000 years with an inclination of 20° to 40° was responsible for the structure of the Kuiper belt. They proposed that the perturbations of this planet excited the eccentricities and inclinations of the trans-Neptunian objects, truncated the planetesimal disk at 48 AU, and detached the orbits of objects like Sedna from Neptune. During Neptune's migration this planet is posited to have been captured in an outer resonance of Neptune and to have evolved into a higher perihelion orbit due to the Kozai mechanism leaving the remaining trans-Neptunian objects on stable orbits.
- In 2012, after analysing the orbits of a group of trans-Neptunian objects with highly elongated orbits, Rodney Gomes of the National Observatory of Brazil proposed that their orbits were due to the existence of an as yet undetected planet. This Neptune-massed planet would be on a distant orbit that would be too far away to influence the motions of the inner planets, yet close enough to cause the perihelia of scattered disc objects with semi-major axes greater than 300 AU to oscillate, delivering them into planet-crossing orbits similar to those of (308933) 2006 SQ372 and (87269) 2000 OO67 or detached orbits like that of Sedna. Alternatively the unusual orbits of these objects could be the result of a Mars-massed planet on an eccentric orbit that occasionally approached within 33 AU. Gomes argued that a new planet was the more probable of the possible explanations but others felt that he could not show real evidence that suggested a new planet. Later in 2015, Rodney Gomes, Jean Soares, and Ramon Brasser proposed that a distant planet was responsible for an excess of centaurs with large semi-major axes.

Searches for Planet Nine

Visibility and location

Due to its extreme distance from the Sun, Planet Nine would reflect little sunlight, potentially evading telescope sightings. It is expected to have an apparent magnitude fainter than 22, making it at least 600 times fainter than Pluto.[643] which is about 100 times dimmer than Planet Nine is expected to be. For comparison, the Hubble Space Telescope has detected objects as faint as 31st magnitude with an exposure of about 2 million seconds (555 hours) during Hubble Ultra Deep Field photography. However, Hubble's field of view is very narrow, as is the Keck Observatory Large Binocular Telescope. Brown hopes to make a request for use of the *Hubble Space Telescope* the day the planet is spotted.</ref> If Planet Nine exists and is close to its perihelion, astronomers could identify it based on existing images. For its aphelion, the largest telescopes would be required. However, if the planet is currently located in between, many observatories could spot Planet Nine. Statistically, the planet is more likely to be closer to its aphelion at a distance greater than 500 AU. This is because objects move more slowly when near their aphelion, in accordance with Kepler's second law.

Searches of existing data

The search in databases of stellar objects performed by Batygin and Brown has already excluded much of the sky the predicted planet could be in, save the direction of its aphelion, or in the difficult to spot backgrounds where the orbit crosses the plane of the Milky Way, where most stars lie. This search included the archival data from the Catalina Sky Survey to magnitude c. 19, Pan-STARRS to magnitude 21.5, and infrared data from WISE.

David Gerdes who helped develop the camera used in the Dark Energy Survey claims that it is quite possible that one of the images taken for his galaxy map may actually contain a picture of Planet Nine, and if so, new software developed recently and used to identify objects such as 2014 UZ224 can help to find it.

Michael Medford and Danny Goldstein, graduate students at the University of California, Berkeley, are also examining archived data using a technique that combines multiple images, taken at different times. Using a supercomputer they will offset the images to account for the calculated motion of Planet Nine, allowing many faint images of a faint moving object to be combined to produce a brighter image.

A search combining multiple images collected by WISE and NEOWISE data has also been conducted. The initial search which covered the region of the sky identified by Holman and Payne using Cassini range data is being expanded to cover other regions away from the galactic plane.

Ongoing searches

Because the planet is predicted to be visible in the Northern Hemisphere, the primary search is expected to be carried out using the Subaru Telescope, which has both an aperture large enough to see faint objects and a wide field of view to shorten the search. Two teams of astronomers—Batygin and Brown, as well as Trujillo and Sheppard—are undertaking this search together, and both teams cooperatively expect the search to take up to five years. Brown and Batygin initially narrowed the search for Planet Nine down to roughly 2,000 square degrees of sky near Orion, a swath of space, that in Batygin's opinion, could be covered in about 20 nights by the Subaru Telescope. Subsequent refinements by Batygin and Brown have reduced the search space to 600–800 square degrees of sky.

A zone around the constellation Cetus, where *Cassini* data suggest Planet Nine may be located, is being searched as of 2016[644] by the Dark Energy Survey—a project in the Southern Hemisphere designed to probe the acceleration of the Universe. DES observes about 105 nights per season, lasting from August to February.

Radiation

Although a distant planet such as Planet Nine would reflect little light, it would still be radiating the heat from its formation as it cools due to its large mass. At its estimated temperature of 47 K, the peak of its emissions would be at infrared wavelengths. This radiation signature could be detected by Earth-based infrared telescopes, such as ALMA, and a search could be conducted by cosmic microwave background experiments operating at mm wavelengths.[645]</ref> Additionally, Jim Green of NASA is optimistic that it could be observed by the *James Webb Space Telescope*, the successor to the *Hubble Space Telescope*, that is expected to be launched in 2018.

Citizen science

Zooniverse Backyard Worlds: Planet 9 project

The Zooniverse Backyard Worlds project, started in February 2017, is using archival data from the WISE spacecraft to search for Planet Nine. The project will additionally search for substellar objects like brown dwarfs in the neighborhood of the Solar System.

Zooniverse SkyMapper project

In April 2017, using data from the SkyMapper telescope at Siding Spring Observatory, citizen scientists on the Zooniverse platform reported four candidates for Planet Nine. These candidates will be followed up on by astronomers to determine their viability. The project, which started on 28 March, completed their goals in less than three days with around five million classifications by more than 60,000 individuals.

Efforts toward indirect detection

Cassini measurements of perturbations of Saturn

An analysis of *Cassini* data on Saturn's orbital residuals was inconsistent with Planet Nine being located with a true anomaly of –130° to –110° or –65° to 85°. The analysis, using Batygin and Brown's orbital parameters for Planet Nine, suggests that the lack of perturbations to Saturn's orbit is best explained if Planet Nine is located at a true anomaly of 117.8°+11°
–10°. At this location, Planet Nine would be approximately 630 AU from the Sun, with right ascension close to 2^h and declination close to –20°, in Cetus. In contrast, if the putative planet is near aphelion it could be moving projected towards the area of the sky with boundaries: right ascension 3.0^h to 5.5^h and declination –1° to 6°.

An improved mathematical analysis of *Cassini* data by astrophysicists Matthew Holman and Matthew Payne tightened the constraints on possible locations of Planet Nine. Holman and Payne developed a more efficient model that allowed them to explore a broader range of parameters than the previous analysis. The parameters identified using this technique to analyze the Cassini data was then intersected with Batygin and Brown's dynamical constraints on Planet Nine's orbit. Holman and Payne concluded that Planet Nine is most likely to be located within 20° of RA = 40°, Dec = –15°, in an area of the sky near the constellation Cetus.

The Jet Propulsion Laboratory has stated that according to their mission managers and orbit determination experts, the *Cassini* spacecraft is not experiencing unexplained deviations in its orbit around Saturn. William Folkner, a planetary scientist at JPL stated, "An undiscovered planet outside the orbit of Neptune, 10 times the mass of Earth, would affect the orbit of Saturn, not *Cassini* ... This could produce a signature in the measurements of *Cassini* while in orbit about Saturn if the planet was close enough to the sun. But we do not see any unexplained signature above the level of the measurement noise in *Cassini* data taken from 2004 to 2016." Observations of Saturn's orbit neither prove nor disprove that Planet Nine exists. Rather, they suggest that Planet Nine could not be in certain sections of its proposed orbit because its gravity would cause a noticeable effect on Saturn's position, inconsistent with actual observations.

Analysis of Pluto's orbit

An analysis of Pluto's orbit by Matthew J. Holman and Matthew J. Payne found perturbations much larger than predicted by Batygin and Brown's proposed orbit for Planet Nine. Holman and Payne suggested three possible explanations: systematic errors in the measurements of Pluto's orbit; an unmodeled mass in the Solar System, such as a small planet in the range of 60–100 AU (potentially explaining the Kuiper cliff); or a planet more massive or closer to the Sun instead of the planet predicted by Batygin and Brown.

Optimal orbit if objects are in strong resonances

An analysis by Sarah Millholland and Gregory Laughlin indicates that the commensurabilities (period ratios consistent with pairs of objects in resonance with each other) of the extreme TNOs are most likely to occur if Planet Nine has a semi-major axis of 654 AU. They used 11 then-known extreme TNOs with their semimajor axis over 200, and perihelion over 30 AU[646], with five bodies close to four simple ratios (5:1, 4:1, 3:1, 3:2) with a 654 AU distance: 2002 GB_{32}, 2000 CR_{105} (5:1), 2001 FP_{185} (5:1), 2012 VP_{113} (4:1), 2014 SR_{349}, 2013 FT_{28}, 2004 VN_{112} (3:1), 2013 RF_{98}, 2010 GB_{174}, 2007 TG_{422}, and (90377) Sedna (3:2). Beginning with this semi-major axis they determine that Planet Nine best maintains the anti-alignment of their orbits and a strong clustering of arguments of perihelion if it is near aphelion and has an eccentricity $e \approx 0.5$, inclination $i \approx 30°$, argument of perihelion $\omega \approx 150°$, and longitude of ascending node $\Omega \approx 50°$ (the last differs from Brown and Batygin's value of 90°).[647]</ref> The favored location of Planet Nine is a right ascension of 30° to 50° and a declination of –20° to 20°. They also note that in their simulations the clustering of arguments of perihelion is almost always smaller than has been observed.

A previous analysis by Carlos and Raul de la Fuente Marcos of commensurabilities among the known ETNOs using Monte Carlo techniques revealed a pattern similar to that of the Kuiper belt, where accidental commensurabilities occur due to objects in resonances with Neptune. They find that this pattern would be best explained if the ETNOs were in resonance with an additional planetary-sized object beyond Pluto and note that a number of these objects may be in 5:3 and 3:1 resonances if that object had semi-major axis of $\approx$700 AU.

Dynamical stability of distant objects

Juliette Becker and colleagues, noting the orbits must remain stable for an extended period for anti-alignment to occur, analyzed the stability of eight extreme trans-Neptunian objects under the influence of Planet Nine and Neptune. They found that many of these objects would remain stable, with semi-major axes varying by less than 100 AU, if Planet Nine's orbital elements fell in a broad region with semi-major axes ranging from 500 to 1200 AU and eccentricities ranging from 0.3 to 0.6 with lower eccentricities being favored at smaller semi-major axes. A Planet Nine with a semi-major axis of 700 AU and eccentricity of 0.4 provided the greatest stability, which contrasts with Batygin and Brown's best-fit orbit with an eccentricity of 0.6. They also determined that anti-alignment became more likely if Planet Nine had a higher eccentricity, with some overlap between favored regions of stability and anti-alignment indicating that the objects may be on the borders of instability. Some objects like 2007 TG422 and 2013 RF98, which were unstable under the influence of Neptune alone, became more stable when Planet Nine was added, although their semi-major axes tended to migrate. A number of objects spent significant periods near a resonance with Planet Nine with some hopping between resonances.

An analysis by Carlos and Raul de la Fuente Marcos revealed that if Planet Nine were on its nominal orbit and in the direction favored by the analysis of Cassini data, some objects like Sedna would remain on stable orbits while the orbits of others would become unstable leading to their ejection from the Solar System. Other possible orbits, however, could leave all six of the objects identified by Batygin and Brown on stable orbits that maintain anti-alignment.

Correlation of arguments and longitudes of perihelion

Trujillo and Sheppard in a paper announcing the discovery of several more distant objects noted a correlation between the longitude of perihelion and the argument of perihelion of these objects. Those with a longitude of perihelion of 0–120° have arguments of perihelion between 280–360°, and those with longitude of perihelion of 180–340° have argument of perihelion 0–40°. The statistical significance of this correlation was 99.99%. They suggest that the correlation is due to the orbits of these objects avoiding close approaches to Planet Nine by passing above or below its orbit. Trujillo and Sheppard also noted that the arguments of perihelion of ETNOs with perihelion less than 35 AU are opposite those with perihelion greater than 35 AU.[648]

Ascending nodes of large semi-major axis objects

In a paper by Carlos and Raul de la Fuente Marcos evidence is shown for a possible bimodal distribution of the nodal distances of the ETNOs. This correlation is unlikely to be the result of observational bias since it also appears in the nodal distribution of large semi-major axis centaurs and comets. If it is due to the extreme TNOs experiencing close approaches to Planet Nine, it is consistent with a planet with a semi-major axis of 300–400 AU.

Secular dynamics of extreme TNOs

Hervé Beust numerically calculated the Hamiltonian describing the secular dynamics of objects perturbed by Planet Nine. Plots of eccentricity vs. longitude of perihelion using these results formed closed curves, or libration islands for aligned and anti-aligned objects. These resemble the plots of Batygin and Brown's original paper that showed the evolution of the orbital elements of ETNOs in simulations under the influence of Planet Nine. Beust also produced similar plots for objects in resonance with a Planet Nine at a semi-major axis of 665 AU, for example Sedna in a 3:2 resonance, as proposed by Malhotra, Volk and Wang. The libration islands in some of these cases included locations other than alignment or anti-alignment. Beust notes that while the phase protection of resonant objects provides additional protection, bodies in the anti-aligned population need not be in resonance with Planet Nine to remain in long-term stable orbits.

Saillenfest, Fouchard, Tommei, and Valsecchi expanded on Beust's work, examining secular dynamics for the non-planar case including the effects of the eccentric Kozai mechanism,[649]</ref> that can result in the flipping of orbits from prograde to retrograde. For a Planet Nine in the plane of the other planets they found that the aligned and anti-aligned configurations identified by Beust extended to inclined objects with semi-major axis under 300 AU. They also identified stable configurations where inclinations oscillated around 90°, that may be related to perpendicular objects observed in simulations by Batygin and Brown, for semi-major axes of 200 to 300 AU. At higher semi-major axes the orbits of these perpendicular objects crossed those of the other planets. Beyond 300 AU the appearance of alignment and anti-alignment was more the result of sticky chaos rather than confinement: orbits wandered everywhere but spent more time in regions of relative stability associated with secular resonances. With an inclined Planet Nine its influence extended as low as 70 AU, generating some orbital chaos, and beyond 150 AU was strong enough to potentially generate flips from prograde to retrograde orbits. Unlike Beust their analysis did not extend to mean-motion resonances. In a previous paper they found that objects objects in resonance with Neptune with semi-major axes as large as 350 AU could undergo significant increases in their perihelia and inclinations due to the Kozai mechanism if there is no Planet Nine. Trujillo and Sheppard noted that in simulations 2013 FT_{28} had significant variations in its orbit and suggested this was due to interactions with Neptune's resonances.

Orbits of nearly parabolic comets

An analysis of the orbits of comets with nearly parabolic orbits identifies five new comets with hyperbolic orbits that approach the nominal orbit of Planet Nine described in Batygin and Brown's initial paper. If these orbits are hyperbolic due to close encounters with Planet Nine the analysis estimates that Planet Nine is currently near aphelion with a right ascension of 83°–90° and a declination of 8°–10°. Scott Sheppard, who is skeptical of this analysis, notes that many different forces influence the orbits of comets.

Possible disrupted binary

Similarities between the orbits of 2013 RF_{98} and (474640) 2004 VN_{112} have led to the suggestion that they were a binary object disrupted near aphelion during an encounter with a distant object. The visible spectra of (474640) 2004 VN_{112} and 2013 RF_{98} are also similar but very different from that of 90377 Sedna. The value of their spectral slopes suggests that the surfaces of (474640) 2004 VN_{112} and 2013 RF_{98} can have pure methane ices (like in the case of Pluto) and highly processed carbons, including some amorphous silicates. The disruption of a binary would require a relatively close encounter with Planet Nine, however, which becomes less likely at large distances from the Sun.

Search for additional extreme trans-Neptunian objects

Finding more objects would allow astronomers to make more accurate predictions about the orbit of the hypothesized planet. The Large Synoptic Survey Telescope, when it is completed in 2023, will be able to map the entire sky in just a few nights, providing more data on distant Kuiper belt objects that could both bolster evidence for Planet Nine and help pinpoint its current location.

New extreme trans-Neptunian objects discovered by Trujillo and Sheppard include:

- 2013 FT_{28}, located on the opposite side of the sky (Longitude of perihelion aligned with Planet Nine) – but well within the proposed orbit of Planet Nine, where computer modeling suggests it would be safe from gravitational kicks.
- 2014 SR_{349}, falling right in line with the earlier six objects.

- 2014 FE_{72}, an object with an orbit so extreme that it reaches about 3000 AU from the Sun in a massively-elongated ellipse – at this distance its orbit is influenced by the galactic tide and other stars.

Other new extreme trans-Neptunian objects discovered by the Outer Solar System Origins Survey include:

- 2013 SY_{99}, which has a lower inclination than many of the objects, and which was discussed by Michele Bannister[650] at a March 2016 lecture hosted by the SETI Institute and later at an October 2016 AAS conference.
- 2015 KG_{163}, which has an orientation similar to 2013 FT_{28} but has a larger semi-major axis that may result in its orbit crossing Planet Nine's.
- 2015 RX_{245}, which fits with the other anti-aligned objects.
- 2015 GT_{50}, which is in neither the anti-aligned nor the aligned groups; instead, its orbit's orientation is at a right angle to that of the proposed Planet Nine. Its argument of perihelion is also outside the cluster of arguments of perihelion.

Batygin and Brown also predict a yet-to-be-discovered population of distant objects. These objects would have semi-major axes greater than 250 AU, but they would have lower eccentricities and orbits that would be aligned with that of Planet Nine. The larger perihelia of these objects would make them fainter and more difficult to detect than the anti-aligned objects.

Commentary

Batygin was cautious in interpreting the results of the simulation developed for his and Brown's paper, saying, "Until Planet Nine is caught on camera it does not count as being real. All we have now is an echo." Batygin's and Brown's work is similar to how Urbain Le Verrier predicted the position of Neptune based on Alexis Bouvard's observations and theory of Uranus' peculiar motion.

Brown put the odds for the existence of Planet Nine at about 90%. Greg Laughlin, one of the few researchers who knew in advance about this paper, gives an estimate of 68.3%. Other skeptical scientists demand more data in terms of additional KBOs to be analysed or final evidence through photographic confirmation. Brown, though conceding the skeptics' point, still thinks that there is enough data to mount a search for a new planet.

Brown is supported by Jim Green, director of NASA's Planetary Science Division, who said that "the evidence is stronger now than it's ever been before".

Tom Levenson concluded that, for now, Planet Nine seems the only satisfactory explanation for everything now known about the outer regions of the Solar

System. Alessandro Morbidelli, who reviewed the paper for *The Astronomical Journal*, concurred, saying, "I don't see any alternative explanation to that offered by Batygin and Brown."

Malhotra remains agnostic about Planet Nine, but noted that she and her colleagues have found that the orbits of extremely distant KBOs seem tilted in a way that is difficult to otherwise explain. "The amount of warp we see is just crazy," she says. "To me, it's the most intriguing evidence for Planet Nine I've run across so far."

Further reading

- Calandrelli, Emily; Escher, Anna (16 December 2016). "The top 15 events that happened in space in 2016"[651]. TechCrunch. Astronomers find evidence for Planet 9. Retrieved 2016-12-16.

External links

	Wikimedia Commons has media related to ***Planet Nine***.

- A New Planet in our Solar System? NASA Takes a Look[652] (*NASA* video, 21 January 2016)
- A new 9th planet for the solar system?[653] (*Science Magazine* video, 20 January 2016)
- The Search for Planet Nine[654] – blog by study authors
- A summary of the history behind the search and claims[655] for a ninth planet
- Could You Live on Planet Nine?[656] – Science article by Rhett Allain
- 'Planet Nine': Facts About the Mysterious Solar System World[657] (space.com infographic)
- Planet Nine May Help Us Slingshot Our Way to the Stars[658]
- Is Planet Nine truly "discovered"?[659] A brief history of discoveries before and leading-up to the predicted planet
- Solar System Survey Casts Doubt on Mysterious "Planet Nine"[660] *Nature* 23 June 2017
- Is There a Giant Planet Lurking Beyond Pluto? A race is on to discover Planet Nine using classical astronomy and new computational techniques[661] *IEEE Spectrum* 31 July 2017 – a good summary as of publication date

Appendix

References

[1]

[2]Comet Space Missions http://spider.seds.org/spider/Comets/c_missions.html, accessed 2007-10-23.

[3]//en.wikipedia.org/w/index.php?title=Template:SolarSummaryLarge&action=edit

[4]An astronomical unit, or AU, is the average distance between the Earth and the Sun, or about 150 million kilometres. It is the standard unit of measurement for interplanetary distances.

[5]Zeilik & Gregory 1998, p. 207.

[6]The combined mass of Jupiter, Saturn, Uranus and Neptune is 445.6 Earth masses. The mass of remaining material is ~5.26 Earth masses or 1.1% (see Solar System#Notes and List of Solar System objects by mass)

[7]The reason that Saturn, Uranus and Neptune all moved outward whereas Jupiter moved inward is that Jupiter is massive enough to eject planetesimals from the Solar System, while the other three outer planets are not. To eject an object from the Solar System, Jupiter transfers energy to it, and so loses some of its own orbital energy and moves inwards. When Neptune, Uranus and Saturn perturb planetesimals outwards, those planetesimals end up in highly eccentric but still bound orbits, and so can return to the perturbing planet and possibly return its lost energy. On the other hand, when Neptune, Uranus and Saturn perturb objects inwards, those planets gain energy by doing so and therefore move outwards. More importantly, an object being perturbed inwards stands a greater chance of encountering Jupiter and being ejected from the Solar System, in which case the energy gains of Neptune, Uranus and Saturn obtained from their inwards deflections of the ejected object become permanent.

[8]

See also

[9]Zeilik & Gregory 1998, pp. 118–120.

[10]In all of these cases of transfer of angular momentum and energy, the angular momentum of the two-body system is conserved. In contrast, the summed energy of the moon's revolution plus the primary's rotation is not conserved, but decreases over time, due to dissipation via frictional heat generated by the movement of the tidal bulge through the body of the primary. If the primary were a frictionless ideal fluid, the tidal bulge would be centered under the satellite, and no transfer would take place. It is the loss of dynamical energy through friction that makes transfer of angular momentum possible.

[11]Duncan & Lissauer 1997.

[12]Zeilik & Gregory 1998, p. 320–321.

[13]Zeilik & Gregory 1998, p. 322.

[14]http://adsabs.harvard.edu/abs/1997Icar..125....1D

[15]//doi.org/10.1006%2Ficar.1996.5568

[16]http://arquivo.pt/wayback/20160520023943/http://media.skyandtelescope.com/video/Solar_System_Sim.mov

[17]http://www.skyandtelescope.com

[18]http://www.cfa.harvard.edu/seuforum/animations/animations/galaxycollision.mpg

[19]http://www.space.com/common/media/video/player.php?videoRef=mm32_SunDeath

[20] Citing

[21]All numbers in this article are short scale. One billion is 10^9, or 1,000,000,000.

[22]Hydrothermal vent communities live so deep under the sea that they have no access to sunlight. Bacteria instead use sulfur compounds as an energy source, via chemosynthesis.

[23]Solar radiation http://curry.eas.gatech.edu/Courses/6140/ency/Chapter3/Ency_Atmos/Radiation_Solar.pdf

[24]1.88 Gcd/m^2 is calculated from the solar illuminance of 128000 lux (see sunlight) times the square of the distance to the center of the Sun, divided by the cross sectional area of the Sun. 1.44 Gcd/m^2 is calculated using 98000 lux.

[25] Iben, I Jnr (1965) "Stellar Evolution. II. The Evolution of a 3 M_{sun} Star from the Main Sequence Through Core Helium Burning". (Astrophysical Journal, vol. 142, p.1447)
[26] Ross and Aller 1976, Withbroe 1976, Hauge and Engvold 1977, cited in Biemont 1978.
[27] Corliss and Bozman (1962 cited in Biemont 1978) and Warner (1967 cited in Biemont 1978)
[28] Smith (1976 cited in Biemont 1978)
[29] Signer and Suess 1963; Manuel 1967; Marti 1969; Kuroda and Manuel 1970; Srinivasan and Manuel 1971, all cited in Manuel and Hwaung 1983
[30] Kuroda and Manuel 1970 cited in Manuel and Hwaung 1983:7
[31] Our Local Galactic Neighborhood, NASA http://interstellar.jpl.nasa.gov/interstellar/probe/introduction/neighborhood.html,
[32] *A short History of scientific ideas to 1900*, C. Singer, Oxford University Press, 1959, p. 151.
[33] The Arabian Science, C. Ronan, pp. 201–244 in *The Cambridge Illustrated History of the World's Science*, Cambridge University Press, 1983; at pp. 213–214.
[34] *"While environmental exposure to UV radiation is known to contribute to the accelerated aging of the outer layers of the eye and the development of cataracts, the concern over improper viewing of the Sun during an eclipse is for the development of "eclipse blindness" or retinal burns."*
[35] http://www.scholarpedia.org/article/Solar_activity
[36] https://doi.org/10.4249%2Fscholarpedia.3967
[37] http://sohowww.nascom.nasa.gov/
[38] http://www.nso.edu/
[39] http://www.astronomycast.com/astronomy/episode-30-the-sun-spots-and-all/
[40] http://www.boston.com/bigpicture/2008/10/the_sun.html
[41] http://www.acrim.com/
[42] http://www.suntrek.org/
[43] https://web.archive.org/web/20050518081349/http://www.solarphysics.kva.se/
[44] http://alienworlds.glam.ac.uk/sunStructure.html
[45] https://www.youtube.com/watch?v=qpMRtvFD8ek&hl=fr
[46] https://web.archive.org/web/20100315111135/https://science.nasa.gov/headlines/y2010/12mar_conveyorbelt.htm
[47] https://www.youtube.com/watch?v=w-41gAPmUG0&feature=youtube_gdata&ab_channel=NASAGoddard
[48] https://www.youtube.com/watch?v=6tmbeLTHC_0&ab_channel=NASAGoddard
[49] http://quake.stanford.edu/~wso/gifs/HCS.html
[50] Kenneth R. Lang, *The Sun from Space* (2000) Springer, . " 2. Discovering Space https://books.google.com/books?id=Sn5gZ6gHKakC&pg=PA17" (page 17)
[51] http://www.nineplanets.org/medium.html
[52] McGRAW-HILL ENCYCLOPEDIA OF Science & Technology, 8th ed., (c)1997, vol. 16, page 685
[53] Kristian Birkeland, "Are the Solar Corpuscular Rays that penetrate the Earth's Atmosphere Negative or Positive Rays?" in *Videnskapsselskapets Skrifter*, I Mat – Naturv. Klasse No.1, Christiania, 1916.
[54] *Philosophical Magazine*, Series 6, Vol. 38, No. 228, December 1919, 674 (on the Solar Wind)
[55] [Brian Harvey, *Russian planetary exploration: history, development, legacy, prospects*. Springer, 2007, p.26.]
[56] David Darling, *Internet Encyclopedia of Science*. http://www.daviddarling.info/encyclopedia/L/Luna.html
[57] Zirker, J. B. (1977), Coronal holes and high-speed wind streams, *Reviews of Geophysics*, *15*(3), 257-269
[58] NASA Study Using Cluster Reveals New Insights Into Solar Wind, NASA, Greenbelt, 2012, p.1
[59] http://www.swpc.noaa.gov/ace/MAG_SWEPAM_24h.html
[60] http://www.swpc.noaa.gov/SWN/sw_dials.gif
[61] http://solarscience.msfc.nasa.gov/SolarWind.shtml
[62] http://www.suntrek.org
[63] http://clusterlaunch.esa.int/science-e/www/object/index.cfm?fobjectid=45273
[64] http://www.physorg.com/news205411403.html

[65]http://news.discovery.com/tech/solar-wind-energy-power.html
[66]http://journals.cambridge.org/action/displayAbstract?fromPage=online&aid=7451968&fulltextType=RA&fileId=S1473550410000066
[67]Carole Haswell, Transiting Exoplanets https://www.amazon.com/dp/0521139384
[68]Michael Perryman, The Exoplanet Handbook https://www.amazon.com/dp/0521765595
[69]Sara Seager, Exoplanets https://www.amazon.com/dp/0816529450
[70]http://www.bbc.com/news/science-environment-17454005
[71]Namely: KOI 326.01 [Rp=0.85], KOI 701.03 [Rp=1.73], KOI 268.01 [Rp=1.75], KOI 1026.01 [Rp=1.77], KOI 854.01 [Rp=1.91], KOI 70.03 [Rp=1.96] – Table 6). A more recent study found that one of these candidates (KOI 326.01) is in fact much larger and hotter than first reported.
[72]http://www.cfa.harvard.edu/news/2014-01
[73]Extract of page 51 https://books.google.com/books?id=bJoYlBWbCAYC&pg=PA51
[74]Gallant, R. 1986. *The National Geographic Picture Atlas of Our Universe*. National Geographic Society, 2nd edition.
[75]If the area of Washington is about 177 km^2 and 2.5 miles is taken to equal 4 km, Solomon's estimate would equal about 700 cubic kilometres of ice, which would have a mass of about 600 billion tons (6×10^{14} kg).
[76]*Popular Astronomy: A Review of Astronomy and Allied Sciences* https://books.google.com/books?id=ePc-AQAAIAAJ, 1896
[77]Mercury Closest Approaches to Earth generated with:
1. Solex 10 http://chemistry.unina.it/~alvitagl/solex/ (Text Output file http://home.surewest.net/kheider/astro/SolexMerc.txt)
2. Gravity Simulator charts http://www.orbitsimulator.com/cgi-bin/yabb/YaBB.pl?num=1235936812
3. JPL Horizons 1950–2200 http://home.surewest.net/kheider/astro/Mercury.txt
[78]U. Le Verrier (1859), (in French), "Lettre de M. Le Verrier à M. Faye sur la théorie de Mercure et sur le mouvement du périhélie de cette planète" https://archive.org/stream/comptesrendusheb49acad#page/378/mode/2up, Comptes rendus hebdomadaires des séances de l'Académie des sciences (Paris), vol. 49 (1859), pp. 379–383. (At p. 383 in the same volume Le Verrier's report is followed by another, from Faye, enthusiastically recommending to astronomers to search for a previously undetected intra-mercurial object.)
[79](look at 1964 and 2013)
[80]– Numbers generated using the Solar System Dynamics Group, Horizons On-Line Ephemeris System http://ssd.jpl.nasa.gov/horizons.cgi?find_body=1&body_group=mb&sstr=1
[81]See Twilight#Astronomical twilight
[82], , .
[83]See also the Greek article about the planet.
[84]at pp. 118–122.
[85]Mercury http://scienceworld.wolfram.com/astronomy/Mercury.html at Eric Weisstein's 'World of Astronomy'
[86]Golden, Leslie M., A Microwave Interferometric Study of the Subsurface of the Planet Mercury (1977). PhD Dissertation, University of California, Berkeley
[87]"NASA extends spacecraft's Mercury mission" http://www.upi.com/Top_News/US/2011/11/15/NASA-extends-spacecrafts-Mercury-mission/UPI-55131321408343/. UPI, 15 November 2011. Retrieved 2011-11-16.
[88]https://history.nasa.gov/SP-423/sp423.htm
[89]http://planetarynames.wr.usgs.gov/Page/MERCURY/target
[90]http://planetarynames.wr.usgs.gov/Page/mercuryQuadMap
[91]http://messenger.jhuapl.edu/
[92]http://messenger-act.actgate.com/msgr_public_released/react_quickmap.html
[93]http://www.solarviews.com/eng/mercury.htm
[94]http://www.astronomycast.com/2007/08/episode-49-mercury/
[95]http://www.geody.com/?world=mercury
[96]http://www.messenger-education.org/Interactives/ANIMATIONS/Day_On_Mercury/day_on_mercury_full.htm

[97] http://www.psrd.hawaii.edu/Archive/Archive-Mercury.html
[98] http://www.esa.int/Our_Activities/Space_Science/BepiColombo_overview2
[99] http://www.sixtysymbols.com/videos/mercury.htm
[100] https://apod.nasa.gov/apod/ap130612.html
[101] https://www.flickr.com/photos/136797589@N04/35846055060/
[102] Defining the rotation as retrograde, as done by NASA space missions and the USGS, puts Ishtar Terra in the northern hemisphere and makes the axial tilt 2.64°. Following the right-hand rule for prograde rotation puts Ishtar Terra in the southern hemisphere and makes the axial tilt 177.36°.
[103] Misstated as "Ganiki Chasma" in the press release and scientific publication. UNIQ-ref-0-3951097dbf03c794-QINU
[104] Numbers generated by Solex https://web.archive.org/web/20150918233453/http://chemistry.unina.it/~alvitagl/solex/
[105]
[106] Several claims of transit observations made by medieval Islamic astronomers have been shown to be sunspots.<ref name="VaqueroVázquez2009">
[107] Redfern, Martin (25 May 2004). "Venus clouds 'might harbour life'" http://news.bbc.co.uk/2/hi/3746583.stm. *BBC News*. Retrieved 30 December 2015.
[108] http://solarsystem.nasa.gov/planets/venus
[109] http://nssdc.gsfc.nasa.gov/planetary/planets/venuspage.html
[110] http://nssdc.gsfc.nasa.gov/imgcat/thumbnail_pages/venus_thumbnails.html
[111] http://www.mentallandscape.com/V_Venus.htm
[112] http://www.mentallandscape.com/C_CatalogVenus.htm
[113] http://www.nineplanets.org/venus.html
[114] http://eclipse.gsfc.nasa.gov/transit/catalog/VenusCatalog.html
[115] http://www.geody.com/?world=venus
[116] http://www.mapaplanet.org/explorer/venus.html
[117] http://planetarynames.wr.usgs.gov/Page/VENUS/target
[118] http://www.lpi.usra.edu/resources/vc/vchome.shtml
[119] http://planetologia.elte.hu/venusz-terkep-elte-ttk-kavucs.pdf
[120]
[121] Earth's circumference is almost exactly 40,000 km because the metre was calibrated on this measurement—more specifically, 1/10-millionth of the distance between the poles and the equator.
[122] Early edition, published online before print.
[123] Including *eorþe*, *erþe*, *erde*, and *erthe*. UNIQ-ref-0-3951097dbf03c794-QINU
[124] Oxford English Dictionary, "earth, *n.*[1]" Oxford University Press (Oxford), 2010.
[125] As in *Beowulf* (1531–33):
Wearp ða wundelmæl wrættum gebunden
yrre oretta, þæt hit on ***eorðan*** *læg,*
stið ond stylecg. UNIQ-ref-1-3951097dbf03c794-QINU UNIQ-ref-2-3951097dbf03c794-QINU
"He threw the artfully-wound sword so that it lay upon the **earth**, firm and sharp-edged."<ref name=beo>*Beowulf*. Trans. Chad Matlick in "*Beowulf*: Lines 1399 to 1799" http://www.as.wvu.edu/english/oeoe/english311/1799.html. West Virginia University. Accessed 5 August 2014. &
[126] As in the Old English glosses of the *Lindisfarne Gospels* (Luke 13:7):
Succidite ergo illam ut quid etiam **terram** occupat: *hrendas* uel *scearfað forðon ðailca* uel *hia to huon uutedlice* ***eorðo*** *gionetað* uel *gemerras*. UNIQ-ref-3-3951097dbf03c794-QINU
"Remove it. Why should it use up the **soil**?"<ref>*Mounce Reverse-Intralinear New Testament*: " Luke 13:7 https://www.biblegateway.com/passage/?search=Luke%2013:7&version=MOUNCE". Hosted at *Bible Gateway*. 2014. Accessed 5 August 2014. &
[127] As in Ælfric's *Heptateuch* (Gen. 1:10):
Ond God gecygde ða drignysse ***eorðan*** *ond ðære wætera gegaderunge he het sæ.*<ref>Ælfric of Eynsham. *Heptateuch*. Reprinted by S.J. Crawford as *The Old English Version of the Heptateuch, Ælfric's Treatise on the Old and New Testament and his Preface to Genesis*. Humphrey Milford (London), 1922. http://wordhord.org/nasb/genesis.html Hosted at *Wordhord*. Accessed 5 August 2014.

[128] King James Version of the Bible: " Genesis 1:10 https://www.biblegateway.com/passage/?search=Genesis%201:10&version=KJV". Hosted at *Bible Gateway*. 2014. Accessed 5 August 2014.

[129] As in the Wessex Gospels (Matt. 28:18):
Me is geseald ælc anweald on heofonan & on ***eorðan***.
"All authority in heaven and on **earth** has been given to me."<ref>*Mounce Reverse-Intralinear New Testament*: " Matthew 28:18 https://www.biblegateway.com/passage/?search=Matthew+28%3A18&version=MOUNCE". Hosted at *Bible Gateway*. 2014. Accessed 5 August 2014. &

[130] As in the Codex Junius's *Genesis* (112–16):
her ærest gesceop ece drihten,
helm eallwihta, heofon and ***eorðan,***
rodor arærde and þis rume land
gestaþelode strangum mihtum,
frea ælmihtig.<ref>" Genesis A http://www.maldura.unipd.it/dllags/brunetti/OE/TESTI/GenesisA/DATI/testo.html". Hosted at the Dept. of Linguistic Studies at the University of Padua. Accessed 5 August 2014.

[131] Killings, Douglas. *Codex Junius 11*, I.ii http://www.gutenberg.org/files/618/618-h/618-h.htm. 1996. Hosted at Project Gutenberg. Accessed 5 August 2014.

[132] As in Ælfric's *On the Seasons of the Year* (Ch. 6, §9):
Seo ***eorðe*** *stent on gelicnysse anre pinnhnyte, & seo sunne glit onbutan be Godes gesetnysse.*
"The **earth** can be compared to a pine cone, and the Sun glides around it by God's decree.<ref>Ælfric, Abbot of Eynsham. "*De temporibus annis*" Trans. as " On the Seasons of the Year http://faculty.virginia.edu/OldEnglish/aelfric/detemp.html". Hosted at Old English at the University of Virginia, 1998. Accessed 6 August 2014.

[133] Tacitus. *Germania*, .

[134] Simek, Rudolf. Trans. Angela Hall as *Dictionary of Northern Mythology*, D.S. Brewer, 2007.

[135] *The New Oxford Dictionary of English*, "earth". Oxford University Press (Oxford), 1998.

[136] Schopf, JW, Kudryavtsev, AB, Czaja, AD, and Tripathi, AB. (2007). *Evidence of Archean life: Stromatolites and microfossils.* Precambrian Research 158:141–155.

[137] Schopf, JW (2006). *Fossil evidence of Archaean life.* Philos Trans R Soc Lond B Biol Sci 29;361(1470) 869-85.

[138] If Earth were shrunk to the size of a billiard ball, some areas of Earth such as large mountain ranges and oceanic trenches would feel like tiny imperfections, whereas much of the planet, including the Great Plains and the abyssal plains, would feel smoother.<ref>

[139] Locally varies between 5 and 200 km.

[140] Locally varies between 5 and 70 km.

[141] https://ngdc.noaa.gov/mgg/global/etopo1_surface_histogram.html

[142] World http://www.nationalgeographic.com/xpeditions/atlas/index.html?Parent=world&Mode=d&SubMode=w at the Xpeditions Atlas http://www.nationalgeographic.com/xpeditions/, *National Geographic Society, Washington D.C., 2006.*

[143] //en.wikipedia.org/w/index.php?title=Earth&action=edit

[144] Lovelock, James. *The Vanishing Face of Gaia*. Basic Books, 2009, p. 255.

[145] https://books.google.com/books?id=xwjlZjFNFlAC#v=onepage&q&f=false

[146] http://adsabs.harvard.edu/abs/2003deu..book.....C

[147] //www.worldcat.org/oclc/52082611

[148] http://education.nationalgeographic.com/education/encyclopedia/earth/?ar_a=1

[149] http://solarsystem.nasa.gov/planets/profile.cfm?Object=Earth

[150] http://solarsystem.nasa.gov/

[151] http://www.nasa.gov/centers/goddard/earthandsun/earthshape.html

[152] http://www.usgs.gov/

[153] https://web.archive.org/web/20090430041323/http://eol.jsc.nasa.gov/Coll/weekly.htm

[154] http://earthobservatory.nasa.gov/

[155] http://www.astronomycast.com/stars/episode-51-earth/

[156] https://www.youtube.com/watch?v=74mhQyuyELQ

[157] https://www.youtube.com/watch?v=l6ahFFFQBZY

[158]This image was taken by the *Rosetta* spacecraft's Optical, Spectroscopic, and Infrared Remote Imaging System (OSIRIS), at a distance of ≈240,000 kilometres (150,000 mi) during its February 2007 encounter. The view is centered on the Aeolis quadrangle, with Gale crater, the landing site of the *Curiosity* rover, prominently visible just left of center. The darker, more heavily cratered terrain in the south, Terra Cimmeria, is composed of older terrain than the much smoother and brighter Elysium Planitia to the north. Geologically recent processes, such as the possible existence of a global ocean in Mars's past, could have helped lower-elevated areas, such as Elysium Planitia, retain a more youthful look.

[159]

[160]//tools.wmflabs.org/geohack/geohack.php?pagename=Mars&params=0_N_180_W_globe: Mars

[161]//tools.wmflabs.org/geohack/geohack.php?pagename=Mars&params=3.34_N_219.38_E_ globe:Mars

[162]//tools.wmflabs.org/geohack/geohack.php?pagename=Mars&params=30_N_260_W_globe: Mars

[163]//tools.wmflabs.org/geohack/geohack.php?pagename=Mars&params=0_N_310_W_globe: Mars

[164]Mustard, Jack (July 9, 2009) MEPAG Report to the Planetary Science Subcommittee http://www.lpi.usra.edu/pss/july2009/presentations/05MEPAG.pdf. lpi.usra.edu. p. 3

[165]//en.wikipedia.org/w/index.php?title=Mars&action=edit

[166]https://dmoztools.net/Science/Astronomy/Solar_System/Planets/Mars/

[167]http://mars.jpl.nasa.gov/

[168]http://www.google.com/mars/

[169]https://www.google.com/maps/space/mars/

[170]http://www.geody.com/?world=mars

[171]http://photojournal.jpl.nasa.gov/targetFamily/Mars

[172]http://mars.jpl.nasa.gov/multimedia/

[173]http://www.msss.com/science-images/

[174]http://hirise.lpl.arizona.edu/

[175]http://sos.noaa.gov/Datasets/dataset.php?id=224

[176]https://www.usgs.gov/media/videos/rotating-globe-mars-geology

[177]https://www.youtube.com/watch?v=Jr1Xu2i-Uc0

[178]http://themis.mars.asu.edu/node/5470

[179]https://www.flickr.com/photos/136797589@N04/32657197761/in/photostream/

[180]https://www.flickr.com/photos/136797589@N04/albums/72157680411399285

[181]http://planetarynames.wr.usgs.gov/Page/MARS/target

[182]http://planetarynames.wr.usgs.gov/Page/mars1to5mTHEMIS

[183]http://pubs.usgs.gov/sim/3292/

[184]http://planetologia.elte.hu/terkep/mars-viking-en.pdf

[185]http://planetologia.elte.hu/terkep/mars-mola-en.pdf

[186]Linda T. Elkins-Tanton, *Asteroids, Meteorites, and Comets*, 2010:10

[187]and 1853, p. 216

[188]: "[Professor Peirce] then observed that the analogy between the ring of Saturn and the belt of the asteroids was worthy of notice."

[189] – search for asteroids in the main belt regions with a diameter >100.

[190]

[191]//en.wikipedia.org/w/index.php?title=Asteroid_belt&action=edit

[192]This value was obtained by a simple count up of all bodies in that region using data for 120437 numbered minor planets from the Minor Planet Center orbit database http://www.minorplanetcenter.org/iau/MPCORB.html, dated February 8, 2006.

[193]Cometary origin of the zodiacal cloud and carbonaceous micrometeorites – Implications for hot debris disk http://arxiv.org/pdf/0909.4322.pdf

[194]http://www.nineplanets.org/asteroids.html

[195]https://web.archive.org/web/20070418220152/http://www.nineplanets.org/asteroids.html

[196]http://solarsystem.nasa.gov/planets/profile.cfm?Object=Asteroids

[197]http://solarsystem.nasa.gov

[198] http://www.astronomycast.com/astronomy/episode-55-the-asteroid-belt/

[199] https://web.archive.org/web/20080307033324/http://www.astronomycast.com/astronomy/episode-55-the-asteroid-belt/

[200] http://www.solstation.com/stars/asteroid.htm

[201] https://web.archive.org/web/20070515163455/http://www.solstation.com/stars/asteroid.htm

[202] https://web.archive.org/web/20070524185042/http://solarsystem.nasa.gov/planets/profile.cfm?Object=Asteroids

[203] https://web.archive.org/web/20111113212032/http://burro.astr.cwru.edu/stu/media/asteroid_all_axisvecc.jpg

[204] https://web.archive.org/web/20111113204941/http://burro.astr.cwru.edu/stu/media/asteroid_all_axisvincl.jpg

[205] http://nssdc.gsfc.nasa.gov/planetary/planets/asteroidpage.html

[206] https://web.archive.org/web/20070411201013/http://nssdc.gsfc.nasa.gov/planetary/planets/asteroidpage.html

[207] http://www.planetary.org/explore/topics/asteroids_and_comets/facts.html

[208] https://web.archive.org/web/20070428143402/http://www.planetary.org/explore/topics/asteroids_and_comets/facts.html

[209] This image was taken by the *Dawn* spacecraft on 2 May 2015, during a "rotation characterization" orbit, above the surface of Ceres. Visible at center and center right are two bright spots, a phenomenon common on Ceres, in Oxo and Haulani craters respectively. Ahuna Mons is also visible in the image as a noticeable, bluff hill, seen just right of bottom.

[210] (produced with Solex 10 http://chemistry.unina.it/~alvitagl/solex/ written by Aldo Vitagliano; see also Invariable plane)

[211] Calculated based on the known parameters

[212] The value of 0.37 is the mean moment of inertia, which is thought to better represent Ceres' interior structure than the polar moment of inertia (0.39), due to the high polar flattening. UNIQ-ref-0-3951097dbf03c794-QINU

[213] APmag and AngSize generated with Horizons http://ssd.jpl.nasa.gov/horizons.cgi?find_body=1&body_group=sb&sstr=1 (Ephemeris: Observer Table: Quantities = 9,13,20,29)

[214] Due to its eccentric orbit, the dwarf planet Pluto was also within the orbit of Neptune from 1979 to 1999, and will be again from approximately 2227 to 2247.

[215] NASA Science News: Water Detected on Dwarf Planet Ceres https://science.nasa.gov/science-news/science-at-nasa/2014/22jan_ceres/, by Production editor: Dr. Tony Phillips | Credit: Science@NASA (22 January 2014)

[216] (Page 335) http://articles.adsabs.harvard.edu/cgi-bin/nph-iarticle_query?1994QJRAS..35..331H&defaultprint=YES&page_ind=4&filetype=.pdf

[217] All other languages but one use a variant of *Ceres/Cerere*: Russian *Tserera*, Persian *Seres*, Japanese *Keresu*. The exception is Chinese, which uses 'grain-god(dess) star' (穀神星 *gǔshénxīng*). Note that this is unlike the goddess Ceres, where Chinese does use the Latin name (刻瑞斯 *kèruìsī*).

[218] Unicode value U+26B3

[219] In 1807 Klaproth tried to change the name to the more etymologically justified "cererium", but it did not catch on.<ref>

[220] Gaherty, Geoff; "How to Spot Giant Asteroid Vesta in Night Sky This Week", 3 August 2011 How to Spot Giant Asteroid Vesta in Night Sky This Week | Asteroid Vesta Skywatching Tips | Amateur Astronomy, Asteroids & Comets | Space.com http://www.space.com/12537-asteroid-vesta-skywatching-tips.html

[221] NASA/JPL, *Dawn Views Vesta*, 2 August 2011 http://www.ustream.tv/recorded/16375687/highlight/191666 ("*Dawn* will orbit two of the largest asteroids in the Main Belt").

[222] Approximately forty percent that of Australia, a third the size of the US or Canada, 12× that of the UK

[223] Preferential formation of sodium salts from frozen sodium-ammonium-chloride-carbonate brines – Implications for Ceres' bright spots. Tuan H. Vu, Robert Hodyss, Paul V. Johnson, Mathieu Choukroun. *Planetary and Space Science*, Volume 141, July 2017, Pages 73-77

[224] 0.72–0.77 anhydrous rock by mass, per McKinnon, William B.; (2008) "On the Possibility of Large KBOs Being Injected Into the Outer Asteroid Belt" http://adsabs.harvard.edu/abs/2008DPS....40.3803M. *American Astronomical Society,* DPS meeting No. 40, #38.03

[225] Ceres: The Smallest and Closest Dwarf Planet http://www.space.com/22891-ceres-dwarf-planet.html. *Space.com* 22 January 2014

[226] Dwarf Planet Ceres, Artist's Impression http://www.nasa.gov/jpl/herschel/dwarf-planet-ceres-pia17830. 21 January 2014. NASA

[227] This emission rate is modest compared to those calculated for the tidally driven plumes of Enceladus (a smaller body) and Europa (a larger body), 200 kg/s<ref name="Hansen2006">

[228] Approximately a 10% chance of the asteroid belt acquiring a Ceres-mass KBO. McKinnon, William B.; (2008) "On the Possibility of Large KBOs Being Injected Into the Outer Asteroid Belt" http://adsabs.harvard.edu/abs/2008DPS....40.3803M. *American Astronomical Society,* DPS meeting No. 40, #38.03

[229] Houtkooper, Joop M.; "Glaciopanspermia: Seeding the Terrestrial Planets with Life?" http://sciwww.esac.esa.int/SB/MARSEXPLORATION/docs/Presentations/Houtkooper.pdf Institute for Psychobiology and Behavioral Medicine, Justus-Liebig-University, Giessen, Germany

[230] Martinez, Patrick, *The Observer's Guide to Astronomy*, page 298. Published 1994 by Cambridge University Press

[231] China's Deep-space Exploration to 2030 by Zou Yongliao Li Wei Ouyang Ziyuan Key Laboratory of Lunar and Deep Space Exploration, National Astronomical Observatories, Chinese Academy of Sciences, Beijing http://english.nssc.cas.cn/ns/NU/201410/W020141016603613379886.pdf

[232] http://photojournal.jpl.nasa.gov/archive/PIA20014_fig2_Kerwan.pdf

[233] http://photojournal.jpl.nasa.gov/archive/PIA20014_fig1_Asari-Zadeni.pdf

[234] http://photojournal.jpl.nasa.gov/archive/PIA20014_fig3_Occator.pdf

[235] https://www.youtube.com/watch?v=XJYfeZoHbzU

[236] http://dawn.jpl.nasa.gov/

[237] http://orbitsimulator.com/gravity/articles/ceres.html

[238] http://ssd.jpl.nasa.gov/horizons.cgi?find_body=1&body_group=sb&sstr=1

[239] http://www.keplersdiscovery.com/Asteroid.html

[240] http://aa.usno.navy.mil/publications/reports/asteroid_ephemerides.html

[241] http://adsabs.harvard.edu/abs/1999AJ....117.1077H

[242] //doi.org/10.1086%2F300728

[243] http://imgur.com/4B1TfHg

[244] https://www.flickr.com/photos/136797589@N04/36035176992/

[245]

[246] Basics of Spaceflight - Chapter 4. Interplanetary Trajectories http://www2.jpl.nasa.gov/basics/bsf4-1.php

[247] Argo - A Voyage Through the Outer Solar System http://www.spacepolicyonline.com/pages/images/stories/PSDS%20GP1%20Hansen_Argo_Neptune%20Mission%20Concept.pdf

[248] AVIATR http://www.lpi.usra.edu/opag/Oct2011/presentations/1_AVIATR_Barnes.pdf

[249] OPAG http://www.lpi.usra.edu/opag/

[250]

[251] G. Maise - Exploration of Jovian Atmosphere Using Nuclear Ramjet Flyer http://www.niac.usra.edu/files/library/meetings/annual/jun02/510Maise.pdf

[252] JET - Cost Capped Titan-Enceladus Proposal http://futureplanets.blogspot.com/2011/03/jet-cost-capped-titan-enceladus.html

[253]

[254] Tsou, et al. - LIFE http://online.liebertpub.com/doi/abs/10.1089/ast.2011.0813?journalCode=ast& link courtesy of Future Planetary Exploration http://futureplanets.blogspot.com/2012/12/life-sample-return-for-enceladus.html

[255] ARPS Enabled Titan Rover Concept with Inflatable Wheels http://trs-new.jpl.nasa.gov/dspace/bitstream/2014/41465/1/05-2366.pdf

[256] Uranus Pathfinder http://www.mssl.ucl.ac.uk/planetary/missions/uranus/

[257] http://www.lpi.usra.edu/opag/nov_2007_meeting/presentations/solar_power.pdf

[258] http://www.nasa-academy.org/soffen/travelgrant/bethke.pdf

[259]http://www.spacepolicyonline.com/pages/images/stories/PSDS-AGU-Spencer.pdf
[260]http://www.lpi.usra.edu/opag/
[261]http://www.lpi.usra.edu/opag/reports.html
[262]http://www.centennialofflight.gov/essay/SPACEFLIGHT/outer_planets/SP30.htm
[263]http://historicspacecraft.com/Probes_Outer_Planets.html
[264]See for example:
[265]The Deuterium-Burning Mass Limit for Brown Dwarfs and Giant Planets http://arxiv.org/abs/1008.5150, David S. Spiegel, Adam Burrows, John A. Milsom
[266]Rocky core solubility in Jupiter and giant exoplanets http://arxiv.org/abs/1111.6309, Hugh F. Wilson, Burkhard Militzer, 2011
[267]The Interior of Jupiter, Guillot et al., in *Jupiter: The Planet, Satellites and Magnetosphere*, Bagenal et al., editors, Cambridge University Press, 2004
[268]http://www.space.com/scienceastronomy/060816_planet_qanda.html
[269]http://news.bbc.co.uk/1/hi/sci/tech/4798205.stm
[270]http://www.sf-encyclopedia.com/entry/gas_giant
[271]
[272]Based on the volume within the level of 1 bar atmospheric pressure
[273]Saturn http://www.universeguide.com/Saturn.aspx. Universe Guide. Retrieved 29 March 2009.
[274]Bizarre geometric shapes that appear at the center of swirling vortices in planetary atmospheres might be explained by a simple experiment with a bucket of water but correlating this to Saturn's pattern is by no means certain.
[275]Laboratory experiment of spinning disks in a liquid solution forms vortices around a stable hexagonal pattern similar to that of Saturn's.
[276]Jean Meeus, *Astronomical Algorithms* (Richmond, VA: Willmann-Bell, 1998). Average of the nine extremes on p 273. All are within 0.02 AU of the averages.
[277]See also the Greek article about the planet.
[278]//www.worldcat.org/issn/0027-9358
[279]//www.worldcat.org/oclc/643483454
[280]http://adsabs.harvard.edu/abs/2007CeMDA..98..155S
[281]//doi.org/10.1007%2Fs10569-007-9072-y
[282]https://books.google.com/books?id=stFpBgAAQBAJ&pg=PA250
[283]http://solarsystem.nasa.gov/planets/profile.cfm?Object=Saturn
[284]http://nssdc.gsfc.nasa.gov/planetary/factsheet/saturnfact.html
[285]http://planetarynames.wr.usgs.gov/?System=Saturn
[286]http://saturn.jpl.nasa.gov/
[287]http://www.solarviews.com/eng/saturn.htm
[288]http://www.astronomycast.com/2007/10/episode-59-saturn/
[289]https://www.flickr.com/photos/vontom/5362248377/
[290]http://www.ciclops.org/view_event/185/Saturn_Rev_175_Raw_Preview?js=1
[291]http://www.sixtysymbols.com/videos/saturn_photpgrapher.htm
[292]http://www.bbc.co.uk/programmes/b06vmr1m
[293] These are the mean elements from VSOP87, together with derived quantities.
[294] Calculated using data from Seidelmann, 2007. UNIQ-ref-0-3951097dbf03c794-QINU
[295] Calculation of He, H_2 and CH_4 molar fractions is based on a 2.3% mixing ratio of methane to hydrogen and the 15/85 He/H_2 proportions measured at the tropopause.
[296]Journal of the Royal Society and Royal Astronomical Society 1, 30, quoted in Miner, p. 8.
[297]Royal Astronomical Society MSS W.2/1.2, 23; quoted in Miner p. 8.
[298]RAS MSS Herschel W.2/1.2, 24, quoted in Miner p. 8.
[299]RAS MSS Herschel W1/13.M, 14 quoted in Miner p. 8.
[300]Johann Elert Bode, Berliner Astronomisches Jahrbuch, p. 210, 1781, quoted in Miner, p. 11.
[301]Miner, p. 11.
[302]Because, in the English-speaking world, the latter sounds like "your anus", the former pronunciation also saves embarrassment: as Pamela Gay, an astronomer at Southern Illinois University Edwardsville, noted on her podcast, to avoid "being made fun of by any small schoolchildren ... when in doubt, don't emphasise anything and just say /ˈjʊərənəs/. And then run, quickly."<ref>
[303]RAS MSS Herschel W.1/12.M, 20, quoted in Miner, p. 12

[304]: [In original German]: "Bereits in der am 12ten März 1782 bei der hiesigen naturforschenden Gesellschaft vorgelesenen Abhandlung, habe ich den Namen des Vaters vom Saturn, nemlich Uranos, oder wie er mit der lateinischen Endung gewöhnlicher ist, Uranus vorgeschlagen, und habe seit dem das Vergnügen gehabt, daß verschiedene Astronomen und Mathematiker in ihren Schriften oder in Briefen an mich, diese Benennung aufgenommen oder gebilligt. Meines Erachtens muß man bei dieser Wahl die Mythologie befolgen, aus welcher die uralten Namen der übrigen Planeten entlehnen worden; denn in der Reihe der bisher bekannten, würde der von einer merkwürdigen Person oder Begebenheit der neuern Zeit wahrgenommene Name eines Planeten sehr auffallen. Diodor von Cicilien erzahlt die Geschichte der Atlanten, eines uralten Volks, welches eine der fruchtbarsten Gegenden in Africa bewohnte, und die Meeresküsten seines Landes als das Vaterland der Götter ansah. Uranus war ihr, erster König, Stifter ihres gesitteter Lebens und Erfinder vieler nützlichen Künste. Zugleich wird er auch als ein fleißiger und geschickter Himmelsforscher des Alterthums beschrieben... Noch mehr: Uranus war der Vater des Saturns und des Atlas, so wie der erstere der Vater des Jupiters."; [Translated]: "Already in the pre-read at the local Natural History Society on 12th March 1782 treatise, I have the father's name from Saturn, namely Uranus, or as it is usually with the Latin suffix, proposed Uranus, and have since had the pleasure that various astronomers and mathematicians, cited in their writings or letters to me approving this designation. In my view, it is necessary to follow the mythology in this election, which had been borrowed from the ancient name of the other planets; because in the series of previously known, perceived by a strange person or event of modern times name of a planet would very noticeable. Diodorus of Cilicia tells the story of Atlas, an ancient people that inhabited one of the most fertile areas in Africa, and looked at the sea shores of his country as the homeland of the gods. Uranus was her first king, founder of their civilized life and inventor of many useful arts. At the same time he is also described as a diligent and skilful astronomers of antiquity ... even more: Uranus was the father of Saturn and the Atlas, as the former is the father of Jupiter."

[305] Jean Meeus, *Astronomical Algorithms* (Richmond, VA: Willmann-Bell, 1998) p 271. From the 1841 aphelion to the 2092 one, perihelia are always 18.28 and aphelia always 20.10 astronomical units

[306] Weird water lurking inside giant planets https://www.newscientist.com/article/mg20727764.500-weird-water-lurking-inside-giant-planets.html, New Scientist,1 September 2010, Magazine issue 2776.

[307] Mixing ratio is defined as the number of molecules of a compound per a molecule of hydrogen.

[308] Michael Schirber – **Missions Proposed to Explore Mysterious Tilted Planet Uranus** (2011) – Astrobiology Magazine http://www.space.com/13248-nasa-uranus-missions-solar-system.html. Space.com. Retrieved on 2 April 2012.

[309] https://books.google.com/books?id=ZqA5AAAAcAAJ

[310] //www.worldcat.org/issn/0027-9358

[311] //www.worldcat.org/oclc/643483454

[312] http://sci.esa.int/science-e/www/object/index.cfm?fobjectid=35653

[313] http://nssdc.gsfc.nasa.gov/planetary/factsheet/uranusfact.html

[314] http://solarsystem.nasa.gov/planets/profile.cfm?Object=Uranus

[315] http://solarsystem.nasa.gov/

[316] http://www.projectshum.org/Planets/uranus.html

[317] http://photojournal.jpl.nasa.gov/targetFamily/Uranus

[318] http://www.ciclops.org/ir_index/81/Voyager_at_Uranus

[319] http://www.astronomycast.com/astronomy/episode-62-uranus/

[320] http://www.solarviews.com/raw/uranus/urfamily.jpg

[321] http://www.sixtysymbols.com/videos/uranus.htm

[322] https://www.youtube.com/watch?v=h3ppbbYXMxE

[323] —Select "Ephemeris Type: Orbital Elements", "Time Span: 2000-01-01 12:00 to 2000-01-02". ("Target Body: Neptune Barycenter" and "Center: Solar System Barycenter (@0)".)

[324] Orbital elements refer to the Neptune barycentre and Solar System barycentre. These are the instantaneous osculating values at the precise J2000 epoch. Barycentre quantities are given because, in contrast to the planetary centre, they do not experience appreciable changes on a day-to-day basis from the motion of the moons.

[325] (produced with Solex 10 http://chemistry.unina.it/~alvitagl/solex/ written by Aldo Vitagliano; see also Invariable plane)

[326]

[327] Based on the volume within the level of 1 bar atmospheric pressure

[328] Neptune is denser and physically smaller than Uranus because Neptune's greater mass gravitationally compresses the atmosphere more.

[329] Moore (2000):206

[330] See also the Greek article about the planet.

[331] "Uranus and Neptune get Hebrew names at last" http://www.haaretz.com/uranus-and-neptune-get-hebrew-names-at-last-1.1053, Haaretz.com

[332] "Hebrew names to Uranus and Neptune" http://www.hayadan.org.il/hebrew-names-to-uranus-and-neptune-3112090 Hayadan.org.il (Hebrew)

[333] "Appendix 5: Planetary Linguistics" http://nineplanets.org/days.html, Nineplanets.org

[334] The mass of Earth is 5.9736 kg, giving a mass ratio

UNIQ-math-0-3951097dbf03c794-QINU

The mass of Uranus is 8.6810 kg, giving a mass ratio

UNIQ-math-1-3951097dbf03c794-QINU

The mass of Jupiter is 1.8986 kg, giving a mass ratio

UNIQ-math-2-3951097dbf03c794-QINU

Mass values from

[335] Weird water lurking inside giant planets https://www.newscientist.com/article/mg20727764.500-weird-water-lurking-inside-giant-planets.html, New Scientist,1 September 2010, Magazine issue 2776.

[336] Burgess (1991):64–70.

[337] Imke de Pater and Jack J. Lissauer (2001), *Planetary Sciences https//books.google.com*, 1st edition, page 224.

[338] Jean Meeus, *Astronomical Algorithms* (Richmond, VA: Willmann-Bell, 1998) 273. Supplemented by further use of VSOP87. The last three aphelia were 30.33 AU, the next is 30.34 AU. The perihelia are even more stable at 29.81 AU

[339] (Bill Folkner at JPL) https://twitter.com/elakdawalla/status/21525820626

[340] —Numbers generated using the Solar System Dynamics Group, Horizons On-Line Ephemeris System.

[341] Hubble Space Telescope discovers fourteenth tiny moon orbiting Neptune | Space, Military and Medicine http://www.news.com.au/technology/sci-tech/hubble-space-telescope-discovers-fourteenth-tiny-moon-orbiting-neptune/story-fn5fsgyc-1226679913807. News.com.au (16 July 2013). Retrieved on 28 July 2013.

[342] Mass of Triton: 2.14 kg. Combined mass of 12 other known moons of Neptune: 7.53 kg, or 0.35%. The mass of the rings is negligible.

[343] UNIQ-math-3-3951097dbf03c794-QINU

[344] See the respective articles for magnitude data.

[345] Moore (2000):207.

[346] In 1977, for example, even the rotation period of Neptune remained uncertain.

[347] First Ground-Based Adaptive Optics Observations of Neptune and Proteus http://citeseerx.ist.psu.edu/viewdoc/download?doi=10.1.1.66.7754&rep=rep1&type=pdf Planetary & Space Science Vol. 45, No. 8, pp. 1031-1036, 1997

[348] Uranus and Neptune https://books.google.com/books?id=8s1JV-TrXacC&pg=PA147#v=onepage&q=Neptune&f=false Reports on Astronomy 2003-2005, p. 147 f.

[349] Burgess (1991):46–55.

[350] Tom Standage (2000). *The Neptune File: A Story of Astronomical Rivalry and the Pioneers of Planet Hunting*. New York: Walker. p. 188.

[351] http://nssdc.gsfc.nasa.gov/planetary/factsheet/neptunefact.html

[352] http://www.nineplanets.org/neptune.html

[353] http://www.astronomycast.com/astronomy/episode-63-neptune/

[354] http://solarsystem.nasa.gov/planets/profile.cfm?Object=Neptune

[355] http://solarsystem.nasa.gov

[356] http://www.projectshum.org/Planets/neptune.html

[357] http://www.sixtysymbols.com/videos/neptune.htm

[358] http://www.planetary.org/blogs/guest-blogs/2013/neptune-the-new-amateur-boundary.html

[359] //en.wikipedia.org/w/index.php?title=Template:Distant_minor_planets_sidebar&action=edit

[360] //en.wikipedia.org/w/index.php?title=Centaur_(minor_planet)&action=edit

[361] This would make Echeclus ($q = 5.8$ AU, $T_J = 3.03$) and Okyrhoe ($q = 5.8$ AU; $T_J = 2.95$), which have traditionally been classified as centaurs, and Hidalgo ($q = 1.95$ AU; $T_J = 2.07$), which has traditionally been considered an asteroid and is classified as a centaur by JPL, Jupiter-family comets, not centaurs.

[362] For the purpose of this diagram, an object is classified as a centaur if its semi-major axis lies between Jupiter and Neptune

[363] (Solex 10) http://chemistry.unina.it/~alvitagl/solex/

[364]

[365] Bauer, J. M., Fernández, Y. R., & Meech, K. J. 2003. " An Optical Survey of the Active Centaur C/NEAT (2001 T4) http://www.journals.uchicago.edu/PASP/journal/issues/v115n810/203101/203101.html", Publication of the Astronomical Society of the Pacific", **115**, 981

[366] Hainaut & Delsanti (2002) *Color of Minor Bodies in the Outer Solar System* Astronomy & Astrophysics, **389**, 641 datasource http://www.sc.eso.org/~ohainaut/MBOSS

[367] A class of Magnesium Iron Silicates $(Mg, Fe)_2SiO_4$, common components of igneous rocks.

[368] //en.wikipedia.org/wiki/Centaur_(minor_planet)#endnote_1

[369] http://www.minorplanetcenter.org/iau/lists/Centaurs.html

[370] http://www.daviddarling.info/encyclopedia/C/Centaur.html

[371] http://adsabs.harvard.edu/abs/2010IJAsB...9..227H

[372] //arxiv.org/abs/1007.2541

[373] //doi.org/10.1017%2FS1473550410000212

[374] http://www.jpl.nasa.gov/news/news.php?release=2013-234

[375] //en.wikipedia.org/w/index.php?title=Comet&action=edit

[376] European Space Agency http://www.esa.int/Our_Activities/Space_Science/Rosetta/Touchdown!_Rosetta_s_Philae_probe_lands_on_comet, Rosetta's Philae probe lands on comet.

[377] Halley: Using the volume of an ellipsoid of $15\times8\times8$ km * a rubble pile density of 0.6 g/cm^3 yields a mass (m=d*v) of 3.02E+14 kg.
Tempel 1: Using a spherical diameter of 6.25 km; volume of a sphere * a density of 0.62 g/cm^3 yields a mass of 7.9E+13 kg.
19P/Borrelly: Using the volume of an ellipsoid of 8x4x4km * a density of 0.3 g/cm^3 yields a mass of 2.0E+13 kg.
81P/Wild: Using the volume of an ellipsoid of 5.5x4.0x3.3km * a density of 0.6 g/cm^3 yields a mass of 2.28E+13 kg.

[378] T. Vogel - Comets and Jets - NASA http://hubblesite.org/hubble_discoveries/comet_ison/blogs/comets-and-jets

[379] Dr. Emily Baldwin - Dry ice fuels comet jets (2010) - Astronomy Now http://www.astronomynow.com/news/n1011/11hartley/

[380] Oregonian (29 Oct 2015), "Comet's oxygen shakes theories on solar system", p. A5

[381] (Solution using the Solar System Barycenter and barycentric coordinates. Select Ephemeris Type:Elements and Center:@0)

[382] On 1758 November 14, Alexis Clairaut announced to the Royal Academy of Sciences in Paris his prediction of the date at which Halley's comet would return: • Clairaut (January 1759) "Mémoire sur la cométe de 1682," https://babel.hathitrust.org/cgi/pt?id=ucm.5324346563;view=1up;seq=44 *Le Journal des Sçavans*, pp. 38–45. On p. 44, Clairaut predicted that Halley's comet would return in mid April 1759. From p. 44: https://babel.hathitrust.org/cgi/pt?id=ucm.5324346563;view=1up;seq=50 *"... il me paroît que la Cométe attendue doit passer à son périhélie vers le milieu du mois d'Avril prochain."* (... it seems to me that the expected comet must pass its perihelion towards the middle of next April.) On p. 40, Clairaut stated that his prediction might be slightly incorrect due to the presence of unknown planets beyond Saturn: *"Un corps qui passe dans des régions aussi éloignées, & qui échappe à nos yeux pendant des intervalles aussi longs, pourroit être soumis à des forces totalement inconnues; telles que l'action d'autres Cométes, ou même de quelque planéte toujours trop distante du Soleil pour*

être jamais apperçue." (A body [i.e., Halley's comet] which passes into regions so remote, and which escapes our eyes during such long intervals, might be subjected to totally unknown forces; such as the action of other comets, or even of some planet always too far from the sun to ever be perceived.) On 1759 April 7, the French astronomer Joseph-Nicolas Delisle announced to the Royal Academy of Sciences in Paris that he and his assistant Charles Messier had observed the return of Halley's comet, as predicted: • de l'Isle (June 1759) "Lettre de M. de l'Isle ... contenant la découverte du retour de la Comète de 1682, ... " https://babel.hathitrust.org/cgi/pt?id=ucm.5324346563;view=1up;seq=360 (Letter from Mr. de l'Isle ... containing the discovery of the return of the comet of 1682), *Le Journal des Sçavans*, pp. 356–364. De l'Isle subsequently admitted that the comet's return had first been seen by a German amateur astronomer and farmer, Georg Palitzsch: • de l'Isle (August 1759) "Seconde lettre de M. de l'Isle," https://babel.hathitrust.org/cgi/pt?id=ucm.5324346563;view=1up;seq=527 *Le Journal des Sçavans*, pp. 523–529. From p. 526: *"... j'ai reçu une Lettre d'Heidelberg le premier Avril au soir, dans laquelle l'on m'écrit que l'on a publié à Leipsick le 24 Janvier de cette année un Mémoire Allemand dans lequel il est dit que cette Comète a été vue en Saxe par un Paysan, nommé Palisch, le 25 & 26 Décembre de l'année dernière; j'ai bien de la peine à concevoir comment ce Paysan aura pû la découvrir, cette Comète, ... "* (... I received a letter from Heidelberg on the first of April in the evening, in which it is written to me that there had been published at Leipzig on the 24th of January of this year a German memoir in which it is said that this comet had been seen in Saxony by a peasant, named Palisch, on the 25th and 26th of December of last year; I can hardly conceive how this peasant could have discovered it, this comet, ...) The story behind the rediscovery of Halley's comet was given by Joseph Lalande in: • Delalande, *Tables astronomiques de M. Halley, ... Et l'Histoire de la Comete de 1759.* [Astronomical tables of Mr. Halley, ... and the history of the comet of 1759.] (Paris, France: Durand, 1759), pp. 91 ff. https://babel.hathitrust.org/cgi/pt?id=mdp.39015006990892;view=1up;seq=105 Lalande acknowledged the contributions of Madame Lepaute to predicting the return of Halley's comet on p. 110. From p. 110: https://babel.hathitrust.org/cgi/pt?id=mdp.39015006990892;view=1up;seq=124 *"... mais il faut convenir que cette suite immense de détails m'eût semblé effrayante, si* Madame LEPAUTE, *appliquée depuis long-temps & avec succès aux calculs Astronomiques, n'en eût partagé le travail."* (... but it must be admitted that this immense series of details would have seemed frightening to me if *Madame LEPAUTE*, [who has] long applied [herself] successfully to astronomical calculations, had not shared in the work.) See also: • Broughton, Peter (1985) "The first predicted return of comet Halley," *Journal for the History of Astronomy*, **16** : 123–132. Available at: Astrophysics Data System http://adsabs.harvard.edu/full/1985JHA....16..123B See also: • Clairaut, *Théorie du mouvement des comètes, ...* https://books.google.com/books?id=WQRbq4SA_zkC&pg=PP7#v=onepage&q&f=false [Theory of the movement of comets, ...] (Paris, France: Michel Lambert, 1760); see especially the preface.

[383] Y-J. Choi, P.R. Weissman, and D. Polishook *(60558) 2000 EC_98*, IAU Circ., **8656** (Jan. 2006), 2.

[384] https://books.google.com/books?id=kC5oQgAACAAJ

[385] https://dmoztools.net/Science/Astronomy/Solar_System/Small_Bodies/Comets/

[386] http://solarsystem.nasa.gov/planets/profile.cfm?Object=Comets

[387] http://solarsystem.nasa.gov/

[388] http://www.icq.eps.harvard.edu/

[389] http://www.magnet.fsu.edu/education/community/slideshows/comet/index.html

[390] http://smallbodies.ru/en/

[391] https://sites.google.com/site/nearearthobjectwatch/

[392] http://www.cfeps.net/

[393] //en.wikipedia.org/w/index.php?title=Template:Distant_minor_planets_sidebar&action=edit

[394] Kuiper belt – oxforddictionaries.com http://www.oxforddictionaries.com/definition/english/Kuiper-belt

[395] Johnson, Torrence V.; and Lunine, Jonathan I.; *Saturn's moon Phoebe as a captured body from the outer Solar System*, Nature, Vol. 435, pp. 69–71

[396] NEW HORIZONS *The PI's Perspective* http://pluto.jhuapl.edu/overview/piPerspective.php?page=piPerspective_08_24_2012

[397]The literature is inconsistent in the usage of the terms *scattered disc* and *Kuiper belt*. For some, they are distinct populations; for others, the scattered disc is part of the Kuiper belt. Authors may even switch between these two uses in one publication.<ref>Weissman and Johnson, 2007, *Encyclopedia of the solar system*, footnote p. 584
[398]Randall 2015, p. 106.
[399]Davies, p. 2
[400]Davies, p. 14
[401]Davies p. 38
[402]Randall 2015, p. 105.
[403]Davies p. 39
[404]Davies p. 191
[405]Davies p. 50
[406]Davies p. 51
[407]Davies pp. 52, 54, 56
[408]Davies pp. 57, 62
[409]Davies p. 65
[410]Davies p. 199
[411]Clyde Tombaugh, "The Last Word", Letters to the Editor, *Sky & Telescope*, December 1994, p. 8
[412]Davies p. 104
[413]Davies p. 107
[414]Davies p. 118
[415]New Horizons' catches a wandering Kuiper Belt Object not far off http://www.spacedaily.com/reports/New_Horizons_catches_a_wandering_Kuiper_Belt_Object_not_far_off_999.html spacedaily.com Laurel MD (SPX). December 7, 2015.
[416]http://www2.ess.ucla.edu/~jewitt/kb.html
[417]http://www2.ess.ucla.edu/~jewitt/kb/gerard.html
[418]http://www.physics.ucf.edu/~yfernandez/cometlist.html
[419]http://solarsystem.nasa.gov/planets/profile.cfm?Object=KBOs
[420]http://solarsystem.nasa.gov
[421]http://www.boulder.swri.edu/ekonews/
[422]http://www.johnstonsarchive.net/astro/tnos.html
[423]http://www.minorplanetcenter.org/iau/lists/OuterPlot.html
[424]http://www.iau.org/
[425]http://www.astronomy2006.com
[426]http://www.nature.com/nature/journal/v424/n6949/fig_tab/nature01725_F1.html
[427]//www.ncbi.nlm.nih.gov/pubmed/12904784
[428]//doi.org/10.1038%2Fnature01725
[429]http://www.space.com/scienceastronomy/060814_tno_found.html
[430]http://www.astronomycast.com/astronomy/episode-64-pluto-and-the-icy-outer-solar-system/
[431]https://web.archive.org/web/20160304083511/http://canadapodcasts.ca/podcasts/TheDays/1358804?start=0
[432]http://nineplanets.org/kboc.html
[433]http://www.minorplanetcenter.net/iau/lists/TNOs.html
[434]This is a composite of four near-true color photographs taken by the *New Horizons* spacecraft on July 14, 2015 from a distance of 720,000 km (450,000 mi). The most prominent feature in the image, the bright, youthful plains of Tombaugh Regio and Sputnik Planitia, can be seen at lower right. It contrasts the darker, more cratered terrain of Cthulhu Regio at lower left. Because of Pluto's 119.591° tilt at its axis, the southern hemisphere is barely visible in this image; the equator runs through Cthulhu and the southern parts of Sputnik Planitia.
[435]The mean elements here are from the Theory of the Outer Planets (TOP2013) solution by the Institut de mécanique céleste et de calcul des éphémérides (IMCCE). They refer to the standard equinox J2000, the barycenter of the Solar System, and the epoch J2000.
[436]Croswell 1997, p. 50.
[437]Croswell 1997, p. 52.
[438]Croswell 1997, pp. 54–55.

[439]UK Consumer Price Index inflation figures are based on data from Gregory Clark (2016), " The Annual RPI and Average Earnings for Britain, 1209 to Present (New Series) https://measuringworth.com/ukearncpi/", MeasuringWorth.com.

[440]The equivalence is less close in languages whose phonology differs widely from Greek's, such as Somali *Buluuto* and Navajo *Tłóotoo*.

[441]Croswell 1997, p. 57.

[442]"Nine Reasons Why Pluto Is a Planet" http://www.philipmetzger.com/blog/nine-reasons-why-pluto-is-a-planet/ by Philip Metzger

[443]http://www.nasa.gov/mission_pages/newhorizons/main/index.html

[444]http://solarsystem.nasa.gov/planets/profile.cfm?Object=Pluto

[445]http://solarsystem.nasa.gov/

[446]http://nssdc.gsfc.nasa.gov/planetary/factsheet/plutofact.html

[447]http://www.lowell.edu/

[448]http://www.astrobio.net/pressrelease/5055/sharpest-ever-views-of-pluto-and-charon

[449]http://www.ifa.hawaii.edu/info/press-releases/PlutoPictures/Pluto-Tholen-10-07.html

[450]http://www.sixtysymbols.com/videos/pluto.htm

[451]http://www.nasa.gov/image-feature/goddard/views-of-pluto-through-the-years

[452]http://photojournal.jpl.nasa.gov/archive/PIA19873_FLYTHROUGH_ANIMATION_V5.mp4

[453]https://www.youtube.com/watch?v=ds_OlZnV9qk

[454]http://www.scientificamerican.com/video/a-day-on-pluto-reconstructed-from-new-horizons-images/

[455]http://pluto.jhuapl.edu/News-Center/News-Article.php?page=20170714-2

[456]https://www.flickr.com/photos/136797589@N04/35260746413/in/photostream/

[457]https://www.flickr.com/photos/136797589@N04/albums/72157686474817595

[458]The latter per the anglicized pronunciation of the .

[459]Calculated on the basis of other parameters.

[460] The Pluto system: Initial results from its exploration by New Horizons http://science.sciencemag.org/content/350/6258/aad1815 Stern et. al. October 16, 2015 Table 1

[461]Charon Geology and imaging Our derived radius of Charon is 606 ± 3 km, similar to ground-based measurements (29); we also determined that Charon is not detectably oblate, with an upper bound on polar flattening of 1%

[462]

[463]Charon http://dictionary.oed.com/cgi/entry/50037018?single=1&query_type=word&queryword=charon&first=1&max_to_show=10 at Oxford English Dictionary

[464]Astronomer Mike Brown can be heard pronouncing it in ordinary conversation on the KCET interview at 42min 48sec. Being a long-time resident of California, he does not distinguish the vowel of the name *Sharon* and the vowel of the classical pronunciation of *Charon*.

[465]

[466]Pronounced 'with a soft "sh" ' per

[467]US Naval Observatory spokesman Jeff Chester, when interviewed on the NPR commentary (at 2min 49sec), says Christy pronounced it rather than classical . In normal conversation, the second vowel is reduced to a schwa: in RP (ref: OED).

[468]Pronounced "Sharon" per and per

[469]Hal Weaver, who led the team that discovered Nix and Hydra, also pronounces it (with a General American accent) on the Discovery Science Channel documentary *Passport to Pluto*, premiered 2006-01-15.

[470]http://solarsystem.nasa.gov/planets/profile.cfm?Object=Plu_Charon

[471]http://solarsystem.nasa.gov

[472]http://adsabs.harvard.edu//full/seri/AJ.../0083//0001005.000.html

[473]http://www.boulder.swri.edu/~buie/pluto/chphases.html

[474]http://www.boulder.swri.edu/~buie/pluto/plutomap1.html

[475]http://news.bbc.co.uk/2/hi/science/nature/4230918.stm

[476]http://www.cbat.eps.harvard.edu/iauc/03200/03241.html

[477]https://web.archive.org/web/20060118020147/http://eso.org/outreach/press-rel/pr-2006/pr-02-06.html

[478]http://arxiv.org/abs/astro-ph/0602082

[479] https://web.archive.org/web/20090320174750/http://spacescience.arc.nasa.gov/agu/abstracts/Deschetal.pdf
[480] http://pluto.jhuapl.edu/news_center/news/20130710.php
[481] http://nasatech.net/ntNewHorizons_PAGE.html
[482] http://pluto.jhuapl.edu/News-Center/News-Article.php?page=20170714-2
[483] https://www.flickr.com/photos/136797589@N04/36101560426/
[484] https://www.flickr.com/photos/136797589@N04/albums/72157686474817595
[485] The Rapa Nui pronunciation is [ˈmakeˈmake], which is approximated in English as US: /ˌmɑːkiˈmɑːki/ ***MAH**-kee-**MAH**-kee*,The Rapa Nui pronunciation is [ˈmakeˈmake], which is approximated in English as US: /ˌmɑːkiˈmɑːki/ ***MAH**-kee-**MAH**-kee*, UK: /ˈmækiˈmæki/ ***MAK**-ee-**MAK**-ee*, or as /ˌmɑːkeɪˈmɑːkeɪ/ ***MAH**-kay-**MAH**-kay*. The first is an anglicized pronunciation; the second is more Polynesian, and is used by Brown and his students.<ref>Podcast *Dwarf Planet Haumea* http://365daysofastronomy.org/2009/03/31/march-31st/ (Darin Ragozzine, at 3′11”) UK: /ˈmækiˈmæki/ ***MAK**-ee-**MAK**-ee*, or as /ˌmɑːkeɪˈmɑːkeɪ/ ***MAH**-kay-**MAH**-kay*. The first is an anglicized pronunciation; the second is more Polynesian, and is used by Brown and his students.<ref>Podcast (Darin Ragozzine, at 3′11”)
[486] Calculated using $(a–b)/a$ and the dimensions from
[487] Calculated using the dimensions from assuming an oblate spheroid.
[488] The Kuiper Belt is a ring of bodies beyond Neptune.
[489]
[490] It has an apparent magnitude in opposition of 16.7 vs. 15 for Pluto.
[491] Based on Minor Planet Center online Minor Planet Ephemeris Service: March 1, 1930: RA: 05h51m, Dec: +29.0.
[492] //en.wikipedia.org/w/index.php?title=Makemake&action=edit
[493] Preliminary simulation of Makemake ()’s orbit http://www.orbitsimulator.com/gravity/articles/newtno.html and the 2009-02-04 nominal (non-librating) rotating frame for Makemake. http://www.orbitsimulator.com/yabbfiles/Attachments/Makemake40000y_13to7.gif See (182294) 2001 KU76 for a proper 11:6 resonance libration.
[494] http://cfa-www.harvard.edu/mpec/K05/K05O42.html
[495] http://hamilton.dm.unipi.it/cgi-bin/astdys/astibo?objects:Makemake;main
[496] http://ssd.jpl.nasa.gov/sbdb.cgi?sstr=136472;orb=1;view=Far
[497] http://ssd.jpl.nasa.gov/horizons.cgi?find_body=1&body_group=sb&sstr=136472
[498] http://www.ing.iac.es:8080/PR/press/2005FY9.html
[499] https://theskylive.com/makemake-info
[500] http://www.klet.org/klenot/html/k05f09y.html
[501] http://www.gps.caltech.edu/~mbrown/makemake.gif
[502] http://antwrp.gsfc.nasa.gov/apod/ap080716.html
[503] http://www.orbitsimulator.com/gravity/articles/newtno.html
[504] *how-**MAY**’-ah*, with three syllables according to the English pronunciation in Hawaii,<ref> *New dwarf planet named for Hawaiian goddess* http://www.heraldnet.com/article/20080919/NEWS02/709199740 (HeraldNet, September 19, 2008)
[505] http://dps08.astro.cornell.edu/AAS_WebcastSchedule_2008.html
[506] Michael E. Brown. *How I Killed Pluto and Why It Had It Coming*, chapter 9: ”The Tenth Planet”
[507] //en.wikipedia.org/w/index.php?title=Haumea&action=edit
[508] Paul Gilster: *Fast Orbiter to Haumea* http://www.centauri-dreams.org/?p=8680. Centauri Dreams—The News of the Tau Zero Foundation. July 14, 2009, retrieved January 15, 2011
[509] http://neo.jpl.nasa.gov/orbits/2003el61.html
[510] http://www.johnstonsarchive.net/astro/astmoons/am-136108.html
[511] http://365daysofastronomy.org/2009/03/31/march-31st/
[512] http://twitpic.com/59rbgj
[513] http://twitpic.com/5acp9q
[514] //en.wikipedia.org/w/index.php?title=Template:Distant_minor_planets_sidebar&action=edit
[515] Maggie Masetti. (2007). *Cosmic Distance Scales – The Solar System http://heasarc.gsfc.nasa.gov/docs/cosmic/solar_system_info.html*. Website of NASA’s High Energy Astrophysics Science Archive Research Center. Retrieved 2008 07-12.

[516]The literature is inconsistent in the use of the phrases "scattered disc" and "Kuiper belt". For some, they are distinct populations; for others, the scattered disc is part of the Kuiper belt. Authors may even switch between these two uses in a single publication.<ref>McFadden, Weissman, & Johnson (2007). *Encyclopedia of the Solar System*, footnote p. 584

[517]Schmadel, Lutz D. (2003); *Dictionary of Minor Planet Names* (5th rev. and enlarged ed. edition). Berlin: Springer. Page 925 (Appendix 10). Also see McFadden, Lucy-Ann; Weissman, Paul & Johnson, Torrence (1999). *Encyclopedia of the Solar System*. San Diego: Academic Press. Page 218.

[518]Remo notes that Cis-Neptunian bodies "include terrestrial and large gaseous planets, planetary moons, asteroids, and main-belt comets within Neptune's orbit." (Remo 2007)

[519](Preprint version (pdf) http://www2.ess.ucla.edu/~jewitt/papers/2006/DJ06.pdf)

[520]Images were taken on October 21, 2003; the object was not detected and identified until 2005.

[521] UNIQ-ref-0-3951097dbf03c794-QINU Both are common; the former is the literary pronunciation of the name, but Brown and his students use the latter. UNIQ-ref-1-3951097dbf03c794-QINU

[522]Calculated from the mean radius

[523]Calculated by dividing the listed mass by the listed volume

[524]Calculated based on the known parameters

[525]Calculated from the listed diameter (*D*) and albedo (*p*) using UNIQ-math-2-3951097dbf03c794-QINU <ref>

[526]//en.wikipedia.org/w/index.php?title=Eris_(dwarf_planet)&action=edit

[527]As of February 2016//en.wikipedia.org/w/index.php?title=Eris_(dwarf_planet)&action=edit UNIQ-nowiki-3-3951097dbf03c794-QINU , Sedna is 85.7 astronomical units (1.282×10^{10} km; 7.97×10^{9} mi) from the Sun,Images were taken on October 21, 2003; the object was not detected and identified until 2005. whereas Eris is 96.3 astronomical units (1.441×10^{10} km; 8.95×10^{9} mi) from the Sun. Both are common; the former is the literary pronunciation of the name, but Brown and his students use the latter. Eris is still close to its 1977 aphelion (furthest distance from the Sun), whereas Sedna is nearing its 2076 perihelion (closest approach to the Sun). Sedna will overtake Eris as the farthest presently known spherical minor planet in 2114.

[528]Brown is joking on this point. It was in fact the Disney character Pluto that was named after the newly discovered "planet", though Venetia Phair, Pluto's christener, had to counter accusations her whole life that she named the planet after a cartoon dog.

[529]For an example of an amateur image of Eris, see Fred Bruenjes' Astronomy http://www.moonglow.net/ccd/pictures/pc/index.html#tnos

[530]//en.wikipedia.org/w/index.php?title=Template:TNO-distance&action=edit

[531]The Resolution of the High Resolution Channel of the ACS is 40 marcsec (milliarcseconds) and the size of 1 pixel is ~25 marcsec i.e. ~1875 km at the distance of Eris.

[532]The reference to 'direct' measurement by the HST should not mislead into thinking that this method is as 'direct' and model-independent as measuring say Neptune's size. Basically, the method consists in finding the statistically best fit to a *smeared* image of the size of less than 2 pixels by comparing it with smeared images of the background stars, using a given computer model of the optics (PSF). A non-technical description of the method is given on Brown's page http://www.gps.caltech.edu/~mbrown/planetlila/index.html#size, a detailed description of this approach and its limitations are discussed in a paper on Quaoar

[533]http://www.minorplanetcenter.net/db_search/show_object?object_id=Eris&commit=Show

[534]http://www.gps.caltech.edu/~mbrown/planetlila/

[535]http://www.gps.caltech.edu/~mbrown/planetlila/moon/

[536]https://web.archive.org/web/20080626220548/http://www.pluggd.tv/audio/channels/kcet_podcast__hammer_conversations/episodes/2h10l

[537]https://web.archive.org/web/20051210231211/http://www.johnstonsarchive.net/astro/astmoons/am2003ub313.html

[538]http://www.minorplanetcenter.net/mpec/K05/K05O41.html

[539]http://neo.jpl.nasa.gov/orbits/2003ub313.html

[540]http://www.spaceflightnow.com/news/n0509/09solarssytem/

[541] https://web.archive.org/web/20051108052322/http://www.slackerastronomy.org/wordpress/index.php/archive/10th-planet/

[542] https://web.archive.org/web/20060426030013/http://www.iau.org/TRANS-NEPTUNIAN_OBJECT_2003_UB.324.0.html

[543] http://www.orbitsimulator.com/gravity/articles/newtno.html

[544] https://web.archive.org/web/20060114202745/http://www2.keck.hawaii.edu/optics/staff/mvandam/gabrielle

[545] http://pr.caltech.edu/media/Press_Releases/PR12724.html

[546] https://web.archive.org/web/20051202081513/http://www.spitzer.caltech.edu/Media/happenings/20050807/

[547] The Solar Wind http://solarscience.msfc.nasa.gov/SolarWind.shtml

[548] "The Unexpected Structure of the Heliotail" http://www.astrobio.net/pressrelease/5553/the-unexpected-structure-of-the-heliotail, *Astrobiology*. July 12, 2013

[549] Cole, Steve. "NASA Satellite Provides First View of the Solar System's Tail" http://www.nasa.gov/press/2013/july/nasa-satellite-provides-first-view-of-the-solar-systems-tail/#.UmxeevmkrJa NASA News Release 12-211, July 10, 2013

[550] Oct.15/09 IBEX team announcement at http://ibex.swri.edu/

[551] Dave McComas, IBEX Principal Investigator at http://ibex.swri.edu/

[552] http://www.universetoday.com/am/publish/size_of_heliopause.html

[553] http://www.cosmosmagazine.com/node/1296

[554] http://photojournal.jpl.nasa.gov/catalog/PIA17460

[555] http://spacefellowship.com/news/art14466/cassini-data-helps-redraw-shape-of-our-solar-system.html

[556] http://www.dartmouth.edu/~hrm/pubabs.html

[557] http://voyager.jpl.nasa.gov/mission/interstellar.html

[558] http://helios.gsfc.nasa.gov/heliosph.html

[559] http://www.galex.caltech.edu/

[560] http://solar-heliospheric.engin.umich.edu

[561] http://www.sciencedaily.com/releases/2010/05/100521191114.htm

[562] http://www.nasa.gov/mission_pages/voyager/heliosphere-surprise.html

[563] http://www.agu.org/journals/eo/eo1136/2011EO360001.pdf#anchor

[564] http://adsabs.harvard.edu/abs/2011EOSTr..92..297S

[565] //doi.org/10.1029%2F2011eo360001

[566] http://web.mit.edu/space/www/helio.review/axford.suess.html

[567] http://www.uiowa.edu/~ournews/2003/december/120803gurnett.html

[568] http://interstellar.jpl.nasa.gov/

[569] http://www.cnn.com/2005/TECH/space/05/25/voyager.space/index.html

[570] https://www.newscientist.com/article.ns?id=mg18625015.000

[571] https://science.nasa.gov/headlines/y2006/21sep_voyager.htm?list33024

[572] http://adsabs.harvard.edu/abs/2004AdSpR..34...66W

[573] http://www.dartmouth.edu/~hrm/R/heliosphere.html

[574] http://www.astronomycast.com/astronomy/episode-65-the-end-of-our-tour-through-the-solar-system/

[575] D.Jewitt, A.Delsanti *The Solar System Beyond The Planets* in *Solar System Update : Topical and Timely Reviews in Solar System Sciences* , Springer-Praxis Ed., (2006) Preprint of the article (pdf) http://www.ifa.hawaii.edu/faculty/jewitt/papers/2006/DJ06.pdf

[576] //en.wikipedia.org/w/index.php?title=Template:Distant_minor_planets_sidebar&action=edit

[577] D.Jewitt, A. Moro-Martın, P.Lacerda *The Kuiper Belt and Other Debris Disks* to appear in *Astrophysics in the Next Decade*, Springer Verlag (2009). Preprint of the article (pdf) http://www.ifa.hawaii.edu/faculty/jewitt/papers/2008/JML08.pdf

[578] (subscription required)

[579] Twelve minor planets with a semi-major axis greater than 150 AU and perihelion greater than 30 AU are known,<ref name=mpcq30a150>

[580] is excluded from the count because it has an observation arc of only 76 days and hence its semi-major axis is not securely known. http://ssd.jpl.nasa.gov/sbdb.cgi?sstr=2003SS422

[581]Objects with a Minor Planet Center designation number have an orbit with more observations taken over a longer period of time, which is therefore better determined and more securely known, than the orbit of objects with only a provisional designation.

[582]Method of diameter calculation: "Assumed" means the albedo of the object is assumed to be 0.04, and the object's diameter is calculated accordingly.

[583]minorplanetcenter.net: q>40, a>50 http://minorplanetcenter.net/db_search/show_by_properties?perihelion_distance_min=40&semimajor_axis_min=50

[584]https://www.pbs.org/wgbh/nova/next/space/a-new-addition-to-the-solar-system-could-hint-at-something-larger/

[585]//en.wikipedia.org/w/index.php?title=Template:Distant_minor_planets_sidebar&action=edit

[586]The Oort cloud's outer limit is difficult to define as it varies over the millennia as different stars pass the Sun and thus is subject to variation. Estimates of its distance range from 50,000 to 200,000 AU.

[587]Absolute magnitude is a measure of how bright an object would be if it were 1 AU from the Sun and Earth; as opposed to apparent magnitude, which measures how bright an object appears from Earth. Because all measurements of absolute magnitude assume the same distance, absolute magnitude is in effect a measurement of an object's brightness. The lower an object's absolute magnitude, the brighter it is.

[588]Scientific American - Kepler Spacecraft May Be Able to Spot Elusive Oort Cloud Objects - 2010 http://www.scientificamerican.com/article/kepler-oort-cloud/

[589]http://solarsystem.nasa.gov/planets/profile.cfm?Object=KBOs

[590]http://solarsystem.nasa.gov/

[591]http://www.nineplanets.org/kboc.html

[592]http://adsabs.harvard.edu/abs/1994QJRAS..35....1M

[593]http://arxiv.org/abs/1411.1844

[594]Tombaugh (1946), p. 73.

[595]//en.wikipedia.org/w/index.php?title=Planets_beyond_Neptune&action=edit

[596]Croswell (1997), p. 43

[597]Croswell (1997), p. 43.

[598]

[599]Littman (1990), p. 70.

[600]Croswell p. 50

[601]Croswell (1997), p. 49.

[602]Croswell (1997), pp. 32–55.

[603]"Percival Lowell's three early searches for Planet X" http://www.astronomy.com/year-of-pluto/2015/05/percival-lowells-three-early-searches-for-planet-x. *Astronomy Magazine*. May 14, 2015.

[604]Tombaugh (1946), p. 79

[605]

[606]Rawlins also took into account Pluto's stellar occultation failure as reported in 1966 by I. Halliday, R. Hardie, O. Franz, and J. Priser in *P. A. S. P.* volume 78 page 113.

[607]Croswell (1997), p. 57.

[608]Croswell (1997), pp. 57–58.

[609]

[610]Croswell (1997), pp. 62–63.

[611]Croswell (1997), p. 63.

[612]Croswell (1997), p. 66.

[613]Littmann (1990), p. 204.

[614]Pluto, Eris, and the Dwarf Planets of the Outer Solar System - talk by Mike Brown - the argument about dwarf planet sizes beyond Neptune is 50 minutes into his talk - on youtube (the WGBH link doesn't work) https://www.youtube.com/watch?v=WHNO079G1i8&feature=youtu.be&t=3135

[615]New Scientist, 23 July 2005, issue 2509, Far-out worlds, just waiting to be found https://www.newscientist.com/article/mg18725091.500-farout-worlds-just-waiting-to-be-found.html

[616]http://adsabs.harvard.edu/abs/1946ASPL....5...73T

[617]https://www.youtube.com/watch?v=xGfv3Oay_pY

[618] http://www.nineplanets.org/hypo.html#planetx

[619] http://adsabs.harvard.edu/abs/1993AJ....105.2000S

[620] //doi.org/10.1086%2F116575

[621] http://adsabs.harvard.edu/abs/1993Natur.363...18Q

[622] //doi.org/10.1038%2F363018b0

[623] http://adsabs.harvard.edu/abs/1985Natur.313...36W

[624] //doi.org/10.1038%2F313036a0

[625] http://www8.nationalacademies.org/astro2010/DetailFileDisplay.aspx?id=288

[626] http://theconversation.com/claims-about-new-planets-that-turned-out-to-be-wrong-and-why-planet-nine-may-be-different-53573

[627] Most news outlets reported the name as *Phattie* (a slang term for "cool" or "awesome"; also, a marijuana cigarette) UNIQ-ref-0-3951097dbf03c794-QINU but *The New Yorker* quote cited above uses "fatty" in what appears to be a nearly unique variation. The apparently correct spelling has been substituted.

[628] *The New Yorker* put the average orbital distance of Planet Nine into perspective with an apparent allusion to one of the magazine's most famous cartoons, *View of the World from 9th Avenue*: "If the Sun were on Fifth Avenue and Earth were one block west, Jupiter would be on the West Side Highway, Pluto would be in Montclair, New Jersey, and the new planet would be somewhere near Cleveland. UNIQ-ref-1-3951097dbf03c794-QINU "

[629] See RA/Dec chart at

[630] See embedded video simulation at

[631] Two types of protection mechanisms are possible:<ref name=KopanyasKozai>

[632] Given the orbital eccentricity of these objects, different epochs can generate quite different heliocentric unperturbed two-body best-fit solutions to the semi-major axis and orbital period. For objects at such high eccentricity, the Sun's barycenter is more stable than heliocentric values. Barycentric values better account for the changing position of Jupiter over Jupiter's 12 year orbit. As an example, 2007 TG$_{422}$ has an epoch 2012 heliocentric period of $\sim$13,500 years,<ref name=TG422-2012>

[633] Batygin and Brown provide an order of magnitude estimate for the mass. • If M were equal to 0.1 $M_{\oplus}$, then the dynamical evolution would proceed at an exceptionally slow rate, and the lifetime of the Solar System would likely be insufficient for the required orbital sculpting to transpire. • If M were equal to 1 $M_{\oplus}$, then long-lived apsidally anti-aligned orbits would indeed occur, but removal of unstable orbits would happen on a much longer timescale than the current evolution of the Solar System. Hence, even though they would show preference for a particular apsidal direction, they would not exhibit true confinement like the data. • They also note that M greater than 10 $M_{\oplus}$ would imply a longer semi-major axis. Hence they estimate that the mass of the object is likely in the range of 5 $M_{\oplus}$ to 15 $M_{\oplus}$.

[634] The average of longitude of the ascending node for the 6 objects is about 102°. In a blog published later, Batygin and Brown constrained their estimate of the longitude of the ascending node to 94°.

[635] A link to the plots of the orbital evolution of all 15 is included in the arxiv version of paper.

[636] This process was slower (or not observed) for the Sedna clones with smaller (or zero) inclinations of Planet Nine.

[637] Of the eight objects with semi-major axis $>$ 150 AU OSSOS found three with arguments of perihelion (ω) outside the cluster previously identified by Trujillo and Sheppard (2014): 2015 GT$_{50}$, 2015 KH$_{163}$, and 2013 UT$_{15}$.

[638] In their paper Brown and Batygin note that "the vast majority of this (primordial planetesimal disk) material was ejected from the system by close encounters with the giant planets during, and immediately following, the transient dynamical instability that shaped the Kuiper Belt in the first place. The characteristic timescale for depletion of the primordial disk is likely to be short compared with the timescale for the onset of the inclination instability (Nesvorný 2015), calling into question whether the inclination instability could have actually proceeded in the outer solar system."

[639] The Search for Planet 9 https://www.youtube.com/watch?v=MptrypvBTag Talk by author Dr. Renu Malhotra, Public talk at TEDxPortland, Published on 17 July 2017

[640] In their paper Brown and Batygin note that "the lack of $\omega \sim 180$ objects suggests that some additional process (such as encounter with a passing star) is required to account for the missing objects at 180°".

[641] Assuming that the orbital elements of these objects have not changed, Jílková et al. proposed an encounter with a passing star might have helped acquire these objects – dubbed sednitos (ETNOs with $q > 30$ and $a > 150$) by them. They also predicted that the *sednitos* region is populated by 930 planetesimals and the inner Oort Cloud acquired ~440 planetesimals through the same encounter.<ref name=JilkovaSednito>

[642] In their paper Brown and Batygin note that the fact that the ratio of the perturbed object to perturber semimajor axis is nearly equal to one "means that trapping all of the distant objects within the known range of semimajor axes into Kozai resonances likely requires multiple planets, finely tuned to explain the particular data set". Hence, they do not favor this theory, as they view it too unwieldy.

[643] The 8-meter Subaru Telescope has achieved a 27.7 magnitude photographic limit with a ten-hour exposure,<ref>

[644] //en.wikipedia.org/w/index.php?title=Planet_Nine&action=edit

[645] It is estimated that to find Planet Nine, telescopes that can resolve a 30 mJy point source are needed, and that can also resolve an annual parallax motion of ~5 arcminutes per year.<ref name=aasnovacmb>

[646] http://minorplanetcenter.net/db_search/show_by_properties?perihelion_distance_min=30&semimajor_axis_min=200

[647] A 3-D version of this orbit and those of several ETNOs is available.<ref name="Millholland_3D_orbit">

[648] Five of the eight objects with perihelion < 35 AU are the perpendicular objects noted by Batygin and Brown

[649] The eccentric Kozai-Lidov mechanism is discussed in a review by Smadar Noaz.<ref name="Noaz_2016">

[650] http://www.astro.uvic.ca/~micheleb/

[651] https://techcrunch.com/timeline/the-top-15-events-that-happened-in-space-in-2016/

[652] https://www.youtube.com/watch?v=ujruGHhHQIk

[653] https://www.youtube.com/watch?v=42GeoCVaZQg

[654] http://www.findplanetnine.com/

[655] http://theconversation.com/claims-about-new-planets-that-turned-out-to-be-wrong-and-why-planet-nine-may-be-different-53573

[656] https://www.wired.com/2016/01/could-you-live-on-planet-nine/

[657] http://www.space.com/31671-planet-nine-discovery-explained-infographic.html

[658] https://www.theatlantic.com/science/archive/2016/01/planet-nine-may-help-us-slingshot-our-way-to-the-stars/425010/

[659] http://mintonsunday.livemint.com/news/is-planet-nine-the-new-x/2.4.3926705692.html

[660] https://www.scientificamerican.com/article/solar-system-survey-casts-doubt-on-mysterious-ldquo-planet-nine-rdquo/

[661] http://spectrum.ieee.org/aerospace/satellites/is-there-a-giant-planet-lurking-beyond-pluto

Article Sources and Contributors

The sources listed for each article provide more detailed licensing information including the copyright status, the copyright owner, and the license conditions.

Discovery and exploration of the Solar System *Source:* https://en.wikipedia.org/w/index.php?oldid=800612879 *License:* Creative Commons Attribution-Share Alike 3.0 *Contributors:* Alinor, Andrewjohnbayles, Armanjafari, Avenue, BD2412, Beland, Bgwhite, BlubberMax, Chezcake1234, Chrisjj, ClueBot NG, Czolgolz, Dane, Dawnseeker2000, Donner60, Drmies, Edknol, EvenGreenerFish, Fjörgynn, Flyer22 Reborn, Fotaun, FraPado86, Frosty, Funandtrvl, Gap9551, IronGargoyle, J 1982, JFG, Jagged 85, Jim.henderson, Jirka.h23, Jopparn, JorisvS, Lasunncty, Materialscientist, Mogism, Nehrams2020, Nergaal, Richie9247w, Rorschach, Rstml, Sbharris, Serendipodous, Shenme, Spideratseds, Srwalden, Tdadamemd sioz, The Tom, Tom.Reding, Tony Mach, Tushar Kuware, UY Scuti, Widr, Wikishovel, 43 anonymous edits 1

Formation and evolution of the Solar System *Source:* https://en.wikipedia.org/w/index.php?oldid=805174690 *License:* Creative Commons Attribution-Share Alike 3.0 *Contributors:* 1963don, Anrnusna, Antoni Barau, Arbnos, Ashill, Astronaut Willis, Audin, BatteryIncluded, Bear-rings, Bender235, Besselfunctions, BirdValiant, Blackninga3216, BlakePlays, Bomb319, CES1596, CLCStudent, CanadianLinuxUser, ChamithN, Charles Edwin Shipp, Ckatz, ClueBot NG, CommonsDelinker, Crzer07, CyanoTex, DVdm, DatGuy, David.moreno72, Davros69999, Dawnseeker2000, Dcirovic, Delphan Gruss EVI, DemocraticLuntz, Dobie80, Dodshe, Dolkungbighead, Download, DrKay, DrStrauss, Drbogdan, Dukebaker, Dwarfpimp, Earthandmoon, Ehrenkater, El C, Elimagic, Eric Kvaalen, Escape Orbit, Excirial, Fama Clamosa, Filiosus's Saga, FlightTime, Friginator, Frinthruit, Gakrivas, Gap9551, GeneralizationsAreBad, GiantSnowman, Gilliam, Gilo1969, Hdjensofjfnen, Headbomb, Heron, Hillbillyholiday, Houjou, Intel4004, Iridescent, IronGargoyle, J 1982, JFG, JaconaFrere, Jarble, Jcpag2012, Jim1138, Jmencisom, Joe Joe S.34567, John, JohnGolt, Jon Kolbert, JorisvS, Josve05a, Jwratner1, Kbh3rd, Kheider, Kianu123456789123456789, Kostas20142, Kozmosis, L1A1 FAL, Lawrence5641, Lchappell, Lugia2453, MarkSewath, Materialscientist, Mchcopl, Messier8, MusikAnimal, NSH002, Newone, Niceguyedc, Niseto, Ode2joy, Orenburg1, Oshwah, Oxcross, PlanetUser, Quenhitran, RA0808, Reatlas, Red Planet X (Hercolubus), Reddi, Rjwilmsi, Rmosler2100, Rod57, Rothorpe, Ruslik0, Seaphoto, Serendipodous, Serols, Sir Cumference, Sithlord Timmy, Slightsmile, Space Infinite, Steven Corder, Tetra quark, Tfr000, The Herald, Tide rolls, ToBeFree, Tom.Reding, Treisijs, Urhixidur, V620 Cephei, Vsmith, Wtmitchell, Xinxinliu, Zezen, Zundark, 166 anonymous edits 16

Sun *Source:* https://en.wikipedia.org/w/index.php?oldid=806164912 *License:* Creative Commons Attribution-Share Alike 3.0 *Contributors:* A2soup, Agge.se, Airplaneman, Allixpeeke, Arado, Archon 2488, Asgrrr, Ashill, Autumn Wind, BD2412, BU Rob13, Bender235, Blanche of King's Lynn, Bob Stein - VisiBone, Bogger, Brandmeister, Brmb101, Captain Cornwall, CommonsDelinker, Cpiral, CuriousMind01, DILNN1, DOwenWilliams, DRAGON BOOSTER, DVdm, Dao1, Daqu, Darsie42, Davemck, David J Johnson, Davidbuddy9, Dawnseeker2000, Dbachmann, Dbfirs, Ddvche, Denimadept, DocWatson42, DoctorKubla, Double sharp, DrKay, Dragons flight, Drbogdan, Earthandmoon, Eco-climber, Edokter, Edward, Eric Kvaalen, Exoplanetaryscience, Felix179793, Feminist, Fkbreitl, FlightTime, Fmadd, Gap9551, GünniX, HalloweenNight, HapHaxion, Headbomb, Heididoerr061, Hertz1888, Hillbillyholiday, Howpper, Hypergeek14, I am. furhan., Iggy the Swan, Inyouchuu shoku, Ironnail, Isambard Kingdom, Izno, Jcpag2012, Jkokavec, Jo-Jo Eumerus, JoeSperrazza, JoeyPknowsalotaboutthat, John D. Croft, Jonesey95, JorisvS, Judist, Jusdafax, Justin15w, Kanjuzi, Katolophyromai, KiloByte, L4tybj, Loraof, LordofPens, MJSS, Macchess, Mark Schierbecker, MartinZ, Materialscientist, MattSucci, Max Paxman, Mazarin07, MeDrewNotYou, Motivação, Nedim Ardoğa, Niceguyedc, Nutster, Nwbeeson, Nyttend, Octoberwoodland, Ost316, Paine Ellsworth, Palnatoke, Patar knight, Pdebee, Phil Boswell, Pintoch, PlanetUser, Pocketthis, Praemonitus, Primefac, RTBoyce, Randy Kryn, Red Planet X (Hercolubus), RichardWeiss, Rjensen, Rjwilmsi, Rolf h nelson, RubenGarciaHernandez, Ruslik0, Sandstein, Sbharris, Sbmeirow, Scwlong, Serendipodous, Sinsearach, Sir Cumference, Skeletoroforange, Smash1gordon, Spacepotato, Stamptrader, Steve03Mills, Steviedman, Stigmatella aurantiaca, Student7, TLT2232, TStein, Thor Dockweiler, Titanium Dragon, Tom.Reding, Tony1, Trappist the monk, U-95, Valuesize, VoxLuna, Wavelength, WiKaffiN, WolfmanSF, Xaxafrad, Yogee23, YuRi YuZi, Zedshort, İnternion 36

Interplanetary medium *Source:* https://en.wikipedia.org/w/index.php?oldid=798218463 *License:* Creative Commons Attribution-Share Alike 3.0 *Contributors:* Ageekgal, Art Carlson, Art LaPella, Ashill, Ask123, BD2412, Beland, Bender235, Braincricket, Brews ohare, CanadianLinuxUser, Chamberlian, Crispmuncher, D. F. Schmidt, DabMachine, Doradus, Drbogdan, Druzhnik, Exodio, Fotaun, Funkjazz, Gaius Cornelius, Gene Nygaard, Headbomb, Iantresman, J 1982, JTN, Jdaloner, Johndric Valdez, Jowa fan, Knick99, Ktr101, Loraof, Mani1, Meshakhtev, Mmeijeri, Mpfiz, Nondistinguished, Novangelis, RJHall, Reventløv, Riana, Ricardogpn, Roberto Mura, Ruslik0, Serendipodous, Shalom Yechiel, Shawn Worthington Laser Plasma, Tassedethe, Tcamps42, Tetra quark, The machine512, Tom.Reding, Trevayne08, Vina-iwbot～enwiki, Who, WikiBotas, William Allen Simpson, WolfmanSF, World-traveller, *Zzzzzzzzzz*, 22 anonymous edits 73

Solar wind *Source:* https://en.wikipedia.org/w/index.php?oldid=805989687 *License:* Creative Commons Attribution-Share Alike 3.0 *Contributors:* Acalamari, Alecbaravik, Alexandermcnabb, Ardaglash, Ashanda, Ashill, AwamerT, Benavidezt, Bender235, Brian.rtl, Buzzzsherman, Cadiomals, Captain Quirk, Chrism, Citation bot 1, ClueBot NG, CogitoErgoSum14, CommonsDelinker, Crossland～enwiki, Crowsnest, Dan100, Dimath, Discospinster, Dododerek, Dondervogel 2, Dynamitecow, Ed Poor, Ekendrick, Elenceq, Entranced98, Fartherred, Favonian, Fign44, Forcimonie, Fotaun, Fæ, Gap9551, Gillespie09, GliderMaven, Gpuica, Graham87, Grutness, HaeB, Hairy Dude, Halmstad, Happy Attack Dog, Headbomb, Heliophysics, Hmains, Hugh Hudson, IVAN3MAN, Ibrahm-07, Icairns, Incnis Mrsi, Isambard Kingdom, J 1982, J.delanoy, JaconaFrere, James McBride, Jauhienij, Jdagius, Jerry Wright, Jim1138, Joel B. Lewis, John Hyams, JorisvS, Jrising, Justin15w, Kadefe, Karlpoppery, KeViNdUrAnT600000, Khazar2, Kingpin13, Kmote, KoshVorlon, Lfstevens, Lithopsian, LordElektroWIKI, LordofPens, Lucas Thoms, Lynnbwilsoniii, MadGeographer, Magioladitis, Malcolmxl5, Mappetop, Marek69, Marissa.English, Materialscientist, Medium69, Michael C Price, Michael H 34, Millennium bug, Mjasfca, Mouse Nightshirt, Mrienstra, MusikAnimal, Myasuda, N2e, Ninney, Orionus, Orphan Wiki, Oshwah, Ost316, Pinethicket, Pleiotrop3, R'n'B, RICHARDS VIEW, Reatlas, Rfassbind, RiskNerd, Rjwilmsi, RockMagnetist, Rothorpe, Rousfo, Ruslik0, Sanam Pakuwal, SergeantHippyZombie, Shawn Worthington Laser Plasma, Shooke, Slightsmile, Some jerk on the Internet, Steve Wise, Tassedethe, Titodutta, Tlettington, Tom.Reding, Vaibhav7101, Vanjagenije, Vegaswikian, Vgy7ujm, Vrenator, Vsmith, WIMwizard, Widr, Wiki guy752, William M. Connolley, Wknight94, Xeno, 170 anonymous edits 76

Terrestrial planet *Source:* https://en.wikipedia.org/w/index.php?oldid=804201328 *License:* Creative Commons Attribution-Share Alike 3.0 *Contributors:* 1cr malkin, 1exec1, 4th-otaku, Alansohn, Allens, Aniten21, Ansh Mathur, Astredita, Avoided, Baconfry, Basilo, BatteryIncluded, Bear-rings, Benbest, Bjankuloski06en～enwiki, Blackbird5555, Blaxthos, Bosteeboy, Braincricket, Cadiomals, CamxxCore, Cazorla, Cgt, Chipotle, Christopher Thomas, Ckatz, Clpo13, ClueBot NG, Cyberbot II, Davidbuddy9, Denisarona, Dna-webmaster, Drbogdan, EvenGreenerFish, Ewigekrieg, Ewulp, Excirial, Filiosus's Saga, Fuddle, Gap9551, GermanJoe, Gniniv, Good Olfactory, Hans Dunkelberg, Harlem Baker Hughes, Hughleesonsmith, Izno, J 1982, J991, Jcpag2012, Jfkduihv, JorisvS, Kheider, Kintetsubuffalo, Kwamikagami, Layla, the remover, Leofil2, Lightmouse, LuK3, Lugia2453, Mann jess, Marikanessa, MartinZ, Matthurricane, Mercurywoodrose, Mike Rosoft, Mpvdm, Narky Blert, Nergaal, Nightscream, PhnomPencil, PlanetStar, Pthag, Quinton Feldberg, QuiteUnusual, Raine12, Reatlas, Rjwilmsi, Roentgenium111, Rothorpe, Roxy the dog, Ruslik0, Scjessey, Scooter20, Shadowjams, Smalljim, Smettems, Stas1995, Super48paul, Swcisel, TLSuda, Tamfang, Terraflorin, Terrestrial111, The High Fin Sperm Whale, The Tom, The Transhumanist, ThePlatypusofDoom, TheWhistleGag, Tide rolls, Tom.Reding, TomS TDotO, Tomásdearg92, Trappist the monk, Vanished user tj23rpoij4tikkd, Vegan11, Viriditas, Vsmith, Waldir, Wavehunter, Wer900, Wiae, Widr, Wieralee, Wing7990, Wknight94, Wtmitchell, Yahia.barie, Yamaguchi先生 , Yurivict, Zedmelon, 110 anonymous edits 88

Mercury (planet) *Source:* https://en.wikipedia.org/w/index.php?oldid=805541261 *License:* Creative Commons Attribution-Share Alike 3.0 *Contributors:* 331dot, A2soup, Ahatmose, Alexandritechrysoberyl, Andyjsmith, Arado, Archon 2488, Arianewiki1, AshLin, Aveh8, BD2412, Bagunceiro, BatteryIncluded, Becky Sayles, Beland, Bencherlite, Bender235, Bomb319, Cablehorn, Cadiomals, ComicsAreJustAllRight, Crisco 1492, CuriousEric, DN-boards1, DOwenWilliams, DVdm, Dawnseeker2000, DePiep, Deeday-UK, DelftUser, Dhtwiki, Dlthewave, DocWatson42, Double sharp, DrKay, Drbogdan, Earthandmoon, Edulovers, Efroimsk, Ehrenkater, Elaqueate, Exoplanetaryscience, FlightTime, GeoffreyT2000, Graham1973, Hedwig in Washington, Hellbus, Hesperian, Hillbillyholiday, Hmains, Hms1103, Hoho24, Huntster, Huritisho, Iggy the Swan, IggyKoopa408, Jcpag2012, JeanLucMargot, Jim.henderson, John, John85, Jon Kolbert, JorisvS, Jpgordon, Kharkiv07, Kheider, Kohran, Kudzu1, Kwamikagami, Lafuzion, Loraof, LouScheffer, MaelstromOfSilence, Manubhatt3, MartinZ, Materialscientist, Mikhail Ryazanov, Mitchumch, Modest Genius, Mousenight, Mr Stephen, Mx. Granger, NSH002, Nedim Ardoğa, Nihiltres, Nijoakim, Nvvchar, Nwbeeson, Oldag07, Pawyilee, Pdebee, PhilipTerryGraham, Piledhigheranddeeper, PlanetUser, Plastikspork, PlayStation 14, Primefac, Purgy Purgatorio, Quindraco, Rfassbind, Rjwilmsi, Rodrigolopes, Rothorpe, Rpot2, Rusl, Ryulong, Sailsbystars, Saros136, Sbmeirow, Schuy B., Signedzzz, Skatebiker, SkoreKeep, Slightsmile, Smk65536, Smkolins, Steve03Mills, SucreRouge, TAnthony, TJRC, Tetra quark, Thanatos666, The Transhumanist, TheGlatiator, TheWhistleGag, Thecodingproject, Thor Dockweiler, Thryduulf, Tom.Reding, Trappist the monk, Twinsday, Tóraí, Ugly Ketchup, Urhixidur, Velho, VirtualDave, W like wiki, White whirlwind, Wieralee, Wikifan2744, WolfmanSF, WurmWoode, YBG, Zundark, НиколаБ 94

Venus *Source:* https://en.wikipedia.org/w/index.php?oldid=806124444 *License:* Creative Commons Attribution-Share Alike 3.0 *Contributors:* A2soup, AkanoToE, Arglebargle79, Arostian, Axl, Barjimoa, BatteryIncluded, Bender235, Bomb319, Coviekiller5, Cpt Wise, DN-boards1, DiScOrD tHe LuNaTiC, Diego Daniel Garcia Gonzalez, Dilidor, Dlthewave, Dondervogel 2, Double sharp, DrKay, Drbogdan, Driftwoodzebulin, Editor2020, Edulovers, Explicit, Finell, Finnusertop, Firth m, Fixuture, Fotaun, Fountains of Bryn Mawr, Frietjes, Gap9551, GentleDjinn, Graeme Bartlett, HalloweenNight, Headbomb, Hellbus, Hillbillyholiday, Huntster, Huritisho, IVORK, IfYouDoIfYouDon't, Iggy the Swan, Isambard Kingdom, J 1982, JeanLucMargot, JerrySa1, Jim1138, John, John of Reading, JohnAM bva, Johnuniq, Jon Kolbert, JorisvS, Josve05a, Justice and low, Justin Stafford, Katolophyromai, Kendall-K1, Kheider, Kraftlos, Krenair, Kwamikagami, Kyle castellan, LL221W, LlywelynII, Loraof, Lylahearts, Magioladitis, Marconi sparks, MartinZ, Martinkunev, Millie Yang, Mz7, Narky Blert, Nihiltres, Pdebee, Penskins, Pinoyaussie, PlanetUser, Praemonitus, Randy Kryn, Reb2016, Rfassbind, Richerman, Rjwilmsi, Robert.Allen, Rod57, Ron Cho, Rothorpe, SLIMHANNYA, Sandhyavempati, Saros136, Saurusaurus, Sbmeirow, Serendipodous, Sir Anon, Steve03Mills,

TAnthony, TaigaBridge, The Anome, The Transhumanist, Tom.Reding, Tomruen, Trackteur, Twinsday, Urhixidur, VQuakr, Vranak, W like wiki, White whirlwind, Wikid77, Wingedsubmariner, Woodlot, XeloriaGrand, Yiyiyiyu, Ywaz, Zaslav, Zedshort, Zimbabweed, Георгий Елизаров 130

Earth *Source:* https://en.wikipedia.org/w/index.php?oldid=806032373 *License:* Creative Commons Attribution-Share Alike 3.0 *Contributors:* 1618033golden, A2soup, AlexanderLevian, AnAwesomeArticleEditor, Archlinux, ArnoldReinhold, Attic Salt, BatteryIncluded, Bear-rings, BeatlesLedTV, Bmakailon, Bobbie73, Bomb319, Bongwarrior, Boorider7, Brandmeister, Brianga, Btornado, CelestialStar14, Cgx8253, Charlesdrakew, CheChe, Club-justin, ClueBot NG, CuriousMind01, DBrant (WMF), DVdm, Dao1, Dawnseeker2000, Dbfirs, Democratics, Dhtwiki, Dlthewave, DocWatson42, Double sharp, Dr.K., DrKay, Drbogdan, Duncan.Hull, EeeveeeFrost, Ermahgerd9, Exoplanetaryscience, Facts707, Farleyeye, Firth m, Floatjon, Fmadd, Fotaun, G0mx, Gorthian, Hawkeye75, Hddty., Headbomb, Huntster, I am. furhan., Iboughttoomanygames, Iggy the Swan, Isambard Kingdom, JEH, JFG, JJBers, JeanLucMargot, Jeppesn, Joeinwiki, Jon Kolbert, Jorge Iani, JorisvS, Just plain Bill, Kajabla, Kellner21, Kind Tennis Fan, Kku, Ldweisberg, Levonscott, Lopifalko, LuciferZH, Magic9mushroom, Mareklug, MarioProtIV, MartinZ, Materialscientist, Mikenorton, Mindbuilder, Mojo0306, Motivação, Mujtaba!, Mykhal, Narky Blert, Navarre0107, NeatNit, Omanyd, Owain Knight, Paine Ellsworth, Permstrump, Piledhigheranddeeper, Pinguinn, Pocketthis, Power~enwiki, Praemonitus, Prokaryotes, Randy Kryn, Red Director, Redom115, Rhombuth, Rich Farmbrough, Rjwilmsi, Robert1947, Rodw, Saurusaurus, Scrossuk, Sijadthelastpoet, Someguy432, Spike-from-NH, Srich32977, SuNaW, Sunrise, Super ninja2, Supertali64, TAnthony, Tdadamemd sioz, Terminator Series 800 Model 101, The Transhumanist, The snare, TheDragonFire, TheFreeWorld, Tomruen, Turkeybutt JC, TwoTwoHello, Ubcule, United Massachusetts, Usernamekiran, W like wiki, Wayne Elgin, Weorthe, Woodstone, Woscafrench, XeloriaGrand, Zefr, 老陳 155

Mars *Source:* https://en.wikipedia.org/w/index.php?oldid=806160076 *License:* Creative Commons Attribution-Share Alike 3.0 *Contributors:* A2soup, Aelon51, Aeonx, Alexandermcnabb, AlphaBetaGamma01, Arado, Ashwinr, BD2412, Barek, Barjimoa, BatteryIncluded, BenRG, Bender235, Bentogoa, Bgwhite, BiHVolim, Blinndsay, Bmags16, Brian Everlasting, CFCF, CV9933, Claudio M Souza, Colonel Wilhelm Klink, CommonsDelinker, CostcoSample, CuriousMind01, DN-boards1, DVdm, DanHobley, Daniel.Cardenas, Dawnseeker2000, Dcirovic, Desildorf, Dissident93, Dlthewave, Dolphin33438, Double sharp, DrKay, Drbogdan, Edulovers, Fgrosshans, Finnusertop, Firth m, Fish567, FlightTime, Fogelmatrix, Fotaun, Frmorrison, G0mx, Gap9551, Garfield Garfield, Gelatinxbox, Geoffrey.landis, Gildir, Glevum, Gob Lofa, Gulumeemee, HalloweenNight, Headbomb, Huntster, Huritisho, Iggy the Swan, Isambard Kingdom, JeanLucMargot, Jerzy, Jmencisom, John, Jon Kolbert, Jondel, JorisvS, Jytdog, K. Badri Vishal, Katolophyromai, Kdammers, Khan singh, Kheider, Kwamikagami, Loraof, LordofPens, Magioladitis, Maranello Prime, Marfinan, Mark Schierbecker, MartinZ, Materialscientist, Menchi, Michael Hardy, Modest Genius, Nemoanon, NewEnglandYankee, Newone, Nfrango, NightShadow23, Nihiltres, Nikkimaria, OeBoe, Omanyd, OmoiEgaite, Originalwana, Oshwah, Ost316, Paine Ellsworth, Pdebee, PeacePeace, PhilipTerryGraham, PlanetUser, Plastikspork, Poiuytrewqvtaatv123321, Praemonitus, Rain drop 45, Red Director, Rfassbind, Rfl0216, RhinoMind, Rivertorch, Rjwilmsi, Rodrigovalle, RomanSpa, Sacchipersempr, SamRathbone, SamuelRiv, SantaWinsAgain, Saros136, Sbmeirow, Scarlettail, SchutteGod, SeoMac, Sepguilherme, Serendipodous, Situphobos, Skepticalgiraffe, Skyfall, Skynxnex, SteepLearningCurve, Steve03Mills, Tex.Bold, The Transhumanist, TheFreeWorld, Thecodingproject, Thevideodrome, Tom.Reding, Tranngocnhatminh, Trappist the monk, TriniGoji, Twinsday, Unbuttered Parsnip, Veedubber86, Vgy7ujm, W like wiki, Wavelength, Weneedwikipedia, Wiki-whisky, WolfmanSF, Yoyozobi, Zedshort 190

Asteroid belt *Source:* https://en.wikipedia.org/w/index.php?oldid=804101696 *License:* Creative Commons Attribution-Share Alike 3.0 *Contributors:* 10metreh, A2soup, AD21147, Adam Hauner, Bear-rings, Bender235, Blacx0, Brockpag, Canoe1967, Capitalismojo, CapnZapp, Charles Matthews, Cherkash, ClueBot NG, Colonies Chris, Culveyhouse, DN-boards1, Dawnseeker2000, Dcirovic, Donner60, Double sharp, Dr bab, DrKay, Drbogdan, Earthandmoon, Elizabeth Linden Rahway, Essa.alserkal, EuroCarGT, Exoplanetaryscience, Finlay McWalter, Fleivium, Flyer22 Reborn, Frietjes, GEOGOZZ, Gani94, Gap9551, Gareth Griffith-Jones, Garethhugs, Gbawden, Gilliam, Gonmrm, Govindaharihari, Graphium, Gruekiller, Headbomb, Hiberniantears, IVAN3MAN, IceDragon64, Improve~enwiki, Isambard Kingdom, J 1982, Jackfork, Jamesx12345, Japs 88, Jclf123, Jehochman, Jiròni, Jmencisom, Jnestorius, John, John Cline, John of Reading, JorisvS, Josethe fatasscow, Joshua Doubek, Jusdafax, Kheider, Krenair, Kwamikagami, KylieTastic, Lambiam, Lexi sioz, Makecat, Malinadams, Materialscientist, Maxwell Verbeek, Mcswell, MinusDR, Monkomics, Mtu, MusikAnimal, Myasuda, N2e, NawlinWiki, Nick4800, Ninney, Oddbodz, Optakeover, Oranjblud, Oshwah, P. S. F. Freitas, Primergrey, Qu Hongcheng, QuantumShadow, RP88, Reatlas, Renerpho, Rfassbind, Rjwilmsi, RobertG, Robertinventor, Rolf h nelson, Roufengyang, RoyGoldsmith, Ruslik0, Satellizer, Secondhand Work, Serendipodous, Shellwood, Sir Cumference, Skizzik, Smashbash2015, Snow Blizzard, Solarra, Squally47, Tbhotch, Tetra quark, That guy 567890, This lousy T-shirt, Tolly4bolly, Tom.Reding, TomK32, Tournesol, Tushar Shrotriya, Ugog Nizdast, Unreliable3361, Vacation9, Vsmith, Weedwhacker128, Weeklyscorpion9, Weirvile, Werddemer, Wikipelli, Winner 42, 158 anonymous edits 229

Ceres (dwarf planet) *Source:* https://en.wikipedia.org/w/index.php?oldid=806069464 *License:* Creative Commons Attribution-Share Alike 3.0 *Contributors:* Abductive, Acroterion, Adam9007, Algmwc5, Amys eye, Ashmoo, BatteryIncluded, Bender235, Caughtacough, ClueBot NG, Coemgenus, CommonsDelinker, CuriousMind01, DVdm, Dcirovic, DerFussi, DinoSlider, Dl2000, Doncurzio, Double sharp, DrKay, Drbogdan, Exoplanetaryscience, Favonian, Ferrofield, Frietjes, Graeme Bartlett, Gulumeemee, HakanIST, Hanif Al Husaini, INeverCry, Icek~enwiki, J 1982, J. 'mach' wust, JFG, JackofOz, Jdcomix, Jeffryfisher, Jennica, Jon Kolbert, Jonathan Chone, JorisvS, Josephskywalker, Kwamikagami, Lithium6ion, Mahmudmasri, Master of Time, Materialscientist, Mevagiss, Michael Barera, Moron4077, Njardarlogar, Non-dropframe, Nyttend, Originalwana, Ornithoptera, Oshwah, Ost316, OwenBlacker, P Aculeius, P199, Paul August, Paul Wil1iams, Pdebee, Phatom87, PhilipTerryGraham, Pinkomega, PlanetUser, PlyrStar93, ProgrammingGeek, RPF2, Randy Kryn, RelaxtJosh, Rfassbind, Rjwilmsi, Ronaldo22161810, Rothorpe, Ruslik0, Sario528, Ser Amantio di Nicolao, Serols, Signedzzz, Soccer.preston10, Starless~enwiki, TAnthony, Tamfang, Tarl N., Tbayboy, TeddieBearFan, The Rambling Man, Titanium Dragon, Tom.Reding, Urhixidur, WadeSimMiser, Whoop whoop pull up, WolfmanSF, XavierGreen, Yeryry, Yikkayaya, Ymblanter, Zefr, 105 anonymous edits 248

Outer planets *Source:* https://en.wikipedia.org/w/index.php?oldid=797499900 *License:* Creative Commons Attribution-Share Alike 3.0 *Contributors:* BatteryIncluded, Beatlesobssessedfreak, ClueBot NG, Consider42, Crystallizedcarbon, Fotaun, Good Olfactory, Icairns, Jcpag2012, JorisvS, Kbdank71, Kwamikagami, LuK3, Mark Arsten, Maver1ck, Novangelis, Peaceduck, Pinethicket, Racerx11, Reatlas, Rfassbind, Rookkey, Serendipodous, Serols, SimonP, Tbayboy, Tom.Reding, TracyRenee, Urhixidur, WOSlinker, Wikid77, Wikielwikingo, Yintan, Zfr, 30 anonymous edits 281

Giant planet *Source:* https://en.wikipedia.org/w/index.php?oldid=768426607 *License:* Creative Commons Attribution-Share Alike 3.0 *Contributors:* .digamma, Abishai 300, Alansohn, Aldaron, Alenkoshy2000, AmaryllisGardener, Araympuertorico, Astredita, Batemqi4382, Begoon, Bennyvk, Bento00, Bigpapa12345, Braincricket, Calabe1992, Ceradon, ClosedEyesSeeing, ClueBot NG, CommonsDelinker, CowboySpartan, Cyberbot II, Cyberpower678, Dan Koehl, Dcirovic, Demomoer, Dna-webmaster, Donner60, Drbogdan, E. Fokker, EagerToddler39, Elijaheac, Erudhalion, EugeneZelenko, Falcon8765, Firetraner, Fotaun, Freakmighty, Freiza667, Funandtrvl, Galloping Moses, Gilliam, Glencthomas, Gobonobo, GregKaye, Grrahnbahr, Gyrobo, Hamiltondaniel, Hans Dunkelberg, Headbomb, Helpsome, Henrykus, Hi13579, Hurunui99, IMAMJESUS, Isaaccates, J 1982, Jab843, JackFrost2121, Jakew, Jianhui67, John spracklen, JohnBlackburne, JorisvS, Josve05a, Juleed, Jusdafax, Katiltrout, Kheider, Kn1467, Kwamikagami, L1A1 FAL, Lazylaces, Lieutenant of Melkor, Limefrost Spiral, Materialscientist, Messier8, Minna Sora no Shita, Monty845, Murgh, MusikAnimal, Nestrs, Oshwah, Paradies, Paulinho28, Phil Boswell, Philg88, PlanetStar, PlanetUser, Pseudomonas, Qurq, R'n'B, RA0808, RJHall, Racerx11, Reatlas, Redrose64, Regoldberg, Richard75, Rjwilmsi, Roentgenium111, Ruslik0, Scotstout, Scottywong, Serendipodous, Sideways713, Some jerk on the Internet, Suckstosuck, SwisterTwister, TallTsar, The High Fin Sperm Whale, The Stickler, Tideflat, Tolly4bolly, Tom.Reding, TwistOfCain, Velella, W. P. Uzer, Webclient101, Weeklyscorpion9, Whoop whoop pull up, Wikipelli, Wildman3386, Will Beback Auto, WolfmanSF, Yahia.barie, Yintan, قیصرانی, ჰ, 150 anonymous edits 286

Saturn *Source:* https://en.wikipedia.org/w/index.php?oldid=806116956 *License:* Creative Commons Attribution-Share Alike 3.0 *Contributors:* 564dude, 72, A Great Catholic Person, A2soup, Adam9007, Ahpook, AlexiusHoratius, Arado, Archeologo, Arjayay, Awesomegamer, BD2412, Barjimoa, BatteryIncluded, Bender235, Bentogoa, Bobogoobo, Brandmeister, BurritoBazooka, C.Fred, Cannolis, Ceoil, ChamithN, Churri101, ClueBot NG, ComicsAreJustAllRight, Crystallizedcarbon, D.C.Rigate, DN-boards1, Dawnseeker2000, Deepanshu1707, Dewritech, Dogman15, Dondervogel 2, Donner60, Double sharp, DrKay, Drbogdan, EP111, ESkog, Edokter, Edulovers, Edwininlondon, Ehrenkater, El C, EricEnfermero, Eurodyne, Excirial, FamAD123, Fbergo, FlightTime, Florian Blaschke, Flyer22 Reborn, Frietjes, Frood, GSS-1987, GreenMeansGo, Gulumeemee, Gymnophoria, HalloweenNight, Hamiltondaniel, Hdjensofjfnen, Headbomb, Hellrider19999, Hillbillyholiday, Hunterm267, Huntster, Iggy the Swan, Isambard Kingdom, JRPG, JamesCox 13, Jimfbleak, JoeHebda, John, John of Reading, Jon Kolbert, JorisvS, Josve05a, JustinTime55, KH-1, Kalimantanas, Kanjuzi, Kheider, Kozan Huseyin, Kwamikagami, L3X1, LC03, Labtek00, Lasunncty, LiberatorG, Loodog, Mahveotm, MartinZ, Materialscientist, Mfb, Miles Edgeworth, Mojoworker, Niceguyedc, Nikkimaria, Noscoper21, Octoberwoodland, Omanyd, Oshwah, Parejkoj, Pdebee, PhilipTerryGraham, Piledhigheranddeeper, PlanetUser, Plastikspork, Praemonitus, Randy Kryn, Redav, Rfassbind, Richard L. Peterson, Rjwilmsi, Ruslik0, Sakura Cartelet, Samsara, Saros136, Sbmeirow, Sebas.318, Sebastian.318, Seleepen, Serols, Sharfooth, Shellwood, Silverhorse, Simplexity22, SkyWarrior in public, Space-Age Meat, SparklingPessimist, Spartanwolf223, SpeedEvil, Steve03Mills, SubSpace, TAnthony, Tbhotch, The Transhumanist, Tom.Reding, TomGeog, Tompop888, Tony Mach, TonyBallioni, Trappist the monk, Twilight Magic, TwoTwoHello, Unbuttered Parsnip, User-duck, V1adis1av, VT0207, VVsuace666, VirtualDave, Vsmith, W like wiki, WOtP, Wedemboyz2088, Whack boy 2.0, Wieralee, William M. Connolley, WolfmanSF, Worldbruce, Zavierpl, Zedshort, 111 anonymous edits 295

Uranus *Source:* https://en.wikipedia.org/w/index.php?oldid=806115030 *License:* Creative Commons Attribution-Share Alike 3.0 *Contributors:* "alyosha", 1Martin33, 564dude, A. Parrot, A.amitkumar, A2soup, Abductive, Abishai 300, Acopyeditor, Albanaco, AlbertBickford, Alcazar84, Alsee, Ambi Valent, Andrew Gray, Andyjsmith, AstroLynx, Axl, B14709, BD2412, Barjimoa, BatteryIncluded, Bender235, Blobbie244, Bob rulz, Boltz6100, Br'er Rabbit, CV9933, Cadiomals, Cancina5645, Chackerian, Chaheel Riens, Cheetahrock21, Chivas112, Clockery, ClueBot NG, Coinmanj, Corinne, Coviekiller5, Cypkerth, DN-boards1, DOSGuy, Dcirovic, Deejaye6, Deepanshu1707, Delusion23, Derschmidt, Deturtlemon1, Dilidor, Dinkytown, Dja1979, Dominicanpapi82, Donutcity, Dotdotdotatsignapostrophe, Double sharp, DrKay, Drbogdan, EP111, Earthandmoon, Egsan Bacon, Ehrenkater, Epicgenius, EricAnthony101, Ewen, Eyreland, FeRD NYC, FerJox, Finell, Finnusertop, FlightTime, Fotaun, Fourthords, Frietjes, Gallina3795, Gob Lofa, Gparker, HalloweenNight, Hamiltpj, Headbomb, Hellbus, Hillbillyholiday, Howcheng, Huntster, Iggy the Swan, Iph, Isambard Kingdom, Jcpag2012, Jguy, Jmencisom, JoeHebda, Joeleoj123, John, Johnuniq, Jomey, Jon Kolbert, Jonathan Barrett, Jonesey95, JorisvS, JustAMuggle, KAP03, Katydidit, Kelisi, Kheider, Koavf, Kwamikagami, Kyng, Kyucasio, LahmacunKebab, LandruBek, Lentower, Lindenhurst Liberty, Lithopsian, Logeion, Loraof, Magioladitis, MarchOrDie, MartinZ, Materialscientist, Mean as custard, Mikhail Ryazanov, Moonraker12, Mortense, Mr Stephen, N Shar, Naddy, NebY, Nedim Ardoğa, Niceguyedc, Nmillerche, Nvvchar, ONUnicorn, OccultZone, Oldag07, Onceinawhile, Onore Baka Sama, Orange-kun, Orion 8, Ospalh, P. S. F. Freitas, Pdebee, PhilipTerryGraham, Plagktos, Plastikspork, PlayStation 14, Primergrey, PulauKakatua19, Q6637p, Rab V, Racerx11, RagingR2, Randy

Kryn, Reatlas, Rfassbind, Rhlozier, Richard L. Peterson, Rivertorch, Rjwilmsi, Robert the Devil, RockMagnetist (DCO visiting scholar), Rondo66, Rosscaldwell, Rothorpe, Ruslik0, Ryulong, Sailsbystars, SamX, Santa Claus of the Future, Saros136, Serendipodous, SheikhTheBaby, SkateTier, SkoreKeep, Smartie2thaMaxXx, Smk65536, Sol2y, SoylentPurple, Spartanwolf223, Spiderjerky, Srich32977, Starcluster, StringTheory11, TAnthony, Teddyktchan, Tetra quark, Thanhtinsaosang, The Transhumanist, Thedrumsmasher, Thewriter6669, Tim!, Toa Nidhiki05, Tom.Reding, Tomeasy, TommyBoy, Trackteur, Transphasic, Unbuttered Parsnip, Vsmith, W like wiki, Wavyinfinity, Whoop whoop pull up, WolfmanSF, Wrxsti fan, ZarhanFastfire, ZerodEgo, הסרפד, 展翅飛翔 319

Neptune *Source:* https://en.wikipedia.org/w/index.php?oldid=804998694 *License:* Creative Commons Attribution-Share Alike 3.0 *Contributors:* 81120906713, A2soup, AaronMP84, Adelson Velsky Landis, Alanobrien, Ambi Valent, Arado, Archon 2488, Arjayay, ArthurianLegend, Asisman, Azmodes, Barjimoa, BatteryIncluded, Bender235, Bensational, Bil1, Bill the Cat 7, BlueMoonlet, Bunkyray5, Cadiomals, Cancina5645, Coladar, CuriousEric, Cuttycuttiercuttiest, DN-boards1, DSmurf, Dawnseeker2000, Deepanshu1707, Delusion23, Dickdock, Dinkytown, Dodshe, Double sharp, DrKay, Drbogdan, Edulovers, Ehrenkater, Et Cum Spiritu Tuo, Exoplanetaryscience, FerJox, Fotaun, Fountains of Bryn Mawr, Gallina3795, Gap9551, Georgelazenby, Glenn L, Go-in, GoPlayerJuggler, Gob Lofa, Gored82, GrundyCamellia, Hbdragon88, Headbomb, Hillbillyholiday, Hipporoo, Huntster, Hz.tiang, Iggy the Swan, InedibleHulk, Inhighspeed, Inoyu49, Irrawaddy, Isambard Kingdom, JPaestpreornJeolhlna, Jcpag2012, Jengelh, Jmostly, John, Jon Kolbert, Jonesey95, JorisvS, Josve05a, Jrrs, JustAMuggle, KAP03, Kelvin13, Kheider, Koavf, Kwamikagami, LL221W, Labtek00, Lightlowemon, Lindenhurst Liberty, Lo Ximiendo, Martin Hogbin, MartinZ, Master of Time, Materialscientist, Mckburton, Me, Myself, and I are Here, Mezigue, Mikhail Ryazanov, Mscuthbert, Naddy, Nedim Ardoğa, Neuralwarp, Nicolas Eynaud, Nihiltres, Ohconfucius, Oldag07, Onore Baka Sama, Orange-kun, Orenburg1, P. S. F. Freitas, Patteroast, Pdebee, PlanetStar, PlayStation 14, PulauKakatua19, RPH, Randy Kryn, Reatlas, Redd Foxx 1991, Rfassbind, Rhatsa26X, RhinoMind, Richard75, Rivertorch, Rjwilmsi, Robert Skyhawk, RobertLovesPi, RockMagnetist (DCO visiting scholar), Roentgenium111, Rothorpe, Ryulong, Sailsbystars, SandyGeorgia, Saros136, Sbmeirow, SeL, SecretAssassin, Serendipodous, SeventeenTurtle, Silenceisgod, SkoreKeep, Skynorth, Smk65536, SoylentPurple, Spiderjerky, Srich32977, Stamptrader, Starcluster, Steel1943, Stub Mandrel, Summstjr, TAnthony, Tetra quark, The Transhumanist, Thor Dockweiler, Tom.Reding, Trackteur, Trappist the monk, Twinsday, Tybomb124, Ufwuct, Virtualphtn, W like wiki, Wasill37, Wcp07, WereSpielChequers, WolfmanSF, YoungDrac 348

Centaur (minor planet) *Source:* https://en.wikipedia.org/w/index.php?oldid=804679286 *License:* Creative Commons Attribution-Share Alike 3.0 *Contributors:* (, Acalamari, Alansohn, Anxietycello, Argo Navis, Auric, Avinash.royyuru, Backin72, Beetstra, Black Walnut, BotKung, BotMultichill, Bota47, Bryan Derksen, CFLeon, Chesnok, Ckatz, CommonsDelinker, CuriousEric, Curps, Dan100, David Eppstein, Deuar, Dna-webmaster, Double sharp, Dread83, Duoduoduo, Egil, Eric Kvaalen, Eurocommuter, Exoplanetaryscience, Filiosus's Saga, Fjörgynn, Fotaun, Gene Nygaard, Geoffrey.landis, Giftlite, Gurch, Gveret Tered, Hike395, Icairns, Ilmari Karonen, Ilvon, Iridia, J 1982, Jandalhandler, Jmg38, John Hyams, John a s, Jon Kolbert, JorisvS, Joseph Solis in Australia, Jscotti, Kbdank71, Kheider, Knowledge Seeker, Kogge, Kwamikagami, Lambiam, Lentower, LeonardoRob0t, Ling.Nut, Loginnigol, Lowe4091, Maxim Razin, Mejor Los Indios, Mercurio2688, Mgmirkin, Michaelmas1957, Mogism, Muma~enwiki, Murgh, NTBot~enwiki, Njardarlogar, Normakku, Novangelis, Originalwana, P. S. F. Freitas, Palica, PamD, PedroPVZ, Piledhigheranddeeper, Pt, Reatlas, Rfassbind, Rich Farmbrough, Rjwilmsi, Rmhermen, RobotE, Roentgenium111, Ronhjones, Ruslik0, Samdacruel, Sanddune777, Sardanaphalus, Satori, Serendipodous, Serpentinite, Sethhater123, ShelfSkewed, SidP, Spartan, Suelru, Tamfang, Template namespace initialisation script, The Singing Badger, The Tom, The shaggy one, Thebiggnome, Tom harrison, Tom.Reding, Tothebarricades.tk, Trafford09, Trappist the monk, Trekphiler, Ufviper, Ulflarsen, Urhixidur, WilyD, WolfmanSF, Zedshort, Δ, عباد مجاهد ديرانية, 54 anonymous edits 372

Comet *Source:* https://en.wikipedia.org/w/index.php?oldid=806029609 *License:* Creative Commons Attribution-Share Alike 3.0 *Contributors:* 72, A Great Catholic Person, AKS.9955, Alonso de Mendoza, Anakimi, AnnaGoFast, Anomalocaris, Avantiputra7, Ayushsha, BatteryIncluded, Beland, Bender235, Bentogoa, Boss801, CLCStudent, CityOfSilver, ClueBot NG, Coltonlololololol, CommonsDelinker, Creepyminerboy, Cyberbot II, David Nelson 3.14, David in DC, Dawnseeker2000, Dcirovic, DelvecchioSimone12 5 96, Dicklyon, DivermanAU, Double sharp, Drbogdan, Dreslough, Drmies, Dschslava, El C, Entranced98, Evensteven, Excirial, Exoplanetaryscience, FletcherD, Fotaun, Frietjes, Gap9551, GeoffreyT2000, Gob Lofa, Graphium, Gustavo1100, Headbomb, Hello123993, Hokuzai, Home Lander, HueSatLum, Huntster, Iking5, IngridC-Curtin, IronGargoyle, Isambard Kingdom, IsraphelMac, Jakcox, Jd22292, Jdean15, Jimroland, Jmencisom, JoJo Anthrax, Jonesey95, JorisvS, Juliancolton, Jytdog, KH-1, Kharkiv07, Kheider, Kwamikagami, Lj, Madly123iamme, Magioladitis, MannIMNJ, Manul, Materialscientist, Mr Stephen, Muadd, Myasuda, NQ, Nabla, Natuur12, NeilN, Ninney, Oshwah, PlanetUser, Plucas58, Potatoswatter, Praemonitus, Red Director, Rfassbind, Ricardo A. Olea, Rico402, Rjwilmsi, Rrh02, Rwessel, Scalhotrod, Sergei Schmalz, Serols, Simplexity22, Snoop Dogg the king of Stoners, Spark802, Spartan198, Strait, Supdiop, Super Spheal, TAnthony, Tetra quark, The Voidwalker, Thor Dockweiler, Titus III, Tom.Reding, Ugog Nizdast, UzzUi, Velella, VexorAbVikipædia, WadeSimMiser, WikiMasterGhibif, WikiWhoYou, Winner 42, Yamaguchi先生, Zedshort, Zppix, Ленка Милић, Юкатан, 123 anonymous edits 382

Kuiper belt *Source:* https://en.wikipedia.org/w/index.php?oldid=805894384 *License:* Creative Commons Attribution-Share Alike 3.0 *Contributors:* 2WR1, A2soup, ATX-NL, Agmartin, Al Legorhythm, Al1549, Aliciamarie9313, Apidium23, Arjayay, Askari Mark, AwesomeAH18, BatteryIncluded, Bear-rings, Beland, Bender235, Bobdog54, ClueBot NG, CommonsDelinker, Continentaleurope, Corinne, Dan100, Daviddaniel37, Dawnseeker2000, Dcirovic, DeadKing99, Deeday-UK, Doanri, DocWatson42, Donner60, DrKay, Drbogdan, EdChem, El C, Eric Kvaalen, Exoplanetaryscience, Finnusertop, Fishyishere, Gadget850, Gah4, Gap9551, Gob Lofa, Gts-tg, HamilcarBarca1, Hamish59, Hddty., Headbomb, Hibernian, Huntster, IronGargoyle, J 1982, JJMC89, JakeR, Jarble, Jehochman, Jhenderson8, Jiten Dhandha, Jmencisom, Jogar2, John, John "Hannibal" Smith, Jonesey95, JorisvS, Kbh3rd, Keith-264, Kheider, Kleuske, Konlon15, KoshVorlon, Krj373, LL221W, Laoris, Leschnei, Llywrch, Lugsciath, Master of Time, Materialscientist, Mild Bill Hiccup, MisfitToys, Modest Genius, Mrjulesd, Mzanon, Nichodon, Oklo99, Omnipaedista, PoqVaUSA, ProloyAdhikary, ProprioMe OW, RP88, Racerx11, Renerpho, Rfassbind, Ricardo A. Olea, RickinBaltimore, RioVerde, Rjwilmsi, Roentgenium111, Rothorpe, SchutteGod, Serendipodous, Serols, Shellwood, Sir Cumference, Slightsmile, SpaceDude777, Stanning, Steinbach, Tbayboy, Tetra quark, This lousy T-shirt, Tom.Reding, Tomstar214, Trappist the monk, TwoTwoHello, Urhixidur, Wieralee, Wiki Guy123098, Wikid77, Wikishovel, WilyD, WolfmanSF, ,62ז 117 anonymous edits 414

Pluto *Source:* https://en.wikipedia.org/w/index.php?oldid=805915686 *License:* Creative Commons Attribution-Share Alike 3.0 *Contributors:* 134340Goat, A2soup, Afernand74, Archon 2488, Arianewiki1, Art LaPella, Ashill, Barjimoa, BatteryIncluded, Bender235, Bgwhite, Blckhawk1234, Bobsmarty, CV9933, Choor monster, Ciroa, ComicsAreJustAllRight, Couillaud, CuriousEric, CuriousMind01, Cush, DN-boards1, Darkdadaah, Dcirovic, Delusion23, Dhtwiki, DocWatson42, Double sharp, DrKay, Drabkikker, Drbogdan, Duff06, EdChem, Edulovers, Ehrenkater, EncMstr, Erenevi1097, Eric Kvaalen, FallingGravity, Fanyavizuri, Frank Klemm, Gap9551, God's Godzilla, Guy vandegrift, Hillbillyholiday, Hohum, Huntster, Huritisho, Isambard Kingdom, JMyrleFuller, JackofOz, Jbeans, JeanLucMargot, John, Jonathan Chone, JorisvS, Kaldari, Kgrad, Kheider, Kind Tennis Fan, Kk2k7it, Kndimov, Kuzwa, Kwamikagami, LFdoR, LL221W, Lexi sioz, Libertyguy, LightandDark2000, Llamadillo, Lugnuts, MJG639, MadeYourReadThis, Madreterra, Mare-Silverus, Mevagiss, Mezigue, Mieliestronk, Moogsi, NDKilla, Naraht, Neil P. Quinn, Nfr-Maat, Nihiltres, Nitrobutane, Njardarlogar, Nwbeeson, Nyttend, OrbitHawk, Pdebee, Peter Chastain, PhilipTerryGraham, Pine, PlanetUser, PulauKakatua19, Quantanew, Renerpho, Rfassbind, Rjwilmsi, Rothorpe, Rottweiler, Rreagan007, Rudy235, Sae1962, Sanddune777, Saros136, Schnapps17, SchutteGod, Serendipodous, Serpinium, SidP, Sigehelmus, Sir Cumference, Smurrayinchester, Sol2y, StarryGrandma, Stas000D, Steve03Mills, StormWillLaugh, Sundaryourfriend, Supuhstar, TAnthony, TJRC, Tbayboy, The Duke of Waltham, The Rambling Man, TheWhistleGag, Theklan, Tim!, Tohma, Tom.Reding, Tomburbine, Trappist the monk, Unbuttered Parsnip, VQuakr, Victorgrigas, VulpesVulpes42, Wahrmund, Wiklol, WolfmanSF, Writtenright, Xover, Y2kcrazyjoker4, Yann, Yinshiyi 439

Charon (moon) *Source:* https://en.wikipedia.org/w/index.php?oldid=802865417 *License:* Creative Commons Attribution-Share Alike 3.0 *Contributors:* 134340Goat, A2soup, Alexconnorbrown, Andrewjohnbayles, AntanO, Archon 2488, Aria1561, Arjayay, Ashill, Ashmoo, Attilios, BatteryIncluded, Bender235, BillC, Bruce Marlin, BrunoHP, Choor monster, ClueBot NG, CommonsDelinker, Connormah, Curedwales, CuriousEric, D Eaketts, DN-boards1, DaftClub, DeNoel, Deepanshu1707, Destro, DoctorGiants, Double sharp, Drbogdan, EP111, El C, Eric, Ericl, Evensteven, Fama Clamosa, FeanorStar7, FlightTime, Florian Blaschke, Flyer22 Reborn, Frmorrison, Gilliam, HMSLavender, Harizotoh9, Headbomb, Hello, I'm Here To Help Wikipedia!, Howcheng, Huritisho, Hydrargyrum, IceKarma, J. 'mach' wust, JFG, Jackfork, JasonAQuest, Jim1138, JorisvS, Jurryaany, Jusdafax, Kaiketsu, Kheider, Khiemisop, Kuchesezik, Kwamikagami, LightandDark2000, LilHelpa, Loadmaster, MarkHardaker, MartinZ, Materialscientist, Maxus96, Maxwell Verbeek, Mean as custard, Medeis, Mevagiss, Michael hefler, Midgley, Mkmillenium, Mpj7, MrCage, Nick-D, Njardarlogar, Nupy2005, Nwbeeson, Ornithoptera, PhilipTerryGraham, PlanetUser, Pluto and Charon, Quintus314, RP88, Renerpho, Rfassbind, Rjwilmsi, Rothorpe, Rudy235, Ruslik0, Sbneelu, Shanno2012, Simfish, Solomonfromfinland, TAnthony, Tarism A2, Tbayboy, Tetra quark, The Big Eye, The Rambling Man, TheWhistleGag, Throbby, ToThAc, Topherleighto, Trappist the monk, Trollguy39, VQuakr, Vencaslac, WhiskeyJuvenile, Winner 42, WolfmanSF, Yoshi24517, Youtryandyoutry, qwertz, 82 anonymous edits 466

Makemake *Source:* https://en.wikipedia.org/w/index.php?oldid=805160215 *License:* Creative Commons Attribution-Share Alike 3.0 *Contributors:* 3 of Diamonds, AgadaUrbanit, AirCombat, Andrewjohnbayles, Ani2164789, BatteryIncluded, Bear-rings, Bender235, Bigbumsrock, Br'er Rabbit, Brucknerj, CAPTAIN RAJU, CLCStudent, Casliber, ClueBot NG, CyanoTex, D Eaketts, DN-boards1, DanielCristofani, Dcirovic, DemocraticLuntz, Dinglebat500, Donner60, Double sharp, DrKay, Dragons flight, Earthandmoon, EdChem, Exoplanetaryscience, Fotaun, Frmorrison, Gadget850, GeoffCapp, Gilliam, Gob Lofa, Good Olfactory, Gunslinger.1970, Guy1890, Hdjensofjfnen, Headbomb, Hevron1998, Hmains, HumphreyW, IVORK, J 1982, JDAWiseman, JackofOz, Jared mcmullin 6969, Jin Zan, Jmencisom, Jmg38, John "Hannibal" Smith, Jonathan Chone, Jonnyhabenero, JorisvS, Kayliwaites, Kheider, Kushan Senevirathna, Kwamikagami, LFdoR, Mahan, MartinZ, Mean as custard, Millennium bug, MusikAnimal, Mvpo666, Mz7, Naddy, Natyayl, Niceguyedc, Nickst, Nicole Sharp, Ninney, Nuwanpushpakumara, Oshwah, P. S. F. Freitas, Peaceray, Pinkyodeler135, RA0808, RP88, Redrose64, Rfassbind, Rjwilmsi, Robinh, Rsrikanth05, Ruslik0, Seafoxlrt616, Seanearlfrog, Serendipodous, ShadowDragon343, Shb124, Sigil VII, Steve03Mills, Szczureq, Tbayboy, ThatRusskiiGuy, Thingg, Tom.Reding, Tomruen, Trekphiler, Tris2000, Vuerqex, Wester, Widr, Wjfox2005, WolfmanSF, XavierGreen, Yamaguchi先生, ליאור, 121 anonymous edits 480

Haumea *Source:* https://en.wikipedia.org/w/index.php?oldid=805689798 *License:* Creative Commons Attribution-Share Alike 3.0 *Contributors:* A Great Catholic Person, Aeusoes1, Arsia Mons, BG Mullen, BOT-Superzerocool, Beethoven, Benbest, Bender235, Br'er Rabbit, Carmichael, Casliber, Catholic Laitinen, CharlesHBennett, Citation bot 1, Ckatz, ClueBot NG, Colinmartin74, CommonsDelinker, Crystallizedcarbon, Curtis23, DN-boards1, DVdm, David Gerard, Dcirovic, Dg011019, Donner60, Double sharp, DrKay, Dragons flight, DrunkBicyclist, Earthandmoon, EdwardH, Eeekster, Excirial, Exoplanetaryscience, Fotaun, FreeRangeFrog, Gadget850, Ganímedes, Gap9551, Gilliam, Gob Lofa, GoldRingChip, Hans Dunkelberg, Headbomb,

Hevron1998, Inventorfrog, Io Herodotus, Isambard Kingdom, J 1982, JDAWiseman, JFG, Jackfork, Jan.Kamenicek, Jayron32, Jin Zan, JorisvS, Juliancolton, KATANAGOD, KAVEBEAR, Kaldari, Kethrus, Kheider, Knowledgekid87, Kwamikagami, LFdoR, Lightmouse, Lipsquid, Little Mountain 5, MartinZ, Mathrick, Mortense, Mouse20080706, Mr. Stradivarius, Naddy, NottNott, Oknazevad, Oshwah, Ozkidzez91, P. S. F. Freitas, Patteroast, Pinkyodeler135, RJGray, RSStockdale, Racerx11, Reatlas, Reaverdrop, Renerpho, Rfassbind, Rjwilmsi, RobertG, Rothorpe, Ruslik0, ScottM84, Seafoxlrt616, Sigil VII, Solomonfromfinland, Sonicology, Stas000D, Steve Hosgood, Steve03Mills, StevenBjerke, Stylteralmaldo, TAnthony, Tbayboy, The High Fin Sperm Whale, The Tom, Tom.Reding, Trulystand700, Tycho Magnetic Anomaly-1, Unknown5000000000000, Urhixidur, Vacio, Wbm1058, Wikipelli, WolfmanSF, X1\, Yintan, Zandperl, 91 anonymous edits 490

Scattered disc *Source:* https://en.wikipedia.org/w/index.php?oldid=803231664 *License:* Creative Commons Attribution-Share Alike 3.0 *Contributors:* 84user, AMN3SiA, ATBS, Abhishekmathur, Anna Lincoln, Anon lynx, Archolman, Azninvasion344, Bender235, BotKung, Br'er Rabbit, C messier, CalRis25, Calor, Ceranthor, Chiswick Chap, Circeus, Citation bot 1, ClueBot NG, Colipon, CommonsDelinker, Computergeek9911, DGG, Dan100, Dana boomer, Dewritech, Dia^, Dna-webmaster, Dogposter, Dominicanpapi82, DrKay, Earthandmoon, East718, Editorofthewiki, Erwin, Ettrig, Eurocommuter, Fenwayguy, Fjörgynn, Florian Blaschke, Frietjes, Frmorrison, GGT, Gap9551, Geometry guy, Gob Lofa, Goodboybad123, GregU, Gruntler, Hans Langschwanz ist wieder da, Headbomb, Herostratus, IceKarma, Ilvon, J 1982, Jan.Kamenicek, Jeffq, Jimfbleak, Jimp, John Hyams, Jonesey95, JorisvS, Kheider, Kobrabones, Kogge, Kwamikagami, Lamp90, Llightex, Mattisse, Mrshaba, Natural RX, Nergaal, Nihiltres, Ohconfucius, Pauli133, Peyre, Piano non troppo, Piledhigheranddeeper, Poppy, Pytom, Quentonamos, RJHall, RandomStringOfCharacters, RexxS, Rfassbind, Rich Farmbrough, Rjwilmsi, RobertG, Robertgreer, Rothorpe, Ruslik0, SandyGeorgia, Serendipodous, SidP, Sideways713, Sir Cumference, Sources said, The Thing That Should Not Be, The Tom, Tom.Reding, Tony Mach, Trappist the monk, Treisijs, Turkeyleg778, UnGreg, Urhixidur, Wavelength, Whyasknow, Willking1979, WolfmanSF, Ziggie456789, ZooFari, 57 anonymous edits 503

Eris (dwarf planet) *Source:* https://en.wikipedia.org/w/index.php?oldid=805815614 *License:* Creative Commons Attribution-Share Alike 3.0 *Contributors:* 10metreh, 23haveblue, 345Kai, 3primetime3, 6d6b73, A2soup, Anaxial, Angry candy, Archon 2488, BD2412, Bagunceiro, Bender235, Bilorv, Blunter Mercury 6, Bongwarrior, CV9933, Casliber, Cavemanbjork, Cbdorsett, Clayhasavirus, ClueBot NG, Colonel Wilhelm Klink, CommonsDelinker, CuriousEric, DN-boards1, DavidBrooks, Dcirovic, DemocraticLuntz, Dewritech, Dinner1, DivineAlpha, DocWatson42, Double sharp, DrKay, Drewmutt, Dspradau, Egsan Bacon, Erisdawrf, Exoplanetaryscience, Favonian, Feminist, Fotaun, Frietjes, Frmorrison, Gah4, GarnetStrider, GeoffCapp, Gilliam, Gogo Dodo, Goodreads5398, Graeme Bartlett, Gwross, Hameltion, Hintswen, HurluGumene, J 1982, J. 'mach' wust, J173299999, JDAWiseman, JFG, JMilty, Jabberwoch, JackofOz, Jdthood, John, Johnuniq, Jonathunder, JorisvS, Jranson, Kheider, Kwamikagami, LAlawMedMBA, LFdoR, Leitmotiv, LightandDark2000, Lisztmacher, Magioladitis, Mandruss, Master of Time, Mpjacko, Neatsfoot, Ninney, Njardarlogar, NotRealDonaldTrump, Nowa, Ost316, Patteroast, Paul August, Pdebee, Philip Trueman, Phinn, Phoogenb, Pinkyodeler135, PotvinSux, RP88, Racerx11, Redandready, Rfassbind, RiverGooding, Rjwilmsi, Robertinventor, Roentgenium111, Rothorpe, Rsrikanth05, Ruslik0, Sametheon, Scogliorenzella, Serendipodous, Shawn81, Shellwood, SidP, Sigil VII, Simplexity22, Skamecrazy123, Solar-Wind, Solomonfromfinland, Steinbach, Steve03Mills, Stikonas, Svick, Syphonist, TAnthony, TYelliot, Tbayboy, Tbhotch, This lousy T-shirt, Tom.Reding, Tomruen, Trappist the monk, Triple plane, Trystan, Unbuttered Parsnip, VQuakr, Vsmith, WolfmanSF, Wtmitchell, Zedshort, Zeryphex, 137 anonymous edits 513

Heliosphere *Source:* https://en.wikipedia.org/w/index.php?oldid=799639976 *License:* Creative Commons Attribution-Share Alike 3.0 *Contributors:* 49oxen, A2soup, Acdx, Alanwpeck, AlbertBickford, AndrewWTaylor, Avinash.royyuru, Awickert, Axemanstan, BD2412, BIL, Banaticus, Basilicofresco, BatteryIncluded, Beland, Bender235, Bgwhite, BillC, Brett, Brown43, Bubba73, Bunnyhop11, Canuck100, Citation bot 1, ClueBot NG, Colinphilipjohnstone, Cpl Syx, Cuaxdon, DSCrowned, Daggerstab, Darkwind, David Gerard, Dcirovic, Deltik, Dino, Discospinster, Dkevanko, Doggy23woggy4, Download, DrDaveHPP, Ego White Tray, Elenceq, EmilTyf, Emptn, EoGuy, Evolauxia, Explicit, Fgine, Filiosus's Saga, Finell, Fotaun, Frendinius, Frietjes, Gap9551, Gavia immer, GliderMaven, Gob Lofa, HMitrova, Hairy Dude, Hans Dunkelberg, Helium4, Hmainsbot1, Hunnjazal, Isambard Kingdom, J 1982, Jdaloner, Jdemp42, Jon Kolbert, JorisvS, Julesd, Justin15w, Keith-264, Kheider, Kiwifist, Kolbasz, Kompar～enwiki, KylieTastic, Lairdwh, MONGO, Mappetop, Marshallsumter, Martarius, Mathwhiz 29, Maurice Carbonaro, Michael Slana, MicroVirus, Mikeeg555, Mjasfca, Mmarre, Mmeijeri, Mogism, Mscuthbert, Mynameisntbob1, Nipsonanomhmata, Originalwana, Owenozier, Paine Ellsworth, Peter Chastain, Phatom87, Pinethicket, Pongox, Qxukhgiels, R'n'B, RJHall, Reatlas, Rfassbind, Rjwilmsi, Roentgenium111, Roughhead, Rrostrom, Ruby Murray, Ruslik0, Sakkura, SchuminWeb, Sean D Martin, Serendipodous, Sir Cumference, Sleddog116, Solar-Wind, Stardust1043, Swpb, Tassedethe, Tehnakki, Tetra quark, Themightyquill, Thnidu, Tom.Reding, Tommy2010, Tot12, Vsmith, Wacgyver, Wingedsubmariner, WolfmanSF, Wtachi, Yanpas, Zedshort, 133 anonymous edits 525

Detached object *Source:* https://en.wikipedia.org/w/index.php?oldid=799727354 *License:* Creative Commons Attribution-Share Alike 3.0 *Contributors:* Art LaPella, CV9933, Cobaltbluetony, DN-boards1, David Newton, Dawnseeker2000, Dcirovic, Double sharp, Eunomiac, Eurocommuter, Fjörgynn, Headbomb, Hevron1998, Igodard, Ilmari Karonen, Ilvon, Iridia, J 1982, J mareeswaran, Jan.Kamenicek, Jehochman, John Hyams, John of Reading, JorisvS, Jusdafax, Kheider, Kobrabones, Kwamikagami, Ling.Nut, Maury Markowitz, Merovingian, Nergaal, Niceguyedc, Olaf Davis, Pauli133, Poppy, Qurq, Rainald62, Reatlas, Rfassbind, Rjwilmsi, Roentgenium111, Rothorpe, Ruslik0, Serendipodous, Solomonfromfinland, Starcluster, The Tom, Tom.Reding, Tomruen, Trappist the monk, TreasuryTag, WolfmanSF, 17 anonymous edits 542

Sednoid *Source:* https://en.wikipedia.org/w/index.php?oldid=802631557 *License:* Creative Commons Attribution-Share Alike 3.0 *Contributors:* Ain92, Allium, Bender235, CuriousEric, DN-boards1, Double sharp, Fdigiuse, Gah4, Headbomb, J 1982, JorisvS, KTo288, Kheider, Kwamikagami, Malik Shabazz, Natsims, Prokaryotes, RP88, Rjwilmsi, Rothorpe, Solar-Wind, Starcluster, Tom.Reding, Urhixidur, Wikid77, WolfmanSF, 17 anonymous edits 549

Oort cloud *Source:* https://en.wikipedia.org/w/index.php?oldid=804018582 *License:* Creative Commons Attribution-Share Alike 3.0 *Contributors:* 7&6=thirteen, Amteixeira, An Sealgair, Attilios, AxelBoldt, Azmodes, BD2412, BatteryIncluded, Bender235, BlueAzulon, Bornsmall, Britannic124, Cherkash, ChrisGualtieri, Cliff12345, ClueBot NG, Coin945, Contraption5000, Cosmicos, Cumbril, DMacks, Davisonio, Dcirovic, Debresser, Degen Earthfast, Deneb in Cygnus, Discospinster, Dishmaster, Doczilla, Donner60, DrKay, Drbogdan, Elisa.arizono, Excirial, Exoplanetaryscience, FlightTime, Fotaun, Funnyfarmofdoom, Giraffedata, Gob Lofa, Goethean, Headbomb, Heron, Ian Macintosh, Icarus of old, Ilyushka88, InternetMeme, Io Herodotus, Iridescent, Isambard Kingdom, J 1982, JakeR, James McBride, Jan Oort, Jandalhandler, Jeanjung212, Jehochman, Jimp, Joeinwiki, John a s, JorisvS, Jusdafax, Kheider, Killieo999, Kiruria281, Kwamikagami, Kwib, Leptus Froggi, Llightex, Mahusha, Malerooster, Materialscientist, Mathematicsstudent, Medium69, MikeLousado, Nealmcb, NeilN, Ninney, Njardarlogar, Omnipaedista, Optakeover, Ost316, PhoBo, Quondum, RA0808, RP88, Racerx11, Raszoo, Rfassbind, RhinoMind, RichardLyre, Rjwilmsi, Ruslik0, Sadads, Serendipodous, Serols, Silverhill, Sionus, Sir Cumference, Smileguy91, Solar-Wind, Solomonfromfinland, Stephenb, Sémaphore, TAnthony, Tamfang, Tetra quark, The Illusive Man, The Proffesor, TimothyRias, Titodutta, To wikepedia, Tom.Reding, TomS TDotO, Trappist the monk, Tlovejoy, TypoBoy, Vgy7ujm, Vsmith, Wagoorand, Wikid77, WilyD, Winston Yamaguchi先生, Yinweichen, Zachariah1978, Zingvin, 老陳, 110 anonymous edits 552

Vulcanoid *Source:* https://en.wikipedia.org/w/index.php?oldid=757858998 *License:* Creative Commons Attribution-Share Alike 3.0 *Contributors:* 8472, Allixpeeke, Altamel, ArglebargleIV, Bender235, Bryan Derksen, Christopher Thomas, Citation bot 1, CuriousEric, Curps, Cyberbot II, DN-boards1, Devil Master, Glenn, Goustien, Headbomb, Hellbus, Hexdragon, Icairns, Icalanise, Joedeshon, Jonesey95, JorisvS, Joseph Solis in Australia, Kbdank71, Ketiltrout, Kheider, Koavf, LechuCzechu, MCB, Matthiaspaul, MeDrewNotYou, Michaelbusch, Mrwuggs, Njardarlogar, PAR, PKT, PedroPVZ, Philg88, PiMaster3, RandomCritic, Reatlas, Rentaferret, Reyk, Rjwilmsi, Rmhermen, RobotG, Roentgenium111, Rothorpe, Ruslik0, Scwlong, Serendipodous, Shawn81, Stirling Newberry, Szczureq, Teabeard～enwiki, The Singing Badger, Tom.Reding, Urhixidur, Vald, Vyznev Xnebara, Wbm1058, WinstonSmith, XJaM, 32 anonymous edits 562

Planets beyond Neptune *Source:* https://en.wikipedia.org/w/index.php?oldid=805760651 *License:* Creative Commons Attribution-Share Alike 3.0 *Contributors:* 65HCA7, A2soup, AJRG, Acroterion, AddWittyNameHere, Alejandrocaro35, Alexei Kopylov, Alpalfour, Anonimski, Another Believer, ArnoldReinhold, Article editor, BD2412, Bantaar, BatteryIncluded, Bender235, BlackGoldKnight, BrineStans, Bubba73, CAPTAIN RAJU, CambridgeBayWeather, ClueBot NG, ComicsAreJustAllRight, CommonsDelinker, DVdm, Davemck, Dcirovic, Dem1995, Donner60, DrKay, Draconicfire, Drbogdan, DwPnt, Englandball123, Favonian, Gah4, Gap9551, Georgejohnthethird, Ghostofnemo, Gilliam, HHEHEHUIUU, Hairy Dude, Headbomb, Home Lander, Hydrargyrum, Io Herodotus, Iridescent, J 1982, J mareeswaran, JakeR, Jaknouse, JasonAQuest, Jehochman, Jersey92, Jerzy, JorisvS, Josve05a, Juliancolton, Julien1978, Junior5a, Kheider, Kinetic37, Laberinto16, MadGuy7023, Malcolmxl5, Marchjuly, Materialscientist, McGeddon, MeDrewNotYou, Nagualdesign, Nealmcb, Newwhist, Nihiltres, Ost316, Phil Boswell, PhilipTerryGraham, Plantdrew, Praemonitus, PranavSNaik, Primefac, Prokaryotes, Ravenpuff, Reyk, Richard Arthur Norton (1958-), Rjwilmsi, Robertinventor, Rothorpe, Rursus, Ruslik0, SDC, SeaPals, Serendipodous, Serols, Simplexity22, Sithian476, SpaceDude777, Tbayboy, Tom.Reding, Trappist the monk, Trekphiler, Troglo, UU, Vsmith, Wnt, Woodroar, 120 anonymous edits 567

Planet Nine *Source:* https://en.wikipedia.org/w/index.php?oldid=806133219 *License:* Creative Commons Attribution-Share Alike 3.0 *Contributors:* 134340Goat, A2soup, Agmartin, Assscroft, Bamyers99, BatteryIncluded, BernardZ, BrayLockBoy, Clpo13, ClueBot NG, ComicsAreJustAllRight, D Eaketts, DadaNeem, David Eppstein, Dewritech, Dl2000, Double sharp, DrunkBicyclist, El C, Gap9551, Hayman30, Headbomb, Huntster, Jehochman, John of Reading, Jonathunder, Laoris, Leitmotiv, LightandDark2000, NewEnglandYankee, Oerjan, PokeCaptain, Primefac, Prokaryotes, Quantanew, Renerpho, Resolutionisim, Rfassbind, Rjwilmsi, Serendipodous, Sir Robert Pollock, Snrns1, Starcluster, TAnthony, Tbayboy, Tomruen, Twang, UU, UncleBigFoot, ZhockDEtoonsworth, 22 anonymous edits 585

Image Sources, Licenses and Contributors

The sources listed for each image provide more detailed licensing information including the copyright status, the copyright owner, and the license conditions.

Figure 1 *Source:* https://en.wikipedia.org/w/index.php?title=File:Solar_system_scale-2.jpg *License:* Public Domain *Contributors:* This image updated by ComputerHotline ... 1
Figure 2 *Source:* https://en.wikipedia.org/w/index.php?title=File:PaleBlueDot.jpg *License:* Public Domain *Contributors:* Alan Liefting, Benzband, Emijrp, Kenyon, Mattflaschen, Nard the Bard, Tulsi Bhagat, V!nc3r, 7 anonymous edits ... 2
Figure 3 *Source:* https://en.wikipedia.org/w/index.php?title=File:Top_of_Atmosphere.jpg *License:* Public Domain *Contributors:* NASA Earth Observatory ... 2
Figure 4 *Source:* https://en.wikipedia.org/w/index.php?title=File:Pioneer_10-11_-_P51b_-_fx.jpg *License:* Public Domain *Contributors:* NASA 3
Figure 5 *Source:* https://en.wikipedia.org/w/index.php?title=File:NewtonsTelescopeReplica.jpg *License:* Creative Commons Attribution-Share Alike 2.0 Generic *Contributors:* User:Solipsist (Andrew Dunn) ... 4
Figure 6 *Source:* https://en.wikipedia.org/w/index.php?title=File:Mercury_transit_2.jpg *License:* Creative Commons Attribution-Share Alike *Contributors:* Brocken Inaglory ... 4
Figure 7 *Source:* https://en.wikipedia.org/w/index.php?title=File:Asteroid2006DP14.jpg *License:* Public Domain *Contributors:* Huntster, Jcpag2012, OgreBot 2, Szczureq ... 6
Figure 8 *Source:* https://en.wikipedia.org/w/index.php?title=File:Europa_g1_true.jpg *License:* Public Domain *Contributors:* Bricktop, Dbenbenn, Jcpag2012, Smartech∼commonswiki ... 7
Figure 9 *Source:* https://en.wikipedia.org/w/index.php?title=File:Timeline_of_Solar_System_exploration.jpg *Contributors:* FraPado86, Mikhail Ryazanov, Mikima ... 7
Figure 10 *Source:* https://en.wikipedia.org/w/index.php?title=File:Pioneer10-11.jpg *License:* Public Domain *Contributors:* NASA ... 8
Figure 11 *Source:* https://en.wikipedia.org/w/index.php?title=File:PIA16239_High-Resolution_Self-Portrait_by_Curiosity_Rover_Arm_Camera.jpg *License:* Public Domain *Contributors:* Chick Bowen, FAEP, Huntster, INeverCry, Jean-Frédéric, Julian Herzog, Nicolas Perrault III, O'Dea, Pleclown, 4 anonymous edits ... 9
Figure 12 *Source:* https://en.wikipedia.org/w/index.php?title=File:Astronaut_Owen_Garriott_Performs_EVA_During_Skylab_3_-_GPN-2002-000065.jpg *License:* Public Domain *Contributors:* NASA ... 12
Image *Source:* https://en.wikipedia.org/w/index.php?title=File:Sun_in_February_(black_version).jpg *Contributors:* HalloweenNight ... 13
Image *Source:* https://en.wikipedia.org/w/index.php?title=File:Jupiter_and_its_shrunken_Great_Red_Spot_(cropped).jpg *License:* Public domain *Contributors:* PlanetUser ... 13
Image *Source:* https://en.wikipedia.org/w/index.php?title=File:Saturn_closeup.jpg *License:* Public Domain *Contributors:* Saturn_during_Equinox.jpg: NASA/JPL/Space Science Institute derivative work: BullBoxerBAB (talk) ... 13
Image *Source:* https://en.wikipedia.org/w/index.php?title=File:Uranus2_(cropped)-1.jpg *License:* Public Domain *Contributors:* NASA/JPL-Caltech 13
Image *Source:* https://en.wikipedia.org/w/index.php?title=File:Neptune_Full_(cropped).jpg *License:* Public Domain *Contributors:* PlanetUser . 13
Image *Source:* https://en.wikipedia.org/w/index.php?title=File:Africa_and_Europe_from_a_Million_Miles_Away_(cropped).png *License:* Public Domain *Contributors:* NASA ... 13
Image *Source:* https://en.wikipedia.org/w/index.php?title=File:Venus-real_color.jpg *License:* Attribution *Contributors:* NASA or Ricardo Nunes 13
Image *Source:* https://en.wikipedia.org/w/index.php?title=File:Mars_23_aug_2003_hubble_(cropped).jpg *License:* Public domain *Contributors:* NASA, ESA, and The Hubble Heritage Team (STScI/AURA) ... 13
Image *Source:* https://en.wikipedia.org/w/index.php?title=File:Ganymede_g1_true-edit1.jpg *License:* Public Domain *Contributors:* PlanetUser 13
Image *Source:* https://en.wikipedia.org/w/index.php?title=File:Two_Halves_of_Titan.png *License:* Public Domain *Contributors:* NASA/JPL/Space Science Institute ... 13
Image *Source:* https://en.wikipedia.org/w/index.php?title=File:Mercury_in_color_-_Prockter07-edit1.jpg *License:* Public Domain *Contributors:* NASA/Johns Hopkins University Applied Physics Laboratory/Carnegie Institution of Washington ... 13
Image *Source:* https://en.wikipedia.org/w/index.php?title=File:Callisto_(cropped)-1.jpg *License:* Public Domain *Contributors:* NASA/JPL/DLR(German Aerospace Center) ... 13
Image *Source:* https://en.wikipedia.org/w/index.php?title=File:Io_highest_resolution_true_color.jpg *License:* Public Domain *Contributors:* NASA / JPL / University of Arizona ... 13
Image *Source:* https://en.wikipedia.org/w/index.php?title=File:FullMoon2010_(cropped)-1.jpg *License:* Creative Commons Attribution-Sharealike 3.0 *Contributors:* Gregory H. Revera ... 13
Image *Source:* https://en.wikipedia.org/w/index.php?title=File:Europa-moon.jpg *License:* Public Domain *Contributors:* NASA/JPL/DLR ... 13
Image *Source:* https://en.wikipedia.org/w/index.php?title=File:Triton_Voyager_2.jpg *License:* Public Domain *Contributors:* BotMultichill, PlanetUser, Rmhermen, Ruslik0 ... 13
Image *Source:* https://en.wikipedia.org/w/index.php?title=File:Nh-pluto-in-true-color_2x_JPEG-edit.jpg *License:* Public Domain *Contributors:* NASA / Johns Hopkins University Applied Physics Laboratory / Southwest Research Institute ... 13
Image *Source:* https://en.wikipedia.org/w/index.php?title=File:Titania_(moon)_color_cropped.jpg *License:* Public Domain *Contributors:* Titania_(moon)_color.jpg: NASA/JPL ... 13
Image *Source:* https://en.wikipedia.org/w/index.php?title=File:PIA07763_Rhea_full_globe5.jpg *License:* Public Domain *Contributors:* NASA/JPL/Space Science Institute ... 13
Image *Source:* https://en.wikipedia.org/w/index.php?title=File:Voyager_2_picture_of_Oberon.jpg *License:* Public Domain *Contributors:* NASA 13
Image *Source:* https://en.wikipedia.org/w/index.php?title=File:Iapetus_as_seen_by_the_Cassini_probe_-_20071008_(cropped).jpg *License:* Public Domain *Contributors:* NASA/JPL/Space Science Institute ... 13
Image *Source:* https://en.wikipedia.org/w/index.php?title=File:Charon_in_Color_(HQ).jpg *License:* Public Domain *Contributors:* Delusion23, JorisvS, PlanetUser ... 13
Image *Source:* https://en.wikipedia.org/w/index.php?title=File:PIA00040_Umbrielx2.47.jpg *License:* Public Domain *Contributors:* NASA / Jet Propulsion Lab ... 13
Image *Source:* https://en.wikipedia.org/w/index.php?title=File:Color_Image_of_Ariel_as_seen_from_Voyager_2.jpg *License:* Public Domain *Contributors:* NASA/JPL (Digital with colortable: Calvin J. Hamilton) ... 13
Image *Source:* https://en.wikipedia.org/w/index.php?title=File:Dione_color_south.jpg *License:* Public Domain *Contributors:* NASA / JPL / SSI / Gordan Ugarkovic ... 13
Image *Source:* https://en.wikipedia.org/w/index.php?title=File:PIA18317-SaturnMoon-Tethys-Cassini-20150411.jpg *License:* Public Domain *Contributors:* Drbogdan, Huntster, JorisvS, Lotse ... 13
Image *Source:* https://en.wikipedia.org/w/index.php?title=File:PIA19562-Ceres-DwarfPlanet-Dawn-RC3-image19-20150506.jpg *License:* Public Domain *Contributors:* Drbogdan, Illustr, JorisvS, PhilipTerryGraham, Ras67, Tony Patt ... 13
Image *Source:* https://en.wikipedia.org/w/index.php?title=File:Vesta_full_mosaic.jpg *License:* Public Domain *Contributors:* Illustr, JorisvS, PhilipTerryGraham, Salamander724, Szczureq, Torsch ... 13
Image *Source:* https://en.wikipedia.org/w/index.php?title=File:PIA17202-SaturnMoon-Enceladus-ApproachingFlyby-20151028-cropped.jpg *License:* Public Domain *Contributors:* PlanetUser ... 13
Image *Source:* https://en.wikipedia.org/w/index.php?title=File:Miranda.jpg *License:* Public Domain *Contributors:* NASA/JPL-Caltech ... 13
Image *Source:* https://en.wikipedia.org/w/index.php?title=File:Proteus_Voyager_2_croped.jpg *Contributors:* - ... 13
Image *Source:* https://en.wikipedia.org/w/index.php?title=File:Mimas_PIA12568.jpg *License:* Public Domain *Contributors:* NASA / Jet Propulsion Lab / Space Science Institute ... 13
Image *Source:* https://en.wikipedia.org/w/index.php?title=File:Hyperion_in_natural_colours.jpg *License:* Public Domain *Contributors:* Uwe W., 1 anonymous edits ... 13
Image *Source:* https://en.wikipedia.org/w/index.php?title=File:Phoebe_cassini.jpg *License:* Public Domain *Contributors:* NASA/JPL/Space Science Institute ... 13
Image *Source:* https://en.wikipedia.org/w/index.php?title=File:PIA12714_Janus_crop.jpg *License:* Public Domain *Contributors:* NASA / Jet Propulsion Laboratory / Space Science Institute ... 13
Image *Source:* https://en.wikipedia.org/w/index.php?title=File:Amalthea_(moon).png *License:* Public Domain *Contributors:* NASA/Calvin J. Hamilton ... 14
Image *Source:* https://en.wikipedia.org/w/index.php?title=File:PIA09813_Epimetheus_S._polar_region.jpg *License:* Public Domain *Contributors:* NASA/JPL/Space Science Institute ... 14

Image *Source:* https://en.wikipedia.org/w/index.php?title=File:Thebe.jpg *License:* Public Domain *Contributors:* NASA/JPL ... 14

Image *Source:* https://en.wikipedia.org/w/index.php?title=File:Rosetta_triumphs_at_asteroid_Lutetia.jpg *License:* Creative Commons Attribution-Sharealike 2.0 *Contributors:* ESA 2010 MPS for OSIRIS Team MPS/UPD/LAM/IAA/RSSD/INTA/UPM/DASP/IDA ... 14

Image *Source:* https://en.wikipedia.org/w/index.php?title=File:Prometheus_12-26-09a.jpg *License:* Public Domain *Contributors:* NASA / JPL / Space Science Institute ... 14

Image *Source:* https://en.wikipedia.org/w/index.php?title=File:PIA21055_-_Pandora_Up_Close_(cropped).jpg *License:* Public Domain *Contributors:* PhilipTerryGraham, PlanetUser ... 14

Image *Source:* https://en.wikipedia.org/w/index.php?title=File:(253)_mathilde_crop.jpg *License:* Public Domain *Contributors:* BokicaK, Clh288∼commonswiki, DragonFire1024, Ilmari Karonen, MECU, Ruslik0, Vesta∼commonswiki ... 14

Image *Source:* https://en.wikipedia.org/w/index.php?title=File:Hydra_(moon)_from_231_000_kilometres.jpg *License:* Public Domain *Contributors:* NASA/JHUAPL/SWRI ... 14

Image *Source:* https://en.wikipedia.org/w/index.php?title=File:Leading_hemisphere_of_Helene_-_20110618.jpg *License:* Public Domain *Contributors:* Jacklee, JorisvS, Torsch, WolfmanSF ... 14

Image *Source:* https://en.wikipedia.org/w/index.php?title=File:Nix_best_view-true_color.jpg *License:* Public Domain *Contributors:* PhilipTerry-Graham, PlanetUser ... 14

Image *Source:* https://en.wikipedia.org/w/index.php?title=File:243_Ida_large.jpg *License:* Public Domain *Contributors:* NASA/JPL ... 14

Image *Source:* https://en.wikipedia.org/w/index.php?title=File:Atlas_(NASA).jpg *License:* Public Domain *Contributors:* ComputerHotline, Joanjoc∼commonswiki, Parpan05, SJuergen, Svdmolen, 1 anonymous edits ... 14

Image *Source:* https://en.wikipedia.org/w/index.php?title=File:Telesto_cassini_closeup.jpg *License:* Public Domain *Contributors:* Ahonc, Bricktop, ComputerHotline ... 14

Image *Source:* https://en.wikipedia.org/w/index.php?title=File:N00151485_Calypso_crop.jpg *License:* Public Domain *Contributors:* NASA / JPL / Space Science Institute ... 14

Image *Source:* https://en.wikipedia.org/w/index.php?title=File:Phobos_colour_2008.jpg *License:* Public Domain *Contributors:* NASA / JPL-Caltech / University of Arizona ... 14

Image *Source:* https://en.wikipedia.org/w/index.php?title=File:Eros_-_PIA02923_(color).jpg *License:* Public Domain *Contributors:* NASA/JPL/JHUAPL ... 14

Image *Source:* https://en.wikipedia.org/w/index.php?title=File:Pan_by_Cassini,_March_2017.jpg *License:* Public Domain *Contributors:* PhilipTer-ryGraham ... 14

Image *Source:* https://en.wikipedia.org/w/index.php?title=File:Deimos-MRO.jpg *License:* Public Domain *Contributors:* NASA/JPL-caltech/ University of Arizona ... 14

Image *Source:* https://en.wikipedia.org/w/index.php?title=File:951_Gaspra.jpg *License:* Public Domain *Contributors:* Alaniaris, BMacZero, Boki-caK, Farsheed.R, Magog the Ogre, Papa November, PlanetUser, Ruslik0, Stas1995, The Evil IP address, Tryphon, Yonidebest, 6 anonymous edits ... 14

Image *Source:* https://en.wikipedia.org/w/index.php?title=File:PIA02127.jpg *Contributors:* - ... 14

Image *Source:* https://en.wikipedia.org/w/index.php?title=File:2867_Šteins_by_Rosetta_(reprocessed)_-_cropped.png *License:* Creative Commons Attribution-Sharealike 2.0 *Contributors:* ESA 2008 MPS for OSIRIS Team MPS/UPD/LAM/IAA/RSSD/INTA/UPM/DASP/IDA; processing by T. Stryk. 14

Image *Source:* https://en.wikipedia.org/w/index.php?title=File:Daphnis_(Saturn's_Moon).jpg *License:* Public Domain *Contributors:* PlanetUser 14

Image *Source:* https://en.wikipedia.org/w/index.php?title=File:Comet_Borrelly_Nucleus_(cropped).jpg *License:* Public Domain *Contributors:* NASA/JPL ... 14

Image *Source:* https://en.wikipedia.org/w/index.php?title=File:Comet_67P_on_19_September_2014_NavCam_mosaic.jpg *Contributors:* Alborza-gros, Amada44, Arjuno3, Brateevsky, Bulat, Capricorn4049, DryominG, Foundling, Huntster, Jan.Kamenicek, LudwigSebastianMicheler, Miya, Neptunia, Sakhalinio, Steinsplitter, Texnik, Torsch, Yann, Ömer Berkay, 5 anonymous edits ... 14

Image *Source:* https://en.wikipedia.org/w/index.php?title=File:Wild2_3_(cropped-square).jpg *License:* Public Domain *Contributors:* NASA .. 14

Image *Source:* https://en.wikipedia.org/w/index.php?title=File:Methone_PIA14633.jpg *License:* Public Domain *Contributors:* Huntster, Sar-danaphalus, WolfmanSF ... 14

Image *Source:* https://en.wikipedia.org/w/index.php?title=File:Comet_Hartley_2_(super_crop).jpg *License:* Public Domain *Contributors:* NASA/ JPL-Caltech/UMD ... 14

Image *Source:* https://en.wikipedia.org/w/index.php?title=File:Dactyl-HiRes.jpg *License:* Public Domain *Contributors:* NASA/JPL ... 14

Figure 13 *Source:* https://en.wikipedia.org/w/index.php?title=File:Lunar_basalt_70017.jpg *License:* Public Domain *Contributors:* Doug Coldwell, GeoWriter ... 15

Figure 14 *Source:* https://en.wikipedia.org/w/index.php?title=File:Protoplanetary-disk.jpg *License:* Public Domain *Contributors:* NASA ... 17

Figure 15 *Source:* https://en.wikipedia.org/w/index.php?title=File:Lhborbits.png *License:* Creative Commons Attribution-Sharealike 3.0,2.5,2.0,1.0 *Contributors:* en:User:AstroMark ... 24

Figure 16 *Source:* https://en.wikipedia.org/w/index.php?title=File:Sun_red_giant.svg *License:* GNU Free Documentation License *Contributors:* Oona Räisänen (User:Mysid), User:Mrsanitazier. ... 30

Image *Source:* https://en.wikipedia.org/w/index.php?title=File:Cscr-featured.svg *License:* GNU Lesser General Public License *Contributors:* Anomie ... 36

Image *Source:* https://en.wikipedia.org/w/index.php?title=File:Padlock-silver.svg *Contributors:* This image file was created by AJ Ashton. Uploaded from English WP by User:Eleassar. Converted by User:AzaToth to a silv ... 36

Image *Source:* https://en.wikipedia.org/w/index.php?title=File:Sun_white.jpg *Contributors:* Huntster, Jcb, Juggler2005, Ulflund ... 36

Figure 17 *Source:* https://en.wikipedia.org/w/index.php?title=File:Sun_poster.svg *License:* Creative Commons Attribution-Sharealike 3.0 *Contributors:* User:Kelvinsong ... 44

Figure 18 *Source:* https://en.wikipedia.org/w/index.php?title=File:EffectiveTemperature_300dpi_e.png *License:* GNU Free Documentation License *Contributors:* Sch ... 46

Figure 19 *Source:* https://en.wikipedia.org/w/index.php?title=File:2017_Solar_Eclipse_Weiser_Idaho.jpg *Contributors:* User:Zombiepedia ... 48

Figure 20 *Source:* https://en.wikipedia.org/w/index.php?title=File:171879main_LimbFlareJan12_lg.jpg *License:* Public Domain *Contributors:* Hinode JAXA/NASA ... 48

Figure 21 *Source:* https://en.wikipedia.org/w/index.php?title=File:172197main_NASA_Flare_Gband_lg-withouttext.jpg *License:* Public Domain *Contributors:* NASA / JAXA ... 50

Image *Source:* https://en.wikipedia.org/w/index.php?title=File:Sunspot_butterfly_diagram.svg *License:* Creative Commons Attribution-Sharealike 3.0,2.5,2.0,1.0 *Contributors:* Con-struct ... 51

Image *Source:* https://en.wikipedia.org/w/index.php?title=File:Sunspot_area_variation.svg *License:* Creative Commons Attribution-Sharealike 3.0,2.5,2.0,1.0 *Contributors:* Con-struct ... 51

Figure 22 *Source:* https://en.wikipedia.org/w/index.php?title=File:Sun_-_August_1,_2010.jpg *License:* Public Domain *Contributors:* NASA/SDO/ AIA ... 51

Figure 23 *Source:* https://en.wikipedia.org/w/index.php?title=File:Heliospheric-current-sheet.gif *License:* Public Domain *Contributors:* Werner Heil (see "other version" below). ... 52

Figure 24 *Source:* https://en.wikipedia.org/w/index.php?title=File:Solar-cycle-data.png *Contributors:* Basilicofresco, Beland, Capricorn4049, Dragons flight, Drdoht, F1jmm, Lampman, Lissajous, Mgc8, Nils Simon, Szczureq, Torsch, WikipediaMaster, Xenoforme∼commonswiki, Xiong Chiamiov, 5 anonymous edits ... 54

Figure 25 *Source:* https://en.wikipedia.org/w/index.php?title=File:Solar_evolution_(English).svg *License:* Creative Commons Attribution-Sharealike 3.0 *Contributors:* User:RJHall ... 56

Figure 26 *Source:* https://en.wikipedia.org/w/index.php?title=File:Sun_red_giant.svg *License:* GNU Free Documentation License *Contributors:* Oona Räisänen (User:Mysid), User:Mrsanitazier. ... 57

Figure 27 *Source:* https://en.wikipedia.org/w/index.php?title=File:Evolution_of_a_Sun-like_star.svg *Contributors:* User:Szczureq ... 58

Figure 28 *Source:* https://en.wikipedia.org/w/index.php?title=File:Map_of_the_full_sun.jpg *License:* Public Domain *Contributors:* ComputerHot-line, Jcpag2012, O'Dea ... 60

Figure 29 *Source:* https://en.wikipedia.org/w/index.php?title=File:Solvognen_DO-6865_2000.jpg *License:* Creative Commons Attribution-Sharealike 2.5 *Contributors:* Palnatoke ... 62

Figure 30 *Source:* https://en.wikipedia.org/w/index.php?title=File:Sun-bonatti.png *License:* Public Domain *Contributors:* Man vyi, Smerdis of Tlön, Torsch, Ykvach ... 64

Figure 31 *Source:* https://en.wikipedia.org/w/index.php?title=File:Sunspots_and_Solar_Flares.jpg *License:* Public Domain *Contributors:* Di-pankan001, Ras67, 2 anonymous edits ... 65

Figure 32 *Source:* https://en.wikipedia.org/w/index.php?title=File:Moon_transit_of_sun_large.ogv *License:* Public Domain *Contributors:* NASA 66

Figure 33 *Source:* https://en.wikipedia.org/w/index.php?title=File:Giant_prominence_on_the_sun_erupted.jpg *License:* Public Domain *Contributors:* ComputerHotline, Stas1995, Szczureq ... 67

Figure 34 *Source:* https://en.wikipedia.org/w/index.php?title=File:Anatomy_of_a_Sunset-2.jpg *License:* Creative Commons Attribution-Sharealike 3.0 *Contributors:* Jessie Eastland ... 68

Figure 75 *Source:* https://en.wikipedia.org/w/index.php?title=File:Venus,_Earth_size_comparison.jpg *License:* Public Domain *Contributors:* Fyunck(click), Jcpag2012, JorisvS, Lotse, Mates ... 133

Figure 76 *Source:* https://en.wikipedia.org/w/index.php?title=File:Maat_Mons_on_Venus.jpg *License:* Public Domain *Contributors:* Bricktop, Bryan Derksen, ComputerHotline, Crux, Evil Monkey, Lotse, TheDJ, 1 anonymous edits ... 135

Figure 77 *Source:* https://en.wikipedia.org/w/index.php?title=File:Mgn_p39146.png *License:* Public Domain *Contributors:* BillC, ComputerHotline, Lotse, Ruslik0 ... 136

Figure 78 *Source:* https://en.wikipedia.org/w/index.php?title=File:Venus_structure.jpg *License:* Public Domain *Contributors:* NASA (original), User:Vzb83∼commonswiki (derivate) ... 137

Image *Source:* https://en.wikipedia.org/w/index.php?title=File:Venuspioneeruv.jpg *License:* Public Domain *Contributors:* NASA ... 138

Image *Source:* https://en.wikipedia.org/w/index.php?title=File:Venus_globe.jpg *License:* Public Domain *Contributors:* NASA ... 138

Image *Source:* https://en.wikipedia.org/w/index.php?title=File:Synthetic_atmosphere_absorption_spectrum.gif *License:* Creative Commons Attribution-Sharealike 3.0 *Contributors:* User:Darekk2 ... 140

Image *Source:* https://en.wikipedia.org/w/index.php?title=File:Synthetic_Venus_atmosphere_absorption_spectrum.gif *License:* Creative Commons Attribution-Sharealike 3.0 *Contributors:* User:Darekk2 ... 140

Figure 79 *Source:* https://en.wikipedia.org/w/index.php?title=File:Venusorbitsolarsystem.gif *License:* Creative Commons Attribution-Sharealike 3.0 *Contributors:* User:Lookang ... 142

Figure 80 *Source:* https://en.wikipedia.org/w/index.php?title=File:Venus-pacific-levelled.jpg *License:* Creative Commons Attribution-Sharealike 3.0 *Contributors:* Brocken Inaglory ... 144

Figure 81 *Source:* https://en.wikipedia.org/w/index.php?title=File:Phases_Venus.jpg *License:* Attribution *Contributors:* Statis Kalyvas - VT-2004 programme ... 145

Figure 82 *Source:* https://en.wikipedia.org/w/index.php?title=File:Venustransit_2004-06-08_07-49.jpg *License:* GNU Free Documentation License *Contributors:* de:Benutzer:Klingon ... 146

Figure 83 *Source:* https://en.wikipedia.org/w/index.php?title=File:Venus_geocentric_orbit_curve_simplified_Line_(pentagram).svg *License:* Public Domain *Contributors:* AnonMoos ... 147

Figure 84 *Source:* https://en.wikipedia.org/w/index.php?title=File:Venus_Drawing.jpg *Contributors:* Athaenara, DrKay, Firsfron, Infrogmation, OgreBot 2, Poulpy, İnfoCan, 1 anonymous edits ... 148

Figure 85 *Source:* https://en.wikipedia.org/w/index.php?title=File:Phases-of-Venus.svg *License:* Public Domain *Contributors:* Nichalp 09:56, 11 June 2006 (UTC) ... 149

Figure 86 *Source:* https://en.wikipedia.org/w/index.php?title=File:Vénus_télescope.jpg *License:* Creative Commons Attribution-Share Alike *Contributors:* Marc Lecleire ... 150

Figure 87 *Source:* https://en.wikipedia.org/w/index.php?title=File:Mariner_2.jpg *License:* Public Domain *Contributors:* NASA Jet Propulsion Laboratory (NASA-JPL) Image ID : PIA04594 ... 151

Figure 88 *Source:* https://en.wikipedia.org/w/index.php?title=File:Venus_as_captured_by_Mariner_10.jpg *License:* Public Domain *Contributors:* NASA ... 152

Image *Source:* https://en.wikipedia.org/w/index.php?title=File:Venus_symbol.svg *Contributors:* Kyle the hacker ... 154

Image *Source:* https://en.wikipedia.org/w/index.php?title=File:Wikiquote-logo.svg *License:* Public Domain *Contributors:* Rei-artur ... 154

Image *Source:* https://en.wikipedia.org/w/index.php?title=File:The_Earth_seen_from_Apollo_17.jpg *License:* Public Domain *Contributors:* NASA/Apollo 17 crew; taken by either Harrison Schmitt or Ron Evans ... 155

Figure 89 *Source:* https://en.wikipedia.org/w/index.php?title=File:Protoplanetary-disk.jpg *License:* Public Domain *Contributors:* NASA ... 159

Figure 90 *Source:* https://en.wikipedia.org/w/index.php?title=File:USA_10654_Bryce_Canyon_Luca_Galuzzi_2007.jpg *License:* Creative Commons Attribution-Sharealike 2.5 *Contributors:* Luca Galuzzi (Lucag) ... 160

Figure 91 *Source:* https://en.wikipedia.org/w/index.php?title=File:PhylogeneticTree,_Woese_1990.PNG *License:* Creative Commons Attribution-Sharealike 3.0 *Contributors:* User:Maulucioni ... 161

Figure 92 *Source:* https://en.wikipedia.org/w/index.php?title=File:Earth2014shape_SouthAmerica_small.jpg *Contributors:* User:Geodesy2000 163

Image *Source:* https://en.wikipedia.org/w/index.php?title=File:Earth-cutaway-schematic-english.svg *License:* Public Domain *Contributors:* derivative work: Anasofiapaixao Earth_internal_structure.png: USGS ... 165

Image *Source:* https://en.wikipedia.org/w/index.php?title=File:Tectonic_plates_(empty).svg *License:* Public Domain *Contributors:* Dbenbenn, Denys, JorisvS, KJG2007, Micheletb, Santosga, Sarang, Sting, Ævar Arnfjörð Bjarmason ... 167

Figure 93 *Source:* https://en.wikipedia.org/w/index.php?title=File:AYool_topography_15min.png *License:* Creative Commons Attribution 2.5 *Contributors:* Plumbago ... 168

Figure 94 *Source:* https://en.wikipedia.org/w/index.php?title=File:Earth_elevation_histogram_2.svg *License:* Public domain *Contributors:* Ciaurlec, MushiHoshiIshi, OgreBot 2, Pitke, Sbelza, Tano4595, باسم, 和平奮鬥救地球 ... 170

Figure 95 *Source:* https://en.wikipedia.org/w/index.php?title=File:MODIS_Map.jpg *License:* Public Domain *Contributors:* NASA ... 171

Image *Source:* https://en.wikipedia.org/w/index.php?title=File:Felix_from_ISS_03_sept_2007_1138Z.jpg *License:* Public Domain *Contributors:* NASA ... 171

Image *Source:* https://en.wikipedia.org/w/index.php?title=File:Pressure_ridges_Scott_Base_lrg.jpg *License:* Public Domain *Contributors:* Michael Studinger ... 171

Image *Source:* https://en.wikipedia.org/w/index.php?title=File:3D-Clouds.jpg *Contributors:* User:Pocketthis ... 172

Figure 96 *Source:* https://en.wikipedia.org/w/index.php?title=File:Full_moon_partially_obscured_by_atmosphere.jpg *License:* Public Domain *Contributors:* NASA ... 174

Figure 97 *Source:* https://en.wikipedia.org/w/index.php?title=File:Geoids_sm.jpg *License:* Public Domain *Contributors:* Angrense, B jonas, Ciaurlec, Denniss, Look2See1, LudwigSebastianMicheler, Maddox2, Mapmarks, Mario1952, Minor edit, Rainald62, RedAndr, Stewi101015, 1 anonymous edits ... 175

Figure 98 *Source:* https://en.wikipedia.org/w/index.php?title=File:Structure_of_the_magnetosphere-en.svg *License:* Public Domain *Contributors:* Original bitmap from NASA. SVG rendering by Aaron Kaase. ... 176

Figure 99 *Source:* https://en.wikipedia.org/w/index.php?title=File:EpicEarth-Globespin(2016May29).gif *License:* Public Domain *Contributors:* User:Tdadamemd ... 177

Figure 100 *Source:* https://en.wikipedia.org/w/index.php?title=File:Pale_Blue_Dot.png *License:* Public Domain *Contributors:* Voyager 1 ... 178

Figure 101 *Source:* https://en.wikipedia.org/w/index.php?title=File:AxialTiltObliquity.png *License:* Creative Commons Attribution 3.0 *Contributors:* Dna-webmaster ... 179

Figure 102 *Source:* https://en.wikipedia.org/w/index.php?title=File:Moraine_Lake_17092005.jpg *Contributors:* Berrucomons, Ernác, Flcelloguy∼commonswiki, Gorgo, Herzi Pinki, Hike395, Ianare, J 1982, JotaCartas, Kasir, MONGO, Mykola Swarnyk, O (bot), Odysseus1479, Palmman, Para, Raphael17, Shizhao, Sultan11, Taluabeg, Thierry Caro, Yuriy75, Überraschungsbilder, [illegible], 1 anonymous edits ... 181

Figure 103 *Source:* https://en.wikipedia.org/w/index.php?title=File:Pavlof2014iss.jpg *License:* Public Domain *Contributors:* NASA ... 183

Figure 104 *Source:* https://en.wikipedia.org/w/index.php?title=File:Continents_vide_couleurs.png *License:* Creative Commons Attribution-ShareAlike 3.0 Unported *Contributors:* User:Cogito ergo sumo ... 184

Image *Source:* https://en.wikipedia.org/w/index.php?title=File:FullMoon2010.jpg *License:* Creative Commons Attribution-Sharealike 3.0 *Contributors:* Gregory H. Revera ... 185

Figure 105 *Source:* https://en.wikipedia.org/w/index.php?title=File:Earth-Moon.svg *License:* Public Domain *Contributors:* Earth-Moon.PNG: Earth-image from NASA; arrangement by brews_ohare derivative work: Cmglee (talk) ... 186

Figure 106 *Source:* https://en.wikipedia.org/w/index.php?title=File:NASA-Apollo8-Dec24-Earthrise.jpg *License:* Public Domain *Contributors:* AKA MBG, Apalsola, BarnacleKB∼commonswiki, Berrucomons, BoringHistoryGuy, Davepape, Edward, Emijrp, Geitost, George Chernilevsky, Howcheng, J 1982, JalalV, Jcpag2012, Jean-Frédéric, Jimmy Xu, KaragouniS∼commonswiki, Ke4roh, Lx 121, Mario modesto, Medconn, Roberta F., Saperaud∼commonswiki, Schekinov Alexey Victorovich, Sevela.p, Sh1019, Simonizer, Starscream, Surya Prakash.S.A., Takabeg, Tangopaso, TheDJ, Wieralee, Wstrwald, Марсианский хро, باسم, 5 anonymous edits ... 188

Image *Source:* https://en.wikipedia.org/w/index.php?title=File:Earth_symbol.svg *Contributors:* OsgoodeLawyer ... 187

Image *Source:* https://en.wikipedia.org/w/index.php?title=File:OSIRIS_Mars_true_color.jpg *Contributors:* A2soup, Arjuno3, Huntster, Mtrova, Sanandros, WolfmanSF ... 190

Image *Source:* https://en.wikipedia.org/w/index.php?title=File:Mars,_Earth_size_comparison.jpg *License:* Public Domain *Contributors:* User:Jcpag2012 ... 193

Image *Source:* https://en.wikipedia.org/w/index.php?title=File:Mars.ogv *License:* Public Domain *Contributors:* NASA/Goddard Space Flight Center 193

Image *Source:* https://en.wikipedia.org/w/index.php?title=File:GMM-3_Mars_Gravity.webm *License:* Public Domain *Contributors:* NASA's Scientific Visualization Studio ... 193

Figure 107 *Source:* https://en.wikipedia.org/w/index.php?title=File:USGS-MarsMap-sim3292-20140714-crop.png *License:* Public Domain *Contributors:* Chmee2, Drbogdan, Kopiersperre, Mapmarks ... 194

Figure 108 *Source:* https://en.wikipedia.org/w/index.php?title=File:Eso1509a_-_Mars_planet.jpg *Contributors:* ESO/M. Kornmesser ... 195

Figure 109 *Source:* https://en.wikipedia.org/w/index.php?title=File:Spirit_Mars_Silica_April_20_2007.jpg *License:* Public Domain *Contributors:* BotMultichill, DragonFire1024, Foroa, Huntster, Kilom691, 1 anonymous edits ... 196

Figure 298 *Source:* https://en.wikipedia.org/w/index.php?title=File:PIA20036-Charon-YoungestCrater-20150714.jpg *License:* Public Domain *Contributors:* Bodhisattwa, Drbogdan, PhilipTerryGraham 474
Image *Source:* https://en.wikipedia.org/w/index.php?title=File:Pluto-picture.jpg *License:* Public Domain *Contributors:* NASA 475
Image *Source:* https://en.wikipedia.org/w/index.php?title=File:Pluto_and_charon.jpg *Contributors:* - 475
Image *Source:* https://en.wikipedia.org/w/index.php?title=File:First_Color_Animated_Images_show_Pluto_and_its_Moon_Charon.gif *License:* Public Domain *Contributors:* Hedwig in Washington, Lotse, RolandyJake, Torsch, とある白い猫 476
Figure 299 *Source:* https://en.wikipedia.org/w/index.php?title=File:Pluto_charon_150709_color_final.png *License:* Public Domain *Contributors:* JPL/NASA 476
Image *Source:* https://en.wikipedia.org/w/index.php?title=File:Cpmap_cyl_PS717_HR_180.jpg *License:* Public Domain *Contributors:* Kwamikagami, Msaynevirta 477
Image *Source:* https://en.wikipedia.org/w/index.php?title=File:Pluto-Charon-v2-10-1-15.jpg *License:* Public Domain *Contributors:* Aineias, Geek3, WolfmanSF 477
Image *Source:* https://en.wikipedia.org/w/index.php?title=File:PIA19967_-_Charon_in_Detail.jpg *License:* Public Domain *Contributors:* WolfmanSF 478
Image *Source:* https://en.wikipedia.org/w/index.php?title=File:NH-PlutoCharon-Color-NewHorizons-20150711.jpg *License:* Public Domain *Contributors:* Drbogdan, Geek3, Huntster, Lysosome, Offnfopt, Rfassbind, Waldir, 1 anonymous edits 478
Image *Source:* https://en.wikipedia.org/w/index.php?title=File:NH-071315-PlutoCharon-FalseColorComposite-20150713.jpg *License:* Public Domain *Contributors:* Drbogdan, Geek3, Offnfopt, Stas000D, Waldir, 1 anonymous edits 478
Image *Source:* https://en.wikipedia.org/w/index.php?title=File:PIA20375-PlutoMoon-Charon-NightSide-20150717.jpg *License:* Public Domain *Contributors:* Drbogdan, PhilipTerryGraham 478
Figure 300 *Source:* https://en.wikipedia.org/w/index.php?title=File:NH-PlutoMoon-Charon-FlyOverAnimation-20151001.webm *License:* Public Domain *Contributors:* Drbogdan, Reguyla 480
Image *Source:* https://en.wikipedia.org/w/index.php?title=File:Makemake_moon_Hubble_image_with_legend_(cropped).jpg *License:* Public domain *Contributors:* PlanetUser, Ruslik0 480
Figure 301 *Source:* https://en.wikipedia.org/w/index.php?title=File:TheKuiperBelt_Orbits_2003EL61_2005FY9.svg *License:* Creative Commons Attribution-ShareAlike 3.0 Unported *Contributors:* User:Eurocommuter 483
Figure 302 *Source:* https://en.wikipedia.org/w/index.php?title=File:Makemake-LB1-2009Nov26-11UT.jpg *License:* Creative Commons Attribution-Sharealike 3.0 *Contributors:* Kevin Heider @ LightBuckets 485
Figure 303 *Source:* https://en.wikipedia.org/w/index.php?title=File:EightTNOs.png *License:* Creative Commons Attribution-ShareAlike 3.0 Unported *Contributors:* Lexicon 486
Image *Source:* https://en.wikipedia.org/w/index.php?title=File:Makemake_with_moon.JPG *License:* Public domain *Contributors:* Ruslik0, Szczureq 487
Figure 304 *Source:* https://en.wikipedia.org/w/index.php?title=File:Haumea.GIF *Contributors:* Tony Dunn (author of Gravity Simulator) from the Gravity Simulator message board. 494
Figure 305 *Source:* https://en.wikipedia.org/w/index.php?title=File:TheKuiperBelt_Orbits_2003EL61.svg *License:* Creative Commons Attribution-Sharealike 2.5 *Contributors:* User:Eurocommuter 494
Figure 306 *Source:* https://en.wikipedia.org/w/index.php?title=File:EightTNOs.png *License:* Creative Commons Attribution-ShareAlike 3.0 Unported *Contributors:* Lexicon 496
Figure 307 *Source:* https://en.wikipedia.org/w/index.php?title=File:Haumea.svg *License:* GNU Free Documentation License *Contributors:* en:User:Kwamikagami 497
Figure 308 *Source:* https://en.wikipedia.org/w/index.php?title=File:Haumea_Rotation.gif *License:* Creative Commons Zero *Contributors:* User:StephHoover 498
Figure 309 *Source:* https://en.wikipedia.org/w/index.php?title=File:2003EL61art.jpg *License:* Public Domain *Contributors:* A. Feild (Space Telescope Science Institute) 500
Figure 310 *Source:* https://en.wikipedia.org/w/index.php?title=File:Eris_and_dysnomia2.jpg *License:* Public Domain *Contributors:* Eunostos, Ford TTT, Huntster, Jan.Kamenicek, Jcpag2012, John800104, Kalki, Krinkle, Laim, Nicole Sharp, Ruslik0, Shizhao, Shorenster, 2 anonymous edits 504
Figure 311 *Source:* https://en.wikipedia.org/w/index.php?title=File:TheKuiperBelt_Projections_100AU_Classical_SDO.svg *License:* Creative Commons Attribution-ShareAlike 3.0 Unported *Contributors:* User:Eurocommuter～commonswiki 505
Figure 312 *Source:* https://en.wikipedia.org/w/index.php?title=File:TheKuiperBelt_100AU_SDO.svg *License:* Creative Commons Attribution-ShareAlike 3.0 Unported *Contributors:* User:Eurocommuter 508
Figure 313 *Source:* https://en.wikipedia.org/w/index.php?title=File:Lhborbits.png *License:* Creative Commons Attribution-Sharealike 3.0,2.5,2.0,1.0 *Contributors:* en:User:AstroMark 509
Figure 314 *Source:* https://en.wikipedia.org/w/index.php?title=File:2003_UB313_near-infrared_spectrum.png *License:* GNU Free Documentation License *Contributors:* 1989, GifTagger, Jeanjung212, Steinsplitter 510
Figure 315 *Source:* https://en.wikipedia.org/w/index.php?title=File:Tempel_1_(PIA02127).jpg *License:* Public Domain *Contributors:* NASA/JPL-Caltech/UMD 512
Figure 316 *Source:* https://en.wikipedia.org/w/index.php?title=File:EightTNOs.png *License:* Creative Commons Attribution-ShareAlike 3.0 Unported *Contributors:* Lexicon 521
Figure 317 *Source:* https://en.wikipedia.org/w/index.php?title=File:2003_UB313_near-infrared_spectrum.png *License:* GNU Free Documentation License *Contributors:* 1989, GifTagger, Jeanjung212, Steinsplitter 522
Image *Source:* https://en.wikipedia.org/w/index.php?title=File:Voyager_1_entering_heliosheath_region.jpg *License:* Public Domain *Contributors:* NASA/Walt Feimer 525
Image *Source:* https://en.wikipedia.org/w/index.php?title=File:Heliosphere-washbasin.svg *Contributors:* User:Yanpas 525
Image *Source:* https://en.wikipedia.org/w/index.php?title=File:Solar_wind_at_Voyager_1_cut_out.png *License:* Public Domain *Contributors:* Rfassbind 525
Image *Source:* https://en.wikipedia.org/w/index.php?title=File:PIA17046_-_Voyager_1_Goes_Interstellar.jpg *License:* Public Domain *Contributors:* Badzil, Cyclopia, WolfmanSF 525
Figure 318 *Source:* https://en.wikipedia.org/w/index.php?title=File:IBEX_all_sky_map.jpg *License:* Public Domain *Contributors:* NASA/Goddard Space Flight Center Scientific Visualization Studio 526
Figure 319 *Source:* https://en.wikipedia.org/w/index.php?title=File:The_Sun_as_seen_in_January_1st_2015.jpg *License:* Public Domain *Contributors:* Huntster, Sigwald, Tetra quark, Tiptoety 527
Figure 320 *Source:* https://en.wikipedia.org/w/index.php?title=File:Heliospheric-current-sheet.gif *License:* Public Domain *Contributors:* Werner Heil (see "other version" below). 528
Figure 321 *Source:* https://en.wikipedia.org/w/index.php?title=File:Heliosphere-washbasin.svg *Contributors:* User:Yanpas 529
Figure 322 *Source:* https://en.wikipedia.org/w/index.php?title=File:Transitional_regions.jpg *License:* Public Domain *Contributors:* NASA/JPL 531
Image *Source:* https://en.wikipedia.org/w/index.php?title=File:Ibexheliosphererevised.jpg *License:* Public Domain *Contributors:* Credits: NASA/IBEX/Adler Planetarium 532
Figure 323 *Source:* https://en.wikipedia.org/w/index.php?title=File:PIA12375.jpg *License:* Public Domain *Contributors:* NASA/JPL/JHUAPL 533
Figure 324 *Source:* https://en.wikipedia.org/w/index.php?title=File:Solarmap.png *License:* Public Domain *Contributors:* ComputerHotline, Magog the Ogre, Narfiol, Sir48, Waldir, باسم, 1 anonymous edits 534
Figure 325 *Source:* https://en.wikipedia.org/w/index.php?title=File:PioneerH2011.jpg *License:* Creative Commons Zero *Contributors:* CamWow 536
Image *Source:* https://en.wikipedia.org/w/index.php?title=File:IBEXmagneticfieldinfluence.jpg *License:* Public Domain *Contributors:* Southwest Research Institute 537
Figure 326 *Source:* https://en.wikipedia.org/w/index.php?title=File:Evolving_Heliophysics_System_Observatory.jpg *License:* Public Domain *Contributors:* NASA 538
Figure 327 *Source:* https://en.wikipedia.org/w/index.php?title=File:72408main_ACD97-0036-1.jpg *License:* Public Domain *Contributors:* 1989, Anakin101, Bricktop, CarolSpears, ClaudiaM, Mdd, RuPassenger, Ruslik0, Thuresson, Torsch 539
Figure 328 *Source:* https://en.wikipedia.org/w/index.php?title=File:Heliopause_diagram.png *License:* Public Domain *Contributors:* RHorning, Ruslik0 539
Figure 329 *Source:* https://en.wikipedia.org/w/index.php?title=File:NewHeliopause_558121.jpg *License:* Public Domain *Contributors:* Hairy Dude, Papg2010, Rehman, 1 anonymous edits 540
Figure 330 *Source:* https://en.wikipedia.org/w/index.php?title=File:Sentinels_of_the_Heliosphere.ogv *License:* Public Domain *Contributors:* NASA/Goddard Space Flight Center 540
Figure 331 *Source:* https://en.wikipedia.org/w/index.php?title=File:TheTransneptunians_500AU.svg *License:* Creative Commons Attribution-Sharealike 3.0 *Contributors:* Eurocommuter 543
Figure 332 *Source:* https://en.wikipedia.org/w/index.php?title=File:Sedna-NASA.JPG *License:* Public Domain *Contributors:* User:Evercat 550
Figure 333 *Source:* https://en.wikipedia.org/w/index.php?title=File:Oort_cloud_Sedna_orbit.svg *License:* Public Domain *Contributors:* Image courtesy of NASA / Jet Propulsion LaboratoryJPL-Caltech / Robert L. HurtR. Hurt Original text courtesy of NASA / 555

License

Index